THE EXCAVATIONS AT WIJNALDUM

REPORTS ON FRISIA IN ROMAN AND MEDIEVAL TIMES

VOLUME 1

The Excavations at Wijnaldum

Reports on Frisia in Roman and Medieval times

J.C. BESTEMAN, J.M. BOS, D.A. GERRETS,
H.A. HEIDINGA & J. DE KONING

VOLUME 1

Frisia- project

A.A. BALKEMA / ROTTERDAM / BROOKFIELD / 1999

GIA
GRONINGEN INSTITUTE OF ARCHAEOLOGY

AAC
AMSTERDAM ARCHAEOLOGICAL CENTRE

The publication of this book was made possible by grants from the Vereniging voor Terpenonderzoek and the Stichting Nederlands Museum voor Antropologie en Praehistorie.

Cover: The disc-on-bow brooch from Wijnaldum-Tjitsma (photograph: Fries Museum, Leeuwarden), sod structures around 1925 from Sumarderheide (photograph from the collection of Museum Smallingerland, Drachten), drawing from the Wijnaldum excavation, various objects from the excavation (photographs: M.Ydo). Design: J. de Koning.

Translation or correction: Christine Jefferis, Zaandam
Illustration: J. de Koning (AAC)
Photography: M. Ydo (AAC)
unless otherwise stated

Published by
A.A.Balkema, P.O.Box 1675, 3000 BR Rotterdam, Netherlands
Fax: +31.10.4135947; E-mail: balkema@balkema.nl; Internet site: http://www.balkema.nl

A.A.Balkema Publishers, Old Post Road, Brookfield, VT 05036-9704, USA
Fax: 802.276.3837; E-mail: info@ashgate.com

ISBN 90 5410 488 0

Printed in the Netherlands

Contents

Preface

On a cool day in August 1990, four people met in a field near Wijnaldum, a small village north of Harlingen in the province of Friesland (the Netherlands). This curious meeting in the middle of nowhere, on what connoisseurs would recognize as a *terp*, a typical settlement mound, would have far-reaching consequences: it led to a big excavation programme, the revival of the research of early medieval Frisia and eventually to this book. So, what was discussed there?

Wijnaldum already had a certain reputation in archaeology. Here, in the same field on the Tjitsma terp where the meeting mentioned took place, the foot plate of the 7th-century gold cloisonné disc-on-bow brooch was by chance brought to light in the 1950s. The brooch is known as one of the finest pieces of early medieval craftsmanship ever found in the Netherlands. No wonder the discovery in 1990 of some fragments of the missing part of the brooch and of several other metal objects from the Roman period and Early Middle Ages by metal detecting on the surface of the Tjitsma terp aroused excitement in archaeological circles, for several reasons. Thanks to Jan Zijlstra, who identified the objects and recognized the importance of the finds, Jurjen Bos of the Biologisch-Archeologisch Instituut (BAI) of the University of Groningen,[1] at that time responsible for the archaeology of the province of Friesland, was informed. He realized that the status of the terp as a monument protected by law did not prevent man-made and natural erosion. Even the archaeological context of the deposition of the famous brooch, which possibly contained crucial information on the background of the political and economic power of the Frisians in the 7th century was obviously affected. Some action was inevitable. Bos consulted his colleagues Anthonie Heidinga and Jan Besteman from the department of Medieval Archaeology of the Institute of Pre- and Protohistorical Archaeology (IPP)[2] of the University of Amsterdam, who had ample experience with settlement archaeology in the Frisian territory to the west of lake IJsselmeer. So the three met at Tjitsma, together with Jan Zijlstra who had rediscovered the site. Here it was decided to combine forces to excavate part of the terp.

On 15 April 1991 the excavation started and it was finished in June 1993, after 18 months of fieldwork. It was carried out by both institutions mentioned, directed by Jurjen Bos, Jan Besteman and Anthonie Heidinga. Jan Schoneveld and Danny Gerrets were responsible for the daily decisions and procedures as site supervisors in respectively 1991-1992 and 1993.

We would go beyond the scope of this preface if we would mention by name all persons who participated or to whom we are otherwise in debt, and if we were to describe fully all logistic, financial and publicitary aspects of this big and complex enterprise. The first exception should be made for Jan Bergsma and his family, the residents of Tjitsma, who not only allowed the excavation in their backyard but also heartily supported it. They were confronted by the presence of dozens of archaeologists, the press and, during a weekend in 1991 for instance, by 3500 curious visitors, which must have been quite a shock when accustomed only to the whistling of the wind! Not to mention the fact that the regional broadcasting service, Radio Fryslân, weekly stressed the international importance of their Tjitsma.

Indeed, many people were involved in the excavation campaigns – the maximum during a few weeks in 1992 was about 100 participants – and quite a lot of effort was made to communicate with the outside world by press, radio, periodical 'open days', an exhibition in nearby Harlingen and by publishing a booklet about the excavation.[3] Apart from the professional staff and the archaeology students, many amateur archaeologists[4] and volunteers, mainly from the vicinity, who made their first steps in archaeology here and ended up as semiprofessionals, contributed to the success of 'Wijnaldum'. So an effective organisation structure had to be created to cope with such a large group of participants and with the complex activities, not only in the field but also in the residence at Harlingen. Here three houses were rented, where most of the participants were lodged and which functioned as the administrative headquarters. Part of the crew worked here daily at cleaning and processing the find material and at computing the field data in order to get most of the documentation finished during the digging campaign. Here Tineke Volkers swayed the sceptre together with Truus Anema.[5]

A few words should be said about the way the excavation was financed. Sponsoring played an important role, which was rather unusual in Dutch archaeology. It soon became obvious that the costs would exceed the funds obtained from both universities, the Province of Friesland and the municipality of Harlingen. A foundation, the Stichting Archeologisch Terpenonderzoek Wijnaldum, was established to support the Wijnaldum research and a sponsor-finding agency, Jan Walburg, was attracted to raise funds. On the other hand, the archaeologists themselves succeeded in arousing public enthusiasm, if not in creating a national Frisian hype. It resulted in financial contributions by almost 40 foundations and private enterprises, not to mention the contributions by private persons who for instance bought the so-called Wijnaldum coin, struck especially for this purpose at the Rijksmunt by the provincial governor, J. Wiegel.

It was a hard job, those campaigns between 1991 and 1993, but a satisfactory one. Of course, many minor things went wrong, moments of stress did occur, but the scientific and social flexibility proved to be sufficient to overcome most problems. However, a long, heavy and lonely path still had to be run: the analysis and publication of the huge amount of data produced by the excavation. Jurjen Bos and Danny Gerrets took the lead in this phase by coordinating the group which was involved in data analysis. The latter and Jan de Koning did a major job by establishing a firm chronology of the habitation phases of the terp, which has been used as a framework by all other researchers. Now the time has come to show the yield of these efforts.

So this book contains the first results of 14 months' digging at Wijnaldum-Tjitsma. The contributions have been written by a heterogeneous group of specialists. Some of them are at the start of their career, often being grasped by the subject by participating in the excavation at Wijnaldum itself, others are experienced scholars, attracted on account of their specialist knowledge. After an introduction on the why of this excavation, the questions that were asked and the expectations we had, they will present the actual evidence about almost ten centuries of habitation on this terp in Westergo, as the region has been called since the early days. The summary at the end will certainly not be the final word about the settlement, for only part of the dataset has been worked out. We are convinced however of the importance of publishing facts from archaeological fieldwork as soon as possible. In this way the discussion on the meaning of the site will be broadened in an early stage and archaeology can profit now by new evidence on artefact categories and other general aspects produced by this site.

Finally, it took more time to finish this volume than we, naive as we are, have predicted. We thank the publisher and all those authors who kept previous deadlines for their patience.

The editors

NOTES

1. Since 1995: Groningen Institute of Archaeology (GIA).
2. Since 1998: Amsterdam Archaeological Centre (AAC).
3. J.C. Besteman, J.M. Bos & H.A. Heidinga 1992. *Graven naar Friese koningen. De opgravingen in Wijnàldum*, Franeker. A second edition was published in 1993.
4. Most of them were members of the *Argeologysk Wurkferbân van de Fryske Akademy* and of the *Archeologische Werkgemeenschap voor Nederland*.
5. Truus Anema was one of the dedicated volunteers from the village of Wijnaldum. She died before the excavation had finished in 1993.

The Wijnaldum excavation: Searching for a central place in Dark Age Frisia

H.A. HEIDINGA

1 INTRODUCTION

Wijnaldum is a small village, built on an raised mound called a *terp*, in the north-western part of the province of Friesland, in the region which was called Westergo in the Middle Ages. It belongs to the municipality of the nearby town of Harlingen, the seaport of the province. Like many villages in the countryside, in the second half of this century Wijnaldum has lost most of the communal and economic functions that such places used to have in the past,[1] a fate which was reinforced by the short distance to Harlingen. The modest church, around which the main part of village is concentrated, symbolizes better days. As far as we know from written sources, the place has never played an outstanding role in the region. So, why introduce Wijnaldum here?

To the east of the village a row of serried, deserted terps, adjacent to a few scattered farms along the Voorrijp road (Fig. 1), suggests however that the historical village as a nucleated hamlet on the most westerly terp was once preceded by 'another' Wijnaldum: an elongated settlement situated parallel to a stream called Ried (nowadays the Sexbierum canal). Indeed, these rather low terps, which for the greater part escaped levelling by commercial quarrying in the 19th and first half of the 20th century but are now threatened by modern agricultural practices, contain valuable archaeological information on a remote past of Wijnaldum from the Roman Period onwards, as was shown recently by the archaeological research of one of them. Although Wijnaldum suffered a period of decline, if not of non-existence, in this 'prehistoric' past as well (in the Migration Period), its wealth and prosperity during the Early Middle Ages, especially during the 6th-8th centuries, do not fit the modest image of the later Wijnaldum at all. This Golden Age of Wijnaldum coincides with the period in which Frisia emerged as an important factor in Northwest-European economy and politics.

In 1991 the village became 'big news' in the regional press: the seat of the legendary Frisian kings of the Dark Ages was claimed to have been discovered here by a team of archaeologists from the universities of Groningen and of Amsterdam.[2] A booklet that was published shortly after by the excavators was less explicit about these kings, but its title: *Digging for Frisian kings* [3] suggested nevertheless that the archaeologists were on their track. Anyway, thousands of enthusiastic people visited the windy terp, called Tjitsma,[4] east of the village where the three-year excavation campaign took place. Also in archaeological circles the name of Wijnaldum aroused great expectations.

Now the time has come to be called to account scientifically: what is the archaeological truth about Wijnaldum? This volume contains the first reports on the excavation. One more volume will follow in the next few years. The articles deal with the sober archaeological reality, with various subjects like landscape reconstruction, the structure and periodicity of the settlement, pottery, glass, leather, brooches, coins, metal work, wood, botanical evidence, human remains etc. Although many exciting new results have been produced, it is too early yet to determine the full meaning of the site. In the present contributions kingship is hardly the issue, although Frisian aristocracy is explicitly or implicitly present in the background, for it is obvious that Wijnaldum was not an ordinary settlement, at least not in the 6th-8th centuries. How extraordinary and special it was cannot yet be said. The main problem is, and will probably be for a long time, the representativity of the site. Only a part of terp Tjitsma was excavated, so a very limited section of the settlement 'Wijnaldum' (i.e. the unit of the nine aligned terps) is known. The stray finds from the other Wijnaldum terps, however interesting they may be, provide insufficient information about supposed social and economic differentiation be-

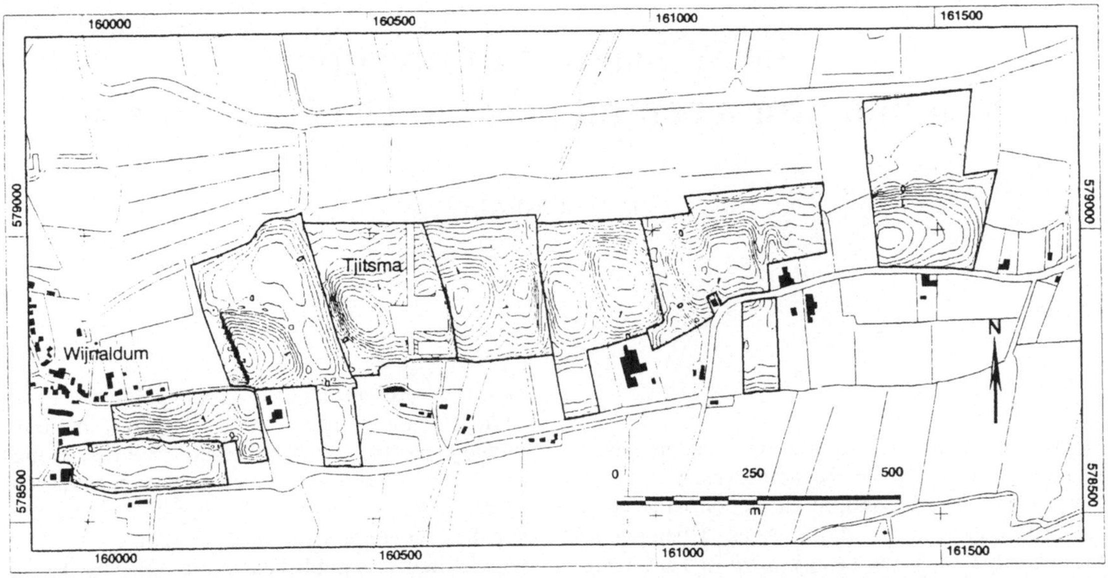

Figure 1. Wijnaldum: the terp row to the east of the village. Only plots of arable land were selected for this survey, so no contour lines of the northern part of the village terp are presented. The location of the excavation at Tjitsma is indicated (after RAAP consultancy, Amsterdam).

tween the various parts of the settlement. And, as will be explained later on, there are hardly any excavated settlements in the Frisian terp region as a whole with which to compare Wijnaldum. Although this lack is compensated to some extent by a huge quantity of finds from many terps that were quarried off in the first half of the century or were browsed recently by metal detectors, the place of Wijnaldum within the settlement system cannot yet be defined with certainty.

However this may be, the archaeological profit from this excavation is enormous. It has not only produced a refined chronology of many artefact categories (which at least made many archaeologists happy), but – more importantly – it also provided a cross-section through the daily life of the people that lived here in the Frisian core area almost continuously from the 2nd till the 11th century. We were informed about their cultural traditions and roots, their surprisingly modest demands on housing in contrast to their wealth, the way they used the environment for a livelihood, the products of their artisans, the way they dealt with their dead and even about the DNA of some of them. Most important is the information on their extra-regional networks which was shown, for example, by coins, metal work, glass and huge amounts of imported pottery. One object, a precious brooch, though made in Friesland (perhaps at Wijnaldum), was manufactured by a top-artisan who probably had his training at some royal court in Anglo-Saxon England.[5] This brooch, the major part of which was already found in the 1950s, was the direct inducement to the excavation and to the rumours about kingship.

In this contribution it will be explained why the excavation took place, which questions were asked and are to be asked, and how this research in the neglected core area of the Frisian realm led to the birth of a major research project on Dark Age Frisia as a whole.

2 WHY THIS EXCAVATION?

The choice for what is euphemistically called 'protection ex-situ' was made for several reasons. In the first place, excavating the site was the best if not the only thing to do, for no law could prevent the farmer from ploughing his fields, and thus from destroying the upper layers of the terp, and no police force could be found to watch Tjitsma effectively. This 'defeat' could be compensated in a way by using this excavation to analyse the process of erosion by agrarian activities in this type of sites in general. In the second place – if not in the first place – research of a site like Wijnaldum was not only desirable but also necessary for scientific reasons. In Dutch archaeology, which nowadays is focused more on protecting sites for future generations than on the past itself, such considerations have become almost illegal. Archaeologists, however, are not in this world for the sole purpose of guarding the archaeological heritage, they are also obliged to rebuild the past and to tell about it. Therefore this opportunity to become informed about a central place in the core area of Frisian achievement could not be neglected. As will be elucidated below, Dutch archaeology has to make up arrears with respect to the Frisians, whose contribution to European history in the Early Middle Ages can hardly be overestimated. The rise of Dorestad, for instance, cannot be understood without some knowledge of what was going on in the terp region, especially in the province of Friesland, in Westergo and Oostergo.

The truth is that archaeological information about this region is astonishing defective. Although the Fries Museum at Leeuwarden contains one of the richest archaeological collections of the country, one of the best monographs ever written in Netherlands: P.C.J.A Boeles, *Friesland tot de elfde eeuw. Zijn vóór- en vroege geschiedenis*, published already in 1927 (2nd revised edition in 1951), deals with the archaeology of the province of Friesland, and modern settlement archaeology was invented, so to speak, in the terp region by A.E. van Giffen in the 1930s,[6] the wave of settlement archaeology of the last decades has almost by-passed this area, especially Westergo and Oostergo. Dutch archaeology can only partly be blamed for that, for the main reason for the archaeological silence in the region is the fact that there appeared to be no necessity to dig in the unchallenged and now legally-protected remains of the terps. As a matter of fact there were some excavations,[7] but the Biological Archaeological Institute at Groningen, which was in practice responsible for the archaeology of the northern provinces, never developed a diachronic research programme for the terp region as it did for the sandy province of Drenthe. Besides, due to the main interests at the time (environmental and subsistence issues, the provenance of the first settlers etc), there was no special preference for the province of Friesland, i.e. Westergo and Oostergo, nor for the most intriguing period of Frisian history: the 'golden age' of the 6th-8th centuries in which these core regions must have contributed greatly to the Frisian achievement of the Dark Ages. Although the volume of the *Berichten van de Rijksdienst voor het Oudheidkundig Bodemonderzoek* from 1965-1966, which contained a synopsis of the prehistory (H.T. Waterbolk), Roman Period (W.A. van Es) and of the Early Middle Ages (H. Halbertsma), was dedicated to Friesland, Dutch archaeology in general seemed somewhat unwilling to explore the enormous potential of the region. An exception was Herre Halbertsma, who started his career during the early 1940s by drawing up an inventory of all terps in the provinces of Friesland and Groningen,[8] and never stopped paying attention to Friesland and Frisian affairs. His contributions in the field however are mainly confined to churches.[9]

Two dissertations in the early 1990s again put the Frisian terp area on the agenda: the study by Gilles de Langen on Oostergo in the Middle Ages and Egge Knol's review of the northern coastal districts in the Early Middle Ages.[10] Especially Knol's study demonstrates the outstanding position of Westergo and Oostergo in the North during the Merovingian Period.[11] The renewed interest in the terp region is also shown by Taayke's recently published dissertation on hand-made pottery from the Iron Age and Roman Period from the area.[12] These works are mainly based on available data from the archaeological archives. However new activities were also undertaken in the field, like the excavation of an early-medieval cemetery at Oosterbeintum in Oostergo in 1987.[13] The time was now ripe for settlement research in the Frisian heartland.

The arguments for excavating the threatened terp of Tjitsma convinced the national board of cultural heritage, so the permission to excavate part of the monument was given.

3 HOW TO EXCAVATE?

Excavating an extremely complex site like a terp is not an easy job. Besides, it was obvious that this particular excavation of a unique monument should be exemplary. The maximum of information should be obtained. It was realized that an effort like this research would transcend nowadays the capacity of any single academic department. So forces were combined: the excavation was to be a joint venture of two institutions, the BAI (now called GIA) of the University of Groningen and the IPP (now AAC) of the University of Amsterdam. At that time Dutch archaeological institutions still operated rather independently and cherished their own traditions and methods, so this 'marriage' was in a way

Figure 2. Work in progress at the Wijnaldum excavation in 1991. Searching for microfinds by means of waterscreen sieving.

an adventure, which worked out very well. It should be stated here that the partnership as such and the fact that it really worked were largely due to the good personal relations of the excavation directors,[14] who had worked together in the past.

The first problem to be solved was: how to excavate the site? Experience and routine in this type of research were actually lacking. Terp-archaeology in the Netherlands, in contrast to the 'flat' settlement archaeology of the sandy soils, had hardly developed methodologically since the days of Van Giffen. This site proved to be even more complex than Van Giffen's Ezinge, for no wooden structures were preserved and hardly any postholes were found: flimsy fragments of turf-walls were the only convincing indications of buildings. So new strategies and methods had to be introduced, as Gerrets amply describes in his contribution about the methodological aspects of the excavation. He explaines for instance how very fruitful the use of metal detectors and several types of sieves has been, but also how the introduction of some advanced computer systems did not actually fulfil the great expectations (as proves often to be the case when naive archaeologists trust that all problems will be solved in this way!).

4 QUESTIONS TO BE ASKED

In 1988 a remarkable volume of thirteen studies on Helgö, the famous rich farm, central place or trade centre, the supposed precursor of Birka from the Migration, Vendel and Viking Period, was published 30 years after archaeological excavations started at this island in Lake Mälaren in central Sweden. In it the results and changing interpretations from the last decades were summed up and a number of scholars, outsiders as well as archaeologists who worked on the project, were invited to give their present view on the meaning of the site.[15] A great variety of opinions about the economic basis, the socio-political and ideological context and the place of the site on the evolutionary ladder (proto-urban or not) passes the review, after which Kristina Lamm concludes: 'This shows that evidence can be used to prove whatever hypothesis one wants. So I am convinced that we shall never reach a definite conclusion on the question of what Helgö really was and we must simply accept that, however rich the material, archaeological evidence cannot give us a true picture of life at that time.' And further on: '...we can only speculate about Helgö's role in central-place development and the growth of political organization in central Sweden.'[16]

Why this depressing introduction when we optimistically start to ask the proper questions about Wijnaldum hoping to find the truth about it? In the first place because Helgö and Wijnaldum have much in common: same period, craft production, many imported goods suggesting international contacts. So the same sort of questions can be asked, although Wijnaldum cannot stand in the shadow of Helgö when one looks at the astonishingly rich finds of the latter place, including a 6th-century statuette of Buddha, a silver bowl, a Mediterranean ladle and an Irish crozier from the 8th century (not to mention the coin hoards). There is another parallel: we too jumped to conclusions about the status of Wijnaldum, dazzled as we were by the splendid gold-cloisonné brooch, as did the excavators of Helgö at first on the grounds of the rich finds from Building Group 2. What arguments are left without such lucky shots? In the same volume Arne Johansen argues that proper concepts and relating criteria, as well as contextual data are indispensable for reaching any valid interpretatation.[17] So the Helgö debate is a useful lesson, a mirror so to speak, for archaeologists who are at the very beginning of interpreting an 'elite-site'. We can only hope that Wijnaldum will eventually be the subject of a similar international debate.

In spite of Johansen's warning not to go into the field without proper concepts, many questions concerning the meaning of the Wijnaldum-Tjitsma site were asked only during and after the excavation (as is mostly the case in archaeology). From the start, however, it was clear that a supra-regional perspective was needed and that not so much the question of how the people succeeded in surviving and keeping dry feet in the salt-marsh should be focused on, but how they were involved in and contributed to the political and economic growth of Frisia in the Early Middle Ages. Dealing with the settlement primarily as an *agrarian* unit, which is the perspective of most settlement archaeology, would not do justice to the real meaning of the site. Farming, cattle-breeding and fishing were interconnected here with activities that went far beyond the scope of a subsistence economy. Obviously the people from Wijnaldum were not wholly dependent on local resources, they did not stand apart (as is the general idea of farming communities) from the great events of their day and they, or at least part of them, participated in networks that extended far beyond the region: directly or indirectly to the Frankish realm, Northern Germany, Scandinavia and England. So, depending on the sort of questions to be asked, different contextual levels should be taken into account: the micro-level of the immediate catchment area or territory of the settlement, the 'nuclear region' of Westergo,[18] the terp region in a wider sense, the 'Frisian' realm and, at the highest level, the North Sea realm and the continental hinterland.

The Wijnaldum research should be considered as a contribution towards solving one of the most intriguing problems of the early history of the Low Countries: how was it possible that this swampy outskirt of civilized Europe produced a *unique group of middlemen in western Europe*,[19] who succeeded in dominating the long-distance trade on the great rivers of Northwest Europe and the North-Sea litoral? In the 8th and 9th centuries, Frisians are attested from Birka in the north, York and London in the west to Strasbourg or even Rome (where a Frisian pilgrim colony was founded in 779) in the south. Although Frisian trade received a lot of attention from historians and archaeologists,[20] very little is known of these people, their organization, socio-political position and their relation with home, i.e. Frisia.[21] Being in the ambiguous position of being protected on the one hand and considered as pariahs on the other for dealing with commerce,[22] and wandering between different cultural systems, thus probably acting in different roles, they obviously did not fit into the world view of our informants about the period, the clergy. But in a way they also escape our historical-anthropological view of early medieval society with its stress on non-commercial gift-exchange. The Frisians and their fellow tradesmen were, so to speak, the representatives of the hidden counterpart of the world as described in Beowulf.[23]

It is a challenge and an obligation to archaeology (especially Dutch archaeology) to investigate the background of the Frisian achievement: which conditions led to it and where and when did it start? It is likely that the outburst of commercial activities in the 8th century had a preamble in Frisian society and economy, probably connected with political growth. In the great days of Frisian trade, Dorestad on the southern border of the Frisian realm was its undisputed centre. As the excavations at Wijk bij Duurstede show, this *emporium* was one of the largest proto-urban conglomerates in Western Europe.[24] The region, called at the time *Fresia citerior*, was also the stage of the Frisian-Frankish struggle for power in the late 7th and early 8th century, and here the only historically known Frisian 'duces' or kings were mentioned (Utrecht is mentioned as their residence). Historical information in general on the Frisians, the weather in the Roman Period or in the heyday of the Frisian expansion, or in the period of the Viking assaults and the rise of the Dutch (then called 'Frisian') counts, is mainly restricted to the western part of their habitat, i.e. the frontier zone with the civilized world. Therefore the misconception could arise that the central part of the Frisian realm, the terp region of Westergo and Oostergo, did play a secondary role in the Frisian achievement. The fact, however, that in the province of Friesland more gold jewellery and coins from the 7th century were found than in any other region in the Netherlands, suggests another story. It seems likely that Central Frisia was the main breeding ground of the Frisian upheaval. It is also likely that Wijnaldum was involved in the rise of Frisian power. The roots of this development are to be sought for in the 5th-6th centuries, for the Frisian society of the Roman Period hardly contributed to the society of early medieval Friesland.

Thus the terps are again on the agenda. Research is resumed now from a new and stimulating perspective. We have to awaken this central coastal district, which was sleeping in a period when, in the adjacent German terp districts, major efforts were being made in the field of settlement and cemetery research.[25]

Some aspects will be dealt with here in more detail: who were the Frisians, in what environmental conditions did they live, how were they organized politically and what was the relation between political structure and economic behaviour? Then we have to look at Wijnaldum to interpret the data against this background.

5 THE BACKGROUND OF WIJNALDUM

5.1 *Who were the Frisians?*

According to the *Lex Frisionum*, the law of the Frisians which was written down probably in 802/3, but anyway after the Frisian realm was incorporated into the Frankish empire, Frisia consisted of three parts. Westergo (the region of Wijnaldum) and Oostergo formed the central part, between the Vlie (a big stream between the present provinces of Friesland and North Holland) and the Lauwers (a river on the border of the present provinces of Friesland and Groningen). It was conquered by the Franks in 734. Eastern Frisia stretched from the Lauwers to the Weser, Western Frisia from the Vlie to the Sincfal, a stream somewhere in the estuary of the river Scheldt. The inland border of this *Frisia Magna* is not indicated, but from other sources we know that at least the central river area in the Netherlands belonged to it.[26] Although it is likely that most inhabitants also named themselves Frisians at that time and shared for instance language and law (to a certain degree), it is not certain what reality in an ethnic or political sense this tripartite entity between Flanders and the Weser represents. It might have been a convenient arbitrary construction of the Franks after conquering the coastal area up to the Weser. It certainly did not include all the people that were called Frisians, for Frisians were also attested outside the Weser border, on the coast of Schleswig-Holstein, so-called North Frisia, which was colonized by them in the 7th century.[27] On the other hand there is ample evidence for non-Frisian elements in law and language, specially in Western Frisia south of the river Rhine, in *Frisia Citerior*. However, one thing is sure: within this Frisia Magna in the 7th and 8th centuries the Franks met a strong 'Frisian' political-military power, which suggests political coherence in the great parter of the Frisian realm at least.

Historical archaeology is favoured but also hampered by the fact that it provides quite different types of sources. If no historical sources were available, there would probably be no Frisian problem, maybe no Frisians at all, for at first sight there are hardly any archaeological arguments for a specific coherent material culture in the area. On archaeological grounds, the northern coastal part of Frisia in the 5th-7th centuries would have been ascribed to the North-Sea or 'Anglo-Saxon' sphere of influence, and the greater part would have been regarded as the periphery of the dominant, 'Frankish' culture. Now, confronted with historically attested Frisians, we are challenged to face the complex and confusing reality behind such a label. We have to tackle the question of how the people we are dealing with defined themselves and how they were labelled by the outside world. We have to be aware of the fact that quite different processes – colonization/migration, long-term acculturation and short term political dominance – could independently or interconnectedly have produced a unity as is suggested by the historical sources. An entity, it should be repeated, of which it is not clear if it should be defined in ethnic, political or other terms. Archaeology however has the potential to trace the material evidence of communication and identity within the Frisian realm and could attribute to the understanding of the nature of proto-historic Frisia, if only the character, duration and context of the transmission of goods and ideas could be established. The problem is that material culture speaks with different tongues, depending on the situation and context, as we have learned from post-processual archaeology. So it will be hard to select 'Frisian' messages from it.

Anyhow, an historical-anthropologically based understanding of the binding and separating forces in those kinds of societies is required, which leads inevitably to the discussion on the tricky subject of ethnicity. It cannot be doubted that there were Frisians in an ethnic sense (i.e. people who considered themselves as belonging to a *gens* called Frisians), but that does not necessarily mean that this identity was articulated visibly for outsiders like archaeologists. Nor did it mean that this identity and the labels and symbols attached to it were durable and inalienable. Wenskus and, more recently, Geary and others stress the fact that early medieval ethnicity was not a given property, but was in fact generated by the leading echelons of society, who needed a god-given legitimacy for power and a common identity to bind their followers.[28] Because ethnicity was a situational construct, as Geary put it, it is very difficult to tackle it archaeologically. As will be described in the contribution by Gerrets and De Koning, the discontinuity of habitation at Wijnaldum and the scarcity of convincing archaeological evidence in the Frisian terp region and in the other parts of Roman Frisia in the 4th early

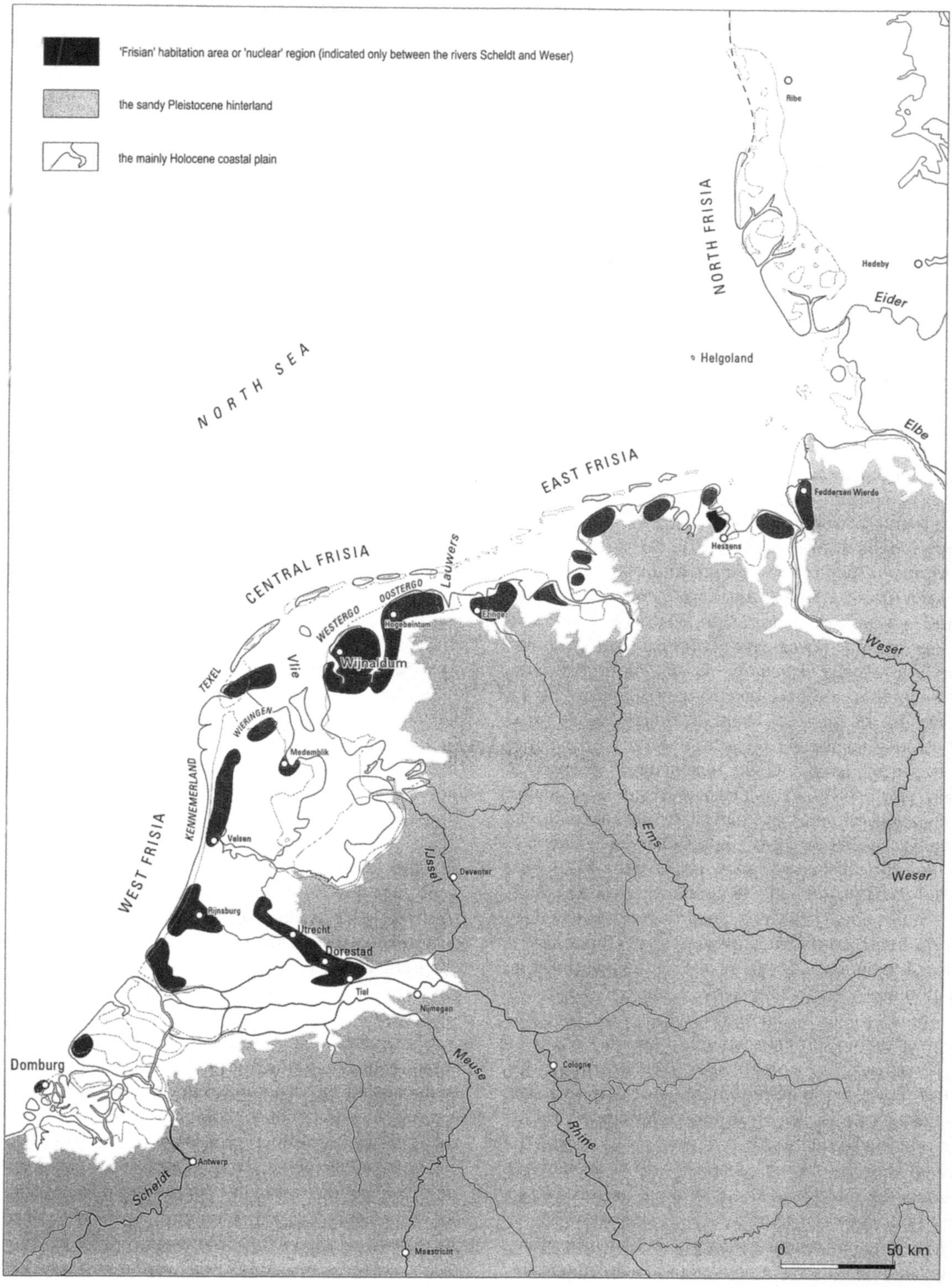

Figure 3. Palaeogeographic reconstruction of Frisia between the river Scheldt and the Danish border in the 7th-8th centurry. The coastal outline between the river Elbe and Ribe is only roughly indicated (drawing AAC).

5th century forces us, in the absence of remaining groups that would have been able to carry on the tribal tradition,[29] to assume that even an ethnic name could be appropriated by newcomers: in the case of the Frisian heartland probably by Anglo-Saxon or Jutish war-leaders from northern Germany or the Cimbric peninsula. Obviously it was still common knowledge that their new habitat was the land of the Frisians. The suggestion is that already in the 5th century the inhabitants of the southern border of the North Sea were predestined to be Frisians by the mere fact that they lived there.

So archaeology shows that the renewed settlement of the Frisian area took place in the 5th century in the terp region. This 'central' Frisia also became the most densely populated part of it (as it was already in the Roman Period). Only in the central river area (the heart of the later *Frisia Citerior)* and on adjacent sandy soils of the central and eastern Netherlands can a thin line from the Roman to the medieval period also be shown. On archaeological and historical grounds it is likely that the Frisians of the terps played an important role in the re-population of West, East and North Frisia in the 6th, 7th and 8th centuries respectively.[30] But also other groups (probably including autochthons) were involved, especially in the West, near the lower courses of the rivers Rhine and Meuse, as the diversity of cemeteries and settlement types from the 6th-7th centuries suggest.[31]

So a major part of the expansion of Frisians in the Early Middle Ages seems to have been a matter of colonization of waste lands, a phenomenon which was repeated on a large scale and under different circumstances during the period of the great reclamations from the 10th century onwards. This type of expansion can explain some shared traditions and a common language, but does not implicate any socio-political coherence and interaction within the Frisian realm.

5.2 *Political power*

Did Frisia exist in a political sense and how and when did political power come into being? Before answering these difficult questions it should be asked how this type of society could have been structured politically on the basis of our knowledge of contemporary societies in Northwest Europe. The society of the terp-Frisians is often described, especially by German scholars, as a sort of *Bauernrepublik.*[32] In our view an anachronistic idea, inspired by the later 'Frisian freedom' and seemingly supported by the 'democratic' chieftains society of Viking Age Iceland. Anyway, the model conflicts with the scarce historical evidence we have. Apart from two Frisian kings in Nero's lifetime and the legendary king Finn from the migration period, mentioned in Anglo-Saxon poetry,[33] there were powerful Frisian leaders at the end of the 7th and early 8th century: Aldgisl (mentioned in 678), his famous successor Radbod (who died in 719), and finally Bubo (defeated in 734 at the river Boorne). For obvious reasons they were called *duces* by the Franks, but the Anglo-Saxon sources speak of kings, which they were without doubt in the eyes of the Frisians. The political and especially the ritual function of kingship must have been indispensable in these kinds of societies.[34] When we meet kingship in all related societies of the period, it is hardly conceivable that the Frisians, even those from the terps, could manage without.

Kingship of this type involves military, political and ritual leadership, a strictly personal relationship with leader and *Gefolgschaft*, sustained by gift-giving, but does not implicate political unification within the Frisian realm or any form of state formation. Possibly the same development took place in Frisia as is assumed for related Anglo-Saxon England: the emergence of minor kingdoms (not exceeding the extent of the *nuclear regions*)[35] in the 6th century and a transformation into larger units in the 7th century.[36] Being dependent on face to face relationship with the followers who came to the king's hall or were visited on the king's circuit, kingship was naturally small-scale.[37] However, Frisia was a maritime society (see below), so the mobility of power, provided by ships, could have generated quite a different geographical dimension of kingship. The description of Viking Age Norse kingship by early Icelandic literature (in particular the Egil Saga), in which the ship played a major role, may be very well applicable to the Frisian situation.

The extent of political power was in the first place dependent on the personal (military) success of the leader. Now it is obvious that the 7th-century leaders like Radbod were highly successful. The fact that their unknown predecessor could conquer *Frisia Citerior*, which was probably all Frankish territory, somewhere in the 7th century (around 650 AD?) and that Radbod could make a stand against the Frankish empire (till 719), is a proof *ex silentio* that their resources could not have been restricted to the central river area around their residence at Utrecht, but encompassed a great part if not the whole of *Frisia Magna*.

The late 'frisification' of *Frisia Citerior* and the paganism of the Frisian kings make it highly improbable that their roots should be sought in the central Netherlands. The conquest of this area with its imperial symbols in the form of Roman *castella* could however have contributed to the consolidation of power and augmenting of status, if not the creation of a new role after the Frankish model, of the successful royal warlords in question.[38] Outside this area, the most likely region where political power could have been bred is the rich and densely populated Frisian terp region. The very beginning of it probably lies in the 5th century, when a new start in Frisia was made, presumably by Anglo-Saxon war-leaders, who left a world of political hierarchy and competition, who had to consolidate their position in a foreign country and who had close networks with elites of England and the northern world. Networks that were maintained by means of shipping. The hiatus of the 4th century implied more than a gap in settlement: not only political power but also the remarkable role of the Frisians on the sea were born in the 5th century.

5.3 *Frisian trade*

In the lengthy debate on the roots, nature and organization of early medieval long-distance trade (which we will not repeat here), there is common agreement about political power as the main force behind it. In the tribal world of northern, Germanic Europe in the Merovingian Period, trade was one of the means of obtaining those luxury goods with which the ruler could perform his indispensable role as gift-giver toward his clients, allies, ancestors and gods. The small-scale exchange performed by clients and middlemen has to be considered as a socio-political network connecting the different Germanic powers, rather than a commercial trade system.[39] In the late 7th or 8th century, the system changed: bulk commodities began to appear, specialized merchants appeared and proto-towns like Dorestad emerged. The role of the king however remained decisive as lord of the *emporia*, protector of the tradesmen and as main profiteer of trade (tolls and taxes). A real market system did not yet exist.

How to fit the Frisians, and especially the terp-Frisians into this picture? The Frisians, who were mentioned on the stage of international trade as early as the late 7th century, soon dominated the traffic from the Frankish empire to Scandinavia and parts of England.[40] It is generally assumed that Dorestad was the focus of this Frisian trade, but most authors on the subject suppose that the skippers and traders had their roots in the terp region.[41] Indeed, this area offered the main prerequisites for being the homeland of long-distance trade, as will be elucidated below. There is every reason to assume that existing socio-political networks and shipping/trading activities by the Frisians on the North Sea and beyond were adapted into the 'second phase' of long-distance trade, now under Frankish patronage.[42] If the dating of the earliest phase of the harbour area at Dorestad (Hoogstraat) is correct – somewhere in the 3rd quarter of the 7th century,[43] then we have to assume that Dorestad developed into a real trading settlement during Frisian hegemony. Thus the Cosmographer of Ravenna could, at the end of the age, rightly describe Dorestad as the '*patria Frisonum*', the capital of the Frisians.[44] The question is, however, if the beginning of Dorestad really heralded a new type of trade. The high percentage of Frankish pottery in the Frisian homeland from the 6th century onward,[45] suggests that the border between socio-political based exchange of luxury goods and some sort of trade of commodities was not so sharp, even in the pre-emporium period.

This picture of Frisian trade raises many questions with regard to the Frisians of the terp region, the people of Wijnaldum and their neighbours. Were they independent local farmers who from time to time organized trading trips to foreign gateway communities to sell their cloth, combs, salt, hides etc. for good gold and other desirable items, and whose 'surplus' sons decided to stay in Dorestad and other Frisian colonies, as Ellmers presumes?[46] Also Van Es sees the traders who visited the maritime fair of Dorestad mainly as free entrepreneurs, although occasionally acting in the service of the king or church.[47] This view, in which kingship is hardly visible at Dorestad and is completely absent in the Frisian homeland, denies however the presumed strong socio-political nature of trade in Germanic society.

Another question is how Frisian was Frisian trade actually: did the Frisians mainly play an intermediate role between the Frankish realm and the emerging polities at the other side of the North Sea, just because they were excellent sailors and lived in the right place to perform such a role, or has the origin of Frisian trade to be sought in Frisia itself? In other words: was production in Frisia the basis of its success?[48] If this was the case, Dorestad had a twofold function in a frontier-zone: it was not only the gateway from the Frankish realm to the north, but also the doorway of the Frisians to the south. The problem is that a North-South stream of secondary cattle products and other software from Frisia will hardly be traceable in the archaeological record, as are the presumed imports like slaves from Scandinavia and England. A major source for understanding these relations is coinage. We can only hint at this complicated subject here.[49] However, a few remarks can be made. In the first place it is striking that so many gold coins (mainly *tremisses)*, produced in northern Gaul, but especially in Dorestad (Madelinus), were found in the province of Friesland and that they are almost missing in the regions down to the Baltic, i.e. in northern Germany and southern Scandinavia.[50] Then, in Frisia itself, coin production probably started as early as the late 6th century, well before similar developments took place in the eastern North-Sea region and the Baltic, the presumed exchange partners. Whatever role coins played in Frisia in the Merovingian Period, commercial or socio-political, they were used here, in contrast to the neighbouring Germanic world. This could mean that Frisia itself was the main partner of Francia (i.e. Austrasia) at the time, not merely the stepping stone to the Baltic. Apart from possible economic relations it is probable that Francia made political investments in this northern neighbour, as the Romans did before in the Germanic trans-limes tribes.

A final question on the subject is if Dorestad became the main focus for trade from Frisia, in particular from the terp region, as it probably did for Austrasia? It is more likely that two systems stood side by side in harmony: the channelling of products via a 'modern' *emporium* and direct long-distance networks overseas, which were maintained in the old fashion, even when Frisian kings had made way for distant Frankish monasteries.[51] The two trading systems and other means of exchange must have been intermingled in some way, as is suggested by the fact that wine and pepper were among the tributes to be delivered from estates in the terp region to far-away ecclesiastical institutions.[52]

After this lengthy detour, it is time to look at the Frisian terp region of Wijnaldum itself, about which so many suggestions have been made here.

5.4 *The Frisian terp region*

'Good subsistence conditions are the precondition for taking part in long-distance exchange. A productive agriculture is the basis of a large, rich population. A large population means power. And power made a Germanic society an attractive partner on the political scene, where luxury goods exchange was important in the transmission of information.'[53] Well, does the Frisian coastal area fit into this profile of a successful chiefdom as described by Ulf Näsman for Scandinavia in the Germanic Iron Age? In many respects, it does. The first scholar to recognise the great potential of the terp region, especially of Westergo (see Fig. 4) and – to a lesser degree – of Oostergo, was the historian B.H. Slicher van Bath, who wrote an important study on the economic and social conditions in the Frisian districts from 900 to 1500.[54] He stresses the vital importance of having good knowledge of the demographic developments of an area in order to understand its social and political role. Now, on the basis of historical and archaeological data, he came to the conclusion that in the 9th-10th centuries the Frisian terp districts were not only the demographic centre of the Netherlands but their population density was surpassed in Western Europe only by very few regions like the Seine district, Saint-Omar and probably the Rhine region near Cologne.[55] The population of Westergo then would have been about 14,500. Although this figure is rather arbitrary, and some of Slicher van Bath's assumptions may be criticized, his presentation of Friesland as an extremely crowded area is on the whole convincing, as are some of the conclusions he drew on the nature of Frisian society and economy which must have differed greatly from the inland districts. 'These districts must have struck contemporaries as urban rather than rural because of the large number of villages and the enormous proportions of the population assumed by shipping, commerce and industry', thus he roughly depicts the region in rather anachronistic terms.[56] Population surplus, the stress on cattle-breeding (which not only provided food, but also desirable export products like hides, leather, wool, parchment, and bone articles), the need for products that were not available in the salt-marshes, and the maritime way of life, shipping, were in his opinion the intermingled factors behind specialized craft production, long-distance trade, the early development of a money economy and expansion. Indeed, the enormous reservoir of people could very well explain the continuous emigration and, in particular, the expansion in the Early Middle Ages to the western and northern coastal regions and to trading places all over west-

Figure 4. Palaeogeographic map of Westergo in the early Middle Ages. Note the island-like nature of the district and the aligned pattern of the terps (in black) in the northern part. Place-names which are mentioned in the text are indicated (after Vos, this volume).

ern Europe, which recalls the colonizing activities of another maritime society of the past: the Greeks.[57] Slicher van Bath also drew some conclusions on the social structure of the terp region, which are less convincing. He is probably right in saying that a feudal system of the type known from the Frankish heartland was missing here, but that does not mean that clientship with related forms of tribute did not exist. If indeed cattle played a central role in society, they could also have served as a tribute item as they did for instance in early-medieval Irish society.[58]

Slicher van Bath describes the Frisian coastal region at the very end of the Early Middle Ages, but most of the demographic and economic conditions must have existed earlier, if not much earlier. Westergo, for instance, was also a densely populated district in the Roman Period[59] and then too tribute in the form of hides was paid (to the Romans), but obviously conditions for developing long-distance trade and large-scale shipping overseas were still missing.[60] Anyway, early-medieval society in the region had to be built up again from the 5th century onward (see above). In this period the ship appeared as new trump of the terp districts, especially when the sail came into use somewhere in the 6th century.

It is generally agreed that the coastal Frisians' sailing skill was the first prerequisite for their success as long-distance traders and that the special landscape, surrounded by water and cut by many gullies, had given rise to the tradition of sailing.[61] The great impact of living and working in a maritime cultural landscape, as for instance the inhabitants of Westergo did, has never been thoroughly conceptualized and investigated in the Netherlands as it has in Denmark.[62] The duality of life on land as well as on the water put its stamp not only on settlement patterns and subsistence economy but must have also have had great effect on communication, social and political conditions, gender and age-determined division of labour, mentality and the perception of time, space and of life after death. The possible political dimension of shipping has already been mentioned. In this sort of society one has to take into account a dichotomy of extreme cultural isolation on the one hand (where life in the remote salt-marshes is concerned) and a high degree of cosmopolitanism among the travelling population on the other. But by now, apart from iron rivets not even one fragment of boat has been found in the Netherlands that can be related to Frisian shipping in the period concerned. The maritime world of the Frisians still awaits discovery.

The reader might have great expectations of the Frisian terp region in the first millennium, of Westergo in particular. So what has archaeology to offer in building up the real regional context of Wijnaldum? Disappointingly little in the field of excavated settlements and cemeteries, as was said before. The archaeological information from accidental finds and incidental observations during the commercial quarrying activities in the past, and from the increasing number of metal detector finds of the last years might be more crucial. Although a number of surveys are available (see above), an up-to-date evaluation is highly necessary. Two Ph.D. students are working now on Westergo: Kristin Bosma deals with the ecology and subsistence, Danny Gerrets with the cultural and socio-political aspects. In this volume aspects of Westergo in a wider sense will also be dealt with, for instance in the contribution by Peter Vos on the landscape. So we shall confine ourselves here to some general observations concerning the Dutch terp region (provinces of Friesland and Groningen) in the first millennium.

On the basis of the available data, which are summarized in the literature mentioned, one gets an impression of the region which in a way corresponds to the supposed cultural dichotomy. The scanty information we have from the few excavated settlements does not at first sight suggest that any of them were connected with elites, long-distance trade or industrial activities. Interpretation is hampered however by the fact that only part of the settlements were excavated, most of the data are awaiting examination and that only preliminary publications are available (or not). So any conclusion on the nature of the sites is premature. Another problem is that the settlements involved are hardly comparable since they are located in different nuclear regions and represent partly different eras. The sites which provided information about the early Middle Ages are Ezinge (Fivelgo, Groningen), Leens (Hunzingo, Friesland), Foudgum and Driesum (Oostergo, Friesland) and Tritsum (Westergo, Friesland). In all, except Leens, the Roman Period is also present (Ezinge started already in the Middle Iron Age).[63] Anyway, the general impression is that the inhabitants were occupied mainly with husbandry and agriculture, and lived and worked in rather modest farmhouses and sunken huts and hardly had any luxury imports (apart from pottery). Nothing reminds one of a Feddersen Wierde, the large and rich terp settlement from the Roman Period in Land Wursten (Germany). Ezinge, however, might be the exception, for rather unusual imported late-Roman pottery was found (including African redware), as well as a precious gold pyramid-shaped knob from the 7th century. The very phases which produced these interesting finds are however hardly represented or were missed by the excavation.[64]

Knol did an admirable job by summarizing and

evaluating the scattered information on burial practices in the terp region. In a way the cemeteries, which are known from the Migration Period onward, mirror the modesty of the settlements: rather poorly furnished inhumation and cremation graves. Weapon-graves are rare, except at the very end of this type of burial practice, in the 8th-9th century. Whatever statement the Frisians made by mortuary ritual and way of housing, they surely had not the intention to demonstrate their wealth to archaeologists, though wealthy they were, in particular in Westergo and Oostergo. This is evidenced not only by the remarkable quantity of imported pottery from the Frankish realm.

We have already referred to the special importance of coins and mintage in Frisia in the 6th and 7th centuries. This line was continued in the 8th century with the production of the typical North Sea coins, the sceattas. Although there is no real proof of the involvement of the terp region in the production, many of these coins ended up here in the ground, as did gold and silver in general. A considerable proportion was deposited as buried treasure or votive offering, especially in Westergo in the late 6th and 7th century: the hoards of Wieuwerd, Dronrijp, Achlum, Midlum and Scharnegoutum for instance. As Mazo Karras has shown,quite a lot of the fine brooches, pendants, bracteates and other jewellery, which belong stylistically to the Anglo-Saxon/Scandinavian sphere of influence, were probably produced in Friesland itself.[65] This might be an indication that coin production also took place in the region. Also in the Carolingian Period much precious metal, mostly silver coins, went into the ground. The peak of wealth deposition lies however in the second and third quarter of the 7th century, the very period of Frisian expansion into the central river area around Dorestad. The suggestion is that there was some connection between the two. It is one of the many questions waiting for an answer.

Figure 5. Ezinge terp excavation 1933: excelently preserved Iron-Age houses (photo GIA).

6 FINAL REMARKS

In this introduction we have explained why the excavation at Wijnaldum-Tjitsma took place and what sort of questions should be asked of the muddy archaeological remains. In order to get the reader (and ourselves!) on the right track we have depicted the socio-political, economic and environmental background of this settlement in the first millennium AD. From this it became clear however that we are at the very beginning of understanding the world in which Wijnaldum functioned. This is why the Wijnaldum research became the instigation for the *Frisia project*, a joint enterprise by the two partners of the Wijnaldum excavations (GIA and AAC/IPP) and the State Service for Archaeological Investigations (ROB, Amersfoort).[66] The Wijnaldum research is now embedded in a broad project which focuses on different parts of early-medieval Frisia: the Dorestad region, the Rhine and Meuse estuaries in the province of South-Holland, the province of North Holland and the Frisian terp district, Westergo. Thus, in the end it will be possible to establish the position of Wijnaldum in its regional and supra-regional context. But now, we will first focus on the data of the Tjitsma site itself.

REFERENCES

BASSETT, S. (ed.) 1989a. *The origins of Anglo-Saxon kingdoms*, Leicester.

BASSETT, S. 1989b. In search of the origins of Anglo-Saxon kingdoms. In: S. Bassett (ed.), *The Origins of Anglo-Saxon Kingdoms*, Leicester, 3-27.

BESTEMAN, J.C. 1990. North-Holland AD 400-1200: turning tide or tide turned? In: J.C. Besteman, J.M. Bos & H.A.

Heidinga (eds), *Medieval archaeology in the Netherlands. Studies presented to H.H. van Regteren Altena*, Assen etc., 91-120.

BESTEMAN, J.C., J.M. Bos & H.A. Heidinga 1992. *Graven naar Friese koningen; de opgravingen in Wijnaldum*, Franeker.

BIERMA, M., A.T. CLASON, E. KRAMER et al. (eds) 1988. *Terpen en wierden in het Fries-Groningse kustgebied*, Groningen.

BLOK, D.P. 1979. *De Franken in Nederland*, 3rd ed., Haarlem.

BOELES, P.C.J.A. 1951. *Friesland tot de 11e eeuw*, 2nd ed., 's-Gravenhage.

BOS, J.M. 1955. *Archeologie van Friesland*, Utrecht.

BULT, E.J. & D.P. HALLEWAS 1990. Archaeological evidence for the early-medieval settlement around the Meuse and Rhine deltas up to ca AD 1000. In: J.C. Besteman, J.M. Bos & H.A. Heidinga (eds), *Medieval archaeology in the Netherlands. Studies presented to H.H. van Regteren Altena*, Assen etc., 1-90.

CARVER, M.O.H. (ed.) 1992. *The age of Sutton Hoo, The seventh century in North-Western Europe*, Woodbridge.

CHARLES-EDWARD, TH. 1989. Early medieval kingships in the British Isles, In: S. Bassett (ed.), *The origins of Anglo-Saxon kingdoms*, Leicester, 28-39.

CRIELAARD, J.P. 1998. Het aristocratisch karakter van de vroege Griekse kolonisatie, *Spiegel Historiael* 33: 115-120.

CRUMLIN-PEDERSEN, O. (ed.) 1991. *Aspects of maritime Scandinavia, AD 200-1200: Proceedings of the Nordic Seminar on Maritime Aspects of Archaeology, Roskilde, 13th-15th March, 1989*, Roskilde.

CRUMLIN-PEDERSEN, O. 1996. *Archaeology and the sea*, Amsterdam (18e Kroon-voordracht).

DELVIGNE, J. 1984. *Herinneringen aan het graven in de terp van Ezinge*, Ezinge (Streekhistorische Vereniging Middagherland).

DÜWEL, K., H. JANKUHN, H. SIEMS, et al. (eds) 1985. *Untersuchungen zu Handel und Verkehr der vor- und frühgeschichtlichen Zeit in Mittel- und Nordeuropa. Teil III: Der Handel des frühen Mittelalters*, Göttingen.

ELLMERS, D. 1972. *Frühmittelalterliche Handelsschiffart im Mittel- und Nordeuropa*, Neumünster.

ELLMERS, D. 1985. Die Bedeutung der Friesen für die Handelsverbindungen des Ostseeraumes bis zur Wikingerzeit. In: S.-O. Lindquist (ed.), *Society and trade in the Baltic during the Viking Age*, Visby (Acta Visbyensia 7), 7-54.

ES, W.A. VAN 1965-1966. Friesland in Roman Times, *Berichten van de Rijksdienst voor het Oudheidkundig Bodemonderzoek* 15-16: 37-68.

ES, W.A. VAN 1990. Dorestad centred. In: J.C. Besteman, J.M. Bos & H.A. Heidinga (eds), *Medieval archaeology in the Netherlands. Studies presented to H.H. van Regteren Altena*, Assen etc., 151-182.

ES, W.A. VAN & W.A.M. HESSING (eds) 1994. *Romeinen, Friezen en Franken in het hart van Nederland, van Trajectum tot Dorestad 50 v. C.-900 n. C.*, Utrecht etc.

GEARY, P.J. 1983. Etnic identity as a situational construct in the early middle ages, *Mitteilungen der Anthropologischen Gesellschaft in Wien*, B. 113: 15-26.

GERRETS, D.A. 1995. The Anglo-Frisian relationship seen from an archaeological point of view. In: V.F. Faltings, A.G.N. Walker & O. Wilts (eds), *Friesische Studien II*, Odense (= *Nowele* 12), 119-129.

GERRIETS, M. 1983. Economy and society: Clientship according to the Irish laws, *Medieval Celtic Studies* 6: 43-61.

GUREVICH, A.J. 1990. The Merchant. In: J. Le Goff (ed.), *The medieval World*, London, 243-284.

HALBERTSMA, H. 1963. *Terpen tussen Vlie en Eems*, Groningen.

HALBERTSMA, H. 1965-1966. The Frisian Kingdom, *Berichten van de Rijksdienst voor het Oudheidkundig Bodemonderzoek* 15-16: 69-108.

HALBERTSMA, H. 1971. Kerkopgravingen in Friesland, I, *Publicatieband Stichting Alde Fryske Tsjerken*, 53-86.

HALBERTSMA, H. 1972. Kerkopgravingen in Friesland, II, *Publicatieband Stichting Alde Fryske Tsjerken*, 99-120.

HALBERTSMA, H. 1982. Frieslands Oudheid, Groningen (thesis university of Groningen).

HALBERTSMA, H. 1989. Herinneringen aan St. Willibrord in Friesland. In: G. Kiesel & J. Schroeder (eds), *Willibrord, Apostel der Niederlande, Gründer der Abtei Echternach*, Luxembourg, 42-68.

HEIDINGA, H.A. 1987. *Medieval settlement and economy north of the Lower Rhine; archeology and history of Kootwijk and the Veluwe*, Assen etc.

HEIDINGA, H.A. 1997. *Frisia in the first millennium, an outline*, Utrecht.

HERSCHEND, F. 1997. Striden i Finnsborg, *Tor* 29: 295-333.

HODGES, R. 1977. Trade and urban origins in Dark Age England: an archaeological critique of the evidence, *Berichten van de Rijksdienst voor het Oudheidkundig Bodemonderzoek* 27: 191-215.

HODGES, R. 1982. Dark Age economics: the origins of towns and trade, 600-1000 AD, London.

JOHANSEN, A.B. 1988. Helgö so far. In: A. Lundström (ed.), *Thirteen studies on Helgö*, Stockholm, 72-76.

KNOL, E. 1993. De Noordnederlandse kustlanden in de Vroege Middeleeuwen, Amsterdam (thesis Vrije Universiteit).

KNOL, E., W. PRUMMEL, H.T. UYTTERSCHAUT, et al. 1995-1996. The early medieval cemetery of Oosterbeintum (Friesland), *Palaeohistoria* 37/38: 245-426.

LAMM, K. 1988. Helgö – international trading centre, local market or farming community? In: A. Lundström (ed.), *Thirteen studies on Helgö*, Stockholm, 89-99.

LANGEN, G.J. DE 1992. *Middeleeuws Friesland. De economische ontwikkeling van het gewest Oostergo in de vroege en volle middeleeuwen*, Groningen.

LANGEN, G.J. DE 1995. Het Fries kweldergebied in de vroege Middeleeuwen, *Madoc* 9: 172-187.

LANGEN, G.J. DE & H.T. WATERBOLK 1982-1988. De archeologie van Ezinge. De nederzettings- en onderzoeksgeschiedenis van een Gronings terpdorp, *Jaarverslagen van de Vereniging voor Terpenonderzoek* 66-72: 78-111.

LEBECQ, S. 1983. *Marchands et navigateurs, Frisons du haut moyen Age*, Lille.

LEBECQ, S. 1992. The Frisian trade in the Dark Ages. A Frisian or a Frankish/Frisian trade? In: A. Carmiggelt (ed.), *Rotterdam Papers* 7: 7-16.

MAK G. 1996. *Hoe God verdween uit Jorwerd*, Amsterdam etc.

MAZO KARRAS, M. 1985. Seventh-century jewellery from Frisia: A re-examination, *Anglo-Saxon Studies in Archaeology and History* 4: 159-177.

McGrail, S. (ed.) 1990. *Maritime Celts, Frisions and Saxons*, London (Council of British Arachaeology Research Report 71).

Nässman, U. 1991. Sea trade during the Scandinavian Iron Age: Its character, commodities, and routes. In: O. Crumlin-Pedersen (ed.), *Aspects of maritime Scandinavia, AD 200-1200: Proceedings of the Nordic Seminar on Maritime Aspects of Archaeology, Roskilde, 13th-15th March, 1989*, Roskilde, 23-40.

Niermeyer, J.F. 1964. In: W.J. Alberts & H.D.H. Jansen (eds), *Welvaart in wording: sociaal-economische geschiedenis van Nederland van de vroegste tijden tot het einde der middeleeuwen*, 's-Gravenhage, Ch. 1-2.

Rijntjes, R. 1995. Van limes-fort naar missiepost. De Romeinse castella langs de benedenloop van de Rijn vóór Willibrord, *Madoc* 9: 157-169.

Russchen, A. 1967. *New light on Dark Age Frisia*, Drachten.

Schmid, P. 1991. Mittelalterliche Besiedlung, Deich- und Landesausbau im niedersächsischen Marschgebiet. In: H.W. Böhme (ed.), *Siedlungen und Landesausbau zur Salierzeit, Teil 1: in den nördlichen Landschaften des Reiches*, Sigmaringen, 9-36.

Slicher van Bath, B.H. 1965. The economic and social conditions in the Frisian districts from 900 to 1500, *A.A.G. Bijdragen* 13: 97-133.

Taayke, E. 1997. *Die einheimische Keramik der nördlichen Niederlande, 600 v. Chr. bis 300 n. Chr.*, Groningen (thesis Groningen). Also published in: *Berichten van de Rijksdienst voor het Oudheidkundig Bodemonderzoek* 40(1990): 109-222 (= Teil I); 41(1995): 9-102 (=Teil II); 42(1996-1997): 9-86 (=Teil III).

Waterbolk, H.T. 1965-1966. The occupation of Friesland in the Prehistoric Period, *Berichten van de Rijksdienst voor het Oudheidkundig Bodemonderzoek* 15-16: 13-36.

Wenskus, R. 1961. *Stammesbildung und Verfassung. Das Werden der frühmittelalterlichen Gentes*, Köln etc.

Zijlstra, J. 1990-1994. *Friese bodemvondsten*, 5 vols., (type-script).

NOTES

1. The decline of village life in the last decades is emphatically described by Geert Mak, who took the Frisian village of Jorwerd as an example (Mak 1996).
2. Although the excavation team did not go as far as Jan Zijlstra in connecting Wijnaldum to the legendary Frisian king Finn, who plays a significant role in early-medieval English poetry (Beowulf, the Finnburg fragment, Widsith) as the treacherous hero who defeated and was defeated by the Danes, this Finn became a sort of mascot of the excavation. Zijlstra assumes that the poets had replaced *Win(wald)*, the real name of a local Frisian 'king' after whom Wijnaldum could have been named, by the more familiar *Finn* for literary reasons (Zijlstra 1990-1994, passim). Frans Herschend however put forward the hypothesis that the whole story should be situated in the Danish-Jutish frontier zone of the 6th century. The real Finnburg could have been an elite settlement like Dankirke in Jutland (Herschend 1997).
3. Besteman, Bos & Heidinga (1992).
4. The name 'Tjitsma' is probably derived from a family that once possessed this piece of land at the Foarryp east of Wijnaldum.
5. Mazzo Karras (1985).
6. A.E. van Giffen's spectacular and exemplary excavations at Ezinge (province of Groningen) in 1923-1934 made, although the site was never fully published by him, a tremendous impression on international archaeology (see for instance Delvigne 1984). A revision of the interpretation of Van Giffen was recently published by De Langen & Waterbolk (1989).
7. The excavations which are relevant here are summarized by Knol (1993, ch. 4) and in the later part of this contribution.
8. Halbertsma (1963, the report was already finished in 1944).
9. See for instance Halbertsma (1971, 1972). Halbertsma's main contributions to our knowledge of Frisia are to be found in his historical studies like his impressive, but unpublished dissertation *Frieslands Oudheid* (Halbertsma 1982). For a long time, from the 1960s onwards, only the provincial archaeologist G. Elzinga seemed to care about the archaeological heritage of Friesland.
10. De Langen (1992) and Knol (1993).
11. Knol (1993) published a complete survey of all available data concerning the Early Middle Ages in the Dutch and adjacent East Frisian region (Knol 1993). It shows a remarkable concentration of gold, silver and imported Frankish pottery in Westergo and Oostergo (Figs 55 and 73). This meritorious dissertation shows however that it was still too early to give a proper image of the sociopolitical and economic development of the region for lack of relevant data (settlements in the first place) and due to the fact that an intellectual discourse on such aspects in Frisia still had to be started. In his popular book on the archaeology of Friesland, J.M. Bos gave a brief survey of the archaeological testimony of the Frisian elite of the Early Middle Ages (Bos 1995, ch. 24). See also De Langen (1995).
12. Taayke (1997).
13. Knol et al. (1995/1996).
14. The general direction was in the hands of the author of this chapter and of Jurjen Bos and Jan Besteman. Jan Schoneveld and Danny Gerrets were the site supervisors (in 1990-1991 and in 1993 respectively), responsible for the daily decisions and procedures.
15. Lundström (1988).
16. Lamm (1988, p. 98).
17. Johansen (1988) criticizes the Helgö project for being too focused on the site itself as a unique phenomenon instead of defining its place within the socio-economic context of the time.
18. A 'nuclear region' is defined by Heidinga (1997, ch. 6) as a cluster of settlements, often surrounded by natural borders, which probably represented a social group, whether or not politically organized.
19. Hodges (1977, p. 209).
20. We only mention here: Jankuhn (1958), Niermeyer (1964), Slicher van Bath (1965), Ellmers (1972), Hodges (1977, 1982), Düwel et al. (1985, 1987), Lebecq (1983a, 1992) and Van Es (1990).

21. These and other questions are dealt with in Heidinga (1997).
22. Gurevich (1990).
23. Bazelmans (1996).
24. Recent publications on Dorestad: Van Es (1990) and Van Es & Hessing [1994, ch. 6 (by W.A. van Es) and ch. 26 (by W.J.H. Verwers)].
25. A recent synopsis is published by P. Schmid (1991).
26. Blok (1979, p. 36).
27. Lebecq (1983a, p. 110).
28. Wenskus (1961) and Geary (1983).
29. If the Frisians of the Roman period did not disappear at all from the face of the earth in the 3rd and 4th centuries, they probably went south, where we encounter them probably as Franks.
30. Lebecq (1983a, pp 107-111).
31. On the archaeological diversity of the province of South Holland (Bult & Hallewas 1990). The Byzantine historian Procopius mentions a tribe called the Varini in the region in the second quarter of the 6th century (see Blok 1979, ch. 2, who also deals with other non-Frisian elements in the southern part of Frisia).
32. Schmid (1991). See also for instance Slicher van Bath (1965).
33. The story is recently discussed by Herschend (1997).
34. The ritual function of early-medieval kingship in the reproduction of the tribe and in the life-cycle of its members is clearly shown by Bazelmans (1996).
35. Heidinga (1987, ch. 9).
36. See Bassett (1989, passim).
37. Charles-Edwards (1989, p. 37).
38. Radbod's daughter Theosinda was married to Grimoald, the son of Pippin. Rijntjes (1995, pp 166-167) supposes that such relationships between the Frankish and Frisian elites already existed much earlier.
39. Näsman (1991, p. 26).
40. Lebecq (1992) gives a critical evaluation of the evidence of the role of the Frisians in Northwest European long-distance trade. His conclusion is that they were indeed rather active if not dominant in this field.
41. See for instance: Niermeyer (1964), Slicher van Bath (1965) and Lebecq (1992).
42. How Frisian trade (mainly North-Sea networks) became incorporated into the Frankish trading system, i.e. the continental networks, is dealt with by Lebecq (1992).
43. Van Es (1990, p. 167).
44. Lebecq (1983b, pp 206-208).
45. Knol (1993, pp 189-194).
46. Ellmers (1977, pp 16-29, 266-270). See also Slicher van Bath (1965).
47. Van Es (1994, p. 107).
48. This view is strongly supported by Slicher van Bath (1965) and Lebecq (1992).
49. Arent Pol (Rijksmusum Het Koninklijk Penningkabinet, Leiden) is now preparing a dissertation on gold coinage in the Netherlands, Belgium and northern Germany in the 6th-7th centuries.
50. Personal communication Arent Pol. The province of Friesland has yielded almost as many gold coins from the 6th and 7th centuries as North and Central Germany, Belgium and Luxemburg together.
51. This type of rural trade declined however when towns attracted most commercial activities (Slicher van Bath 1965, p. 114). In the German part of Frisia a special type of trade settlement developed in the 9th century: the Langwurt (Brandt 1985).
52. Wine and pepper from a property of the monastery of Werden in East Friesland (Schmid 1991, p. 15) and wine from several places in the province of Groningen (Slicher van Bath 1965, p. 110). Although these deliveries date from the 10th-11th centuries, they are illustrative for the complicated connections which probably also existed earlier.
53. Näsman (1991, p. 28).
54. Slicher van Bath (1965).
55. Slicher van Bath (1965) and Table 3.
56. Slicher van Bath (1965, p. 104).
57. The Greek colonization of the 8th-7th century BC has recently been explained by Jan Paul Crielaard as a manifestation of power and heroism by the aristocracy rather than a economic solution for overpopulation (Crielaard 1988). Maybe this aspect also played a role in the Frisian colonization.
58. See for instance Gerriets (1983).
59. Halbertsma has estimated that the population of Westergo in this period (1th-3rd century AD) was at least 20,000, which is probably too high (see Slicher van Bath 1965, p. 100).
60. Friesland in the Roman period: see Van Es (1965-1966).
61. See for instance Niermeyer (1964), Slicher van Bath (1965) and Lebecq (1983a, ch.6; 1992).
62. See Crumlin-Pedersen (1991, 1996) and Crumlin-Pedersen & Munch (1995).
63. We make use here of the survey given by Knol (1993, ch. 4).
64. Knol (1993, p. 134).
65. Mazzo Karras (1985).
66. The Frisia project, which started in 1996, is directed by Dr J.A. Bazelmans (secretary, GIA/AAC), Dr J.M. Bos (GIA), J. van Doesburg (ROB), D.A. Gerrets (GIA), Prof. Dr H.A. Heidinga (chairman, AAC), Dr W. Prummel (GIA) and Dr W.J.H. Verwers (ROB). An introductory publication on the project, which explains the research aims of the project and which gives a provisional view of the Frisian area in the Roman and early-medieval period, was published in 1997 by H.A. Heidinga: *Frisia in the First Millennium. An Outline*, Utrecht.

Excavation method

D.A. GERRETS

1 INTRODUCTION

With the excavation at the terp of Wijnaldum-Tjitsma two main research questions had to be answered. The first question was a scientific one: what was the meaning of this settlement in the historical process of political centralization in early medieval Frisia.[1] To be able to answer this question, the emphasis of the research was on the 6th/7th century AD. The second question was related to cultural heritage management: what was the measure of erosion on the terp and which factors were responsible for this.[2] Wijnaldum-Tjitsma was and is an archaeological monument, protected by the National Monuments Act. To excavate the site, permission was needed from the Minister of Culture. The excavation licence of 1991 initially stated that it was only permitted to excavate the main trenches as far as the natural subsoil. In all other trenches excavation had to be limited to one or two levels. This strongly influenced excavation strategy during the first year. In 1992 it became clear that the former State Committee for Archaeological Monuments allowed us to dig without any further restrictions. As little was known about the measure of erosion of terps, the large-scale excavation offered a good opportunity to study terp erosion in more detail.

For this reason and to get a better insight into the stratigraphy of the site it was decided to excavate two main trenches, laid out crosswise over the eastern part of the terp, down to the natural subsoil. As it was expected that erosion would be most severe on the highest part of the terp, it was decided to lay out the north-south oriented main trench at the west side of the east summit of the terp, while the second east-west main trench was laid out over this summit (see Fig. 1).[3] Outside these main trenches a number of trenches would be excavated to get a better understanding of the spatial lay-out of the settlement on the terp. By excavating the top levels we hoped to be able to record the most severely threatened features and finds. Those present in the deeper subsoil would not be excavated.

Initially, the intention was to excavate a substantial part of the east terp during the first campaign between April and October 1991, to continue with the excavation of a substantial part of the west half during the same period in 1992. The results of the first excavation campaign were so rewarding, and circumstances appeared to be so complicated that it was decided to continue the excavation on the same parcel. Finally, the farmer allowed us to return for a third time to continue the excavation from the beginning of April until the beginning of August 1993. This last campaign made it possible to finish the excavation of the main trenches which were crucial for the understanding of the chronological structure of the site.

2 SURVEY BY BORING AND INTENSIVE SYSTEM WALKING

Preceding the first excavation campaign, a number of borings offered a first impression of the stratigraphical structure of the site. The borings showed that dung was only rarely represented as is always assumed for Westergo.[4] This is a clear difference with famous terps excavated in the Netherlands and Germany, e.g. Ezinge, Feddersen Wierde and Elisenhof, where occupation layers are separated from each other by substantial dung layers. It was sometimes very hard to distinguish natural sedimentation layers from raised sod layers, because they originate from the same material. Sods were cut on the surrounding salt-marsh. In most of these sods, which were used for all kinds of construction purposes, storm-surge levels intercalated in more clayey layers were generally very clearly visible. Other sods from more clayey salt-marsh deposits were less well visible. The borings gave a good im-

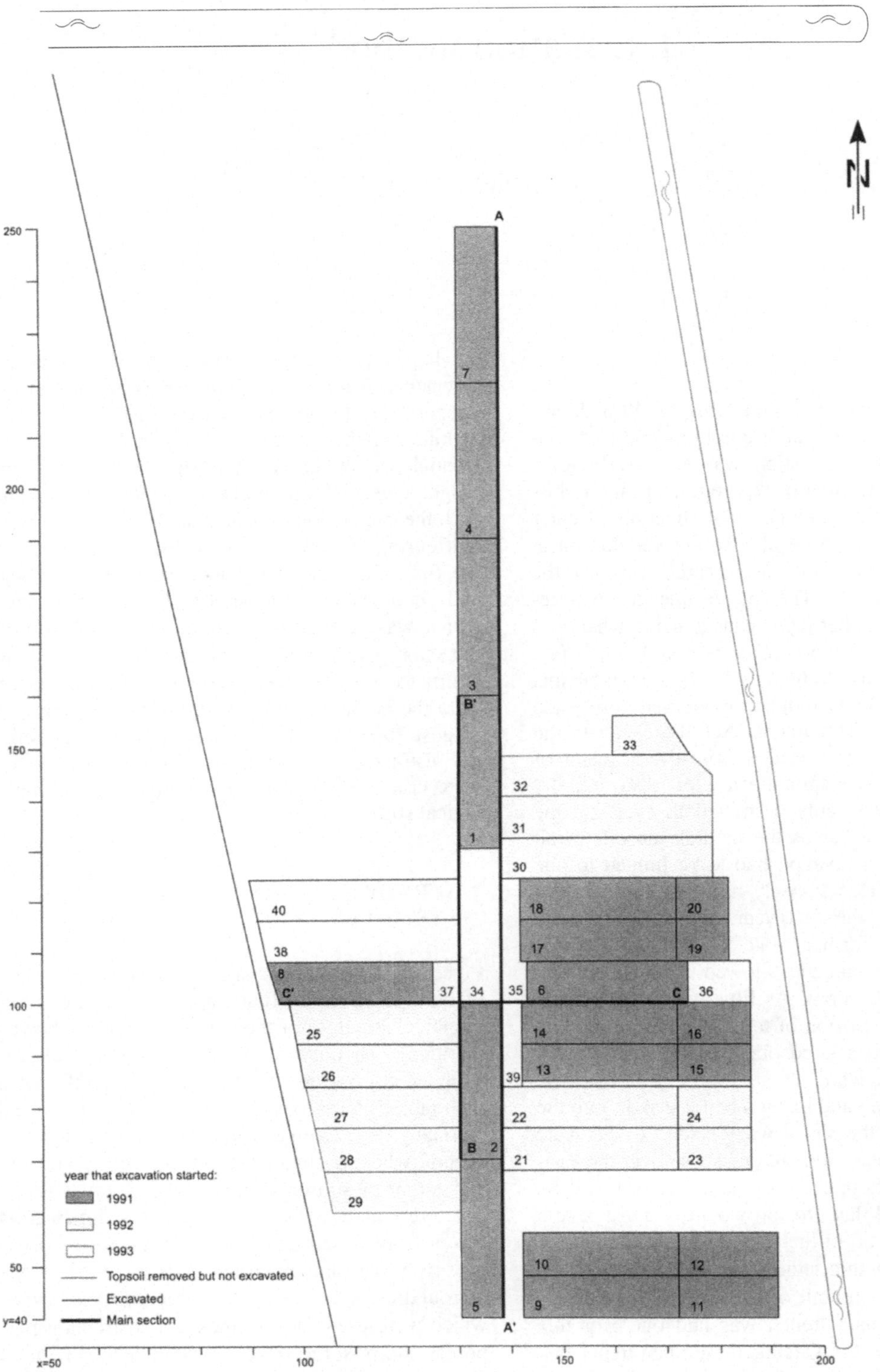

Figure 1. Plan of excavation trenches (numbers) and main sections (characters) with the excavation coordinates, present ditches and year that the excavation started.

pression of the numerous occupation layers, which are more organic with small charcoal particles and small lumps of burned loam. Because of their number it was often impossible to connect the singular occupation layers in two different borings with each other. In one boring at the northern side of the terp an arable soil with a thickness of 1.16 m was observed. This was the first indication for the erosion of terp layers.

To learn more about the relation between the finds in the top soil and in the undisturbed archaeological strata in the subsoil part of the excavated area was preceded by system walking in a restricted area of 16 × 30 m. This area was subdivided in 28 squares of 4 × 4 m and 4 rectangles of 4 × 2 m in which all visible stray finds on the surface were collected. As the numerous metal finds from the top soil could be regarded as a possible indication for eroded graves the whole excavation area was surveyed by physical anthropologists but finally not a single human bone was collected from the surface.

3 THE MAIN SECTIONS, EXCAVATION TRENCHES AND LEVELS

The usual Dutch excavation method in Holocene areas with soft soils and an almost horizontal archaeological stratigraphy is to lay out the excavation levels horizontally as well as to 'follow' the level of the settlement layer(s). The Wijnaldum terp, however, with its numerous settlement layers and raised soil offered a complex and capricious stratigraphy. The choice of the archaeological levels prepared for drawing depended upon the observation of a more or less coherent settlement pattern within 25 cm depth from the former level. If the layered structures made a clear choice impossible the layer was laid following the normal contours of the terp structures as naturally as possible.

The north-south oriented main trench (see Fig. 1) had a total length of 210 m and a width of 8 m and was excavated down to the natural subsoil, which varied in depth from 3/3.5 m in the centre of the terp to 1.80 m at the utmost north of the excavation. To be able to connect the west and the east sections with each other, the main trench was subdivided into smaller trenches of 30 m long, which were excavated in 3 to 10 levels. The east-west oriented cross section had a total length of 76 m and was also 8 m in width and was excavated as far as the natural subsoil. Outside the two main trenches larger trenches were excavated in one or more levels, which were 8 to 16 m wide.

A level was documented as soon as the features changed considerably from those in the preceding levels. Generally levels were at a distance of 20-25 cm from each other, but at greater depth the distance between levels was sometimes more. The first level was generally at a depth between 40 and 60 cm under the surface level, depending on the recognizability of the features.

Initially it was intended to remove the topsoil over the whole surface of the terp to protect finds and features in the subsoil from destruction, but due to a lack of financial means this aim was not completely achieved. The excavation trenches where only the topsoil was removed, but which were not extensively excavated are depicted in Figure 1 with the thin lines.

4 MECHANICAL DIGGER AND METAL DETECTION

For the first time the mechanical digger and metal detectors were introduced on a large-scale terp-excavation. To get a better idea of the spatial layout of the settlement, it was necessary to excavate a large surface. Therefore, a mechanical digger equipped with an extended arm was working almost continuously on the site making new levels or moving soil heaps. During the three excavation campaigns about 8000 m^3 were excavated. The experienced driver was able to dig millimetrewise, while somebody was collecting the finds, visible at the surface and within reach of the mechanical digger, in squares of 4 × 4 m.

Immediately behind the mechanical digger a specialist followed with a metal-detector. All metal finds were measured three-dimensionally. To be sure that all metal finds were located, the mechanical digger generally dug in layers of 5 cm and less at a time. Experiments during the first stage of the excavation showed that metal finds were missed if the layers removed by the mechanical digger became thicker than 5 cm. Eventually about 5000 metal finds were located. The mass of metal finds offers a new source of information about the terp region since Wijnaldum-Tjitsma was the first terp excavation where metal detection was used in such a systematic way. At the same time this poses a problem of comparability with the traditionally excavated terps.

It was remarkable that different people working

Figure 2. A metal detector specialist followed immediately behind the mechanical digger.

with different metal detectors produced very different results, depending on the skill of the users and on the sensitivity of the metal detectors. Some metal detectors are more sensitive to precious metals or alloys like bronze while others are more sensitive to iron. Not all finds were found with the metal detector. Dense concentrations of charcoal meant that metal objects were first found during water screen sieving.

5 DOCUMENTATION OF LEVELS AND SECTIONS

All levels were drawn and naturalistically coloured as were most of the important sections. In the north-south oriented main trench the east section was drawn completely ($A\text{-}A^1$, see Fig. 1) and the main part of the west section ($B\text{-}B^1$). In the east-west main trench the main part of the south section was drawn ($C\text{-}C^1$). Finds in the sections were collected stratigraphically. All stratigraphical units in the main sections were described on overlays of the original drawings. During the 1991 and 1992 campaigns no smaller transections were made with respect to the complicated stratigraphy to prevent unintended damage to underlying features. Only deeper settlement structures such as wells and sunken huts were transected. Especially the remains of sod buildings appeared to be so complicated that, in 1993, it was thought necessary to dig small trial trenches to study the features in their direct stratigraphical context. After the sections in these trial trenches had been studied, the different layers were excavated stratigraphically. Sod structures posed special problems when the soil dried out, because they were hard to recognize then. As the texture of the soil was very sandy, some sods and sod walls were hardly recognizable any more only a few hours after heavy rain. Sometimes, however, under damp weather circumstances, sod structures became visible which had not been observed before. A special problem was posed by the clayey sods. They were especially hard to recognize. In wells the sod lining was only recognized near the groundwater table where the former vegetation layers of the sods were still visible.

6 FINDS AND SAMPLES

As said before metal finds were located with the metal detector and measured three-dimensionally. Finds found with the mechanical digger were collected in squares of 4 × 4 m, while finds in the sections were collected in their stratigraphical position. As sod layers were poor in find material these layers were removed with the shovel to collect more finds. Soil excavated from all other features was separately sieved on water screens with a mesh width of 4 mm. The water screen sieving resulted in all kinds of new sources of information like sherds of glass vessels, beads, charcoal, numerous fish-bones and bones of fowl etc. From the content samples were taken for flotation of palaeobotanical remains. Other samples were sieved on screens with a width of 1 mm for the smallest archaeozoological remains. The results also offer interesting statistical observations on the different methods of find collecting and sampling. The 12,000 find numbers range from a single object (metal objects, stratigraphical finds) to find numbers containing thousands of finds and weighing 25 kg or more (from occupation layers, wells, a metal working dump etc.).

The finds were sorted manually by a large group of students, volunteers from the region and from elsewhere in the country and abroad. During the summer of 1992 it was attempted to spead up the excavation by increasing the workforce. On certain days over a hundred people (staff, students and volunteers) were involved, but this proved to be ineffective because of a lack of experienced pit supervisors in this archaeologically difficult terp site. Thanks to the cooperation of the large number of students and volunteers all finds were washed, dried, sorted, numbered and available for further analysis at the end of the last campaign in 1993. This took place in the excavation support centre in Harlingen. Determination, however, especially of pottery in the field gave a first insight in the periodization of the settlement and must be considered necessary.

7 AUTOMATIC DATA REGISTRATION

To cope with the large amount of data all finds were entered in a DBase IV database while all drawings were digitized in Autocad release 11 as soon as possible in Harlingen. Features and other stratigraphical units did not recieve any numbers in the field because it was thought that it would be possible to make three-dimensional reconstructions of every possible feature with the computer and to combine automatically these find numbers with each separate feature. The expectation that features could be reconstructed three-dimensionally, however, caused some people to become less aware of the stratigraphical relationships in the field and excavating became an industrial process. Finally, possibilities for three-dimensional reconstructions of features with the computer appeared to be very restricted, because there are too many human decisions involved in establishing stratigraphical relationships (colour, texture, intersecting features, erosion etc.). Fortunately the aim of a computer engineered reconstruction did not allow us to drop the traditional drawing skills. Therefore the detailed description of the stratigraphical units in the field and the naturalistically coloured drawings of sections and layers formed the basis of the analysis leading to the reconstruction of the stratigraphical relationships in the years following the end of the excavation. This time with traditional means like coloured pencils and human brains.

The computer data will remain available for possible further analysis in the future.

8 CONCLUSIONS

The systematic introduction of excavation methods like metal detection and water screen sieving produced many find categories in great quantities, which were so far relatively scarce at terp excavations, like ferrous and non-ferrous metal objects, beads, glass fragments and bones of fish and water fowl. This confronts us also with the problem of comparability of the site.

The involvement of large numbers of volunteers made it possible to finish find-processing together with the field work. Thus all find material was immediately accessible for further analysis by the different specialists.

The computer proved to be an important tool in database management. A lot of time can be gained by analysing the large amount of data generally coming from terp excavations with the help of queries from linked databases. During the final analysis all specialists designed their own databases with the only condition that one key variable made it possible to link their database with the main databases. This way they very quickly had the datings of their material at their disposal together with their reliability.

After a lot of money and time was invested in digitizing drawings it appeared very difficult to analyse a complex stratified site like a terp with the help of the computer as we had in mind. In the analysis too many human decisions were involved, which could not be solved by way of automatization. We learned that in relation to the automatization of the excavation documentation it is very important to know from the start which research problems can be solved more easily with the help of the computer and how they can be solved. Only under these conditions will it be clear whether time and money can be gained with the computer. Because the automatization of the excavation documentation of large excavations is the work of experts most of the time, intensive communication between the archaeologist(s) and the computer expert is of decisive importance for the final success of the analysis.

Wijnaldum-Tjitsma also showed the danger of automatization, that one tends to concentrate on the input of data and this postponed the analysis. In this way excavation becomes an industrial procedure.

At Wijnaldum-Tjitsma this development was strengthened by the fact that we believed in the possibility of analysing the stratigraphy of the site afterwards with the computer. Finally we became convinced that at very stratified sites like Wijnaldum-Tjitsma it is necessary to analyse stratigraphy in the field. Making Harris matrixes after every day of excavation can be a very useful tool.[5]

REFERENCES

WATERBOLK, H.T. 1961. Beschouwingen naar aanleiding van de opgravingen te Tritsum, gem. Franekeradeel, *It Beaken* 23/24: 216-226.

NOTES

1. See Heidinga (this volume).
2. See Bos & Gerrets (this volume).
3. At the start of the excavation it was still assumed that the Wijnaldum-Tjitsma terp consisted of the excavated part and the parcel at the west side of the excavation. During the excavation it became clear, however, that the parcels at the east side of the terp also belonged to the same terp.
4. E.g. Waterbolk (1961, pp 217-218).
5. At the start of the excavation it was still assumed that the Wijnaldum-Tjitsma terp consisted of the excavated part and the parcel at the west side of the excavation. During the excavation it became clear, however, that the parcels at the east side of the terp also belonged to the same terp.

The erosion of the Wijnaldum-Tjitsma terp: Aspects of cultural heritage management

J.M. BOS & D.A. GERRETS

1 INTRODUCTION

One of the main research problems at Wijnaldum-Tjitsma was the study of the state of conservation of the terp settlement. It was long thought that the Ancient Monuments Act provided an effective protection for our cultural heritage. Numerous metal objects found by amateur archaeologists with a metal detector at Wijnaldum-Tjitsma made it clear that remains of the former occupation were regularly ploughed to the surface. With the excavation we hoped to find answers to many questions relating to the state of conservation of the terp. To what extent was the terp affected by erosion? Is agrarian exploitation the main factor in the erosion process? Can the erosion process be dated? Before trying to answer these questions, we will first look at the general condition of the Frisian terps.

2 COMMERCIAL QUARRYING OF THE TERPS

The *Friesch Genootschap voor Geschiedenis, Oudheidkunde en Filologie* (Frisian Society for History, Antiquities and Philology; here called the Frisian Society) was established in 1827, one of its departments being devoted to 'Antiquities'. This department concerned itself with, among other things, archaeological finds. Mr. Wopke Eekhoff joined the Frisian Society in 1835, drawing attention to archaeology. In 1839 he proposed to the members 'to sent a request to all private landowners, to notify the Society where they had been digging'. This suggestion was ignored, and Eekhoff started to explore the Frisian terps with some colleagues himself. When he informed the Society on their findings in 1845, he managed to rouse the interest of the other members. The Frisian Society decided to set up an excavation committee with a budget. Furthermore, the Society sent over 150 letters to all borough councils, to owners of terps and other important land owners, asking them to report any digging or collected artifacts. One assumes that Eekhoff must have been disappointed, since one year later no reports or such like had come back.

What gave rise to the idea that there would be any digging, or that there would be any artifacts to collect? From about 1840 systematic quarrying of the terps started. The terp soil, which is rich in phosphates, was sold as fertilizer for the poor sandy soils in the reclaimed peat-moors. Friesland started to exploit its terps as South America has exploited her guano. The soil of a terp of one hectare was sufficient to fertilize 375 hectares of poor sandy soil. For more than a century hundreds of terps were quarried for commercial reasons, which was absolutely disastrous from the viewpoint of cultural heritage management.

After a hesitant start, many Frisian terps were 'opened' in the second half of the 19th century because of an increasing demand for fertilizer. Commercial quarrying of terps reached its peak in the 1920s, followed by a slow but steady decrease. Commercial exploitation stopped immediately after the Second World War.

Practically no larger Frisian terp escaped commercial quarrying. The newspaper *Nieuwe Rotterdamsche Courant* of November 3rd, 1905 reported that the director of the National Museum of Antiquities had written to the curators of Leiden University: 'Slowly the so-called terps in Friesland are disappearing. They are, as reported, being quarried, during which process many antiquities come to light, which may enrich the museums, but none of the terps has been subjected to an archaeological excavation, as would be necessary to throw light on our earlier history'.

'The terps', he continues, 'belong to our most prominent ancient monuments, of great international importance for archaeology. The loss would be irreparable and our reputation in the civilized world would suffer, if they were all to disappear without one of them being systematically excavated!' Finally the readers are told that the Home Secretary invited the *Commissaris van de Koningin* (Governor) of the Province of Friesland to enquire of all mayors whether there were still undamaged terps in their municipalities.

This enquiry showed that more than half of the terps were already partly or completely quarried. In 1944 this was over 75%.[1] One should realise that the enquiry involved all kinds of terps, including small house terps. When we restrict ourselves to the larger terps, the situation is devastating. The fact that in the 1930s and 1940s hardly any new terps were opened was not because of a decrease in the demand for terp soil, but was due to the fact that *all* large terps, as far as they were not inhabited, had in fact been or were being levelled.[2]

The Frisian Society appointed observaters at the quarrying of some important terps like Hoogebeintum and the Burmania terp at Ferwerd. This provided us with the first written reports, but this was not what the director of the National Museum of Antiquities had in mind.

3 EXCAVATING TERPS

While in the Province of Groningen archaeological research of terps had already started before the Second World War, Friesland had to wait until the late 1950s, when the first plans were developed for the excavation of a Frisian terp. After a series of trial excavations, the choice was made for the terp of Tritsum (municipality of Franekeradeel, Westergo) where Waterbolk conducted campaigns from 1959 to 1961.[3] The results were disappointing in the light of the questions asked at the time and no new large-scale excavations were undertaken in Friesland.[4]

The following decennia archaeological interest moved to cultural heritage. It was now considered better to invest the limited manpower in mapping and conservation of the last remnants, than in the destruction of the remaining terps by new excavations. The instrument enforcing this new approach was the national *Monumentenwet* (Ancient Monuments Act) of 1961, which offered the opportunity to protect sites legally. Some of the best preserved terp sites became legally protected sites. The Wijnaldum terps are also an example of these newly created legally protected sites.

During the following years it became evident that the legally protected status did not always mean that sites were indeed well preserved for the future. Sometimes there were indications that archaeological monuments were threatened. However, this seldom resulted in proper excavations. An exception was the find of an early-medieval urn with human cremation material from a recently widened ditch at the remains of the Oosterbeintum terp (municipality of Ferwerderadeel, Oostergo) in 1987. It seemed that an early-medieval cemetery was gradually being destroyed and furthermore the widening and deepening of the ditch could be expected to cause desiccation of the remaining features. Together with the need for a modern excavation of a Frisian early-medieval cemetery, this was the main reason for the partial excavation of the cemetery in 1988-1989.[5] It now appeared that questions might be asked about the efficiency of protection under the Ancient Monuments Act.

The increasing use of metal detectors by amateur archaeologists offered a new possibility to verify the effectiveness of the legal protection of terp monuments. Their existence, however, was generally ignored by professional archaeologists. Jan Zijlstra, a private collector in Leeuwarden, was often visited by these amateur archaeologists. Jan Zijlstra had not only a wide expertise, but he also acknowledged the importance of collecting and recording the metal finds which were less spectectular at first sight. He acquired by far the largest collection of Frisian metal finds and registered their provenance. He was able to point out the abundance of metal finds from the Frisian terps, mostly from Roman times onwards. The information thus collected proved invaluable, and he did not hesitate to communicate his findings to the professional archaeologists active in the area.

The richest find from Friesland had been recovered at the terp Wijnaldum-Tjitsma in the 1950s.[6] What fascinated Zijlstra was that the material recovered from this site in recent years was the greatest in Friesland, both in quality and quantity. Most qualitatively rich finds seemed to belong to the 6th-7th centuries. The destructive impact of 'common agricultural use' on the protected site seemed obvious. Had a cemetery been ploughed here? Or should it be feared that an entire terp was systematically being ploughed down? Large-scale excavation of the Wijnaldum-Tjitsma terp offered the possibility to gain better insight in the destructive processes by natural erosion and agricultural use, affecting many similar sites in the region.

4 THE EROSION OF WIJNALDUM-TJITSMA BY PLOUGHING

The farmer of Wijnaldum-Tjitsma, Mr Bergmans, assured us that ploughing into the subsoil was absolutely undesirable: on terps like this, mixing the subsoil with the rich topsoil would decrease the fertility of the fields. However, this is what is unintentionally happening. Terps are, by definition, artificial mounds. Nature tends to level mounds: with each storm or cloudburst some soil is blown away or washed down and this was what we regularly observed during the excavation campaigns. By consequence, if the plough is adjusted to the same depth at all times, it actually reaches deeper into the terp each time ploughing takes place. This process could be inferred in the working pits 25/26: the brick drains that had been laid in the 1950s, were in 1991 occasionally being ploughed to pieces, the ploughed topsoil having 'reached' their level in the intervening years. Other indications for recent erosion by ploughing are finds from the surface. As said before, Jan Zijlstra collected from amateur archaeologists many metal finds found at Wijnaldum-Tjitsma during the last decade. During the excavation campaigns the whole surface of the terp was intensively surveyed with the metal detector. Most metal finds were found near the present summit of the terp and in the southeast corner of the excavation with a relatively empty zone in between. This resulted in hundreds of metal finds, some of which were apparently affected by their long stay in the topsoil, while many others looked as if they were new. The same picture was true of the pottery. Some sherds were rounded by intensive weathering, indicating their long stay in the topsoil, while other sherds looked as if they had been ploughed up yesterday. The numerous finds from the topsoil were a clear indication of the erosion of the terp. The excavation strategy was adapted to this new knowledge; it was decided to try to document the first level, immediately beneath the ploughed topsoil, of the entire terp.[7]

Another indication for the destructive impact of ploughing is the disc-on-bow brooch from Wijnaldum-Tjitsma.[8] The footplate of this brooch was found in the 1950s when drains were still laid by hand with a special spade. The brooch was probably cut into pieces and the central disc and head plate overseen by the worker who found the footplate. The central disc has not yet been found, while the exact location of the pieces found by amateur archaeologists is not known. But the pieces found during the excavation were all found in the topsoil from an area of 300 m^2. All pieces, but most clearly the large head plate fragment, showed that the remaining part must have been torn apart with a lot of violence, probably by the plough.

Although there are many indications for the destruction of finds by ploughing and the displacement of topsoil by rain and wind, indications for these processes are less clear than one would expect. While the north-south oriented main section shows that the topsoil near the summit of the terp varies in thickness between 22/25 and 30 cm, the topsoil at the northern slope of the terp varies in thickness between 30 and 35 cm. The topsoil in the east-west section varies in thickness between 28 and 32 cm.

5 THE EFFECT OF PLOUGHING ON DISTRIBUTION PATTERNS OF CERAMICS

Before the excavation started, the surface of the later excavation trench 17/18 was intensively surveyed and all finds were collected within 28 squares of 4 × 4 m and 4 rectangles of 2 × 4 m. Then the topsoil was removed by the mechanical digger, and during digging visible finds were collected within the same 28 squares and 4 rectangles. The topsoil was removed in two layers. To reach the first level, which was interpretable, i.e. where features could clearly be distinguished, the subsoil was removed in two more strata, while finds were collected in the same squares and rectangles. Then, the soil was removed by the mechanical digger millimetre after millimetre until the next level was reached, generally 10-15 cm deeper. At the first documented level (at ca 45-60 cm below the surface) ceramics were collected by waterscreen sieving of the contents of the different features. Except for one post-medieval sherd all sherds from this level are considered to be in the right, undisturbed context. By comparing the results from this undisturbed level with the sherds collected earlier from the higher lying strata, three groups of ceramics can be distinguished:

1. The sherds which have a certain relation with the sherds from the same square collected at the first documented level,
2. Sherds which have a possible relation and
3. Sherds which have certainly no relation with the finds from the same square at the first documented level (because they are too young).

The results are presented in Table 1.

Table 1. Relation between finds from the topsoil and from the first documented level.

	No relation with doc. level	Uncertain relation with doc. level	Certain relation with doc. level	Total
Surface	119	136	6	261
1st level (topsoil)	38	25	19	82
2nd level (topsoil)	20	15	32	67
3rd level (subsoil)	3	9	25	37
4th level (subsoil)	2	1	61	64
1st documented level	1	0	–	1
Total	183	186	143	512

It is striking that sherds at the surface are far more numerous than the number of sherds from underlying levels. At least 119 of these sherds (but probably more) do not show any relationship with the sherds from the first documented level. This large number of sherds from the surface can partly be explained by fragmentation. Of the 261 sherds from the surface only 8 weighed more than 5 g. The fracture edges of most sherds were rounded, while most sherds showed traces of heavy weathering. Another reason for this large number of sherds is erosion by wind and rain, which move soil to the lower parts of the surrounding terrain, while the heavier sherds stay in place. The majority of these sherds will have been brought to the site in (post-) medieval times together with other domestic refuse, serving as fertilizer.

Downward transportation of (post-) medieval sherds occurs on a relatively large scale: 183 of the 552 sherds collected were with certainty moved downward, but this phenomenon is almost completely restricted to the topsoil. Only 6 sherds from the undisturbed subsoil were too young compared with the first documented level, where nevertheless one post-medieval sherd was excavated. On the other hand transportation of material from the undisturbed subsoil to the topsoil occurs on a considerable scale (57 sherds, 9.6%).

6 DECAPITATION OF THE TERP BEFORE THE 1950S (Fig.1)

During the boring campaign preceding the excavation in April 1991 a topsoil was observed with a maximum thickness of 1.16 m at the northern slope of the terp. In the main north-south section the maximum thickness of this agricultural layer was 1.26 m. The post 1950s drainage topsoil had a thickness of 36 cm which shows that before the 1950s 90 cm of topsoil from elsewhere on the terp must have been moved here. A detailed study of this agricultural layer showed that three strata could be distinguished: one post-1950s topsoil and two pre-1950s agricultural layers slightly lighter in colour. These layers can probably be related with the filling up of two ENE-WSW orientated ditches. These ditches are at right angles to another subrecently filled up ditch. This last ditch subdivided the excavated field into two plots. Material from all three ditches, especially bricks and glazed red wares, show that they are all post-medieval. Figure 2 shows that the north slope of the east summit of the terp is more gentle than the slope of the west unexcavated summit of Wijnaldum-Tjitsma. It will be clear that this more gentle relief must be explained by the filling up of these ditches.

It is hard to say whether the filling material came from higher up the terp, but this seems very likely. The main section showed that many early medieval layers on the north and south slope of the former Roman house terp were truncated (see Fig. 2). Of the former occupation layers of the Roman Period only dug in features like postholes, the underpart of a hearth, wells etc. were still preserved. There are several indications that early-medieval layers originally overlay this former house terp. A well in the main section intersecting this platform dates from the 5th century while on a southern extension of this platform only the remaining floor level of a 5th-century sunken hut was preserved. Deeply dug in features like wells from the Carolingian Period and even from the late 9th/early 10th century show that occupation must have extended much further north than is suggested by other features still remaining. It is hard to say whether 'late' material from the topsoil in the 'early' zones derives from layers now disappeared on the same location or that they were moved here by ploughing. Nevertheless, all this shows that the reconstruction of the former, but now disappeared occupation layers in Figure 2 is not exaggerated. Therefore, it seems very likely

Figure 1. Excavation trenches and the present relief of Tjitsma terp at Wijnaldum. 1 = Agricultural layer of 1.26 m, 2 = Agricultural layer with a maximum thickness of 1.16 m, 3 = Post-medieval ditches, 4 = Former summit of the terp, 5 = Present field division, and 6 = Subrecent ditch.

that the former early-medieval summit of the terp was located more to the northwest, somewhere above or near the earliest Roman house terp, compared with the present summit.

7 LEACHING

Terp excavations have impressed the world because of their excellently preserved wooden remains of

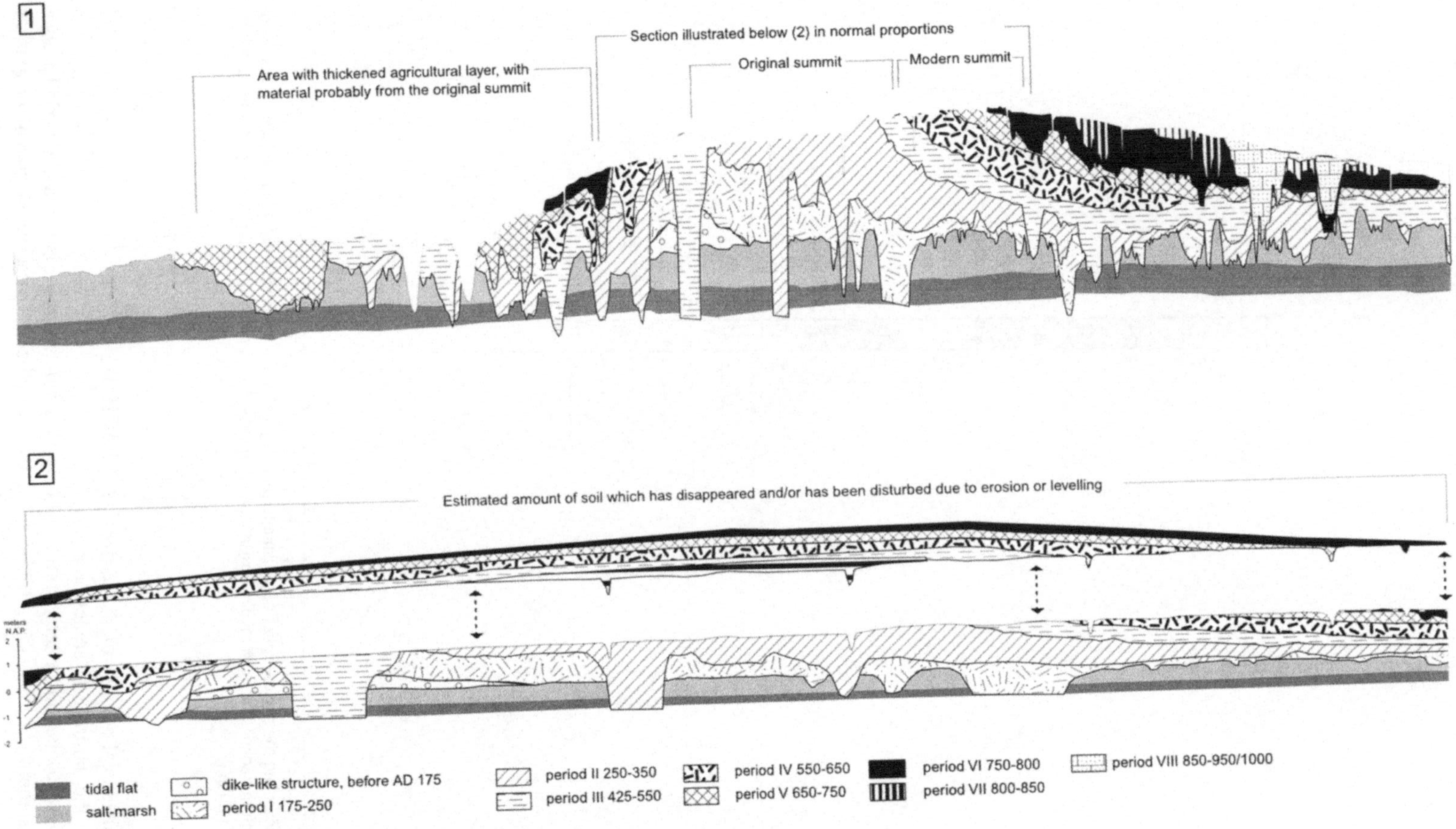

Figure 2. Section A-A′ vertically exaggerated: 1) Showing the modern and the original summit and the part of section A-A′ in normal proportions, and 2) Where the original summit must have been located, the estimated amount of soil which has disappeared or which has been disturbed due to erosion and/or levelling.

houses and other organic finds, as for example, at Ezinge, Feddersen Wierde and Elisenhof. At Wijnaldum-Tjitsma preservation circumstances are less favourable. One reason is the almost complete absence of dung. Massive dung layers close off underlying features from oxygen. This absence of dung seems to be characteristic of terps in the whole Westergo region. Waterbolk has suggested that the dung was used as fuel, because of a general lack of firewood in the region.[9] This assumption seems to be underlined by the research on wood and charcoal by Stuijts, which shows that much construction wood has been used secondarily as firewood.[10] Wood had to be transported to Westergo over a large distance, and people were apparently very economical in their use of raw materials.

A second reason for the relatively poor preservation circumstances is related to the use of the region for agriculture. We have already indicated that drains were laid by hand in small trenches in the 1950s. At the south of the agricultural field these drains are lying in a east-west direction, while at the north side of the field they are lying both in a north-south direction and in an east-west direction. Some additional drains were laid mechanically in an east-west direction at the southernmost part of the agricultural field during the last decade. The first kind of drains lie at a depth varying between 34 cm at the base of the terp to 74 cm near the summit at an interval of 9 to 13 m. The last kind of drains are drawn by a cable at a depth of 1.30 m under the surface, while two huge knives lift up the soil. At Wijnaldum-Tjitsma the impact of these drains differs from location to location. The section above a first drain of this type was still intact, while the section above a second one was completely destroyed.

Agricultural use demands a low ground water table. On September 6th 1991 the ground water table was at a depth of 3.30 m under the surface at 1.46 m under Dutch Ordinance Level (*Nieuw Amsterdams Peil*: NAP). The texture of most terp layers is rather sandy, while dung is generally absent. This means that organic materials like wood and leather are only preserved very near to the ground water table in features dug deep into the natural subsoil like wells and ditches. It is hard to quantify the process of leaching, because the original organic content is unknown.[11] It was, however, clear that processes of leaching must have had a considerable impact at certain, if not most locations. Sods were cut on the salt marsh, but the original vegetation layer was only visible in the zone of the fluctuating ground water table. Also pollen analysis suggests that considerable oxidation took place as pollen are nearly absent in salt marsh deposits, compared with wells and ditches, which are generally much closer to the ground water.[12]

Preservation circumstances seemed to differ considerably as shown by the following data. Dung was preserved in a Carolingian pit, dug into the salt marsh at a depth of 60 cm under NAP in excavation trench 2, at 50 cm above NAP in a former Merovingian drinking place for cattle in excavation trench 13/14 and at 1.00 m above NAP in a pit under a sod house in excavation trench 18, near the summit of the terp. Even closer to the present summit, however, in secondary deposits in a sunken hut in excavation trenches 6/36 a brown organic layer, probably dung, was badly preserved at 1.00 m +NAP.[13] On the contrary, in a sunken hut far to the north in excavation trench 33, wattle was still beautifully visible in a pit at a height of 1.18 m +NAP. Sometimes features were already visible immediately under the topsoil at a depth of 35 cm. Most levels, however, were at a depth of 60 cm under the topsoil, while the organic content of the features at that level was generally poor. The reasons for these local differences in preservation of organic material are not well understood.

Also other find categories showed considerable differences in preservation. Bone was very well preserved as is generally the case on terps, but in some cases bones were falling apart. Copper alloyed objects were generally well preserved, sometimes even with a nice patina. Some objects, however, were desintegrating pieces because of heavy corrosion. Some metal finds from the topsoil were torn into pieces by the plough and/or showed traces of chemical weathering. Iron objects generally suffer from the high salinity of the soil.

8 WIJNALDUM-TJITSMA IN THE PERSPECTIVE OF SOME OTHER EXCAVATED TERPS

It will be clear that terps are affected from the first moment of their existence both by natural processes and by human agricultural activities and that the impact of these last processes increased with the introduction of modern agriculture. Earlier terp excavations like Ezinge and Leens-Tuinsterwierde (both Prov. of Groningen, the Netherlands) and Feddersen Wierde and Elisenhof in Germany impressed the international archaeological community because of their preservation circumstances. We have to keep in mind that these terps all contained substantial dung layers and that erosion had its impact on the upper strata of these terps. At Ezinge the first metre with the traces of Merovingian and Carolingian occupation was considered topsoil and removed without any documentation.[14] The summit of the terp at Leens-Tuisterwierde was at 5.00 m +NAP according to Van Giffen. And although the sandy terp of Leens-Tuinsterwierde contained dung layers the first level was at a height of 3.10-3.30 m +NAP, which means that a considerable amount of soil must have been removed without documentation.[15]

At the beginning of the Tritsum terp excavation, the excavators expected a terp with preservation circumstances like those at Ezinge. Leaching appeared to have affected the terp to such an extent that the results, apart from a preliminary publication, were never published.

9 NEW POLICIES

The results of the excavation at Wijnaldum-Tjitsma improved our understanding about the state of conservation of terps and terp remains in general. It showed that the process of erosion must have been in progress as long as this kind of terps were being ploughed. The intensification of agricultural use since World War II must have hastened it considerably.

Suspicions that this must be happening on all terps used as arable land were soon confirmed by reports from amateur archaeologists using metal detectors. Several 'rich' terps yielded large quantities of metal finds, the result of at least decades of intensive ploughing. Most of these terps 'dried up' after dozens of amateurs had systematically collected the metal from the ploughed topsoil. Yet the amateurs return after each ploughing: it appears new finds are ploughed up each time. The 'old' finds had often suffered from having been ploughed up and down for some time and chemical weathering; the 'fresh' finds are often in fine condition.[16]

The provincial government took immediate action when it became apparent that the legal protection of archaeological sites is inadequate. It commissioned RAAP Archaeological Consultancy Agency (RAAP) in 1992 to start a programme of stocktaking and monitoring as a basis for new policies.[17]

10 CONCLUSION

The Frisian terps are a unique part of Dutch national heritage because of their well developed stratigraphy, their richness in finds, their often favourable preservation circumstances and the crucial role they play in an important part of our history. Unfortunately the importance of the terps was not well understood by former generations and for this reason less than 25% of the original number of these sites still remains.

It would be a serious misunderstanding to believe that the remaining 25% is well preserved, because of the growing awareness of the cultural importance of these monuments. The excavation of the Wijnaldum-Tjitsma terp has made it clear that the quality of the archaeological remains at this site is seriously threatened by a combination of the relief of the site as an artificial dwelling mound, the sandy texture and, most of all, modern agricultural use. Wind and rain cause a gradual erosion of the topsoil as illustrated by the large numbers of sherds lying at the surface and differences in thickness of the topsoil. This process is unintentionally intensified by ploughing. The plough causes transportation of finds from the undisturbed subsoil to the topsoil where they are subject to further destruction by the plough and natural processes like frost, rain and chemical weathering.

Probably the most serious threats, however, as they take place invisibly in the subsoil are the processes of leaching and chemical weathering because of the extremely low ground water table. Together with the sandy structure of the soil and the almost complete absence of dung (which are both generally the case at terps in the Westergo region), they are responsible for the gradual disappearance of the last organic remains. Little is known about the speed of these kinds of processes, but at sites like Wijnaldum-Tjitsma and Tritsum, where the reconstruction of structures must be derived from slight differences in colour, the impact of these processes must be disastrous. They can only be stopped by far reaching measures like the artificial raising of the ground water table and a different agricultural exploitation.

Decapitation seemed to be a phenomenon which is typical of the Wijnaldum-Tjitsma terp and especially the excavated middle part of this terp. Although decapitation is a regularly observed phenomenon at terps, the serious extent of truncation of layers at Wijnaldum-Tjitsma seemed to be caused by the filling up of two ditches somewhere between 1400 and the 1950s. Recent excavations of the terp remnant at Dongjum (Westergo, Friesland) have shown that this process of decapitation is probably a much more general phenomenon. Also at Dongjum, a considerable part of the former occupation layers must have been ploughed to the foot of the mound already in the Middle Ages.

Therefore, most processes of erosion which were observed at Wijnaldum-Tjitsma are probably not unique for this terp but will take place at many other terps in the region and beyond. It makes clear that these sites of great cultural importance are not sufficiently protected by the present Monument Act. We hope that the research carried out at Wijnaldum-Tjitsma together with the results of the inventory carried out by RAAP Consultancy in the terp region will contribute to a new debate about the necessity of additional protective measures.

REFERENCES

ARJAANS, J. 1991. Terpafgravingen in Friesland, *Jaarverslagen van de Vereniging voor terpenonderzoek* 75: 45-56.

BOS, J.M. 1995. *Archeologie van Friesland*, Utrecht.

GEURTS, J.P.A.M. & J.G. VAN NIFTRIK (eds) 1997. Monumentenwet van 1988: wet van 3 december 1988, Stb. 683, tot vervanging van de Monumentenwet, 6th rev. ed., Zwolle (Nederlandse wetgeving: editie Schuurmans & Jordens, 52).

GIFFEN, A.E. VAN 1936. Der Warf in Ezinge, Provinz Groningen, Holland, und seine westgermanischen Häuser, *Germania* 20: 40-47.

GIFFEN, A.E. VAN 1940. Een systematisch onderzoek in een der Tuinster Wierden te Leens, *Jaarverslagen van de Vereniging voor terpenonderzoek* 20-24: 26-117.

HALBERTSMA, H. 1963. Terpen tussen Vlie en Eems, een geografisch-historische benadering, Groningen.

INVENTARISATIE VAN ARCHEOLOGISCHE MONUMENTEN IN FRIESLAND 1984. Overzicht van archeologisch belangrijke terreinen in Friesland 1984, Leeuwarden etc.

KNOL, E., E. KRAMER & G.J. DE LANGEN 1989. De verkenning van het vroeg-middeleeuwse grafveld van Oosterbeintum (Fr.), *Paleoaktueel* 1: 73-80.

KNOL, E., W. PRUMMEL, H.T. UYTTERSCHAUT et al. 1996. Het vroeg-Middeleeuwse grafveld van Oosterbeintum, *Jaarverslagen van de Vereniging voor terpenonderzoek* 73-74.

KNOL, E., W. PRUMMEL, H..T. UYTTERSCHAUT et al. 1995/1996. The early medieval cemetary of Oosterbeintum (Friesland), *Palaeohistoria* 37/38: 245-416.

LANGEN, G.J. DE, H. HOMMES, T.M. PERGER et al. 1993. Het project bodembeschermingsgebieden Friesland, *Paleoaktueel* 4: 80-84.

LANGEN, G.J. DE, H. HOMMES, T.M. PERGER et al. 1995. Tussen beleid en wetenschap. Tussentijds bericht over het project Bescherming op Maat in Friesland. *Paleoaktueel* 6: 60-63.

WATERBOLK, H.T. 1961. Beschouwingen naar aanleiding van de opgravingen te Tritsum, gem. Franekeradeel, *It Beaken* 23/24: 216-226.

ZIJLSTRA, J. 1991. Finns fibula? Belangwekkende Vroeg-Middeleeuwse vondsten te Wijnaldum, *Westerheem* 40: 51-62.

NOTES

1. Arjaans (1991, p. 51).
2. Halbertsma (1963, p. 5) and Arjaans (1991, p. 51).
3. Waterbolk (1961).
4. Ibidem (1961, p. 223).
5. Knol et al. (1989, 1996, 1995/1996).
6. Nijboer & Van Reekum (this volume) and Schoneveld & Zijlstra (this volume).
7. Gerrets (this volume).
8. See also Nijboer & Van Reekum (this volume).
9. Waterbolk (1961, p. 223). Waterbolk refers to the extensive surface of the Westergo region, the high population density and the large distance to the higher pleistocene sandy soils, compared with other parts of the saltmarsh in the northern Netherlands.
10. Stuijts, forthcoming.
11. We discussed this problem with J. Delvigne pedologist of the Department of Biology, University of Groningen.
12. See also contribution of Groenman-van Waateringe (this volume).
13. Gerrets, Excavation method (this volume, Fig. 1.).
14. De Langen & Waterbolk (1989, p. 92).
15. Van Giffen (1940, p. 32).
16. Cf. De Langen et al. (1993, 1995).
17. De Langen et al. (1993).

The Subatlantic evolution of the coastal area around the Wijnaldum-Tjitsma terp

P.C. VOS *with a contribution by* B.A.M. BAARDMAN

1 INTRODUCTION

1.1 *Study area and objectives*

This study about the Subatlantic coastal evolution of Westergo, the area in northwest Friesland between the former estuaries of Middelzee and Marne (Fig. 1), is carried out within the framework of the archaeological research at the terp Wijnaldum-Tjitsma. The terp Wijnaldum-Tjitsma is located in the northwestern part of Westergo, 600 m east of the small village of Wijnaldum. The archaeological excavations that were carried out on this terp between 1991 and 1993 offered an opportunity to study the evolution of salt-marsh deposits. Terp layers and salt-marsh deposits were investigated in two north-south cross-sections. The stratigraphy of the natural and artificial layers are recorded in drawings and photographs, and six lacquer-peels were taken to study the salt-marsh deposits in further detail. Moreover, diatom and pollen samples were taken for palaeo-environmental research and molluscs were sampled in the top of the sand-flat deposits and at the base of the salt-marsh deposits in order to date the changes in these environments at the site. Further, borings were made in the line of the main cross-section in northern and southern direction in order to study the lateral extension of the salt-marsh and tidal-flat deposits.

In Westergo, the Holocene coastal deposits attain thicknesses of 4-12 m, and in the silted up tidal channels thicknesses of even more than 25 m. South and east of Westergo, the Holocene coastal deposits wedge out against the topographically higher Pleistocene grounds of Gaasterland and the Drents Plateau.[1] The presence of the salt-marsh ridges is one of the most characteristic morphological features in Westergo. The configurations of the ridges are expressed on the surface contour map, but also on the soil map of the area because the soil texture of the ridges is lighter than the soil in between the ridges. Salt-marsh ridges are also found in other parts of the northern Netherlands, e.g. in the former Fivel tidal basin in northeast Groningen. The salt-marsh ridges in the northern Netherlands run more or less parallel to the present-day coastline of the Waddenzee. The ridges are generally ca 200-500 m wide and ca 0.5-1.5 m elevated above the surrounding surface, if the terp elevations are not taken into account.

The salt-marsh ridges of Westergo are of special archaeological interest, because the earliest terps are situated on these natural heights. In this way linear settlement patterns (terp rows) developed in this area. The salt-marshes of Westergo have been inhabited from the Late Bronze Age/Early Iron Age up to the present-day (geological and archaeological stratigraphic tables, see Fig. 2). For a clear understanding of the interaction between man and landscape, insight into the depositional history of the ridges is most relevant. However, as yet the genesis of the salt-marsh ridges in the northern Netherlands is uncertain.[2]

The aim of this study is to reconstruct the landscape development and habitation history at the site of Wijnaldum-Tjitsma and the surrounding area (Westergo). Based on the geological and archaeological research at the site of Wijnaldum-Tjitsma, the depositional and habitation history is reconstructed with the help of 6 cross-sections. The representation of the coastal evolution of Westergo, in 8 palaeogeographic maps, has been based primarily on the configuration of the salt-marsh ridges on the morphological map and the soil map, and on the dates of the terp settlements. Before we discuss the landscape reconstruction, first, the geological setting of Westergo, the literature about the salt-marsh ridges and the data used in the environmental reconstructions will be described.

In the appendix, compaction calculations for depositional and anthropogenic layers at the Wijnaldum-Tjitsma site made by Baardman are presented.

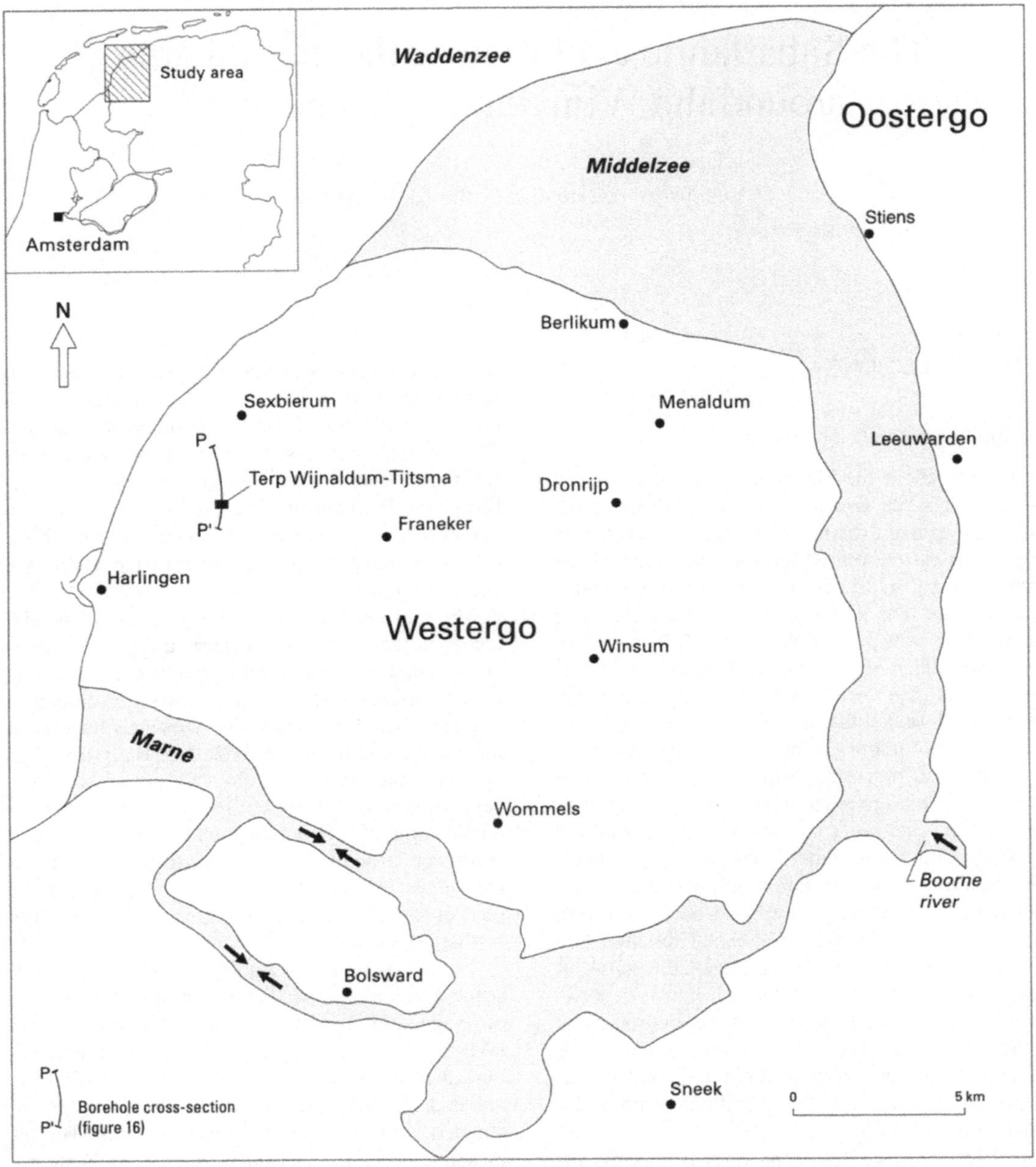

Figure 1. The study area, Westergo located between the former tidal systems of the Marne and Middelzee. Represented are the configuration of the salt-marsh ridges and the datings of the oldest habitation found on these ridges (drawing TNO-NITG).

These calculations were used by the reconstruction of palaeo-tidal levels which could be recognized in the deposits at the site.

1.2 *Geological setting*

During the early Atlantic, northwestern Friesland was drowned by the sea as a result of the world-wide, Holocene sea-level rise. The marine inundation followed the Weichselian Boorne Valley, the main drainage system of the area (Fig. 3). The coastline was still far up north, and northwestern Friesland formed the distal, landward part of the tidal basin. Similar to the Holland tidal basin,

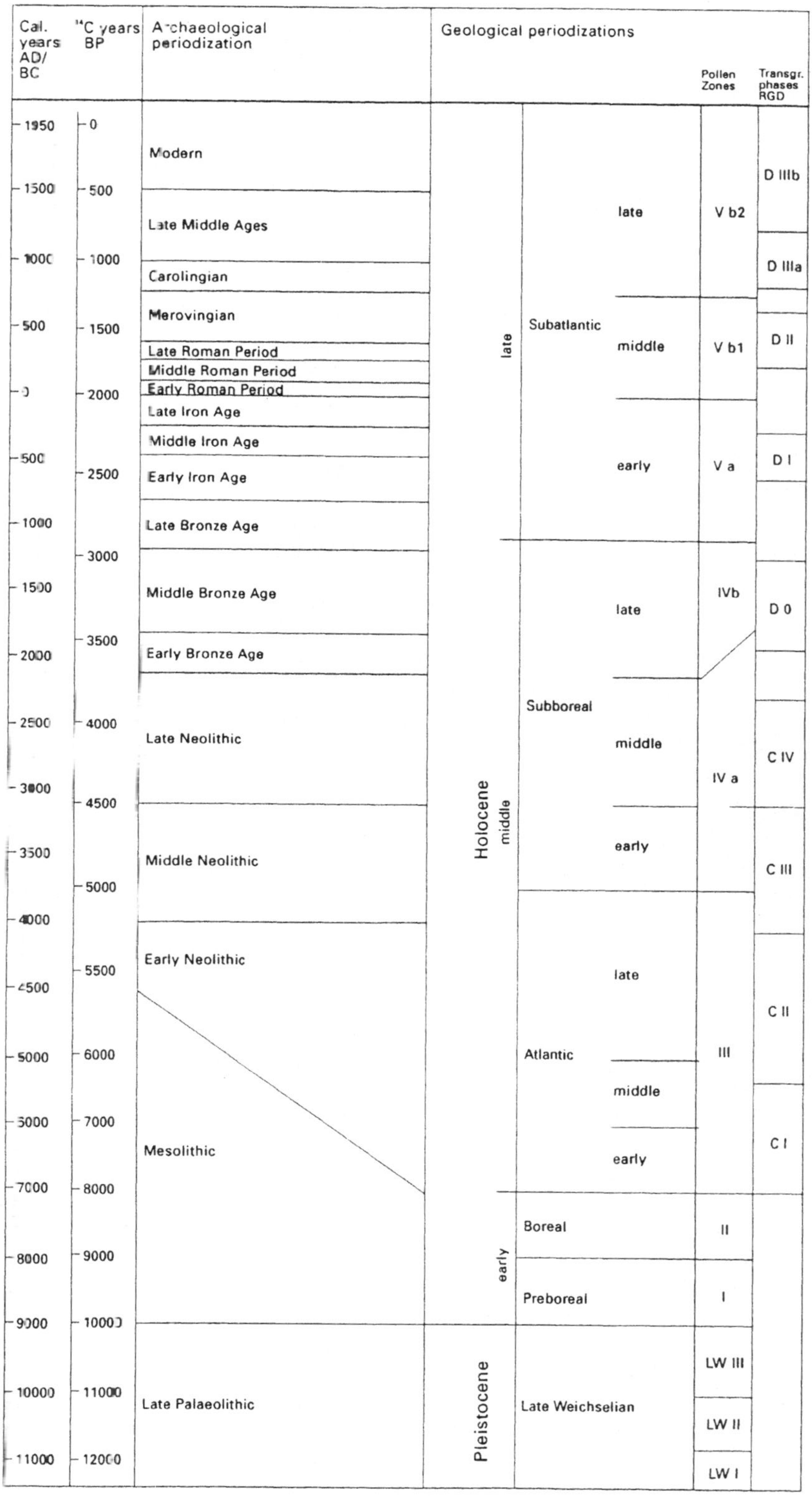

Figure 2. Geological and archaeological periodizations of the Holocene.

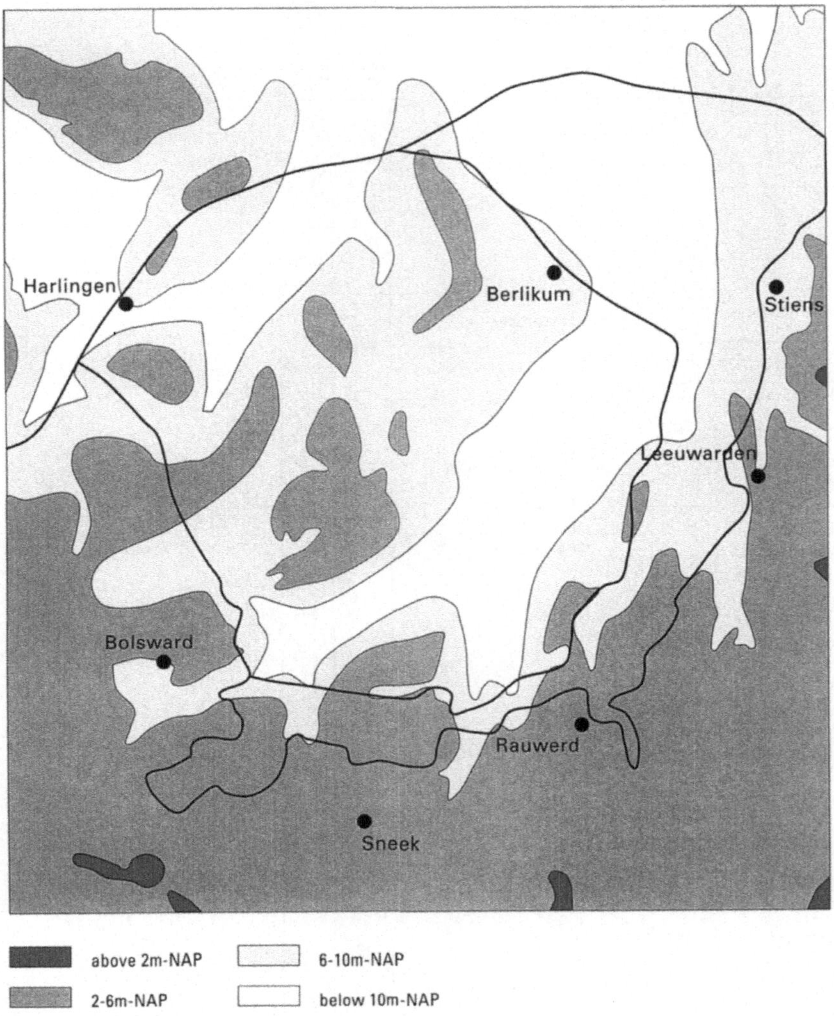

Figure 3. Base Holocene map of NW Friesland. Depth contours are in metres below NAP (drawing TNO-NITG; after Van Gijssel & De Gans 1993, Van der Spek 1994).

sedimentation in Westergo was characterized by low energy conditions during the early Atlantic. The energy of tidal waves was diminished, because of the extensive flood tidal deltas and other intertidal flats in the coast-proximal zone.[3] A brackish water environment with the deposition of clays prevailed. Often the clays contain roots of reed ('rooted clay'), indicating influxes of fresh water.[4] In between the brackish clay area and the southern high Pleistocene grounds, extensive peat swamps developed and formed the so called 'Basal Peat' layer.

Due to the sea-level rise during the Atlantic and early Subboreal the coastal zone and back barrier moved southward. Consequently, the marine facies belts consisting of channel deposits and sand- and mud-flat deposits also shifted landward over the Pleistocene subsurface. In Westergo, sandy channel deposits and intertidal-flat sediments were deposited on the rooted clay and locally tidal channels incised in the older Holocene clays and Pleistocene subsurface (Fig. 4). Around 4000 BP the landward boundary of the tidal basin in northwestern Friesland became more or less fixed as a result of the declining sea-level rise. As sediment supply continued in the area, a net infilling of the basin occurred up to the supratidal level.[5]

From 3000 BP onwards the salt-marshes progressed northward over the sandy tidal flat deposits. At the seaside of the salt-marshes, the typical salt-marsh clays with crinkly sandy laminations were formed.[6] Landward, in the salt-marsh basins,

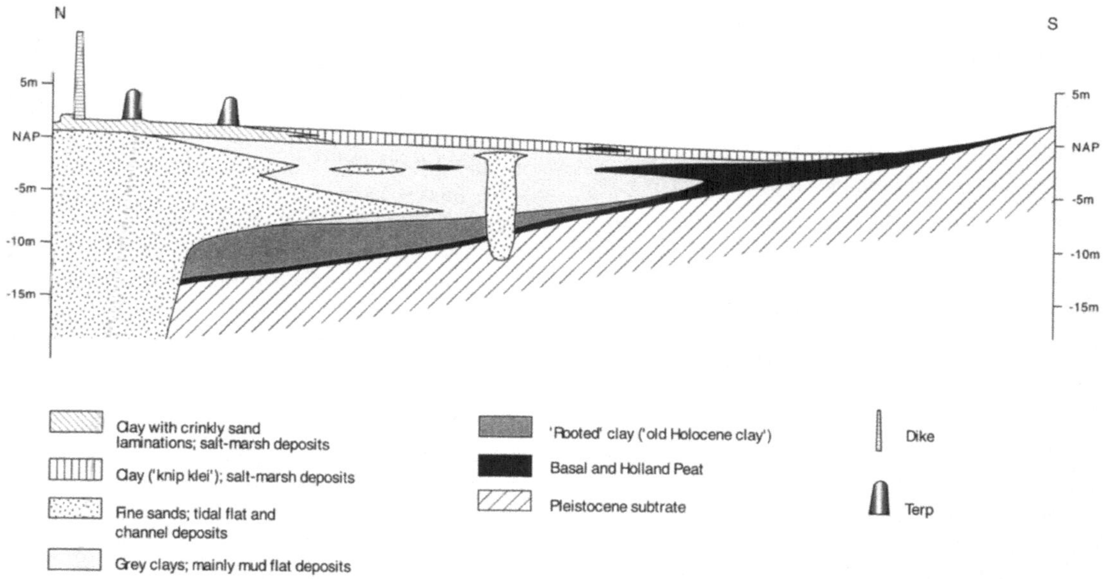

Figure 4. Schematic N-S cross-section through the Holocene coastal deposits of NW Friesland (drawing TNO-NITG).

heavy, sticky clays with a low carbon content were deposited, the so-called '*knip klei*'.[7]

About 700 BC, man started to occupy[8] the higher silted-up salt marshes in the southernmost part of Westergo (the so-called 'Old land'). The salt-marsh ridges, which were formed north of the 'Old land', were colonised during the Late Iron Age and Roman Period, indicating that the northern part of Westergo was silted up to the supra-tidal level and that the area was suitable for habitation (Fig. 1). In the Early Middle Ages the whole area of Westergo had changed into salt-marshes and the area took approximately its present shape.

Due to the filling-up of the channels in Westergo after the Late-Roman Period, the hinterland south of Westergo started to drain via the Middelzee and Marne systems.[9] With the continuing sea-level rise, tidal action penetrated in these systems. The penetration of the sea was possibly enhanced by artificial drainage and subsequent lowering of the peat/clay area south of Westergo.[10] At the end of the 9th century the Middelzee extended rapidly and tidal channels scoured deep into the subsurface.[11] In the 10th century seawater reached as far inland as Sneek and made contact with the Marne system which drained the southwestern side of the Westergo area. The watershed between both systems developed near Bolsward.[12]

The oldest dikes were constructed in the beginning of the 10th century on the levees and salt-

Figure 5. Embankment history of Middelzee (drawing TNO-NITG; from Van der Spek 1994, following mainly Rienks & Walther 1954).

marsh ridges of Westergo and Oostergo.[13] The Middelzee started to silt up and during the next centuries the Middelzee area was reclaimed stepwise from the south to the north (Fig. 5). With the construction of the dike of 1505 in Het Bildt, an almost linear coastline was formed between Westergo and Oostergo. Salt-marshes that accreted at the toes of these dikes were reclaimed in the following centuries.

1.3 *Salt-marsh ridges in the Dutch literature*

Different opinions exist in the Dutch literature about the terminology and genesis of the salt-marsh ridges in the northern Netherlands. Van Veen was one of the first who recognized salt-marsh ridges in the area of the former Fivel tidal basin (Groningen) and he described them as 'shore ridges' or 'levees' (*oeverwallen*). Van Veen[14] attributed the formation of ridges to wave action (Fig. 6) and compared the origin of the ridges with the formation of the coastal barriers *(schoorwallen)* in the Waddenzee. At the southeastern lee-side of the former basin, Van Veen recognized low 'head ridges'. The Fivel basin was closed by W-E oriented 'tail ridges', which developed at the northwestern margin of the basin by the dominant NW winds (Fig. 6). Roeleveld,[15] in his study of the Groningen coastal area, only mentioned the 'marsh-bar' system of the Fivel tidal basin; he called the system a 'secondary complex spit'.

Bakker[16] described ridges in the Westergo area (Friesland) and he explained them by his 'channel-ridge hypothesis': tidal channels of the 'first generation' silted up between AD 0-900 and were transformed into ridges between 1000-1200 due to differences in compaction. The sandy channel fills subsided less than the surrounding clayey subsurface. The relief differences were enlarged by channel incisions in the swales between the ridges around 1200/1300.[17] In later publications,[18] however, Bakker largely rejected his channel-ridge hypothesis and instead attributed the origin of the ridges to the topography of tidal sandflats in the subsurface *(wadzandrelief)* and to the incision of the tidal channels of 1200/1300 AD.

According to the soil scientists and geomorphologists of the former STIBOKA[19] (the present Staring Centre), the formation of the salt-marsh ridges was the result of erosion of the salt-marsh shore face during a transgression and/or storm-surge periods. The eroded sandy sediments were redeposited directly over the salt-marsh edge and due to this deposition a low ridge was formed (Fig. 7). The clayey particles settled during these erosive events further inland in the 'salt-marsh basin'. Roeleveld, following the transgression hypothesis of the salt-marsh ridges of De Smet,[20] de-

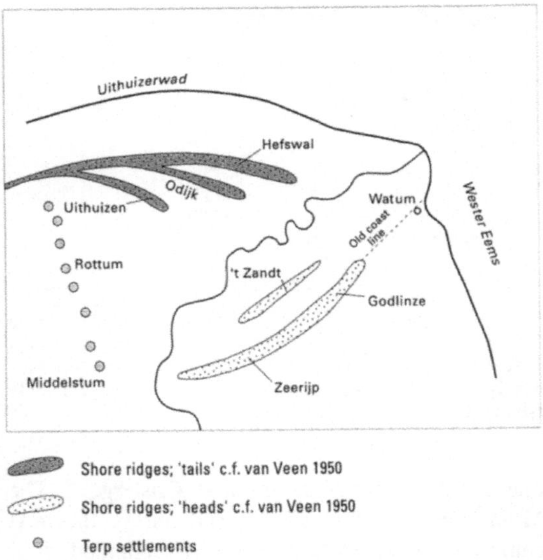

Figure 6. 'Shore ridges' in the silted-up mouth of the Fivel tidal basin (NE-Groningen) (drawing TNO-NITG; after Van Veen 1950).

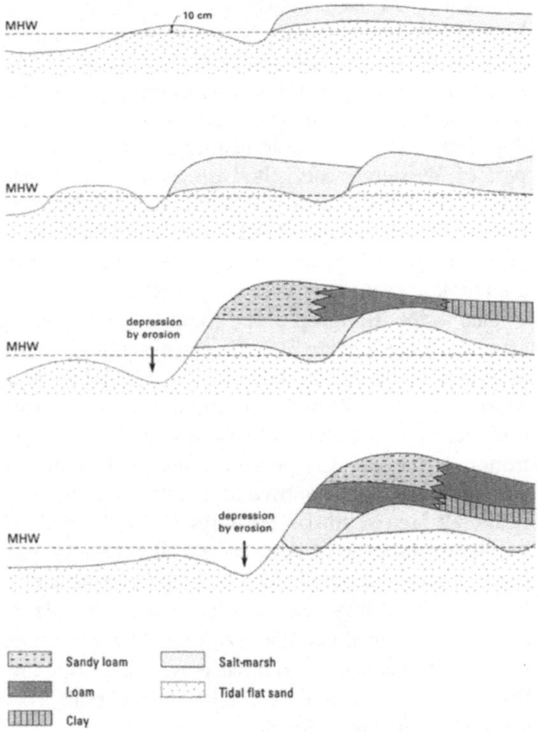

Figure 7. Schematic cross-section, representing the evolution of a salt-marsh ridge (drawing TNO-NITG; after De Smet 1965).

fined the 'marsh bars' of the Marne area in NW Groningen as 'cheniers'.[21]

The ridges in NE Friesland, e.g. Holwerd-Waaxens-Brantgum and Metslawier-Jouswier-Ee, studied by Griede,[22] are not similar to the salt-marsh ridges as described by De Smet (Fig. 7). These ridges in NE Friesland have a different orientation. Rather than parallel to the Waddenzee coastline, they are running more or less perpendicular. Genetically they have proven to be channel-inversion ridges: sandy tidal-channel fills which have a higher elevation than the surrounding clay/peat area due to the differential compaction of the subsoil in the region.

2 CONFIGURATION OF THE SALT-MARSH RIDGES IN WESTERGO

The configurations of the ridges are displayed on the morphological map (surface contour map; Fig. 8), and on the soil map of the area[23] (Fig. 9). The ridges are shown on the soil map because the soil texture of the ridges is lighter than the soil in between the ridges. Man used these low natural heights to construct dwelling mounds (terps). Therefore the salt-marsh ridges can also be detected from the rows of terps on the terp location map (Fig. 10). In the summarizing Figure 11, the names, estimated ages, and stages of development of the salt-marsh ridges and terp rows are given.

2.1 *Morphological map*

Based on the surface contour map, which was derived from the topographical maps (scale 1:10,000) of the area, Westergo can be subdivided into the following morphological sub-areas (Fig. 8):

- 'Old land':
  a) Low land,
  b) Tidal levees of the Marne and Middelzee tidal systems.
- Funnel-shaped palaeo-basin:
  a) Fan-shaped ridges,
  b) Western depression,
  c) Cap ridge.

The 'Old land' is the low area in between the tidal levees of the Marne and the Middelzee, which is located south of the salt-marsh ridges (Fig. 1). The natural surface of the Old land is generally below 0.5 m above NAP (Nieuw Amsterdams Peil, Dutch Datum Level)[24] and no pronounced ridges are found in this low area. Since the oldest settlements found in the Old land date from the Late Bronze Age/Early Iron Age,[25] the Old land must have silted up and changed into a salt-marsh before the 6th century.

Within the funnel shaped palaeo-basin, north of the Old land, three morphological subareas can be recognized: a series of fan-shaped ridges at the eastern side of the basin, a depression at the western side of the basin and a closing cap ridge in the north.

The series of fan-shaped ridges are rather complex. Rather than elongated bars, these ridges are fan-shaped rows consisting of patchy elevations. The 'turning point of the fan' can be postulated eastward of Berlikum. The elevations of the rows vary in height between 0.5-1.2 m above NAP. Because of the patchy character of the fan-shaped ridges the width of the ridges is difficult to determine. At a rough estimate, the ridges are 100 m to 750 m broad.

West of the funnel-shaped basin a depression is located. In general, the depression's surface is below 0.5 m above NAP. In the northwestern part of the depression one ridge is found: Wijnaldum-Voorrijp. The Wijnaldum-Voorrijp ridge is situated in line with the Dongjum-Ried-Berlikum ridge. They are separated from each other by the low land, which is part of the western depression, around the Sexbierumer Vaart. The fan-shaped ridges and the western depression are separated from the Waddenzee by the cap ridge, a large ridge along the present-day sea dike, which is about 1 to 2 km broad. It is remarkable that the northeastern part of the surface of the cap ridge is silted up to a higher level (maximum ca 1.5 m above NAP) than the surface of the southwestern part (maximum ca 1.0 m above NAP). On the northeastern part of the cap ridge also several smaller ridges are found.

2.2 *Soil map*

The configuration of salt-marsh ridges is also expressed by textural differences on the soil map (Fig. 9). The configuration on the soil map shows a pattern more or less comparable with the pattern shown on the morphological map: the Old land, consisting of silty clay or silty clay loam (*zware of lichte klei*), the eastern fan-shaped ridges, consisting of sandy loam (*lichte zavel*), the western depression and swales in between the fan-shaped ridges, consisting of silty clay loam or loam (*lichte klei of zware zavel*) and the cap ridge, consisting of sandy loam (*lichte zavel*).

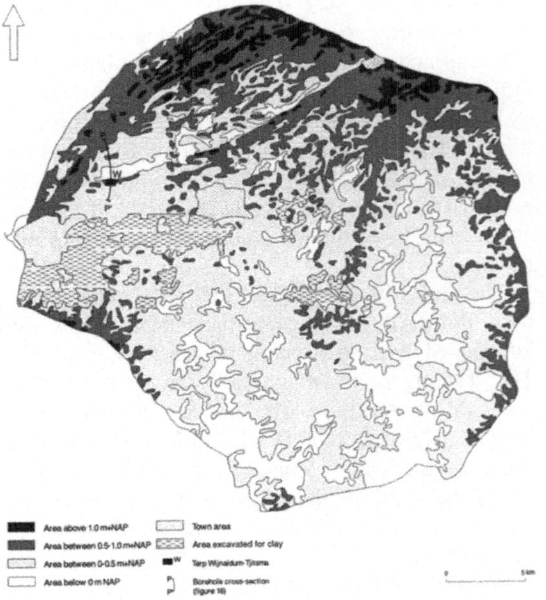

Figure 8. Morphological map of Westergo, surface contour lines based on the topographic map, sheets 5 and 10, scale 1:10,000, 1961, Topographic Survey of the Netherlands, Emmen (drawing TNO-NITG).

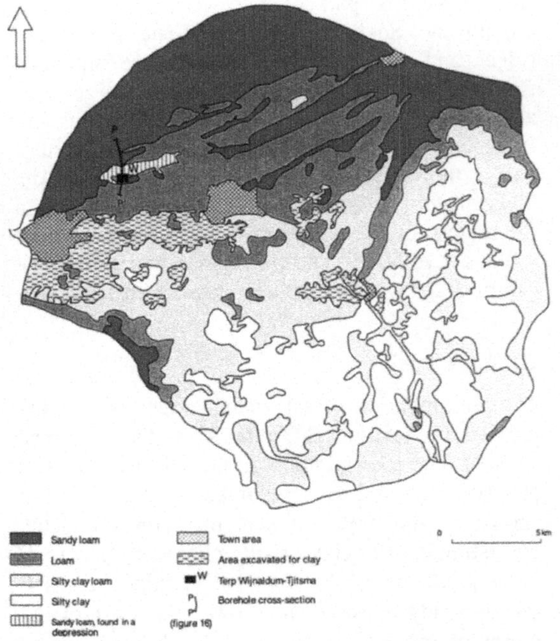

Figure 9. Regional soil map of the Westergo area compiled from the soil maps 1:50,000, sheets 5 (Harlingen) and 10 (Sneek), STIBOKA 1976 and STIBOKA 1974 (drawing TNO-NITG).

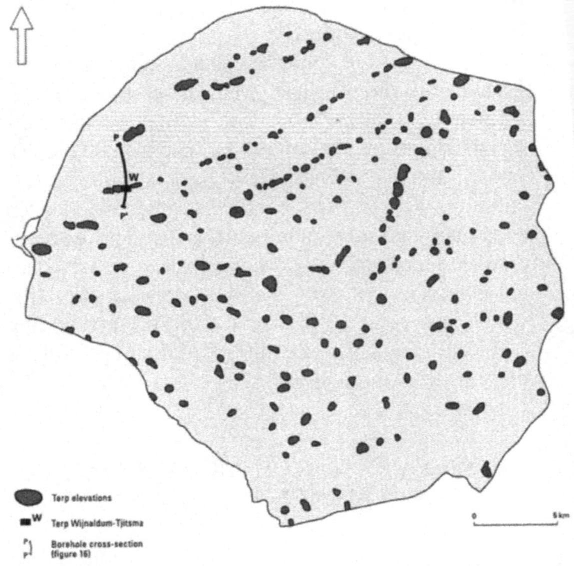

Figure 10. Map with the location of the terps in Westergo (drawing TNO-NITG).

The main difference between the soil map and the morphological map is the more linear configuration of the ridges (Baijum-Dronrijp-Menaldum; Schalsum-Peins-Beetgum; Dongjum-Ried-Berlikum) on the soil map compared with the pattern on the morphological map (Fig. 8). The latter map gives a more complicated impression.

2.3 *The configuration of the terp-rows*

The salt-marsh ridge pattern, shown on the morphological and soil maps, is also clearly expressed by the configuration of the dwelling mounds on the terp location map of Figure 10. The terps in the funnel-shaped basin are largely located on the ridges, and therefore the terp rows clearly show the orientation of the ridges (Figs 10 and 11). This is not the case in the 'Old land' where no salt-marsh ridges occur and where the oldest terps date from the 8th-6th centuries BC.[26] In this area, the distribution of the terps is more or less scattered.

Not all terp rows in the funnel-shaped basin are found on ridges. Especially, in the western part of the funnel-shaped basin where in general the salt-marsh ridges are less well developed, terp rows occur which are not situated on well-defined ridges (Fig. 11). This is e.g. the case with the terp row Sopsum-Hitzum-Tzum, at the western margin of

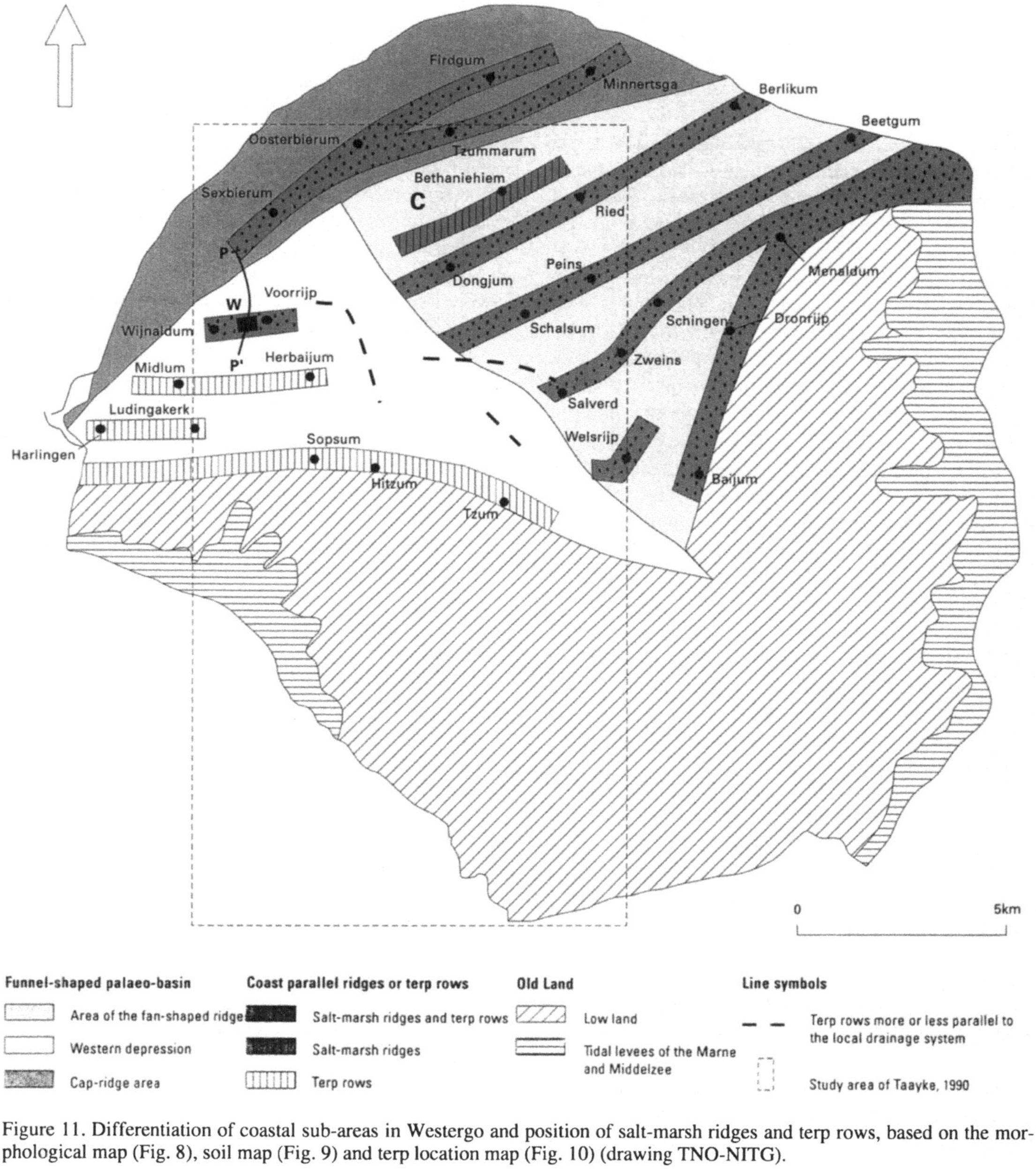

Figure 11. Differentiation of coastal sub-areas in Westergo and position of salt-marsh ridges and terp rows, based on the morphological map (Fig. 8), soil map (Fig. 9) and terp location map (Fig. 10) (drawing TNO-NITG).

the funnel-shaped basin, which, like the terp row on the eastern margin, can be dated in the 3rd/4th centuries BC. On one salt-marsh ridge of the fan-shaped ridges, the ridge of Bethaniënhiem just south of the cap ridge (Fig. 11), no terp row has developed. The absence of the terp settlements on this ridge might be related with the hiatus in the occupation of the Westergo area during the 4th century AD.

3 RESULTS OF THE GEOLOGICAL INVESTIGATIONS AT THE TERP WIJNALDUM-TJITSMA

3.1 *Lithological units*

The archaeological excavations in the terp Wijnaldum-Tjitsma, situated on the salt-marsh ridge of

Wijnaldum-Voorrijp, provided an excellent opportunity to study the natural deposits under the terp layer (Fig. 1). The so-called 'main cross-section', consisting of the eastern walls of the north-south oriented excavation pits WP VII, IV, III, I, XXXIV, II and V (Fig. 12), had a total length of 210 m and had a maximum height of 4 m (Fig. 13). In the central part of the terp, the shorter 'west cross-section' consisting of the west wall of the excavation pit WP I was documented (Figs 14 and 15).

In line with the north-south oriented main cross-section of the archaeological site, boreholes were made with an auger between highway A31 (Harlingen-Leeuwarden) and the provincial road *Hearewei* (Harlingen-Sexbierum) over a distance of about 2.5 km (Fig. 16). The borehole interval was 50 to 150 m and the depth was maximum 5 m.

In the auger bore-hole cross-section and in the main- and western cross-sections of the excavation pits the following natural and/or anthropogenic lithological units were distinguished from the bottom to the top:

- Sand with clay layers,
- Extremely fine sands,
- Clay with crinkly sand laminations,
- 'Wall' deposits, and
- Terp layer.

The lithology, sedimentary environment and dates of these units will now be discussed.

3.1.1 *Sand with clay layers*

Lithology. The unit 'sand with clay layers' was, because of its depth, only visible in the bore-holes. The top of the layer is in general at a depth of about 2 m below NAP, except for the bore-holes in the depression near the Sexbierumer Vaart, where the layer reaches a level up to 1.25 m below NAP. The unit consists of grey, very fine to medium fine sands (classification names cf. NEN 5104[27], Table 1). The medium grain size of the sand varies generally between 105 and 165 µm. The unit contains thin clay layers (3 mm up to 2 cm). The number of clay layers varies from a few to many clay layers. In some zones the amount of clay is so large that the unit should be classified as 'clay with sand layers'. Thin layers of very fine peat detritus, consisting of eroded and rounded peat fragments, and small nails of sea-urchins are generally present in the unit. The unit contains marine molluscs such as *Macoma baltica, Cerastoderma edule* and *Hydrobia ulvae.*

Environment and age. The environmental reconstruction of the unit 'sand with clay layers' is difficult because the deposits were not exposed in the excavation pits. For the reconstruction we have to rely on the information from the auger bore-holes with a diameter of 2.5 cm. Most of the specific sedimentological characteristics of subtidal- and intertidal deposits[28] are not registered in auger bore-holes. A characteristic criterion of an intertidal environment, which is in most cases detectable in an auger core, is bioturbation. Bioturbation is not recognized in the sand with clay layers, which suggests that this unit was largely formed in a tidal channel environment.

Radiocarbon, pollen or archaeological dates are not available for the unit sand with clay layers. Anyway, the unit is older than the overlaying extremely fine sands, of which the formation of the upper part has been radiocarbon dated at 300-200 BC (Fig. 17). We suppose that the major part of the unit 'sand with clay layers', bored in the auger cross-section, was formed during the period 1000-600 BC.

3.1.2 *Extremely fine sands*

Lithology. The unit 'extremely fine sands' (NEN 5104, Table 1) is mostly found in the borings between 0.8 and 2.5 m below NAP. The average level of the top of the extremely fine sands is south of the terp Wijnaldum-Tjitsma somewhat lower than at the northern side (Fig. 16). In the borings south of the terp Wijnaldum-Tjitsma (577-160-7 to 578-160-5) the average top of the extremely fine sands is 108 cm below NAP, whereas the average top in the borings north of the Sexbierumer Vaart (597-160-15 to 580-160- 5) is 97 cm below NAP.

In the main cross-section and the western cross-section of the excavation, the top of the extremely fine sands forms the base of the cross-sections. No more than 5-40 cm of this layer was exposed. In the cross-sections, the top of the layer is found at a

Table 1. Classification of sand (mean grain size) according to the NEN 5104.

Name in Dutch	Name in English	Mean grain fraction between
Uiterst fijn	Extremely fine	63-105 µm
Zeer fijn	Very fine	105-150 µm
Matig fijn	Medium fine	150-210 µm
Matig grof	Medium coarse	210-300 µm
Zeer grof	Very coarse	300-420 µm
Uiterst grof	Extremely coarse	420-2000 µm

Figure 12. Location map of excavation trenches and cross-sections at the site Wijnaldum-Tjitsma near Wijnaldum (drawing TNO-NITG).

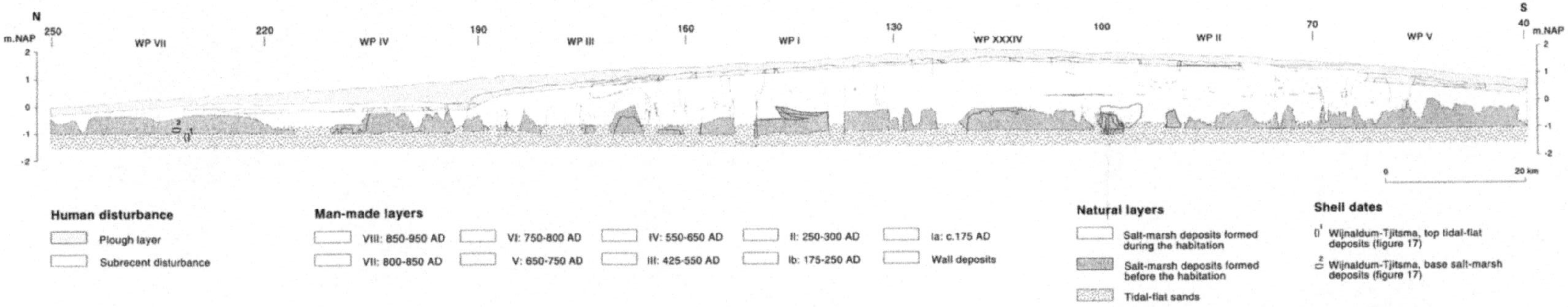

Figure 13. Main cross-section (for location see Fig. 12; drawing TNO-NITG).

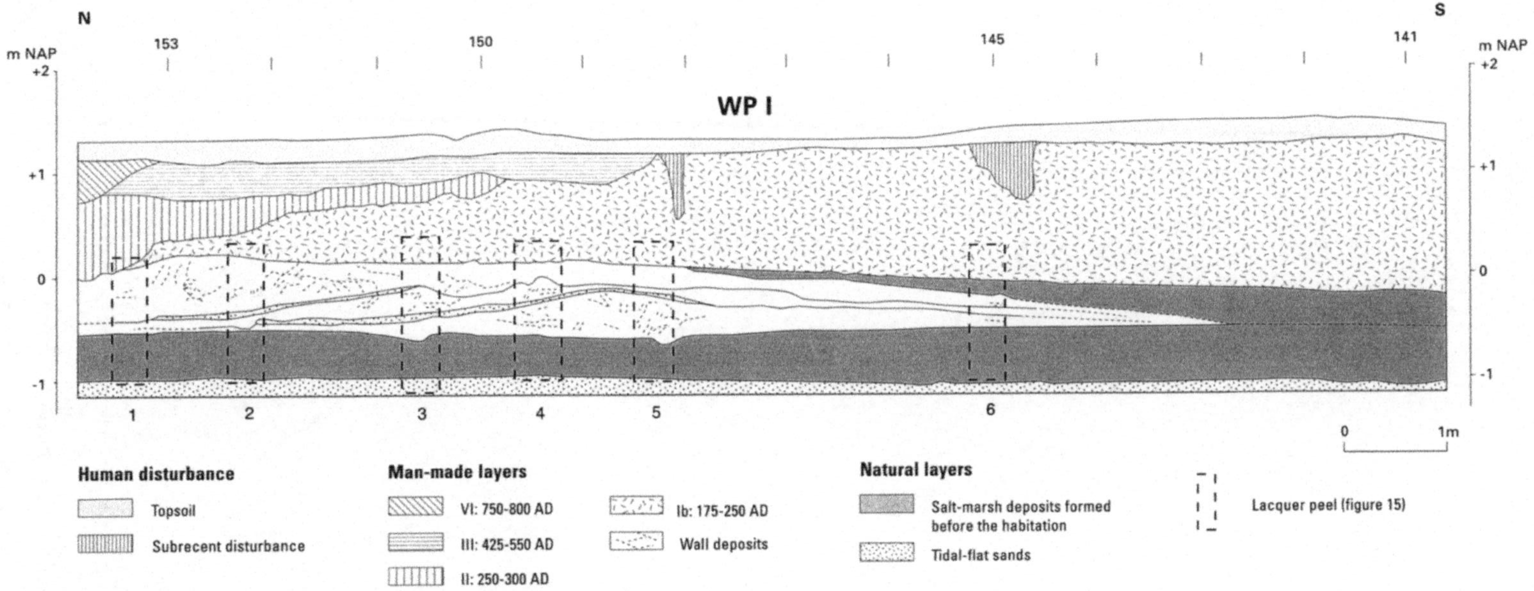

Figure 14. West cross-section (for location see Fig. 12; drawing TNO-NITG). 1. Senaka Welidenya (1996, p. 2). 2. See CUR-commissie (1993) and Locher & De Bakker (1990, part 1: 306-310). 3. See Tysma et al. (1994, F2/34 and F2/8).

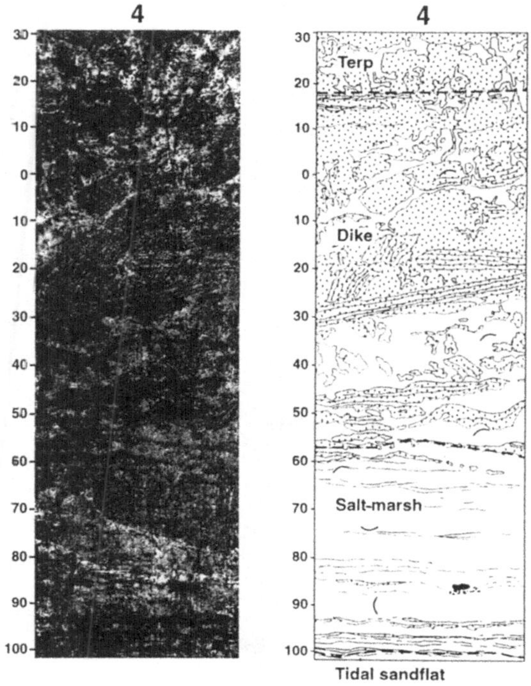

Figure 15. Lacquer-peels of the west cross-section (drawing TNO-NITG).

depth of 95-108 cm –NAP. The boundary with the overlaying clay layer is sharp, but not erosive and in the cross-sections (Figs 13 and 14) the boundary is almost a straight horizontal line.

The medium grain size in the unit varies between 80-100 µm. The sands are grey and generally clayey with a lutum content between 2-8%. Sometimes the clay content of the unit exceeds the 8% lutum limit, which implicates that the unit should be classified as very sandy clay (cf. NEN 5104). Because of the shell particles in the sands, the carbonate content of the unit is high. Sedimentary structures in the clayey sands and sandy clays are not found or are strongly disturbed, as the layer was completely or almost completely homogenized by bioturbation. Double-valved shells of *Scrobicularia plana* in life position are found until about 10 cm from the top of the layer.

Environment and age. The extremely fine sands were formed in an intertidal flat environment. This is evidenced by the occurrence of shells of *Scrobicularia plana* and *Macoma baltica* in 'life position' and by the presence of an *epipsammic* diatom assemblage (diatom zone A, Fig. 18). The deposits

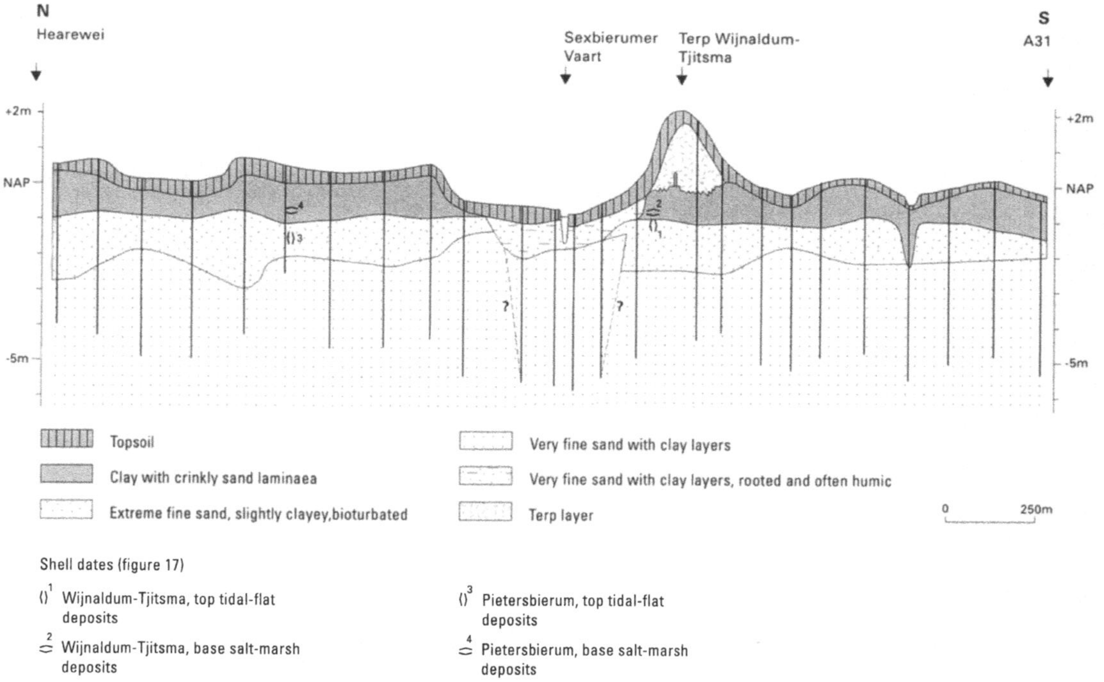

Figure 16. Borehole cross-section Hearewei-A31, crossing the Wijnaldum-Tjitsma terp (for location see Fig. 1; drawing TNO-NITG).

Terp Wijnaldum-Tjitsma: top tidal-flat deposits

Lab.no	Conventional radiocarbonate age (yr BP)	Dated material	Sample attitude (m-NAP)	Cal.years BC/AD 800 700 600 500 400 300 200 100 0 100
GrN 19581	2250±45	Scrobicularia plana	-1.10	

c. 275 BC

Pietersbierum: top tidal-flat deposits

Lab.no	Conventional radiocarbonate age (yr BP)	Dated material	Sample attitude (m-NAP)	Cal.years BC/AD 800 700 600 500 400 300 200 100 0 100
GrA 359	2250±30	Macoma baltica	-1.57	
GrA 360	2270±40	Scrobicularia plana	-1.57	

c. 250 BC

Terp Wijnaldum-Tjitsma: base salt-marsh deposits

Lab.no	Conventional radiocarbonate age (yr BP)	Dated material	Sample attitude (m-NAP)	Cal.years BC/AD 800 700 600 500 400 300 200 100 0 100
GrN 19582	2140±35	Macoma baltica	-0.85	
GrN 19587	2125±40	Cerastoderma edule	-0.85	
GrN 19589	2055±35	Mytilus edule	-0.85	

c. 100 BC

Pietersbierum: base salt-marsh deposits

Lab.no	Conventional radiocarbonate age (yr BP)	Dated material	Sample attitude m-NAP)	Cal.years BC/AD 800 700 600 500 400 300 200 100 0 100
GrA 354	2590±40	Cerastoderma edule	-0.67	
GrA 356	2120±60	Spicula	-0.67	
GrA 357	2260±40	Cerastoderma edule	-0.67	
GvA 358	2140±40	Macoma baltica	-0.67	
GrN 20406	2020±35	Mytilus edule	-0.67	

Cal. 2s-age range — Estimate age

Cal. 1s-age range

Figure 17. Overview of the calibrated, geological radiocarbon dates from the ditch exposure Pietersburum (location, see Fig. 15) and the main cross-section of the Wijnaldum-Tjitsma terp (for location, see Fig. 13). Radiocarbon ages were converted to a sidereal time scale using CALIB 3.0.3 (Stuiver & Reimer 1993). Thick line = 1σ-age range (68.3%) and thin line = 2σ-age range (95.4%) (drawing TNO-NITG).

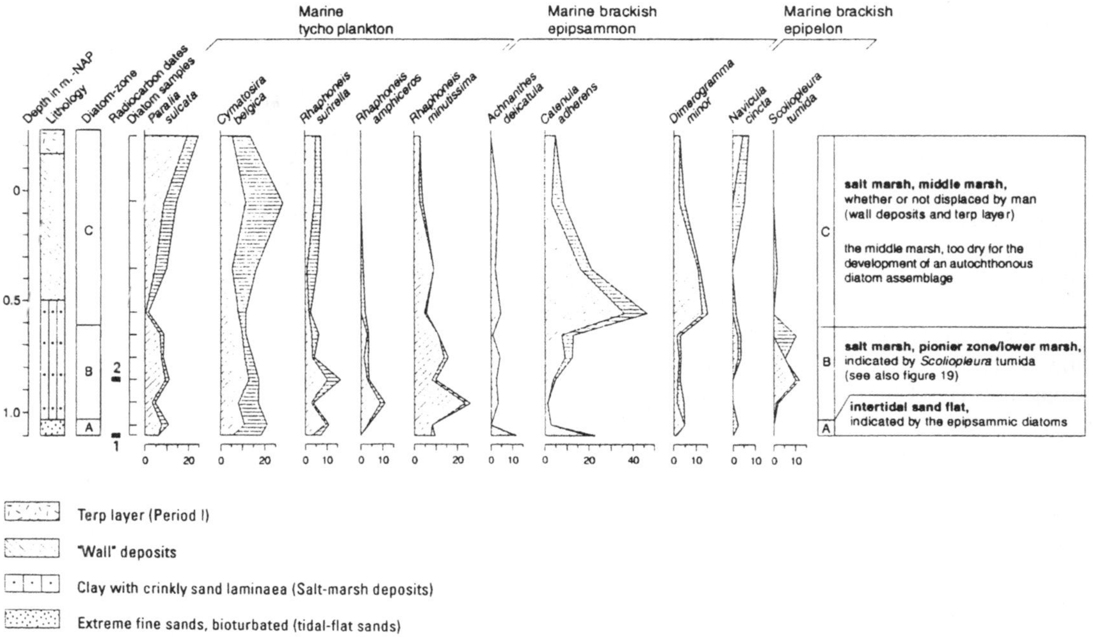

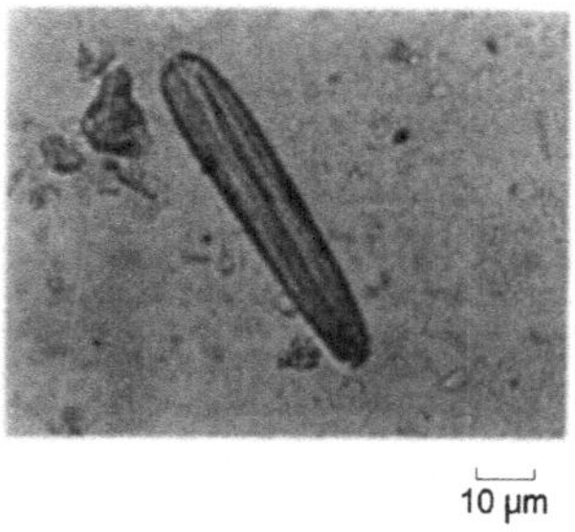

Sampled molluscs (figure 17)		Lab. no	Conv. radioc. age (yr BP)	Cal. 2σ-age range (yr cal. BC)
2	Macoma baltica	GrN 19582	2140± 35	356 - 102
	Cerastoderma edule	GrN 19687	2125± 40	356 - 52
	Mytilus edule	GrN 19689	2055± 35	168 - 8
1	Scrobicularia plana	GrN 19581	2250± 45	398 - 200

Figure 18. Diatom diagram Wijnaldum-Tjitsma. Diagram representing the relative abundances of the individual species and the environmental interpretation. Slanting lines = % underbroken diatoms, horizontal lines = % broken diatoms. Photograph diatom *Scoliopleura tumida* (drawing and photograph TNO-NITG).

are strongly bioturbated, which is an other indication of an intertidal environment.

Shells in life position from the top of the extremely fine sands in the main cross-section of Wijnaldum-Tjitsma and from the upper part of these sands in the borehole Pietersbierum (Fig. 16) have been radiocarbon dated. All the calibrated dates of shell samples from these locations have about the same 2σ-age range: 400-200 BC (Fig. 17). Because of the broad and overlapping 2σ-age ranges, the radiocarbon dates do not give an answer to the question whether the top of the tidal-flat deposits at borehole Pietersbierum is younger than the top of the same deposit under the terp Wijnaldum-Tjitsma. The radiocarbon dated molluscs of the borehole Pietersbierum were derived from a slightly lower level in the extremely fine sands (1.57 m –NAP) than those at Wijnaldum-Tjitsma (1.10 m –NAP), which makes the age comparison of the top of tidal flat deposits on both locations more complicated. Anyway, a younger age of the top of the extremely fine tidal flat sands at borehole Pietersbierum should be expected since the borehole is located 1125 m closer to the Waddenzee than the terp Wijnaldum-Tjitsma (Fig. 16).

3.1.3 *Clay with sand laminations*

Lithology. The clay layer on top of the very fine sands is laminated with thin sand layers with a thickness varying between 1 mm and 1 cm. The lutum content of the clay layer varies between 25 and 35% (Ks3: *klei sterk siltig*, according to the NEN 5104 classification).[27] The medium grain size of the sand in the laminations is 80-90 µm. In most cases the top of the layer has been excavated by man; the sods were used for the elevation of the terp and as construction material. Where not disturbed, mainly under the first house platform from the Middle Roman Period, the clay layer with sand laminations reaches a thickness of 85 cm. The top of the non-disturbed unit lies at a level of 10-15 cm below NAP. The concentrations of sand laminae in the upper 40 cm of the clay layer (sandy zone) are generally higher than those in the basal 35 cm of the unit (clayey zone).

All the sand laminations in the sandy and the clayey zone have a horizontal position and have a rather crinkly shape. The crinkly sand laminations generally can be followed over tens of meters. The sand laminations are only truncated by man-made ditches.

At the base of the unit, a 1 to 3 cm thick humic layer occurs. Locally in the layer thin shell layers are found which contain e.g. doublets of *Cerasotoderma edule*, *Macoma baltica* and *Mytilus edule*. None of these doublets are found in living position. In de oxidation zone above the lowest level of the ground-water table the clay unit is mottled by brown iron spots and nodules and greenish phosphate spots.

Environment and age. The 'clay layer with sand laminations' was formed in a salt-marsh environment, as indicated by the crinkly shape of the sand laminations.[29] In the lower part of these salt-marsh deposits under the terp Wijnaldum-Tjitsma bivalved skeletons of the diatom *Scoliopleura tumida* are found (diatom zone B: 70-105 cm below NAP, Fig. 18), indicating that this species was autochthonous. *Scolipleura tumida* is a species which can only live in a marine to brackish environment that is frequently flooded (below or around MHW). Thus, the occurrence of *Scolipleura tumida* indicates that the lower part of the clay layer with crinkly sand laminations was formed in the pioneer zone or the lowest part of the lower marsh (Fig. 19).[30]

The molluscs from the shell layers found in the lower part of the salt-marsh deposits, in the main cross-section of Wijnaldum-Tjitsma (Fig. 13, WP VII) and in borehole Pietersbierum (Fig. 16), have been radiocarbon dated (Fig. 17). These molluscs are allochthonous as these molluscs do not live in a salt-marsh environment. The presence of doublets of *Cerastoderma edule*, *Macoma baltica* and *Mytilus edule* indicates that during storms living species from the nearby creeks and the tidal-flat environment were picked up and deposited secondarily on the salt-marsh. Therefore, these replaced ('living') shells are reliable time markers.

As was the case with the radiocarbon dates in the top of the tidal flat sands, the 2σ-age ranges of the molluscs from the lower part of the salt-marsh deposits at Wijnaldum-Tjitsma and at borehole Pietersbierum do not differ significantly from each other. According to the dates, *Mytilus edule* at borehole Pietersbierum tends to be somewhat younger than *Mytilus edule* date at Wijnaldum-Tjitsma. Therefore, the best estimated ages of the shell storm layers, are interpreted somewhat younger at borehole Pietersbierum: ca 100 BC at Wijnaldum-Tjitsma and ca 50 BC at borehole Pietersbierum (Fig. 17); and also the age of the base of the salt-marsh deposits (start of the formation of the pioneer zone) are taken to be younger: ca 175-200 BC at Wijnaldum-Tjitsma and ca 125-150 BC at borehole Pietersbierum (Fig. 17).

The amorphous peat material, found in the humic layer at the base of the salt-marsh deposits under the terp Wijnaldum-Tjitsma, appeared to be eroded older peat from the surrounding area. Peat material, sampled in the main cross-section (Fig. 13, WP VII, 225 m) was dated 4855 ± 40 (GrN-19814).

3.1.4 *Wall deposits*

Composition and age. The wall deposits consist of sod layers and naturally deposited cross-stratified sand layers. In the west cross-section and in the main cross-section (Figs 13 and 14, WP I), the wall deposits are found immediately under the first house platform. The construction of this platform preceded the first permanent occupation at the site.[31] They are intercalated in the top of the unit clay with sand laminations and they have a lens-shaped form. The wall deposits were excavated in WP I only. The adjacent area was not excavated deeply enough to expose these deposits. In the west cross-section, the base level of the wall deposits is ca 60 cm –NAP and the top level is ca 15 cm +NAP. The dike-like structure is thus ca 75 cm high and in the cross-section about 10 m wide. The orientation of the dike-like structure in the excavation WP I is WSW-ENE.

The wall deposits are exposed best in the west cross-section (Fig. 14). In the main cross-section, a great part the unit was disturbed by a well (Fig. 13, WP I, 150 m; Period IIIA/B: 425-550 AD). The sods of the wall deposits, which were laid down in regular squares, consisted of clay with many crinkly sand laminations (Figs 14 and 15), which implicates that the sod material was derived from the sandy zone of the 'clay unit with sand laminations'. The two natural, cross-stratified, very fine sand layers found in the northern part of the wall deposits are erosion layers indicating that the wall deposits were truncated two times during storms. The sods on these cross-stratified sand layers show that the dike-like structure was renewed after each storm truncation (Fig. 14).

The wall deposits can not be dated precisely because no datable finds like sherds or bones were discovered. Anyhow, the dike-like structure was constructed shortly before the oldest occupation phase (Period Ia) which started about 175 AD.

3.1.5 *Terp layer*

Composition, age and purpose. The terp layer is a man-made layer. The sods which were used for the terp construction were generally cut from the upper sandy zone of the unit 'clay with sand laminations'. The terp material is often strongly mottled by fragments of charcoal and sherds and by patches of iron and phosphate. The top of the terp layer was decalcified by soil processes (leaching).

The terp Wijnaldum-Tjitsma was occupied from ca AD 175 until 900/950. After the 10th century the inhabitants presumably moved to a terp more to the south on which the actual farm is situated to-day. On the basis of the stratigraphy of the terp and the find material (mainly pottery-assemblages) eight occupation phases could be recognized: Period Ia (ca AD 175), Period Ib (AD 175-250), Period II (AD 250-300), archaeological hiatus (AD 300-425), Period III (AD 425-550), Period IV (550-650, Period V (650-750), Period VI (750-800), Period VII (800-850) and Period VIII (850-950) (Fig. 13).[32]

A house-platform had to be constructed, because the salt-marsh was still flooded during spring/storm tides, as indicated by the storm layers in the dike-like structure. The heightening of an artificial dwelling mound (periods Ia and b) was thus for safety purposes. During the later occupational phases the terp was further raised and extended by the inhabitants of the terp (Period II-VIII).[33] Thin natural sand and clay layers (1 mm up to 2 cm) are intercalated in the lower lateral parts (below 80 cm above NAP, Fig. 13) of the terp and in the infill of ditches.[34] This indicates that the lower lateral sides and the terp ditches (connected with tidal creeks) were also heightened by natural deposition.

After the latest occupation Period VIII, the top of the terp at the site has been flattened by ploughing and wind erosion (Figs 13 and 14). It is estimated that ca 50-75 cm has been removed from the top. Probably, most of the material disappeared during the last century.

3.2 *Reconstruction of palaeo-tidal levels*

3.2.1 *Time-depth assessment of the palaeo-MHW level during the Late Iron Age*

The boundary between the tidal-flat sands and the crinkly laminated salt-marsh deposits at the site Wijnaldum-Tjitsma is found under the first constructed house-platform of Period I at 100 cm –NAP. The occurrence of the diatom *Scolipleura tumida* in the lower part of the salt-marsh deposits (diatom zone B: 105-70 cm –NAP, Fig. 18) indicates that this lower part was formed below or around the palaeo-MHW level (pioneer zone and lowest part of the lower marsh, cf. Fig. 19). This observation is in line with the opinion of Roep and Van Regteren Alterna that the palaeo-MHW level lay up to 20 cm above the lowest level of the crinkly laminated salt-marsh deposits.[35] Assuming a palaeo-MHW of 20 ± 5 cm above the lowest level of the crinkly lamination, the palaeo MHW-level at the site of Wijnaldum-Tjitsma should be at 80 ± 5 cm –NAP, without correction for the subsidence due to compaction of the sediment. The subsidence of the natural tidal flat deposits, caused by the construction of the terp, has been calculated by Baardman (see Appendix). Baardman concluded that a post-sedimentary subsidence had reduced the palaeo-MHW level by 2 to 8 cm. This means that the corrected palaeo-MHW level under the terp Wijnaldum-Tjitsma must have been between 67 and 83 cm –NAP.

The age of the palaeo-MHW level can be reconstructed with the help of radiocarbon dated shells (Fig. 17), found in a storm layer in the salt-marsh deposits, just above the tidal flat deposits (85 cm –NAP; not corrected for compaction; Fig. 13). The date for the shell sample indicates an age of the storm layer of about 100 cal. years BC. The age of the palaeo-MHW level is estimated at 150 ± 50 cal. years BC (ca 2100 ± 50 ^{14}C years BP). The time-

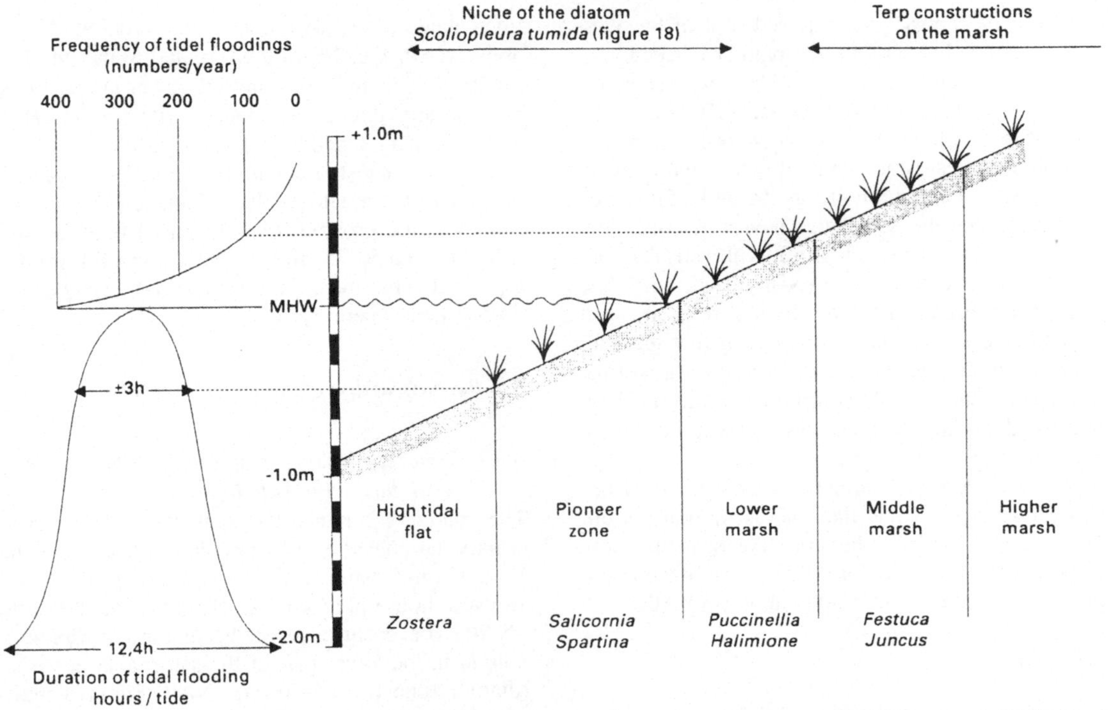

Figure 19. Zonation of the salt marshes in relation to the duration and frequency of floodings (after Dijkema et al. 1990; drawing TNO-NITG).

depth interval of the palaeo-MHW level (or palaeo MHW range box) at the site of Wijnaldum-Tjitsma is thus: 67-83 cm –NAP and 200-100 cal. years BC.

The palaeo MHW range box of the site of Wijnaldum-Tjitsma is plotted in the time-depth diagram of the palaeo MHW trend over the last 2000 years on the Frisian Islands (Fig. 20).[36] Extending the palaeo-MHW bandwidth of De Groot et al. to 2250 BP (Fig. 20), the palaeo MHW range box of the site of Wijnaldum-Tjitsma fits right in the middle of the palaeo MHW bandwidth of the Frisian Islands.

3.2.2 *Top of the salt-marsh deposits under the central podium of the terp*

The top of the undisturbed salt-marsh deposits, under the platform of Period I and lateral of the wall deposits (Figs 13 and 14) can be dated archaeologically around AD 175.[37] The top of the salt-marsh deposits has an height of 15 cm below NAP. The amount of subsidence, because of the construction of a house platform for this level, calculated by Baardman (see Appendix), is 5-15 cm. This means that the surface of the salt-marsh before the construction of the terp was at about 10-0 cm below NAP.

3.2.3 *Maximum storm surge level*

During the habitation of the terp, the storm-surge levels in the area will not have been higher than the lowest permanent habitation on the terp. The height of the permanent habitation during the different occupation phases can be reconstructed with the help of the hearth-plates (Table 2). Most of these structures are related to (possible) main buildings.

The hearth-plate levels, given in Table 2, are in the oldest periods I-IV higher than in the younger periods V-VII. There is, however, one exception (*in Table 2) in the range of hearth-plate levels: a hearth plate at 10 cm +NAP of Period III, found at the southern side of the terp.

Table 2. Lowest levels of the hearth-plates for the periods I-VII at the site of Wijnaldum-Tjitsma.

Occupation phase	Lowest hearth-plate in cm +NAP	Corrected
Period I/II(AD 175-300)	125	133-174
Period III (425-550)	140*	148-189*
Period IV (550-650)	112	120-161
Period V (650-750)	63	71-112
Period VI (750-800)	87	95-135
Period VII (800-850)	80	88-129
Period VIII (850-950)	–	–

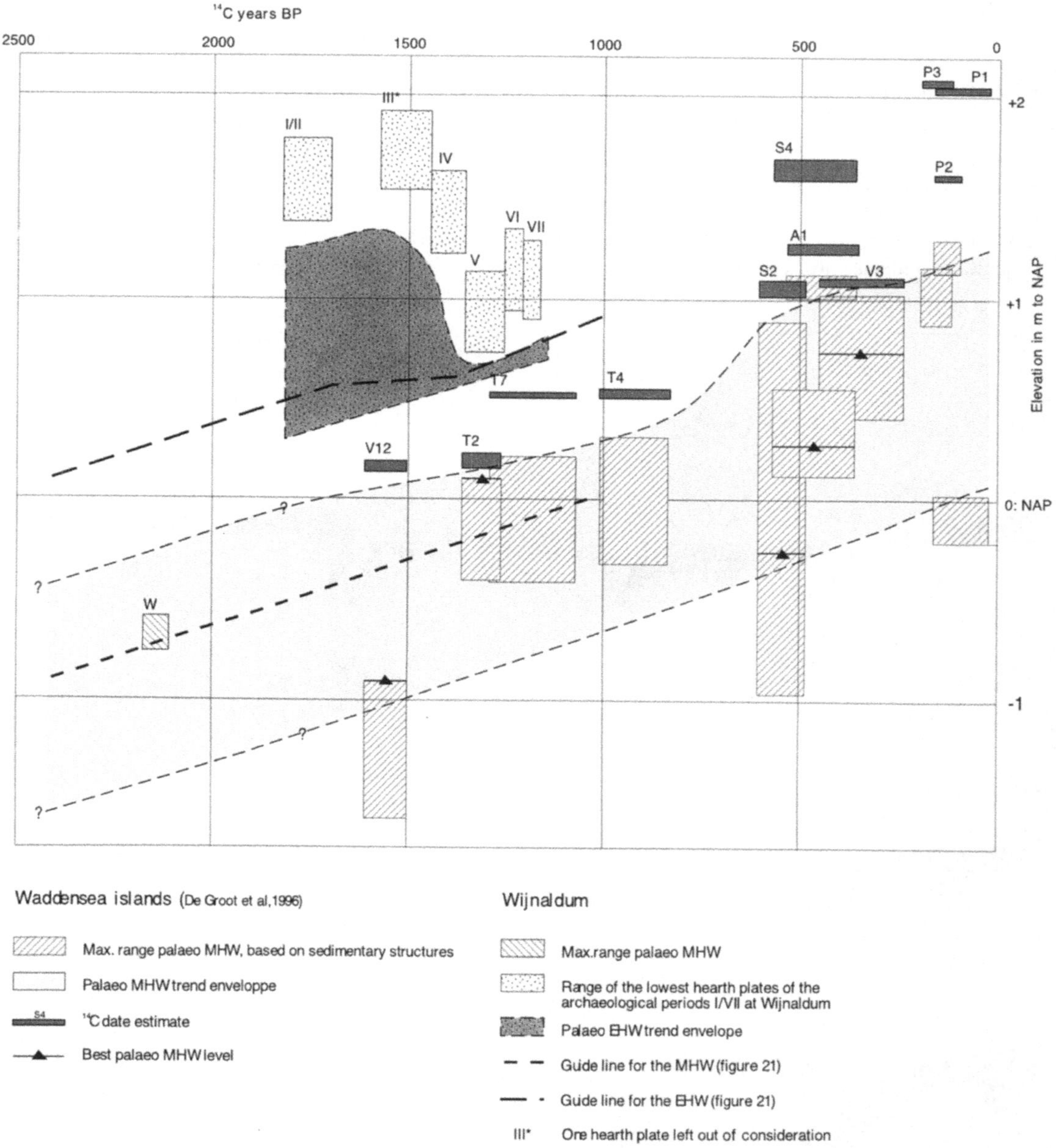

Figure 20. Reconstructed palaeo-tidal levels near the Wijnaldum-Tjitsma terp plotted in the sea-level, time-depth diagram of the Waddenzee area (after De Groot et al. 1996; drawing TNO-NITG).

The higher elevation of the main buildings during the oldest Periods I-IV might be explained by:

- The maximum spring/storm surge levels or Extreme High Water (EHW) during the Periods I-IV were decimeters higher than during the later Periods V-VII, or
- The occupants did not take any risks during the first settlements periods and they preferred to live considerably elevated above the EHW level.

Anyhow, the lower settlement levels during the periods V-VII indicate that during the habitation of the terp (AD 175-850) the EHW level was not significantly raising; possibly the EHW level even subsided.

4 SYNTHESIS

4.1 *Reconstruction of the salt-marsh development and terp construction at Wijnaldum-Tjitsma*

4.1.1 *Method and assumptions*

The reconstructions of the environmental changes and terp constructions between 400 BC and 1000 at Wijnaldum-Tjitsma are presented in the north-south cross-sections of Figure 21. The cross-section depicted in Figure 16 has been used as starting point (stencil plate) for the reconstructions in Figure 21. The lithological units in Figure 16 are related with the following environments in Figure 21:

- Sand with clay layers: tidal channels,
- Extremely fine sands: tidal flats,
- Clay with crinkly sand laminations: salt-marshes.

The time-depth assessments of the different envi-

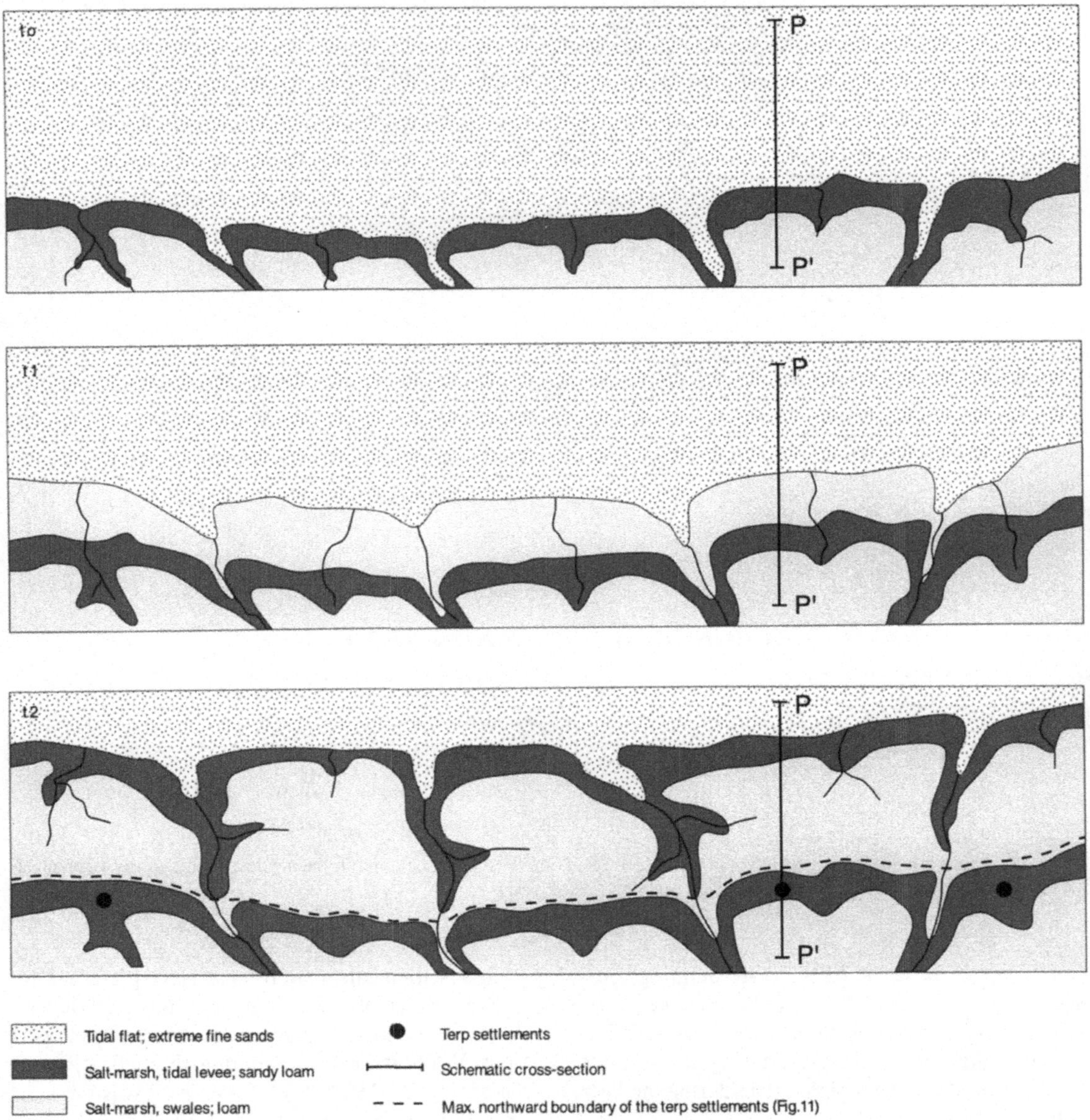

Figure 21. Hypothetic schematization of the seaward expanding salt marsh of Westergo at the time points t0, t1 and t2. a) On maps.

ronments in the cross-sections have been based on radiocarbon datings (Fig. 17) and archaeologically dated terp layers (periods I-VIII, Figs 13 and 14). The palaeo-MHW levels have been reconstructed with the help of the palaeo-MHW trend of the Waddenzee (Fig. 20) and the known palaeo-MHW range box at the site of Wijnaldum-Tjitsma (62-78 cm below NAP at 150-100 cal. years BC, cross-section 3.2.). In the diagram of Figure 20, a guide line has been drawn running through the palaeo-MHW range box and parallel to the basal line of the palaeo-MHW trend envelope. The time-depth positions derived from this guide line have been used in the reconstructions of Figure 21. The EHW levels (maximum storm surge level) in the reconstructions have been placed (arbitrary) 1 m above the palaeo-MHW level.

In the reconstructions, we assume that the Ried was one of the main tidal channels of the western depression. Radiocarbon dates from molluscs in the

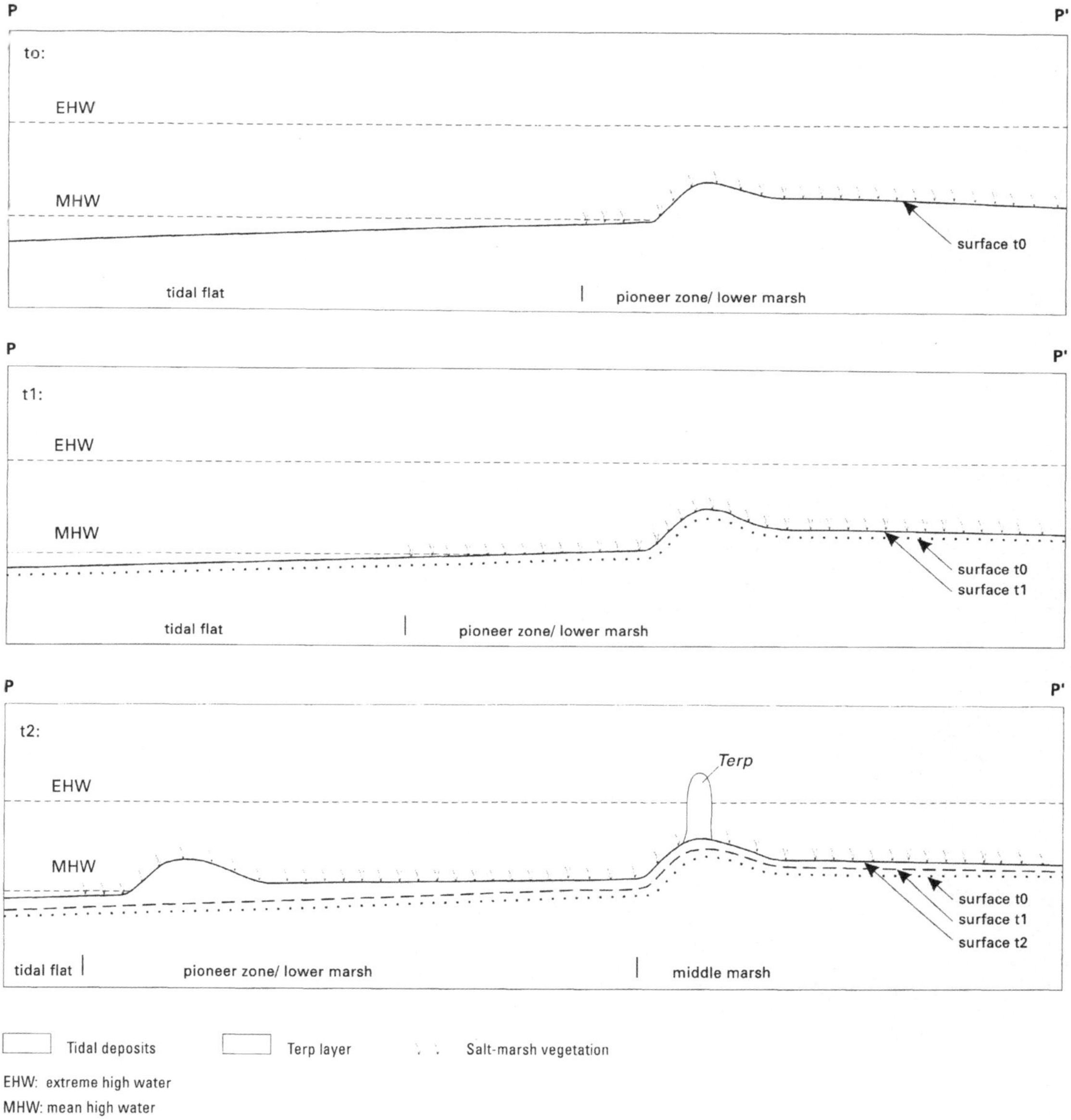

Figure 21. Continued. b) In cross-sections (drawing TNO-NITG).

top of the sand flat deposits and from molluscs at the base of the salt-marsh deposits, both north and south of the Ried (Fig. 17), indicate that the areas north and south of the Ried changed approximately simultaneously from a high tidal sand flat environment into a salt-marsh environment. Therefore, we consider the ridge of Wijnaldum-Voorrijp primarily as a levee of the Ried tidal channel and not as a levee along the seaward margin of the salt marsh.

In general, it is very difficult to assess the palaeo-surface (palaeo-relief) around a terp during the several stages of development. Lack of datable material at a certain distance from the terp and human disturbances (e.g. excavation of material related with the terp construction and ploughing) make it often very difficult to detect the palaeo-salt-marsh surfaces during the different periods of terp construction.

4.1.2 *Reconstruction*

400 BC, the area around the site of Wijnaldum-Tjitsma was still a sand flat environment, which was inundated during high water. The radiocarbon dates of the molluscs found in the upper part of the sand flat deposits (Fig. 17) indicate that around 250 BC the sand flat near the site of Wijnaldum-Tjitsma and the sand flat north of the Ried (site Pietersbierum) reached the high tidal zone during the same

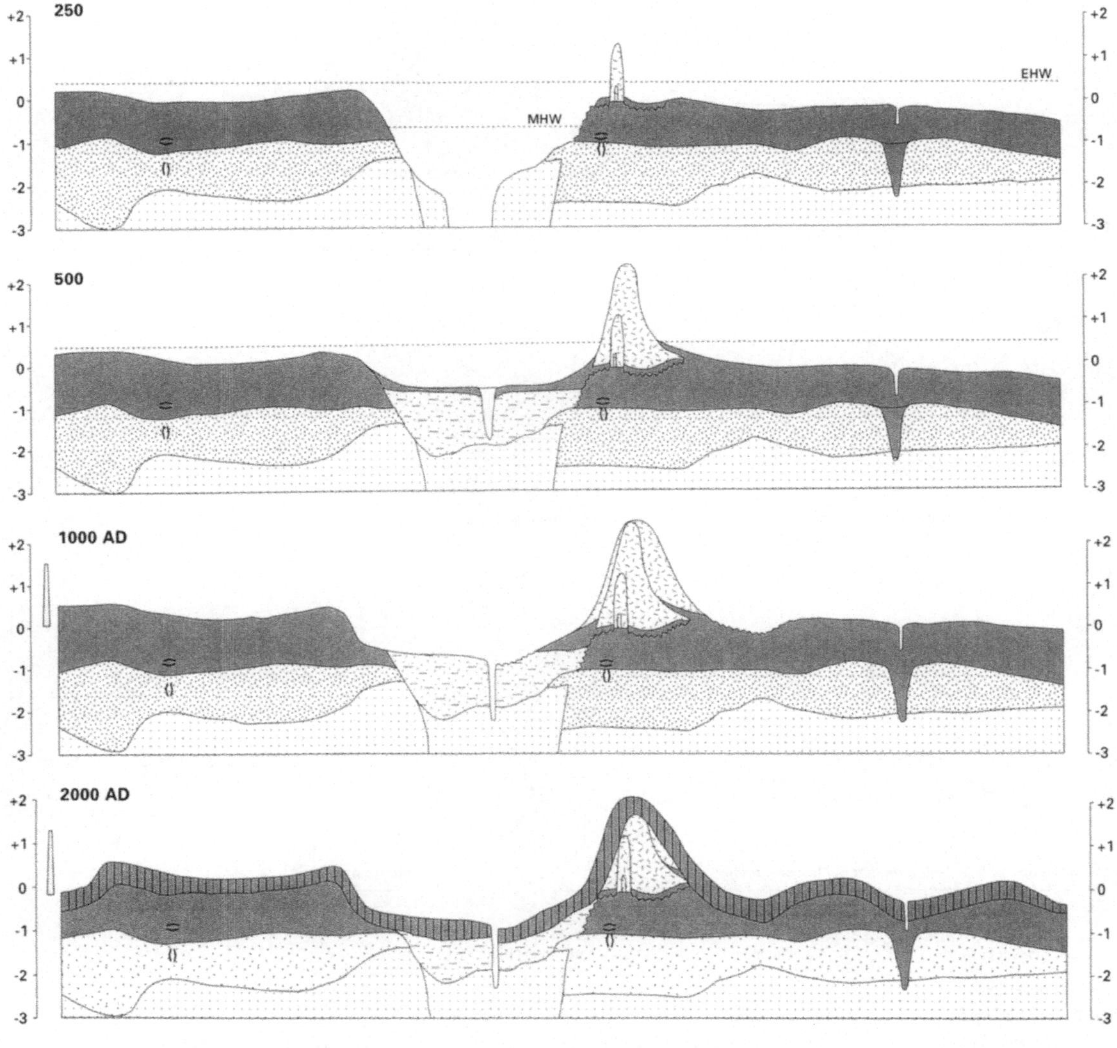

Figure 22. Cross-section representation of the Subatlantic evolution of the coastal area near the site Wijnaldum-Tjitsma (drawing TNO-NITG).

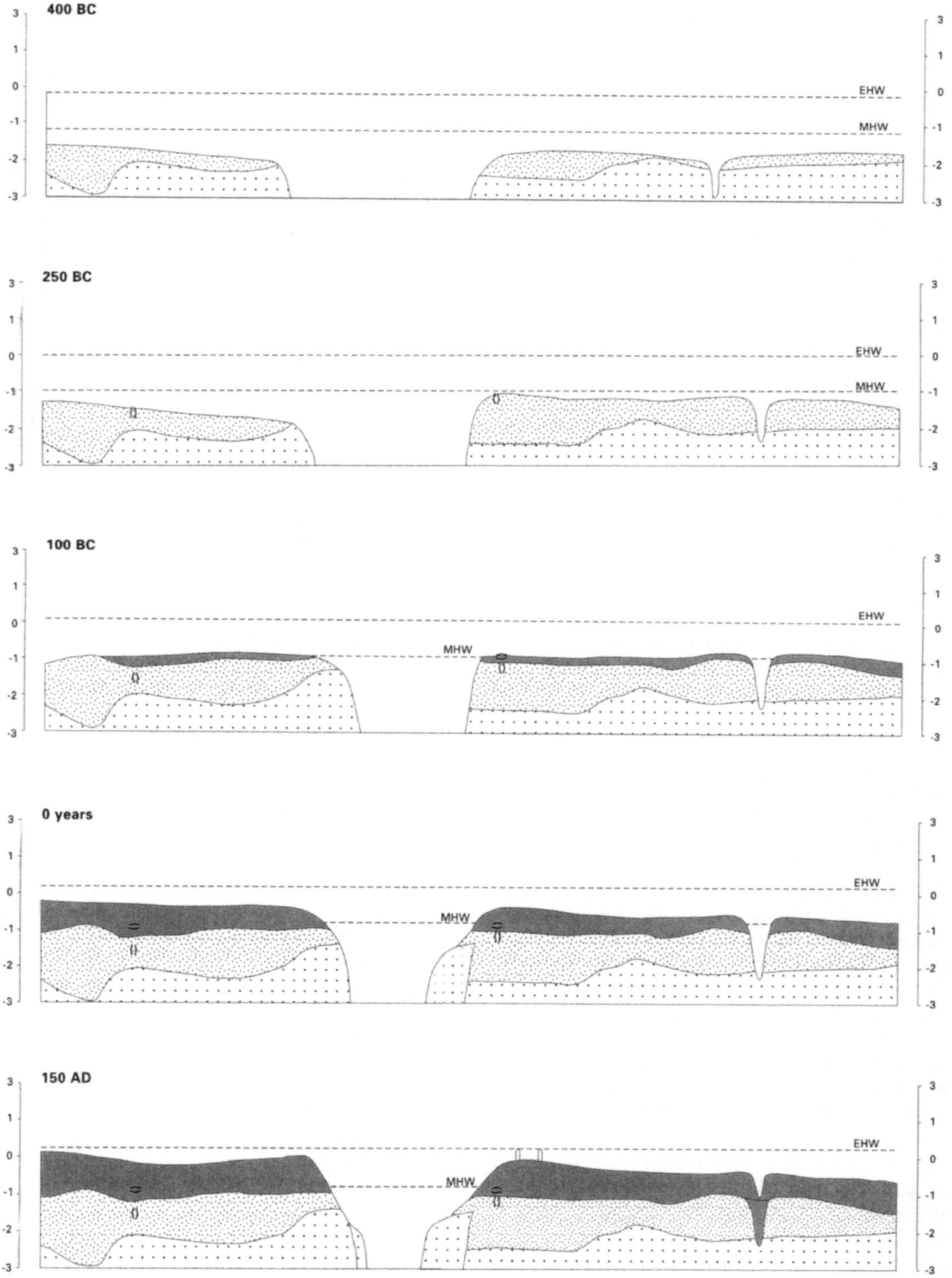

Figure 22. Continued.

period. In all the cross-sections, a tidal channel has been projected in the present-day depression of the Ried (Sexbierumervaart). This is based on the assumption that the Ried is a remnant of a former tidal channel.

Around 150-100 BC, the areas at the site of Wijnaldum-Tjitsma south of the Ried and at the borehole Pietersbierum north of the Ried silted up to the supratidal level. The 2σ-age ranges of the radiocarbon dated shells, found in the storm layers at the base of the salt-marsh deposits at the site of Wijnaldum-Tjitsma and at the borehole Pietersbierum, do not significantly differ from each other (Fig. 17). Possibly, the salt-marsh development at the southern side of the cross-section (site of Wijnaldum-Tjitsma) started earlier than at the northern side (borehole Pietersbierum). The dates of the molluscs (best estimated dates based on e.g. *Mytilus edule*, Fig. 17) at Pietersbierum tend to a slightly younger age of the beginning of the salt-marsh development. Best estimated dates used in the reconstructions (Figs 22 and 23):

- Wijnaldum-Tjitsma, shell storm layer: 100 BC,
- Wijnaldum-Tjitsma, start of the marsh development: 150 BC,
- Pietersbierum, shell storm layer: 50 BC,
- Pietersbierum, start of the marsh development 100 BC.

After ca 50-0 BC, the salt marsh altered in the area from the pioneer zone into the middle marsh. This is well documented by the diatoms at the site of Wijnaldum-Tjitsma (Fig. 18). In the lower part of the salt-marsh deposits the autochthonous species *Scoliopleurea tumida* is found. This diatom species indicates that the marsh was frequently flooded (near the palaeo-MHW level). In the upper part of the marsh deposits, however, autochthonous species are completely absent; only allochthonous marine (tycho-)planktonic species from the North Sea and epipsammic diatoms from the sandflats are found. The absence of autochthonous diatoms indicates that the salt marsh was not frequently flooded anymore and well drained by tidal creeks; the marsh became too dry for the growth of diatoms.

Around AD 150 the first human interference took place in the area: a dike-like structure (wall deposits below the central podium of the terp) was constructed on the marsh. The function of the dike-like structure is not clear. One suggestion, which was made in the field during the excavations, was that the dike was constructed to stimulate the natural sedimentation with the aim of reducing the necessary labour for the final construction of the site. However, this was not very likely to be the case since the dike was only flooded during a few extreme storm surges. Sediment-trap constructions are successful in the regularly flooded intertidal areas, see for instance the present-day reclamation works in the Waddenzee.

Another possibility is that the structure was part of a low square or circular dike to keep the salt-marsh dry during normal high waters (spring tides). Such an embanked part of the marsh could, for instance, have been used for agricultural purposes or to house the cattle of a nearby settlement. Anyhow, the dike was not high enough to resist extreme floodings, in view of the two storm-sand layers found in the wall deposits of the west cross-section. Because of the floodings, one had to construct terps to live safely in the area. Around AD 175 inhabitants from the region built the central podium of the terp on top of the wall deposits (Period I). The top of the central podium was at a level of 133-174 cm above NAP (corrected for settlement, see Appendix). This level is more than 180 cm above the reconstructed palaeo-MHW level represented in Figure 20 (between 50-90 cm below NAP during the Middle Roman Period). After the construction of the central podium of Period I the area around the terp was still flooded during extreme high water levels. This is indicated by a natural sediment layer (clay with crinkly laminations) intercalated in the lateral sides of the terp and in the terp ditches (Period II-VIII). Thus, around the terp, the middle marsh environment continued till the embankment of the area, about 1000.[38] In a brackish marsh environment, which is still flooded a few times a year, trees can not grow. The grassland of the middle marsh area was pre-eminently suitable for cattle-breeding.

Freshwater conditions existed only on the terp. After the construction, a freshwater lens developed under the terp because the water-supply by rainwater exceeded the loss by evaporation (temperate climate conditions). The groundwater was used for drinking-water, as indicated by the wells of the different habitation periods. The freshwater conditions on the terp allowed agricultural activities. Also trees and shrubs could be planted on top of the terp for domestic use.

Probably up to AD 250, the tidal channel of the Ried was draining the marsh area further inland. It is likely that during high water the Ried tidal channel was navigable, which implicates that the site of Wijnaldum-Tjitsma had a good connection with the sea. There were probably also navigable water connections with the hinterland in that period since the marshes and peat area in the back drained via the western depression. This situation changed after the Roman Period, when the formation of the cap ridge northwest of Wijnaldum started to block the opening of the western depression. It is not clear whether the cap ridge completely closed the opening to the sea. Nowadays the cap ridge completely blocks off the depression of the Ried. But, it might be possible that during the medieval period the region's inhabitants artificially maintained a connection to the sea by canalizing the remnant channel. Also the connections to the hinterland could have been artificially improved during the Middle Ages.

Today, the depression of the former Ried north of Wijnaldum-Tjitsma is quite large and has a rather artificial rectangular form (Figs 8 and 9). Possibly the depression has already been enlarged by quarrying clays for e.g. terp and later dike constructions since the Roman Period. The sediments could be transported via the (canalized) channel systems.

Because of the formation of the cap ridge, the hinterland of Westergo started to drain via the tidal systems Marne and Middelzee during the Early Middle Ages. These systems made contact near Bolsward, where a tidal watershed developed.

4.2 *Landscape reconstruction and habitation history of Westergo*

4.2.1 *Palaeo-environment assessment: method and assumptions*

Palaeo-environments. The palaeo-environments which occurred in the former coastal landscape of Westergo were tidal channels, tidal flats and salt marshes. These environments can be differentiated with respect to subsurface deposits on the basis of the sediment composure: tidal flat and channel environments consist of very fine sands whereas the salt-marshes consist of clayey deposits (clay with crinkly sandy laminations, clays and heavy clays or *knip klei*). A sub-environment within the salt-marsh environment are the salt-marsh ridges, higher areas characterized by a more sandy sediment composition than the surrounding lower clay area. The configuration of the salt-marsh ridges is expressed by differences in micro-relief (Fig. 8), differences in soil texture (Fig. 9) and the terp rows (Fig. 10).

Archaeological datings. The time/place assessment of the different tidal environments is largely based on archaeological information. The terp settlements, for the greater part constructed on the salt-marsh ridges, supply information about the age of the ridges, since the ridges were formed before the earliest habitation. The dating of the earliest habitation on the ridges, presented in Figure 11, is primarily based on the pottery investigations of Taayke[39] which were carried out in the central part of Westergo. Besides, the datings of Early Medieval sites by Knol[40] and new data compiled by the *RAAP Archaeological Consultancy*, who have made an archaeological field inventory of the area since 1992,[41] are used.

4.2.2 *Start of terp settlements related to the depositional stage of the salt-marsh ridges*

The ridges became occupied when they were silted up to the level of the middle marsh (Fig. 19). Observations from boreholes show that the terps in the northwest of Friesland were constructed on a layer of salt-marsh deposits (clay with crinkly laminations). In the excavation pit of the terp Wijnaldum-Tjitsma the undisturbed salt-marsh deposits below the central podium of the terp were about 85 cm thick. Bivalved shells in the top of the tidal flat deposits and in the base of the salt-marsh deposits (Figs 13 and 16) indicated that the salt-marsh formation started ca 150 BC and that the sedimentation stopped at this location when the terp habitation started ca AD 175.[42] Thus, in a ca 325-year period a salt-marsh layer of 85 cm was deposited, implying an average sedimentation rate of ca 2.6 mm per year.

The presence of *Scoliopleura tumida* (an autochthonous diatom species which lives in the range between high tidal flat and lower marsh) in the lower part of the salt-marsh deposits at Wijnaldum-Tjitsma (Fig. 17) demonstrates that the lower part of the salt-marsh deposits was frequently flooded, almost during every flood. In the upper part of the salt-marsh deposits (Fig. 17) only marine allochthonous species were found, derived from the sandflats and coastal zone and brought in largely during storms. During the formation of the upper part of the marsh deposits, the environment was too dry for the development of autochthonous diatoms. The thickness of the salt-marsh deposits and the diatom content of the salt-marsh deposits at Wijnaldum-Tjitsma indicate that terp settlements were constructed on the higher silted-up salt marshes, which were only inundated during occasional high floods (spring tides/storm surges). Thus, new inhabitants would settle on terps when the ridge reached approximately the stage of the middle marsh; the area flooded less frequently than 100 times/year (Fig. 18).

Genesis of the salt-marsh ridges. According to the soil scientists and geomorphologists[43] the formation of the salt-marsh ridges resulted from erosion of the salt-marsh shore face during transgression and/or storm-surge periods. The eroded sandy sediments were redeposited directly over the salt-marsh edge and as a result of this deposition a low ridge was formed. The clayey particles settled during these erosive events further inland in the 'salt-marsh basin'. In the coastal area of the Marne in NW Groningen, Roeleveld[44] described these ridges ('bars') as 'cheniers'.

However, the salt-marsh ridges of the northern Netherlands are neither cheniers nor beach ridges.[45] Cheniers and beach ridges consist of rather coarse materials such as sand and shells, whereas the salt-marsh ridges consist of rather sandy clays (clay with crinkly sand laminations). Cheniers are formed in coastal areas where the influx of river sediment is fluctuating.[46] In the whole Dutch Holocene coastal record rivers do not contribute more than 11% of the total sediment budget.[47] In the Westergo coastal area the sediment input by the small local rivers from the Pleistocene hinterland (e.g. Boorne) was negligible, which implies that chenier type ridges can not have been formed in the area. Just as little as cheniers, salt-marsh ridges are beach ridges since the latter are formed in the foreshore of a sandy beach profile, i.e. in an open marine (or large lacustrine) environment.[48] Such a beach environment was not present in the former tidal basin of Westergo (part of the Waddenzee).

We consider the salt-marsh ridges of Westergo as former levees along the seaward margin of the salt-marsh (Fig. 22). As demonstrated at the site of Wijnaldum-Tjitsma, the levees (the ridge of Wijnaldum-Voorrijp; Fig. 11) were formed during a long period (hundreds of years) of marsh inundations (storm floods and spring tides) and they were not set back by erosion. In the 200 m long north-south cross-section through the ridge of Wijnaldum-Voorrijp (Fig. 13), no truncations were found in the marsh deposits; instead, the marsh deposits showed a horizontal 'pan-cake' layering of the crinkly sand laminations.

We attribute the development of a series of ridges of the 'levee type' in Westergo to variations in sediment supply (Fig. 22). However, the causes for the supposed palaeo-variations in sediment supply are still unsure. In theory, the sediment supply on the margins of the salt marsh depends on variations in flooding frequency and/or fluctuations in the availability of sediment from the coastal area. The number of floodings depends on:

- The length of the period during which the margin is exposed to the sea,
- The number of storm surges, and
- The fluctuations in the MHW level (caused by changes in e.g. the sea-level and the configuration of the tidal basin).

The Waddenzee tidal basin was the source area of the salt-marsh deposits and changes in the configuration of the source area influenced the sediment transport to the salt-marsh area. During the Late Iron Age and Roman Period, the former peat area in the western Waddenzee was inundated and the area changed into a tidal area[49] where tidal channels indented the Pleistocene sediments. Part of the eroded material was transported to the salt marshes of Westergo by (storm) waves and tidal currents. The sediment supply to the marsh might have fluctuated during periods of intensified erosion.

Beach ridge formation along the coast of Lake Michigan could be correlated with climate-related lake-level fluctuations.[50] Climate oscillations may as well have played a role in the formation of the salt-marsh ridges in Westergo (e.g. frequency of storm surges; sea-level fluctuations). However, for meso-tidal coastal environments (Westergo), the reconstruction of ridges is more complex than for non-tidal environments (Lake Michigan). As yet, our data set is too limited to postulate a relation between salt-marsh ridge formation and palaeo-climate oscillations.

Assessment of the palaeogeographic maps. In the palaeogeographic reconstructions (Fig. 23a-g), the salt-marsh environment has been subdivided into two legend units: higher (rather) sandy salt-marsh ridges and lower clayey salt-marsh areas. The shapes of the higher sandy ridges have been compiled by joining the relief data (area above 0.5 m +NAP, Fig. 8) and the soil data (loam and sandy loam, Fig. 9). However, in the reconstructions the tidal channels and tidal flats have been combined into one legend unit, since we are not able (yet) to assess the age/place of these different environments.

A major problem in using the relief map for the ridge reconstructions is that the terp dwellers influenced the relief by quarrying the Holocene deposits for the brick industry and for terp and dike constructions during many centuries. With respect to the assessment of the locations of the ridges these areas are the 'white spots' which have to be filled up in the map reconstruction by the author, who, for that purpose, has to make use of information from the direct surroundings.

The seaward expansion of the middle stage of the salt marsh is dated by the earliest habitation of the terp settlements (Figs 10 and 11). In between the archaeologically dated middle marsh and the intertidal sand flat/tidal channel area, a seaward migrating pioneer zone/lower marsh area was present. In the reconstructions, we projected a pioneer zone/lower marsh area in front of the 'archaeologically known' middle marsh area. The boundary between the pioneer zone and the intertidal sand flats has been projected just north of a seaward developing salt-marsh ridge (cf. schematic representation of Fig. 22).

4.2.3 *Landscape reconstruction and habitation history of Westergo*

600 BC. The Late Bronze Age/Early Iron Age settlements found in the Old land, indicate that this area had already been silted up to the middle stage of the salt marsh during the early Subatlantic. The seaward margins of the Old land salt marsh have been drawn at the locations of the rows Sopsum-Hitzum-Tzum and Baijum-Dronrijp-Menaldum (location names, see Fig. 11). The earliest terps on these rows date from the 3rd/4th centuries BC and it is likely that here, around 600 BC, the salt-marsh sedimentation started (pioneer stage). The configuration of the 3rd/4th centuries BC terp rows points to an funnel-shaped, estuarine-like, tidal basin. Small local rivers, draining the hinterland of southern Friesland, discharged into this funnel-shaped tidal basin. Probably the Boorne was one of these local rivers. The courses of the local rivers which date from the Iron Age and Roman Period are difficult to detect, because their channel fills do not show obvious differences compared with the surrounding sediments. In the palaeogeographic maps, we have reconstructed the local rivers and creeks in the Old land on the relatively higher areas (between 0 and 0.5 m below NAP, Fig. 8) and along the terp settlements (Fig. 10), assuming that:

- Along the small rivers and creeks the area silted up to a slightly higher level, and
- The terp settlements were located at strategic points along waterways.

At the location of the eastern terp row of Baijum-Dronrijp-Menaldum a salt-marsh ridge developed of which the surface sediments (soil) consist of loam or sandy loam. At the western side, at the terp row of Sopsum-Hitzum-Tzum, however, this was not the case. Here, no ridge was formed and the surface sediments consist mainly of clay. Also during the later periods the eastern side of the funnel-shaped Westergo basin silted up to a higher level and, consequently, the surface deposits on the ridges are more sandy than the sediments at the western part. The sandy character and higher sedimentation levels of the ridges in the eastern part of the basin can be attributed to larger sediment transports caused by a stronger exposure to waves. The western part of the basin was protected from strong wave action because of the presence of the Grienderwaard (Fig. 24). The Grienderwaard is a remnant of the Pleistocene headland of Texel which was inundated by the sea between 1250 and 100 BC[51] and which is still present in the subsurface of the western Waddenzee. In contrast, the eastern part of the Westergo tidal basin was more strongly affected by waves generated during northwestern and northern storms in the North Sea area. Via the mouth of the palaeo-Boorne Valley, which was located at the present island of Terschelling, the storm waves influenced the sedimentation process of the eastern part of the study area. During storms sandy material was eroded from channels and tidal flat areas and transported to the salt-marsh hinterland. Near the salt marsh the energy was reduced by frictional effects and sandy material was deposited along the margins of the salt marsh and along the creeks.

Because of the lower energetic conditions in the sheltered western part of the Westergo basin (Western depression, Fig. 11), the sediments along the creeks and at the seaward salt-marsh margin in this part of Westergo were primarily clayey. Probably, in this area also low clayey levees (ca 10-20 cm) were formed. However, these levees were too low to be detected from our reconstructed surface contour map (interval 50 cm; Fig. 8).

400 BC. After 600 BC, the Westergo tidal basin silted up stepwise to the supratidal level as indicated by the terp settlements on the salt-marsh ridges in this area. Around 400 BC, the southernmost terp rows Sopsum-Hitzum-Tzum and Baijum-Dronrijp-Menaldum were inhabited. Seaward, new salt-marsh areas developed, but near Wijnaldum the tidal flat environment continued to exist for a few centuries (Fig. 17). The salt marsh-tidal flat

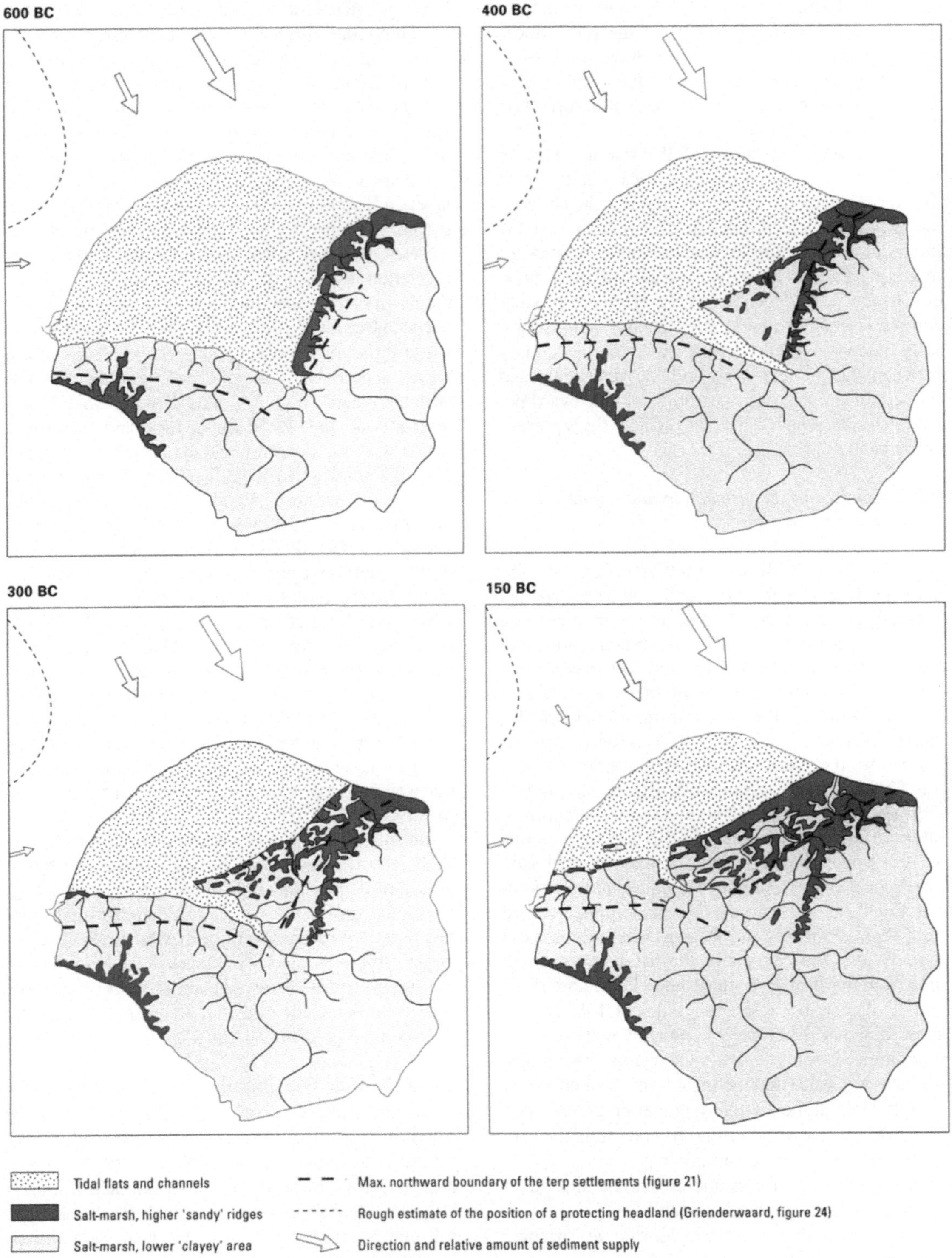

Figure 23. Palaeo-geographic reconstructions of the Westergo coastal area. a) 600 BC, b) 400 BC, c) 300 BC, d) 150 BC.

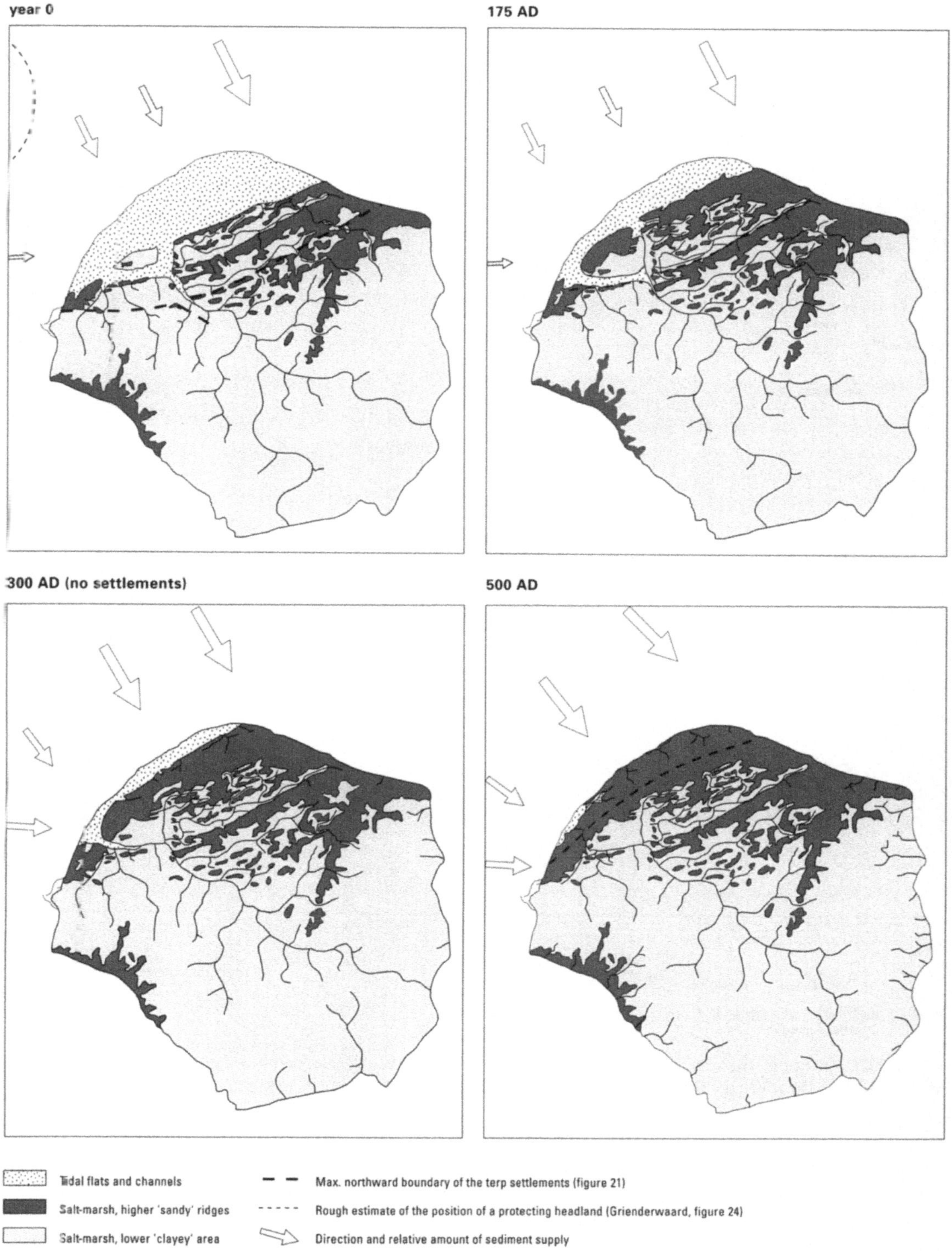

Figure 23. Continued. e) 0 BC/AD, f) AD 150, g) AD 250, and h) 500 (drawing TNO-NITG).

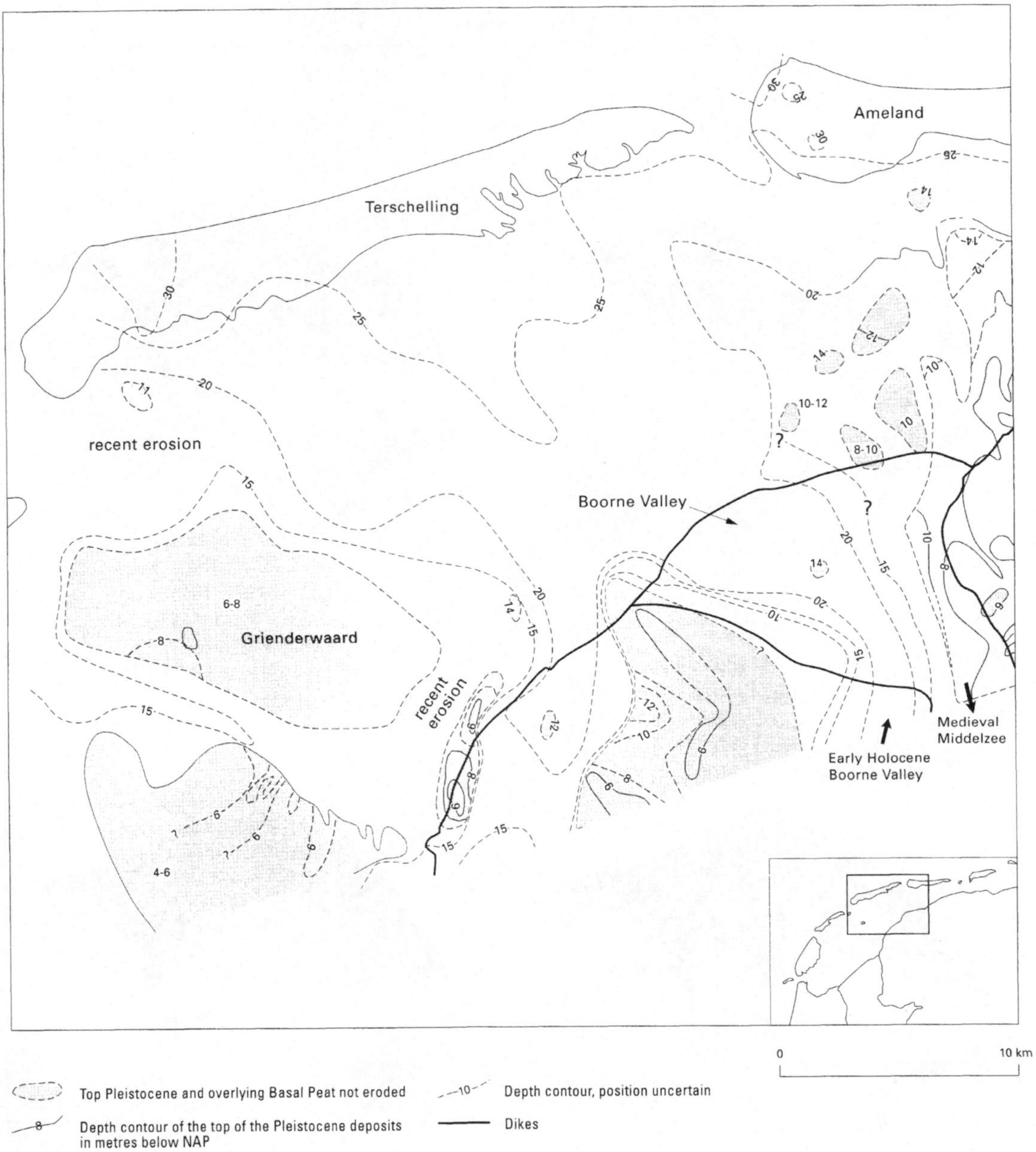

Figure 24. Depth of the base of the base of the Holocene deposits in the Waddenzee, northwest of the study area (after Van Gijssel & De Gans 1996, Van der Spek 1974; drawing TNO-NITG).

margin in the 400 BC reconstruction has been drawn north of the line Salverd-Schingen. Because of the larger sand supplies in the eastern part of the funnel-shaped basin, the salt marsh expanded faster seaward and silted up to a higher level in the eastern part than in the western part. In the western part a depression was formed and the hinterland drained largely via this depression.

300 BC. About 300 BC, terp settlements were built near Welsrijp in the new salt-marsh area in the funnel-shaped basin of the former Boorne estuary.

Seaward of the occupied salt-marsh areas the sandy fan-shaped ridge of Schalsum-Peins-Beetgum started to develop. In the western depression possibly the formation of the ridge of Harlingen-Ludingakerk started. The radiocarbon dated diatoms of the species *Scrobicularia plana* in the top of the extremely fine sands near Wijnaldum (Fig. 17) indicate that this area reached the stage of a high tidal flat.

150 BC. Around 150 BC, the salt-marsh area was probably occupied up to the line Harlingen-Ludingakerk and Salverd-Zweins-Schingen. However, this is only an estimate because the pottery types of this period (G4, V3 and K3a, Fig. 11) do not allow the assessment of an exact date of the terp settlements – the dating of this pottery has a range of 300-400 years.

The salt marsh development (pioneer zone) started in the area near the site of Wijnaldum-Tjitsma (radiocarbon dated bivalved molluscs, Fig. 17) and probably eastward, in line, near the ridge Dongjum-Ried-Berlikum. The western depression remained a small tidal area in between the ridges of Wijnaldum-Voorrijp and Dongjum-Ried-Berlikum. The creeks in the western depression were probably important shipping routes, connecting the occupied hinterland with the sea.

0 BC/AD. Around 0 BC/AD, the occupation extended to the line Midlum-Herbaijum and Schalsum-Peins-Beetgum. The area in line with the site of Wijnaldum-Tjitsma was flooded too frequently to allow terp settlements (lower marsh, Fig. 17). The Grienderwaard (Fig. 24) was still protecting the western part of the (largely silted-up) Westergo tidal basin, but as indicated by the more sandy salt-marsh deposits near Sexbierum, the energy conditions along the western salt-marsh margin had become somewhat higher.

The higher energy conditions were due to the erosion of the Grienderwaard which, in turn, was related with the inundation of the peat area of the western Waddenzee (peat areas on the Texel high grounds) and the enlargement of the opening of the Zuiderzee.[52] Because of the erosion of the Grienderwaard, the influence of wave energy during storms in the western part of the Westergo tidal basin increased and the formation of a sandy cap ridge started along the whole northwestern part of Westergo.

AD 150. During the next centuries the cap ridge gradually closed the Westergo area from the sea. However, the influence of the Grienderwaard on the sedimentation process did not disappear completely since the eastern part of the cap ridge silted up to a significantly higher level (1-2 m above NAP) than the western part (0.5-1 m above NAP).

During the 2nd century AD the ridge of Wijnaldum-Voorrijp and in line the ridge of Dongjum-Ried-Berlikum were occupied. The terp construction has been dated ca AD 175,[53] but man already interfered in the area as indicated by the dike-like structure below the central podium of the terp (wall deposits; Fig. 14). The wall was possibly part of a low dike constructed for agricultural purposes of a nearby settlement.

AD 250. About AD 250, the Westergo tidal basin had been silted up for the greater part to the supratidal level. Middle Roman finds, found on the cap ridge, indicate that in those days also this area was used by man. It is not likely that during the Roman Period the cap ridge already completely closed the silted-up Westergo area from the sea. Probably, the area still drained via the creeks formed in the western depression. The tidal systems like the Middelzee and Marne, which drained the hinterland of Westergo in the Early and Late Middle Ages, had not yet been (re-)activated.[54] The Ried was the major drainage creek intersecting the newly formed cap ridge northwest of Wijnaldum. The strategic position at the creek offered the inhabitants of the terp Wijnaldum-Tjitsma all the advantages related to trade and fishery.

500 AD. Most inhabitants abandoned the Frisian coastal area between AD 300 and 425.[55] The hiatus in the occupation history might explain the fact that on the salt-marsh ridge of Bethaniëhiem, clearly detectable on the morphological map and soil map (Figs 8 and 9), no terp settlements (Figs 10 and 11) were built. During the 5th century, the cap ridge had been silted up to such a level that this area was a better location for the new settlers. Also old terp locations, such as Wijnaldum-Tjitsma (Period III) were occupied again during the 5th century. It is not clear whether the site of Wijnaldum-Tjitsma still had a direct connection with the sea via the Ried about 500. The formation of the cap ridge probably blocked the connection with the sea, but it can not be excluded that the settlers maintained an artificial opening to the sea. Anyway, the main part of the Westergo area drained via the Middelzee and Marne tidal systems.

The Westergo area remained a salt-marsh area until the embankments in the 10th/11th centuries (Fig. 5). This is demonstrated e.g. at the terp Wijnaldum-Tjitsma by the natural salt-marsh intercalations in the terp deposits and ditch fills.[56]

5 CONCLUDING REMARKS

The salt-marsh ridges of Westergo are neither cheniers nor beach ridges.[57] Cheniers and beach ridges are formed along the shoreline of an open marine environment and consist of sand and shells, while the salt-marsh ridges in the Westergo area developed in the back of a tidal basin and their sediment composure is rather clayey (clay with sand layers/laminae). The salt-marsh ridges of Westergo are considered levees along the former seaward margin of the salt marsh (Fig. 22).

The salt-marsh ridge of Wijnaldum-Voorrijp, investigated at the site of Wijnaldum-Tjitsma was not the optimal location to study salt-marsh ridges. The site is located on the relatively small ridge, which developed in the western depression, the former mouth of a local river draining the hinterland of NW Friesland (Figs 11 and 23). For the study of salt-marsh ridges, it should be better to have an exposure in a ridge in the northeastern part of Westergo, since these ridges are more pronounced (fan-shaped ridges, Figs 8 and 9).

At the location of the site of Wijnaldum-Tjitsma, the salt-marsh ridge could not even be detected. The relief on the site is artificial, the differences in height are caused by the man-made terp. The top of the undisturbed salt-marsh deposits below the central podium of the terp (15 cm below NAP, Fig. 13) is lower than the present-day surrounding surface (ca 0-100 cm above NAP, except for the depression of the Ried, Fig. 14). The reason why the present surface around the terp is higher than the undisturbed top of the salt marsh below the central podium of the terp is that below the podium the sedimentation stopped after the construction (ca AD 175) while in the surrounding area the sedimentation continued till the embankment of the area (ca 1000).[58] The sedimentation may have raised the surrounding area by many decimeters; e.g. assuming a low sedimentation rate of 1 mm/year, during the ca 825 years up to the embankments, the area around the terp will have been raised by ca 82.5 cm. Such a sediment layer is thick enough to cover up the original relief which existed before the terp construction.

The natural salt-marsh deposits below the terp layer of Wijnaldum-Tjitsma were formed between ca 150 BC-AD 175 and are ca 85 cm thick. This implies an average sedimentation rate of ca 2.6 mm per year.

The reconstructed palaeo-MHW level at Wijnaldum-Tjitsma (75 ± 8 cm below NAP at 2100 ± 50 ^{14}C years BP) is well in line with the palaeo-MHW trend reconstructed over the last 2000 years on the Frisian Islands (Fig. 20). The terp settlements (level of the lowest hearth-plates) on the terp Wijnaldum-Tjitsma indicate that during the Roman Period and Early Middle Ages the maximum storm-flood level (EHW) was not significantly raising, possibly the EHW level even subsided (Fig. 20).

Based on the habitation succession found on the salt-marsh ridges of Westergo, the middle marsh in the central part of the Westergo region (funnel-shaped palaeo-basin, Fig. 11) expanded in 1000 years (500 BC-AD 500) from about 10 km in the central part of the funnel-shaped ridges to about 5 km at the lateral sides. This implicates a mean progradation of the middle marsh of ca 5-10 m per year.

The rapid expansion of the salt-marsh area in Westergo was possibly enhanced by the Subatlantic erosion of the former peat area in the western Waddenzee. Reworked sediments from this area could be transported by waves and currents via the major channel system northwest of Westergo (relict of the former Boorne valley, Figs 3 and 24). The effects of the northwestern winds were larger in the eastern part of Westergo than in the western part. The western part was sheltered from the northwestern winds by the Grienderwaard, whereas the eastern part was located in line with the Subatlantic channel system which had developed in the central part of the present Terschelling (Fig. 24). Therefore, the eastern part was silted up to a higher level and salt-marsh sediments were in general more sandy here.

In the northern part of the Netherlands the clastic coastal sediments are often regarded in relation with the so-called 'Calais and Duinkerke transgression phases'(Fig. 5). The 'trans-/regression model' is based on the hypothesis that the trans- and regressions were caused by sea-level fluctuations due to supra-regional climatic events. The cycle of the trans- and regression phases was about 300-600 years.[59] The cyclicity of the archaeologically dated salt-marsh ridges (ca 150-250 years) does not correlate with the supposed cyclicity of the Duinkerke transgression phases.

The 'classic' trans-/regression model is also not valid in explaining the evolution of the salt-marshes in the different regions of the northern Netherlands. The salt-marsh developments in the different regions appear to be diachronous. During the Iron Age, the salt marshes in the palaeo-tidal basins of Westergo (this paper) and the Hunze[60] expanded (regressive trend) whereas other areas such as the marsh and peat areas of northeast Friesland[61] and the western Waddenzee[62] disappeared by marine inundations contemporaneously (transgressive trend). The coastal evolution in the northern Netherlands has to be considered a long-term marine infilling process of the Pleistocene palaeo-valley systems which was a consequence of the Holocene sea-level rise. Regionally, the infilling process may have been interrupted or enhanced by coastal particularities caused by tidal inlet/channel displacements, erosion of Pleistocene headlands and peat areas and, from 1000 onwards, by human influences.

Salt-marsh ridges will be studied in the nearby future within the framework of the geological mapping program of the northern Netherlands by the NITG-TNO and, hopefully, these investigations will further improve the understanding of the origin of the salt-marsh ridges in this part of the Netherlands.

ACKNOWLEDGEMENTS

I am indebted to G.J. de Langen and E. Taayke for their advice in dating the terp rows of Westergo and to D.A. Gerrets for stimulating discussions and constructive comments on the manuscript. Thanks to the NITG-TNO section *Vormgeving*, in particular J.J.M. Houkes, A. Koers, B.I. Kromkamp, R.R. Van der Kraan, D. Lepelaar-Moeliker and B. Vos for their contributions to the illustrations and to J.T.E.M. Smit for improving the English text.

APPENDIX

Consolidation in coastal deposits for the reconstruction of palaeo tidal levels at the Wijnaldum-Tjitsma terp (B.A.M. Baardman)

Introduction

The objective of the study is to reconstruct coastal wetlands morphology in Northwest Friesland (Westergo), the Netherlands, for the archaeological investigations of the terp Wijnaldum-Tjitsma.

The present position of specific sediment levels does not coincide with their levels at deposition, due to consolidation of underlying sediments and of the terp itself. Three sediment levels have been considered:

1. The palaeo Mean High Water (MHW) sediment level as deposited in 100-120 BC. The present-day level has been reconstructed at 0.80 ± 0.05 m –NAP,
2. The top of salt-marsh sediments as deposited in AD 150-175. The present-day level is located at 0.15 m –NAP,
3. The hearth of house after the first terp raising in AD 150-175. The present-day level is located at 1.25 m above NAP.

This study gives the calculations of the approximate amount of subsidence due to settlement of the three levels. The corrected levels with respect to present Mean Sea Level (NAP) are summarised below.

Results

Depending on uncertainties in values of geotechnical parameters and the determination of present levels, the following ranges for amount of subsidence and palaeo level with respect to NAP can be given:

Sediment	Present level	Amount of subsidence	Palaeo level	Time of deposition
Palaeo MHW	0.80 ± 0.05 m –NAP	0.02-0.08 m	0.67-0.83 m –NAP	100-120 BC
Top of salt-marsh	0.15 m –NAP	0.05-0.15 m	0.00-0.10 m +NAP	AD 150-175
Hearth of house	1.25 m +NAP	0.08-0.49 m	1.33-1.74 m +NAP	AD 150-175

Data

The situation is schematised as follows:

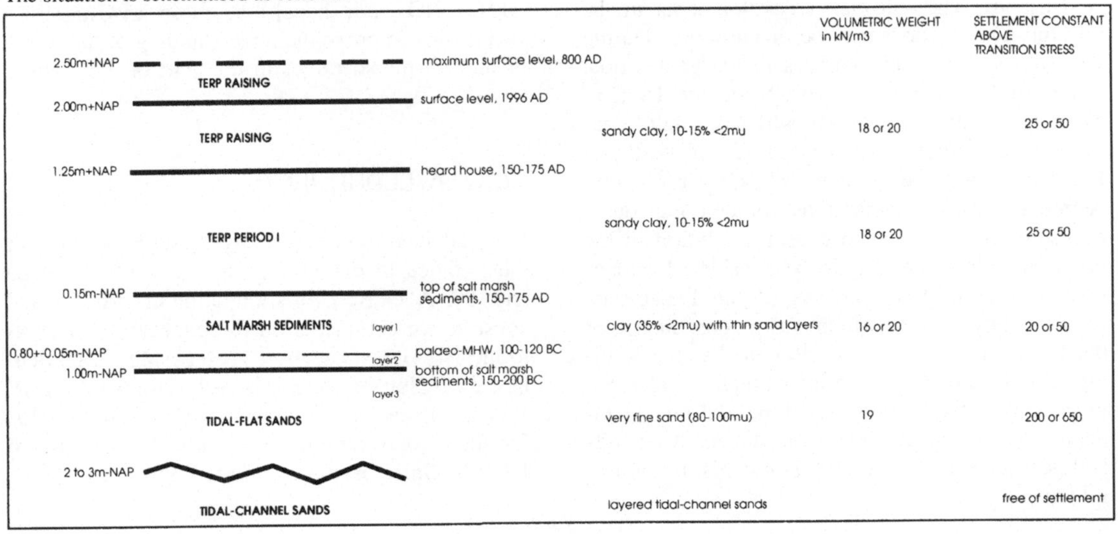

Causes of settlement

Palaeo-MHW sediment lies presently lower than in 100-120 BC due to consolidation of the underlying part of the salt-marsh sediments and the tidal-flat sands.

The top of salt-marsh sediments lies presently lower than in AD 150-175 due to consolidation of all of the salt-marsh sediments and the tidal-flat sands.

The hearth of the house lies presently lower than in AD 150-175 due to consolidation of the terp-period-I sediments, the salt-marsh sediments and the tidal-flat sands. In all mentioned layers consolidation is caused by four loads:

1. Weight of the terp-period-I sediments,
2. Weight of the house,
3. Weight of the (partly removed) dwelling mound (the terp was raised from 1.25 m +NAP to 2.50 m +NAP),
4. Lowering of the ground-water level, assumed to be located at 1.00 m –NAP at lowest.

For the palaeo MHW level an additional load of the upper 0.60-0.70 m of the salt-marsh sediments has to be considered.

As the weight of the dwelling mound above house level (1.25 m +NAP to 2.50 m +NAP, leading to a load of minimum 18 kN/m^3 × 1.25 m = 22.5 kN/m^2) is comparable to or higher than the weight of the house (maximum 20 kN/m^2, value for a contemporary small house according to Senaka Welidenya,[63] the weight of the house is not entered in the calculations of maximum load.

Method

Consolidation has been calculated using Koppejan's formula for a settlement period of more than 27 years:

$$z = \frac{d}{C} \ln \left(\frac{\overline{\sigma}'_{\text{before loading}} + \Delta\sigma'}{\overline{\sigma}'_{\text{before loading}}} \right)$$

in which

z = consolidation of compressible layer, in metres

d = thickness of compressible layer before consolidation, in metres. Approximated in an iterative way

C = Koppejan settlement constant above transition stress for both primary and secondary settlement

$\overline{\sigma}$ = stress in the middle of a layer, in kN/m^2. Is calculated from $\overline{\sigma} = \dfrac{\sigma_{\text{top}} + \sigma_{\text{bottom}}}{2}$

$\overline{\sigma}'_{\text{before loading}}$ = effective stress in the middle of a compressible layer before loading, in kN/m^2

$\Delta\sigma'$ = increment of effective stress in a compressible layer, in kN/m^2

Settlement of each level has been calculated twice, using minimum and maximum values for volumetric weight and settlement constants. For determination of the effective stress before loading, a value between the minimum and maximum of volumetric weight of overlying layers has been entered in the calculations. The level of the ground-water table has been varied from 1.00 m –NAP to top of the salt-marsh sediments (0.15 m –NAP).

Assumptions
- Underlying tidal-channel sands do not settle.
- All occurring stresses are higher than the transition stress. This assumption seems reasonable as the sediments have never been overconsolidated before terp construction.
- The settlement parameters entered in the calculations are literature values.[64] For volumetric weights and settlement constants minimum and maximum values have been used; the real values probably lie between the assumed values. For more exact determination of the geotechnical parameters local expertise or laboratory testing is required.
- The difference in value of the Koppejan settlement constant for settlement during a 27-year period and the value for a 2200-year period can be neglected.

Palaeo level of palaeo-mhw

The present-day level of the palaeo-MHW sediments has been reconstructed at 0.80 ± 0.05 m –NAP; since the deposition of these sediments in 100-120 BC they have subsided due to consolidation of the underlying salt-marsh sediments (layer 2) of 0.20 ± 0.05 m thickness and tidal-flat sands (layer 3) of about 2 m thickness.

Minimum and maximum values of consolidation have been found by varying the present-day level of palaeo-MHW sediments from 0.75 m –NAP to 0.85 m –NAP, and by entering minimum and maximum values for ground-water table levels, volumetric weights and settlement constants.

The maximum load consisted of salt-marsh sediments of 0.65 ± 0.05 m thickness, plus terp raising of 2.65 m thickness. In addition ground-water table lowering to 1.00 m –NAP may have occurred at the time of maximum loading (in AD 800), so the ground-water table may have been 0.20m lower than during deposition of the palaeo-MHW sediments.

Additional assumptions
- Palaeo-MHW sediments do not subside due to consolidation of underlying tidal-flat sands before additional loading is applied (implying that at the moment the salt-marsh sedimentation reaches the palaeo-MHW level, the tidal-flat sands are completely consolidated for that loading).
- The thickness of layer 2 before loading amounts to $d_{\text{before loading, layer 2}} = 0.20$ or 0.30 m, depending on the exact present-day level of palaeo-MHW sediments (reconstructed as 0.80 ± 0.05 m –NAP).
- The thickness of layer 3 before loading amounts to $d_{\text{before loading, layer 3}} = 2.01$ m.

Load

$$p_{\text{salt marsh}} = h_{\text{overlying salt marsh}} \times \tilde{a}_{\text{salt marsh}} = 0.60 \times 20 = 12.0 \text{ kN/m}^2 \quad \text{or} \quad 0.70 \times 16 = 11.2 \text{ kN/m}^2$$

$$p_{\text{terp}} = h_{\text{terp}} \times \gamma_{\text{terp}} = 2.65 \times 20 = 53.0 \text{ kN/m}^2 \quad \text{or} \quad 2.65 \times 18 = 47.7 \text{ kN/m}^2$$

After 30 years p_{terp} and $p_{\text{salt marsh}}$ are completely transferred to the effective stresses, so maximum loading amounts (assumed palaeo-MHW presently located at 0.75 m –NAP):

$$\Delta\sigma' = p_{\text{terp}} + p_{\text{salt marsh}} + p_{\text{gr-wt-lowering}} = (p_{\text{terp}} + p_{\text{salt marsh}}) + \gamma_{\text{water}} \times \Delta h_{\text{water}} =$$

$$(53.0 + 12.0) + 10 \times 0.20 = 67.0 \text{ kN/m}^2$$

Minimum loading, in case no ground-water table lowering has occurred, amounts (assumed palaeo-MHW presently located at 0.85 m –NAP):

$$\Delta\sigma' = p_{\text{terp}} + p_{\text{salt marsh}} = 47.7 + 11.2 = 58.9 \text{ kN/m}^2$$

Consolidation layer 2

Effective stress in the middle of layer 2 before loading:

$$\overline{\sigma}'_{\text{before loading}} = \overline{\sigma}_{\text{total, before loading}} - \overline{\sigma}_{w\text{, before loading}} =$$

$$\left(\frac{(\gamma_{\text{salt marsh}} \times h_{\text{salt marsh}})_{\text{top}} + (\gamma_{\text{salt marsh}} \times h_{\text{salt marsh}})_{\text{bottom}}}{2} \right) - \left(\gamma_{\text{water}} \times \frac{1}{2} h_{\text{water}} \right) =$$

$$\frac{0 + (18 \times (0.20 \text{ or } 0.30))}{2} - 10 \times (0.10 \quad \text{or} \quad 0.15 = 0.80 \text{ kN/m}^2 \text{ or } 1.20 \text{ kN/m}^2$$

$$z_{\text{layer 2}} = \frac{d_{\text{before loading, layer 2}}}{C_{\text{salt marsh}}} \ln\left(\frac{\overline{\sigma}'_{\text{before loading}} + \Delta\sigma'}{\overline{\sigma}'_{\text{before loading}}}\right) = \frac{0.30}{20} \ln \frac{1.20 + 67.0}{1.20} = 0.060 \text{ m}$$

$$\text{or} \quad \frac{0.20}{50} \ln \frac{0.80 + 58.9}{0.80} = 0.017 \text{ m}$$

Consolidation layer 3
Effective stress in the middle of layer 3 before loading:

$$\overline{\sigma}'_{\text{before loading}} = \overline{\sigma}_{\text{total, before loading}} - \overline{\sigma}_{w,\text{ before loading}}$$

in which

$$\overline{\sigma}_{\text{total, before loading}} = \sigma_{\text{layer 2}} + \overline{\sigma}_{\text{tidal – flat sands}} =$$

$$(\gamma_{\text{salt marsh}} \times h_{\text{layer 2}}) + \frac{1}{2}(\gamma_{\text{tidal – flat sands}} \times h_{\text{tidal – flat sands}}) =$$

$$18 \times 0.20 + \frac{1}{2}(19 \times 2.0) = 22.6 \text{ kN/m}^2$$

$$\overline{\sigma}_{w,\text{ before loading}} = \gamma_w \times (h_{\text{layer 2}} + \frac{1}{2} h_{\text{layer 3}}) = 10 \times (0.2 + \frac{1}{2} \times 2.0) = 12 \text{ kN/m}^2$$

$$\Rightarrow \overline{\sigma}'_{\text{before loading}} = 10.6 \text{ kN/m}^2$$

$$z_{\text{layer 3}} = \frac{d}{C_{\text{tidal – flat sands}}} \ln\left(\frac{\overline{\sigma}'_{\text{before loading}} + \Delta\sigma'}{\overline{\sigma}'_{\text{before loading}}}\right) = \frac{2.01}{200} \ln \frac{10.6 + 67}{10.6} = 0.020 \text{ m}$$

$$\text{or} \quad \frac{2.01}{650} \ln \frac{10.6 + 58.9}{10.6} = 0.006 \text{ m}$$

Total subsidence
Total subsidence of the palaeo-MHW sediments due to loading by younger sediments, raisings and possible ground-water table lowering amounts minimum 0.017 + 0.006 = 0.023 m and maximum 0.060 + 0.020 = 0.080 m.

So at the time of deposition the level of the palaeo-MHW sediments was located somewhere between 0.83 m –NAP and 0.67 m –NAP; this range includes the reconstruction uncertainty of the present-day level.

Level of top salt-marsh sediments in AD 150-175

Since the deposition of these sediments in AD 150-175 they have subsided due to consolidation of the underlying salt-marsh sediments (layer 1 + layer 2) of 0.85 m thickness and tidal-flat sands (layer 3) of about 2 m thickness.

The load consisted of terp raising of 2.65 m thickness. In addition a lowering of the ground-water table to 1.00 m –NAP may have occurred at the time of maximum loading (in AD 800).

Additional assumptions

- During deposition of the top of the salt-marsh sediments the ground-water table was located at surface level; the ground-water table has never been lower than 1.00 m –NAP. Also the case has been considered that the ground-water table has never been lower than the top of the salt-marsh sediments in the period AD 150 to AD 1996.
- Thickness of salt-marsh sediments before loading amounts to $d_{\text{before loading, salt marsh}} = 0.95$ m.
- Thickness of tidal-flat sands before loading amounts to $d_{\text{before loading, tidal – flat sands}} = 0.95$ m.

Consolidation salt-marsh sediments
Effective stress in salt-marsh (layer 1 + layer 2) before loading:

$$\overline{\sigma}'_{\text{before loading}} = \overline{\sigma}_{\text{total, before loading}} - \overline{\sigma}_{w,\text{ before loading}} =$$

$$\left(\frac{(\gamma_{\text{salt marsh}} \times h_{\text{salt marsh}})_{\text{top}} + (\gamma_{\text{salt marsh}} \times h_{\text{salt marsh}})_{\text{bottom}}}{2}\right) - \left(\gamma_{\text{water}} \times \frac{1}{2} h_{\text{water}}\right) =$$

$$\frac{0 + (18 \times 0.95)}{2} - 10 \times 0.425 = 4.3 \text{ kN/m}^2$$

Total load (with and without lowering of ground-water table):

$$\Delta\sigma' = p_{terp} + p_{gr-wt-lowering} = p_{terp} + \gamma_{water} \times \Delta h_{water} = 53.0 + 10 \times 0.85 = 61.5 \text{ kN/m}^2$$

$$\text{or} \quad 47.7 + 0 = 47.7 \text{ kN/m}^2$$

$$z_{salt\ marsh} = \frac{d_{before\ loading}}{C_{salt\ marsh}} \ln\left(\frac{\bar{\sigma}'_{before\ loading} + \Delta\sigma'}{\bar{\sigma}'_{before\ loading}}\right) = \frac{0.95}{20} \ln \frac{4.3 + 61.5}{4.3} = 0.130 \text{ m}$$

$$\text{or} \quad \frac{0.95}{50} \ln \frac{4.3 + 47.7}{4.3} = 0.047 \text{ m}$$

Consolidation layer 3
Effective stress in layer 3 before loading:

$$\bar{\sigma}'_{before\ loading} = \bar{\sigma}_{total,\ before\ loading} - \bar{\sigma}_{w,\ before\ loading} =$$

$$\gamma_{salt\ marsh} \times h_{salt\ marsh} + \frac{1}{2}\gamma_{layer\ 3} \times h_{layer\ 3} - \left(\gamma_{water} \times \frac{1}{2} h_{water}\right) =$$

$$\left(18 \times 0.95 + \frac{1}{2}(19 \times 2.00)\right) - 10 \times 1.85 = 17.6 \text{ kN/m}^2$$

$$z_{layer\ 3} = \frac{d_{before\ loading,\ layer\ 3}}{C_{tidal-flat sands}} \ln\left(\frac{\bar{\sigma}'_{before\ loading} + \Delta\sigma'}{\bar{\sigma}'_{before\ loading}}\right) = \frac{2.01}{200} \ln \frac{17.6 + 61.5}{17.6} = 0.015 \text{ m}$$

$$\text{or} \quad \frac{2.01}{650} \ln \frac{17.6 + 47.7}{17.6} = 0.003 \text{ m}$$

Total subsidence
The top of the salt-marsh sediments has subsided at minimum 0.047 + 0.003 = 0.05 m and at maximum 0.130 + 0.015 = 0.15 m. So the top of the salt-marsh sediments was located somewhere between 0.10 –NAP and 0.00 m –NAP in AD 150-175.

Level of hearth of house in AD 150-175

The additional subsidence by the hearth of the house has undergone with respect to the top of the salt-marsh sediments, is the amount of consolidation of the layer terp-period-I.

It has been assumed that consolidation of this layer due to its own weight amounts to one sixth part of its thickness at maximum,[65] and zero at minimum (in case the terp soil was densified perfectly during raising). Further consolidation was caused by the load of the terp raising from 1.25 m +NAP to 2.50 m +NAP.

Additional assumptions
- The ground-water table always remained below the bottom of the terp.
- The thickness of layer terp-period-I before loading amounts to $d_{before\ loading,\ tpl} = 1.63$ m.

Consolidation of the layer terp-period-I
Consolidation due to own weight:

$$z_{due\ to\ own\ weight} = \frac{1}{6} \times 1.63 \text{ m} = 0.271 \text{ m}$$

Effective stress in layer terp-period-I before loading:

$$\bar{\sigma}_{terp} = \bar{a} \times \frac{1}{2} h = 19 \times \frac{1.40}{2} = 13.3 \text{ kN/m}^2$$

Effective stress increment in layer terp-period-I due to terp raising to 2.50 m +NAP:

$$\Delta\sigma' = \gamma_{raising} \times h_{raising} = 1.25 \times 18 = 22.5 \text{ kN/m}^2 \quad \text{or} \quad 1.25 \times 20 = 25 \text{ kN/m}^2$$

$$z_{\text{terp due to raising}} = \frac{1.63}{25} \ln \frac{13.3 + 25}{13.3} = 0.069 \text{ m} \quad \text{or} \quad \frac{1.63}{50} \ln \frac{13.3 + 22.5}{13.3} = 0.032 \text{ m}$$

$$z_{\text{terp, total}} = 0.271 + 0.069 = 0.340 \text{ m} \quad \text{or} \quad 0.032 + 0 = 0.032 \text{ m}$$

Total subsidence of hearth house

Total subsidence of hearth of house is $z_{\text{heard house}} = z_{\text{top salt marsh}} + z_{\text{terp, total}}$. This amounts to minimum 0.05 + 0.032 = 0.082 m and maximum 0.15 + 0.340 = 0.49 m.

So in AD 150-175 the hearth of the house was located somewhere in between 1.33 m +NAP and 1.74 m +NAP. The large range mainly depends on the method of densification during terp construction.

REFERENCES

CUR-COMMISSIE (C68) 1993. Construeren met grond – Grondconstructie op en in weinig draagkrachtige en sterk samendrukbare ondergrond, *Bouwen op slappe grond*, Gouda (Civieltechnisch Centrum Uitvoering Research en Regelgeving, CUR-publicatie 162).

LOCHER, W.P. & H. DE BAKKER 1990. *Bodemkunde van Nederland*, part 1, 2nd ed., Den Bosch.

SENAKA WELIDENYA, H. 1996. GIS application in foundation cost zonation for an area in Sliedrecht, the Netherlands, MSc-thesis ITC Delft.

TYSMA, S., F. VAN HERWIJNEN & P.H.H. LEIJENDECKERS 1994. *Polytechnisch Zakboek*, 46th ed., Arnhem.

REFERENCES

BAKKER, J.P. 1949. Morfologisch onderzoek van Barradeel en zijn betekenis voor het inzicht in de subatlantische transgressie, *Verhandelingen van de Koninklijke Nederlandse Akademie van Wetenschappen, Akademiedagen* (Leeuwarden), Deel II: 121-143.

BAKKER, J.P. 1957. Enige opmerkingen over waterwegen in N.W. Friesland omstreeks 1000 na Chr. In: A. Martin (ed.), *Vroeg-middeleeuwse zandstenen sarcophagen in Friesland en elders in Nederland*, Drachten, 153-163.

BAKKER, J.P. 1958a. Küstenentwicklung und prähistorische Besiedlung in den Niederlanden, *Sitzungsberichte der Physikalische-medizinischen Sozietät zu Erlangen* 78: 102-127.

BAKKER, J.P. 1958b. Landschaps- en sedimentatieontwikkeling in Westelijk Barradeel sinds prehistorische tijd, *It Beaken* 20: 219-239.

BAKKER, J.P. & J.J. WENSINK 1955. Overzicht van de Holocene reliefgeneraties en sedimentopvolging in Barradeel, *Barradeel, Rapport betreffende het onderzoek van het Lânskip-genetysk Wurkforbân fan de Fryske Akademy*, Drachten, 1-27.

BEETS, D.J., A. VAN DER SPEK & L. VAN DER VALK 1994. Holocene ontwikkeling van de Nederlandse kust, Haarlem (Rijks Geologische Dienst, Rapport 40.016).

BEETS, D.J. & P.C. VOS 1995. Quarternary field trips in Central Europe, 16, North Sea Coast, Stops 15-19: The Netherlands. In: W. Schirmer (ed.), INQUA, XIV International Congress, Aug. 3-10, 1995, Berlin, Germany, *Quarternary field trips in Central Europe, 2, Field trips on special topics*, Berlin, 1031-1034.

CATE, J.A.M. TEN & G.C. MAARLEVELD 1977. *Geomorfologische kaart van Nederland 1:50.000, Toelichting op de legenda*,Wageningen etc.

CNOSSEN, J. 1958. Enige opmerkingen omtrent het ontstaan van het Beneden-Boornegebied en de Middelzee in verband met subatlantische transgressie, *Boor en Spade* 9: 24-38.

DELCOURT, P.A., W.H. PETTY & H.R. DELCOURT 1996. Late-Holocene formation of Lake Michigan beach ridges correlated with 70-yr oscillation in global climate, *Quaternary Research* 45: 321-326.

DIJKEMA, K.S., J.H. BOSSINADE, P. BOUWSEMA et al. 1990. Salt marshes in the Netherlands Wadden Sea: rising high-tide levels and accretion enhancement. Expected effects of climatic change on marine coastal ecosystems, *Developments in Hydrology* 57: 173-188.

GIJSSEL, K. & W. DE GANS 1993. *General map of the top of the Pleistocene deposits in the Netherlands*, Haarlem.

GRIEDE, J.W. 1978. *Het ontstaan van Frieslands Noordhoek*, Amsterdam (thesis Vrije Universiteit).

GROOT, T.A.M. DE, H.A. VAN ADRICHEM BOOGAERT, M.M. FISCHER et al. 1987. *Toelichtingen bij de Geologische Kaart van Nederland 1:50.000, Blad Heerenveen West en Oost (10 W, 10 O)*, Haarlem.

HOYT, J.H. 1969. Chenier versus barrier, genetic and stratigraphic distinction. *Bulletin of the American Association of Petroleum Geologists* 53: 299-306.

KING, C.A.M. 1972. *Beaches and coast*, London.

KNOL, E. 1993. *De Noordnederlandse kustlanden in de Vroege Middeleeuwen*, Groningen (thesis Vrije Universiteit Amsterdam).

LICHTER, J. 1995. Lake Michigan beach ridge and dune development, lake level, and variability in regional water balance, *Quaternary Research* 44: 181-189.

NEDERLANDS NORMALISATIE INSTITUUT (NNI) 1989. *Geotechniek: Classificatie van onverharde grondmonsters, NEN 5104*, Delft.

REINECK, H.E. & I.B. SINGH 1980. *Depositional sedimentary environments*, 2nd edition, Berlin etc.

RIENKS, K.A. & G.L. WALTHER 1954. *Binnendiken en slieperdiken yn Fryslân*, Bolsward.

ROELEVELD, W. 1974. *The Holocene evolution of the Groningen marine clay district*, Amsterdam (thesis Vrije Universiteit).

ROEP, T.B. & J.F. VAN REGTEREN ALTENA 1988. Paleotidal levels in tidal sediments (3800-3635 BP); compaction, sea-level rise and human occupation (3275-2620 BP) at

Bovenkarspel, NW Netherlands. In: P.L. de Boer, A. van Gelder, & S.D. Nio (eds), *Tide-influenced sedimentary environments and facies*, Dordrecht, 215-231.

SMET, L.A.H. DE 1965. *De bodem van Groningen*, Wageningen.

SMET, L.A.H. DE & J.J. VLEESHOUWER 1956. De Bodemkundige opbouw van het Groninger kleigebied, *Boor en Spade* 8: 142-159.

SPEK, A. VAN DER 1994. Large-scale evolution of Holocene tidal basins in the Netherlands, Utrecht (thesis University of Utrecht).

STRAATEN, L.M.J.U. VAN 1954. Composition and structure of recent marine sediments in the Netherlands, *Leidse Geologische Mededelingen* 19: 1-110.

STIBOKA 1974. *Bodemkaart van Nederland Schaal 1:50.000, Toelichtingen bij de kaartbladen 10 W Sneek en 10 O Sneek*, Wageningen.

STIBOKA 1976. *Bodemkaart van Nederland Schaal 1:50.000, Toelichtingen bij de kaartbladen 5 W Harlingen en 5 O Harlingen*, Wageningen.

STUIVER, M. & P.J. REIMER 1993. Extended ^{14}C database and revised CALIB 3.0 ^{14}C age calibration program, *Radiocarbon* 35: 215-230.

TAAYKE, E. 1990 and 1997. Die einheimische Keramik der nördlichen Niederlande, 600 v Chr. bis 300 N. Chr., Teil I: Westergo, *Berichten van de Rijksdienst voor het Oudheidkundig Bodemonderzoek* 40: 101-222; Teil V: Übersicht und Schlußfolgerungen, 42: 163-208.

TAYLOR, M. & G.W. STONE 1996. Beach-ridges: a review, *Journal of Coastal Research* 12: 612-621.

TERWINDT, J.H.J. 1988. Palaeo-tidal reconstructions of inshore tidal depositional environments. In: P.L. de Boer, A. van Gelder & S.D. Nio (eds), *Tide-influenced sedimentary Environments and facies*, Dordrecht, 233-263.

THOMPSON, T.A. & S.J. BAEDKE 1995. Beach ridge development in Lake Michigan: shoreline behavior in response to quasi-periodic lake-level events, *Marine Geology* 129: 163-174.

VEEN, J. VAN 1950. Eb- en vloedschaar systemen in de Nederlandse getijdewateren, *Tijdschrift Koninklijk Nederlands Aardrijkskundig Gennootschap* 67: 303-325.

VEENENBOS, J.S. 1949. De bodemkartering van de Friese knipkleigronden, *Boor en Spade* 3: 76-86.

VOS, P.C. 1992. *Paleogeografische reconstructie van het Lauwersmeergebied*, Oosterwolde (Rijks Geologische Dienst- District Noord, project 40009) 1-22.

WEE, M.W. TER 1976. *Toelichtingen bij de Geologische Kaart van Nederland 1:50.000, Blad Sneek (10 W, 10 O)*, Haarlem.

WENSINK, J.J. 1958. De Jongpleistocene en Holocene ontwikkeling van een deel van Westergo, *Geologie en Mijnbouw* 20: 73-87.

ZAGWIJN, W.H. 1986. *Nederland in het Holoceen*, Haarlem (Geologie van Nederland, 1) 1-46.

NOTES

1. Ter Wee (1976, pp 68-82), Groot et al. (1987, pp 10-17 and pp 142-155).
2. In the Dutch literature, there is no consensus about the way these ridges developed, see above section: Salt-marshes in the Dutch literature.
3. Beets & Vos (1995, 1031 pp).
4. Van der Spek (1994, pp 43-45).
5. Beets & Vos (1995, 1031 pp).
6. Van Straaten (1964), Roep & Van Regteren Altena (1988, pp 222-223).
7. Veenenbos (1949).
8. Taayke (1990, 1997).
9. Cnossen (1958, p. 30), Knol (1993, pp 34-35).
10. Van der Spek (1994, p. 102).
11. STIBOKA (1974, p. 37), Ter Wee (1976, p. 84 and p. 89).
12. STIBOKA (1974, p. 35), Van der Spek (1994, p. 98 and p. 102).
13. Van der Spek (1994, fig. 5.3, p. 103 – mainly after Rienks & Walther 1954).
14. Van Veen (1950, pp 303-305).
15. Roeleveld (1974, pp 66-67).
16. Bakker (1949, pp 127-131).
17. The so-called tidal channels of the 'second generation', Bakker (1949, p. 130).
18. Bakker (1957, p. 155; 1958a, pp 115-117; 1958b, pp 228-231), Bakker & Wensink (1955, pp 17-21).
19. De Smet & Vleeshouwer (1956, pp 145-146), De Smet (1965, pp 14-16), Ten Cate & Maarleveld (1977, p. 58.)
20. De Smet (1965, p. 16).
21. Roeleveld (1974, p. 83 – cheniers, King 1972).
22. Griede (1978, p. 107).
23. STIBOKA (1974, 1976).
24. NAP = Normaal Amsterdams Peil, Dutch ordnance level, which is appoximately the present mean sea level.
25. Taayke (1990, 187 ff; 1996, 189 ff).
26. Taayke (1990, 1997).
27. NEDERLANDS NORMALISATIE INSTITUUT (NNI), Geotechniek: Classificatie van onverharde grondmonsters, NEN 5104: 13-14.
28. Terwindt (1988, pp 234-235).
29. Van Straaten (1954), Roep & Van Regteren Altena (1988, pp 222-223).
30. Dijkema (1990 – Bossinade et al. pp 173-188).
31. Period I; see Gerrets & De Koning (this volume).
32. Gerrets & De Koning (this volume).
33. Gerrets & De Koning (this volume).
34. For the construction and stratigraphy of the terp, see Gerrets & De Koning (this volume).
35. Roep & Van Regteren Altena (1988, p. 226).
36. De Groot et al. (1996, p. 76).
37. Gerrets & De Koning (this volume).
38. Van der Spek (1994, fig. 5.3, p. 103 – mainly after Rienks & Walther 1954).
39. Taayke (1990, 1997).
40. Knol (1993, p. 94, Table 6).
41. De Langen, personal communication.
42. Gerrets & De Koning (this volume).
43. De Smet & Vleeshouwer (1956, pp 145-146), De Smet (1965, pp 14-16), Ten Cate & Maarleveld (1977, p. 58).
44. Roeleveld (1974, p. 83).
45. Cf. the descriptions given by Taylor & Stone (1996) in their review of these morphological phenomena, p. 612 and further.

46. Reineck & Singh (1980, pp 352-354 cf. Hoyt 1969).
47. Beets & Vos (1994, p. 29).
48. E.g. Taylor & Stone (1996, pp 612-615), Thompson & Baedke (1995, pp 164-167).
49. Zagwijn (1986, palaeogeographic maps 7-9).
50. Lichter (1995, p. 188), Delcourt et al. (1996, pp 324-325).
51. Palaeogeographic maps 6 and 7 in Zagwijn (1986).
52. Zagwijn (1986, palaegeographic maps 7-9).
53. Gerrets & De Koning (this volume).
54. Van der Spek (1974, p. 101).
55. Gerrets & De Koning (this volume).
56. Gerrets & De Koning (this volume).
57. Cf. the descriptions given by Taylor & Stone (1996) in their review of these morphological phenomena, p. 612 and further.
58. Roeleveld (1974, pp 13-15), Griede (1978, pp 37-40).
59. Zagwijn (1986, palaeogeographic map 7).
60. Vos (1992, pp 9-12).
61. Griede (1978, pp 90-92).
62. Zagwijn (1986, palaeogeographic map 7).
63. Senaka Welidenya (1996, p. 2).
64. See CUR-commissie (1993) and Locher & De Bakker (1990, part 1: 306-310).
65. See Tysma et al. (1994, F2/34 and F2/8).

Settlement development on the Wijnaldum-Tjitsma terp

D.A. GERRETS & J. DE KONING

1 INTRODUCTION

The terp Wijnaldum-Tjitsma is one of the bigger terps of a row of at least five terps near the village of Wijnaldum, about 10 km northeast of the city of Harlingen (Province of Friesland). The present-day village of Wijnaldum itself is situated on the most western terp of this row. The terp Wijnaldum-Tjitsma is the terp in the middle of this cluster measuring 11.2 ha with a height of 2.2 m. By comparison, the highest terp of the Netherlands, Hoogebeintum, with its height of 8.8 m. measured 9.5 ha before quarrying and the large terp of Ezinge measured about 15 ha and was ca 5.5 m high before quarrying/excavation. Of the terp Wijnaldum-Tjitsma slightly more than 8000 m^2 was excavated, which is 7.1% of its total surface. The excavation trenches varied in depth between 0.5 and 4 m.

As we will see below, the settlement started in the Middle Roman Period with the construction of at least two house platforms. Finds collected on the terp outside the excavation area suggest that several more of these house platforms were constructed on the site, which gradually grew together into one dwelling mound. The terp as we know it today is the result of almost 800 years of occupation and complex site formation processes. Part of the top of the terp has disappeared, due to erosion.[1] The restricted area excavated, together with the disappearance of remains of human occupation because of erosion and intensive habitation and the first house platforms gradually growing together into one terp make it hard to define the settlement during the different occupation phases. It is important to bear this in mind when reading the following description of settlement development on the Wijnaldum-Tjitsma terp.

In this chapter a general outline of the development of the settlement at Wijnaldum-Tjitsma is described based on the analysis of the plans and sections and on the analysis of find material, as far as they are now available. The reader should not forget that the analyses of several categories of find material is still going on and that they may alter the picture as it is presented here. It is reasonable to expect, however, that these changes will be refinements, rather than fundamental alterations, now that the analysis of most large find categories is for the greater part finished or at an advanced stage. The main issue of this chapter is the presentation of the development of settlement structure as a whole.

2 THE ANALYSIS OF A TERP LIKE WIJNALDUM-TJITSMA: POSSIBILITIES AND IMPOSSIBILITIES

The excavation of a terp entails a number of special methodological problems, unknown for example to the research of 'two-dimensional' settlements on the higher sandy soils, where accumulation of settlement debris is lacking. Whereas in the latter settlements even the former culture layer is generally absent, terps are an accumulation of several occupation layers on top of each other. Settlements on the sandy soils often moved over a large surface, while occupation on terps tended to concentrate within a very restricted area. Terps did not only develop because of intentional heightening, but also as a result of occupation itself in the accumulation of settlement debris. This offers the advantage that in time a stratigraphy develops in which different occupation phases are represented by separated levels. At the same time the concentration of the occupation on a restricted surface causes a disturbance of older occupation phases. On the terp Wijnaldum-Tjitsma digging of features like ditches and wells caused a considerable transport of soil and therefore of old material to younger layers. This formed an important problem in the dating and interpretation of features. Before the construction of new buildings, remains of former buildings were generally levelled. Therefore, only few structures were (partially) preserved.

The terp Wijnaldum-Tjitsma was occupied from about AD 175 until ca 950; in the second half of the 10th century the occupation had presumably moved to a terp more to the south on which the actual farm stands. This farm burnt down in the 1970s. According to the farmer living in the farmhouse, a great number of bricks was found during the rebuilding of the farm which can be dated in the 13th/14th century; this can be regarded as an indication that there was a brick building standing here.

3 DATING AND ITS LIMITATIONS

During the analysis of the complex stratigraphy of Wijnaldum-Tjitsma, the application of the Harris matrix has proved to be of great value.[2] In the Harris matrix all stratigraphical relationships between features are depicted in an abstract way (see Fig. 1). The Harris matrix forces to follow strict reasoning in establishing the stratigraphical sequence of features and as such helps to get a better understanding of the relative chronology of a site. The collection of finds per feature is an absolute condition for the application of the Harris matrix.

In order not to be influenced by the data of the finds, they were left out of consideration until the final stage of the chronological analysis. The two main cross-sections formed the backbone of the Harris matrix. The datings of the finds were first included after the stratigraphical sequence of all 3713 features had been established.

After the Harris matrix was finished and all sherds were typologically classified separately from each other, we tried to make datings more absolute by combining both kinds of data. Especially the dating of the pottery played an important role in the dating of the features, because of its quantity. In about one third of all features there was datable material, in most cases pottery, very seldom well datable finds like coins, brooches etc. A golden tremissis struck in ca 640 offers a post quem dating, but one can never be certain that it is in the context where it belongs. It can for example have been transported to other younger or older layers by human or natural activities. Only by comparing the dating of the coin with the other find material from the same context in combination with its stratigraphical position, will it be clear whether the post quem dating represents an 'absolute' date in the Harris matrix or not. If this is so, then all features standing in a younger stratigraphical relation to the feature with the tremissis will be younger than 640. In general, the few well datable finds are seldom found in features which mark the transition of one period to an other. As such they only confirm or contradict a dating of the feature in which they were found. Therefore establishing 'absolute' datings for features in a well stratified site as Wijnaldum-Tjitsma is a complicated process of logical reasoning, and careful consideration of many sometimes mutually contradicting variables. Sometimes we had to conclude that a feature had a certain dating only to find shortly after that this dating was contradicted by a stratigraphically somewhat younger feature and the whole process had to start again.

Much pottery, especially from the Migration Period is still little studied and because of this badly dated. Typochronological studies of for example Merovingian wheelthrown ware are mainly based on pottery from cemeteries.[3] In these studies typological characteristics take into account the whole pot and are not very useful for the analysis of large numbers of pottery from settlement excavations, where the archaeologist has to deal only with sherds. Here we were forced to develop a new typology which can be used in settlement research in the future. The Tjitsma excavation will provide us with the first datings for this new typology for Merovingian wheelthrown ware, which are mainly based on the shape of the rim and the fabric.

Therefore typochronological classification of the Merovingian wheelthrown ware and of the indigenous handmade pottery from the Migration Period did not contribute much to the absolute chronology of the site. However, as they are represented in large numbers in pottery assemblages at the site, their percentage in relation to the presence of other finds is of chronological importance. Especially these general quantitative changes in combination with the interrelation of different kinds of pottery in assemblages has proved to be of great chronological value in relation to the Harris matrix.

They helped to produce a refined relative chronology which could be given a more absolute foundation with the help of the proportionally small number of well datable finds.

A problem we encountered in the dating of features is the transport of material on the site caused by human activities. Already at an early stage of the analysis it became clear that the transport of material played an important role in the composition of the find assemblages. In some cases features

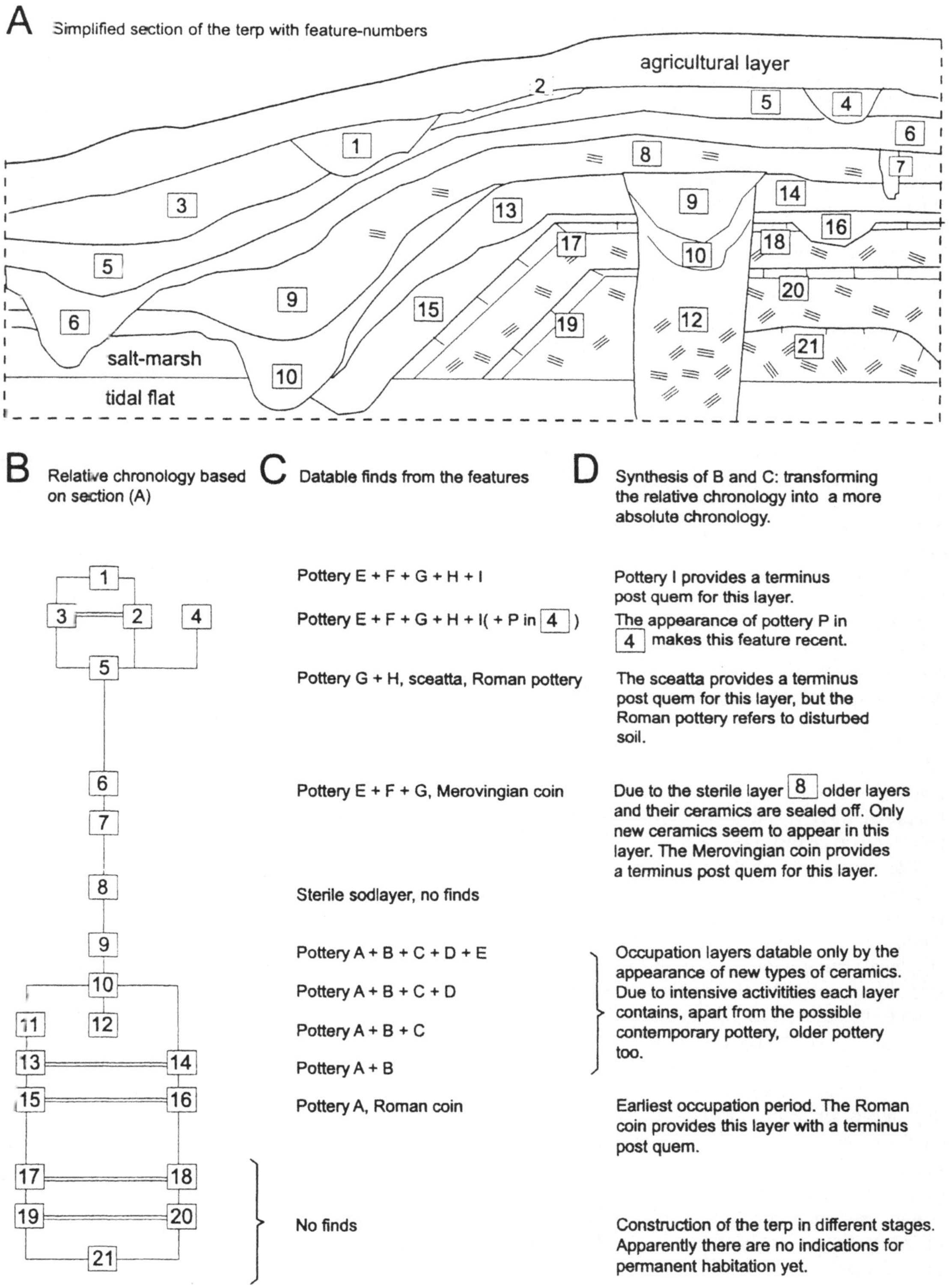

Figure 1. Schematic representation of the dating process based on the situation as met in the Wijnaldum excavation.

were excavated, which regarding their stratigraphical context belonged to the Carolingian Period, but according to the material found in them would have to be dated to the Roman Period. With the help of the Harris matrix transportation of material could be determined with a high degree of certainty, in the case of much older material. This is, however, much more difficult for layers lying immediately on top of each other. As most of the time it is unknown how long certain artefact types stayed in use, transport of material could only on rare occasions be determined for find categories which rapidly succeeded each other. While transport from older, deeper layers to younger, higher lying layers occurred on a large scale (341 features, 9.2%) the reverse e.g. by the activity of animals like moles has hardly occurred (52 features, 1.4%). The find-material appeared consistent with the stratigraphical dating in the case of 1253 features (33.7%). 2067 features (55.7%) without datable find material were able to be dated with the help of the Harris matrix.

The excavation took place by digging artificial levels, more or less following the relief of the topsoil. This meant that levels were sometimes very heterogeneous in material. In some excavation levels remains of occupation representing more than 600 years were excavated. Especially where strata were sloping steeply or where intensive digging had taken place in the past, several occupation phases could be represented in one excavation level. It was only because material was collected featurewise in the field that the different occupation phases could be distinguished.[4]

The specialists involved in the analysis of the different categories of find material other than pottery were provided with datings and related to these datings a valuation of the reliability. If the stratigraphical dating with the help of the Harris matrix was consistent with the material content of the feature the dating was given value 1. If downward transportation of material had taken place the dating got value 2 and must be regarded as unreliable, because in most cases this will mean that a feature dug into older levels was not observed during excavation. When it was clear that upward transportation had taken place the dating got value 3. In cases where no material was excavated from the feature but where a dating could be given because of stratigraphical analysis the dating of the feature got value 4.

Two dendrochronological and eleven radiocarbon datings (see Fig. 2) had an important impact during the analysis. They forced us to critically reconsider our datings and caused some important movements in the absolute datings of the matrix. This kind of absolute dating, however, in an excavation such as Wijnaldum-Tjitsma is less absolute than we thought. A felling-date for an oak, used for a frame at the bottom of a well in the summer of AD 263 seems to be very precise but says little about the date of the well. The wood could have been (re-)used much later.[5] Radiocarbon datings pose the same kind of problem. Although we have consciously selected big fragments of animal bones, because the chance of them being transported is smaller and the dating is more reliable, we will never know whether the dating is representative of the feature.

This does not mean that all datings are built on quicksand. The laws as prescribed by the Harris matrix, that is to say the stratigraphical sequence of layers and features, impose absolute time limits. The features are dated by all the different materials which were found in them and as far as they have been analysed up to now (radiocarbon, dendrochronological datings, pottery, glass, Roman metal finds, coins and fibulae) and by the dating of the feature above and under it.

For some find categories the Harris matrix in combination with the large numbers of finds from Wijnaldum-Tjitsma offers the possibility for a more precise dating of the find material. Unfortunately this analysis has to be omitted because of a lack of funds and a lack of time. Regarding all uncertainties in archaeological dating, the datings are not 'historical' dates, but estimates with a varying degree of reliability and precision. The precision of all datings presented in this volume have, generally speaking, a maximum deviation of fifty years, but in many cases they will be more precise, in only a few cases less. As upward movements occurred on a much larger scale than downward movements, the termini post quem will generally be more reliable than the termini ante quem.

This method of stratigraphical analysis offers new possibilities, but also new problems. It is now possible to say something about the relationship between find material and context, for example whether the material is in the context one would expect it to be, whether transport of material has taken place, or whether material has been brought into the settlement secondarily. This has produced interesting data especially for the material from the Roman Period.[6] How to explain the great number of finds from the Roman Period in an early medie-

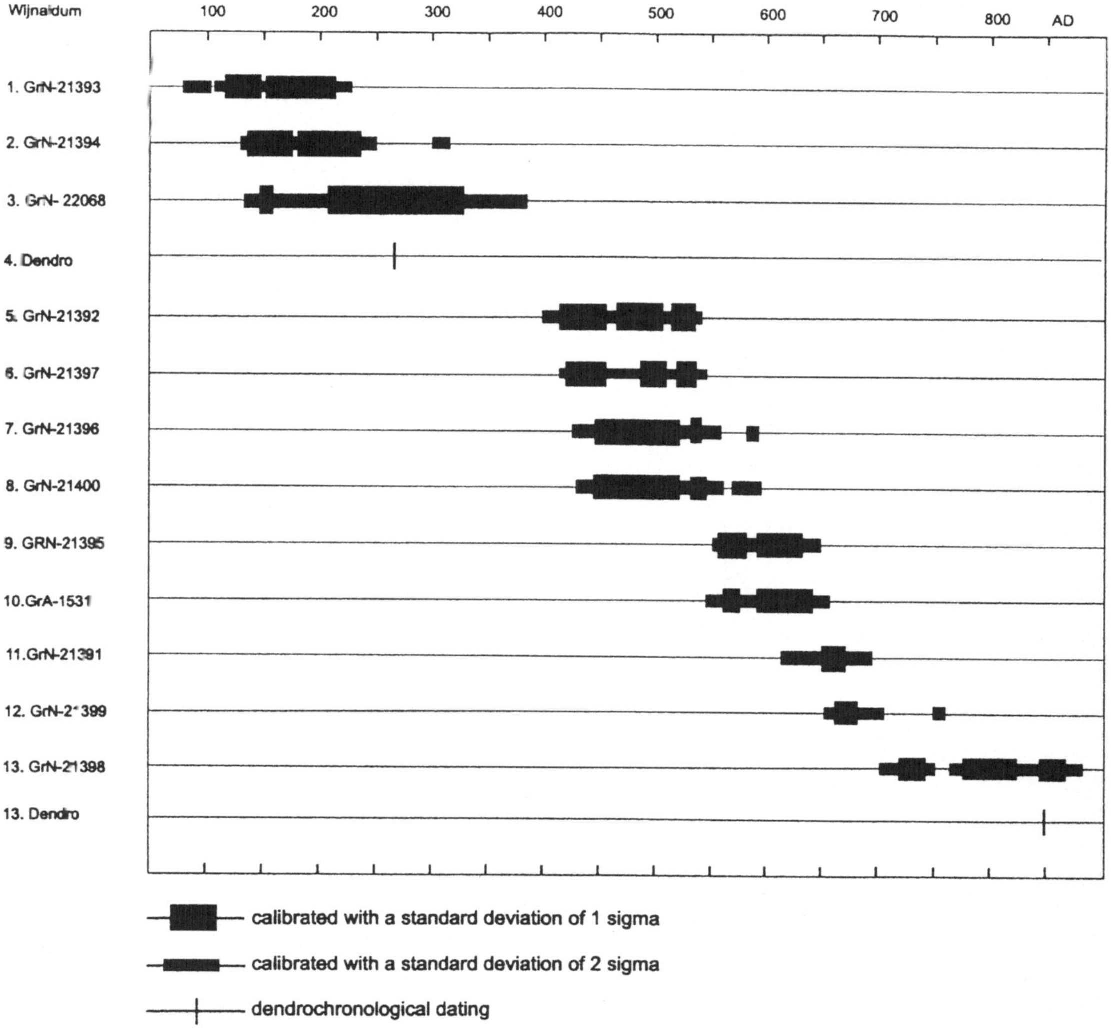

Figure 2. Radiocarbon and dendrochronological datings.

val context? Is this because of transport of material on the terp or were these finds brought into the settlement at a later stage to be reused for another purpose?

For some find categories, like glass, metal and small animal bone, it is the first time in the coastal area of the northern Netherlands that they come from a controlled, stratified context. It will be clear that the new data from the Wijnaldum-Tjitsma terp offer great possibilities for the interpretation of material coming from a less well controlled context.

The precision of the newly acquired chronological data creates serious problems for typochronological classification of the material and for the periodization of the settlement. Changes in settlement structure will seldom coincide with changes in ceramics or changes in types of fibula. The more precisely the material is dated, the more difficult it gets to draw absolute time lines. This shows once more that classification is a descriptive tool of the archaeologist, but it does not always reflect important events or changes in the past.

Figures 3 to 20 represent the eight occupation phases and two sub-phases (IIIa and IIIb) on the Wijnaldum-Tjitsma terp. On the left page the features are represented as they were found in the excavation while the right page shows a reconstruc-

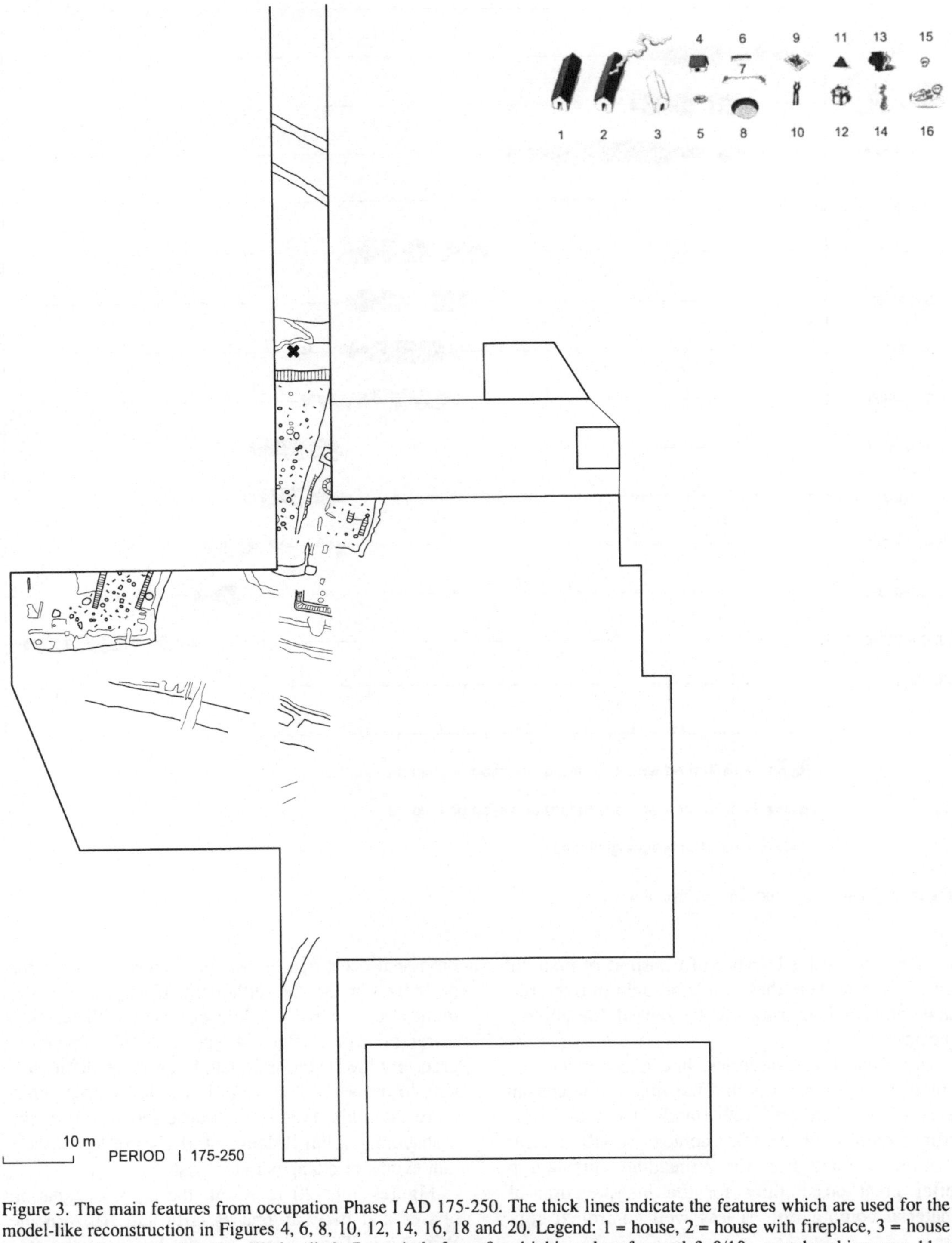

Figure 3. The main features from occupation Phase I AD 175-250. The thick lines indicate the features which are used for the model-like reconstruction in Figures 4, 6, 8, 10, 12, 14, 16, 18 and 20. Legend: 1 = house, 2 = house with fireplace, 3 = house (uncertain), 4 = sunken hut, 5 well, 6 = ditch, 7 = sod platform, 8 = drinking place for cattle?, 9/10 = metalworking area, 11 = fireplace, 12 = granary, 13 = cremation burial, 14 = inhumation burial, 15 = infant inhumation, 16 = horse foetus and infant inhumation. × (thick) = location of 14C sample, × (thin) = location of dendrochronological sample.

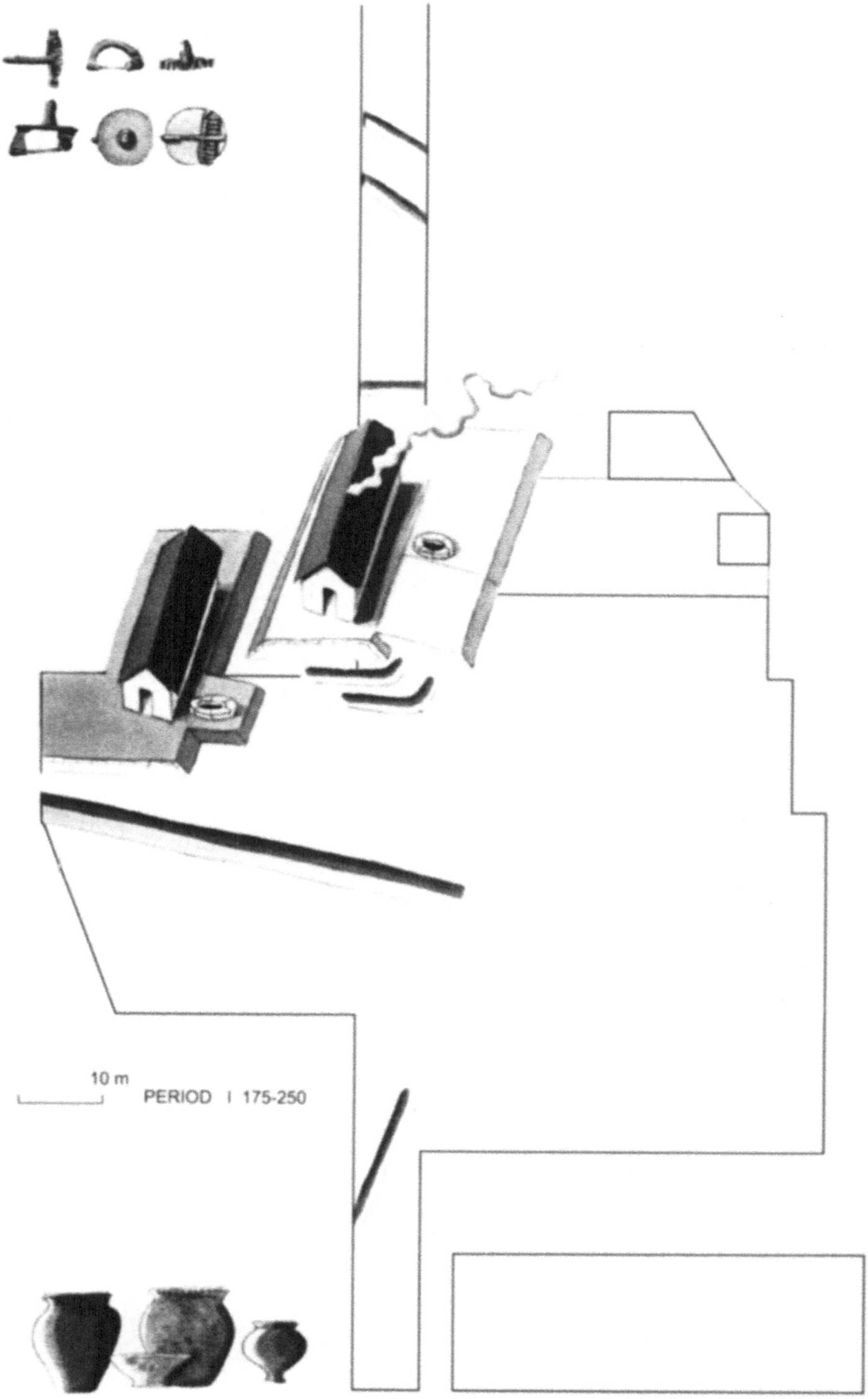

Figure 4. Model-like reconstruction of the settlement between AD 175 and 250. For explanation of symbols of this figure see Figure 3.

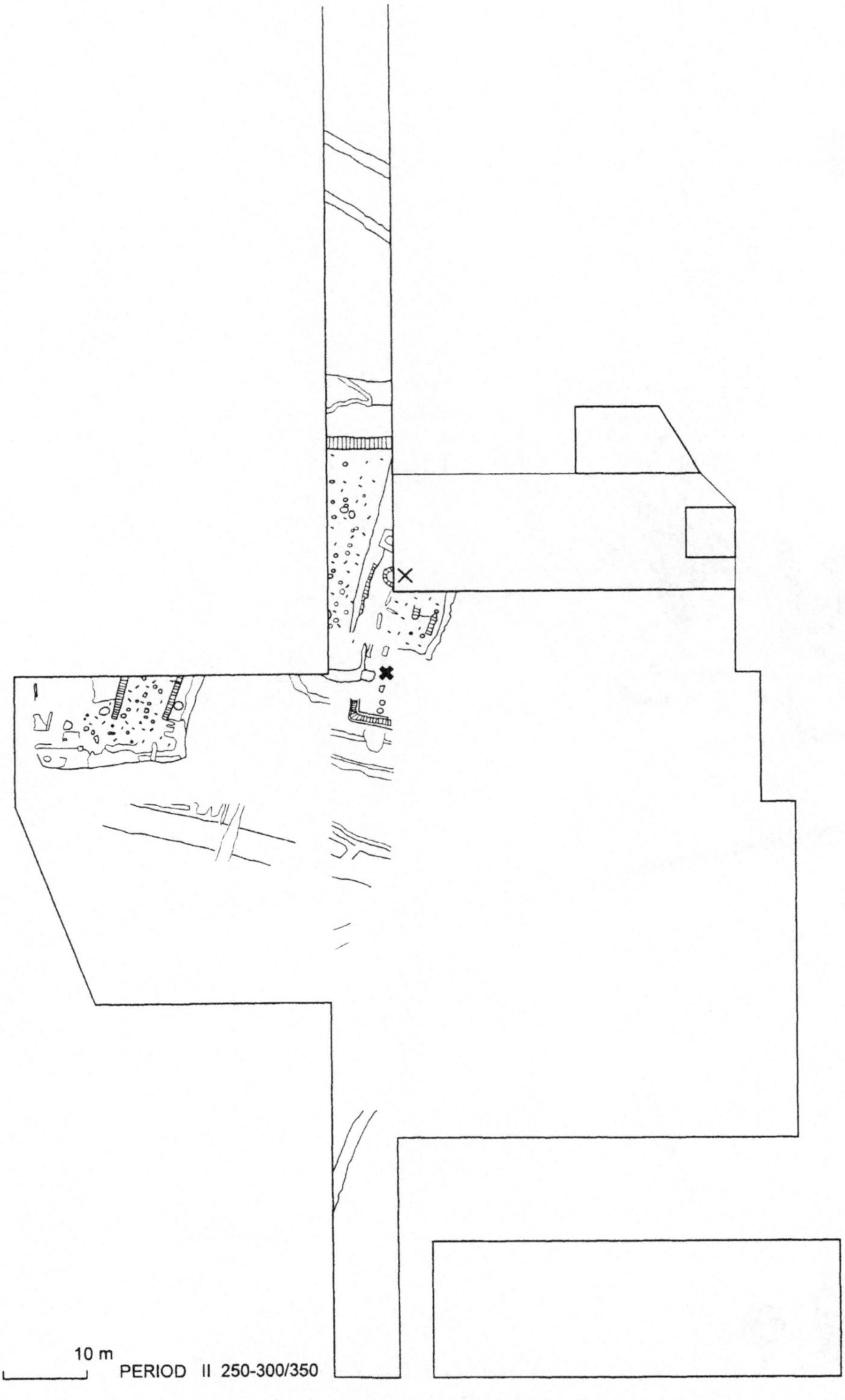

Figure 5. The main features from occupation Phase II AD 250-300/350. For explanation of symbols of this figure see Figure 3.

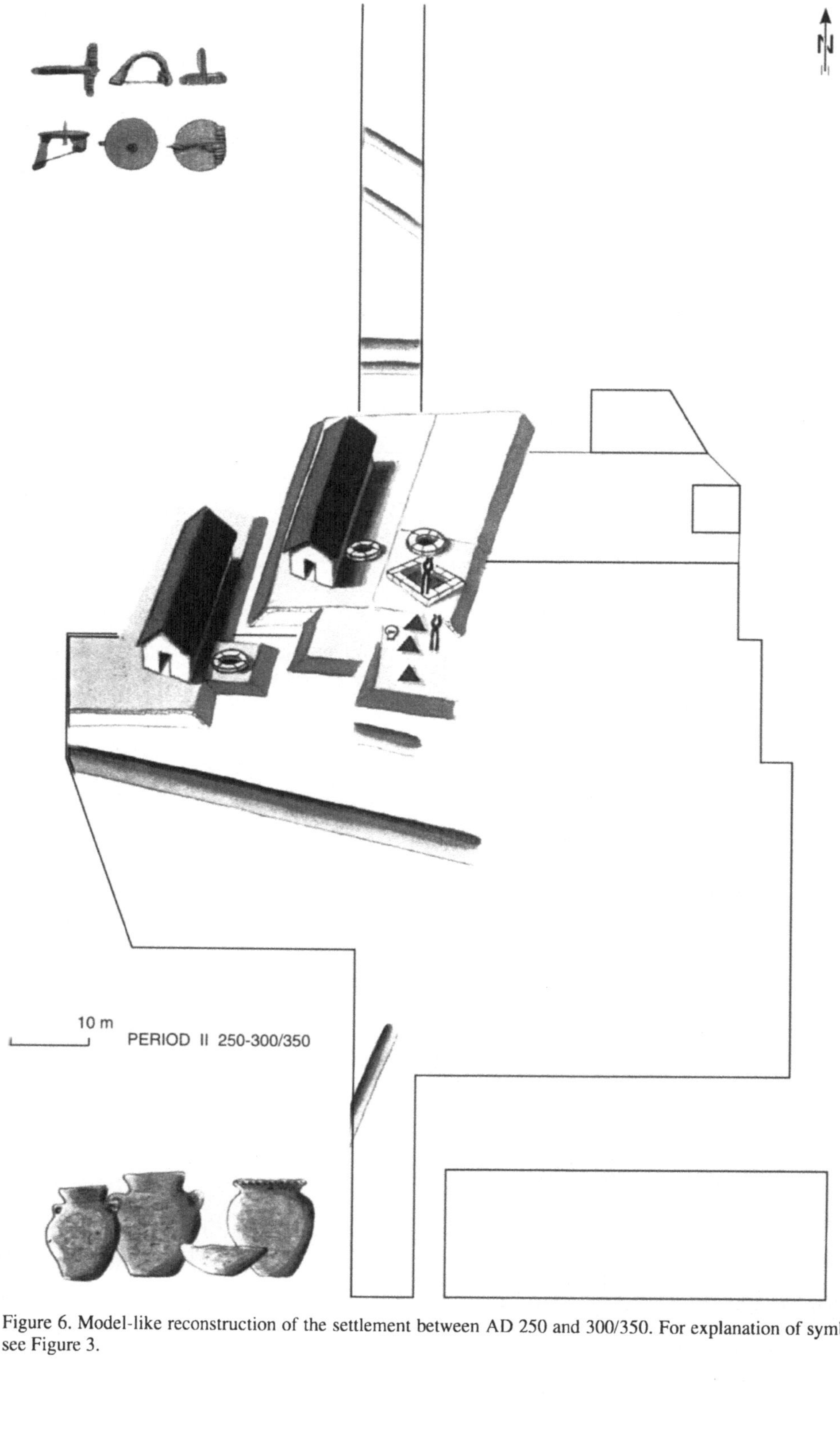

Figure 6. Model-like reconstruction of the settlement between AD 250 and 300/350. For explanation of symbols of this figure see Figure 3.

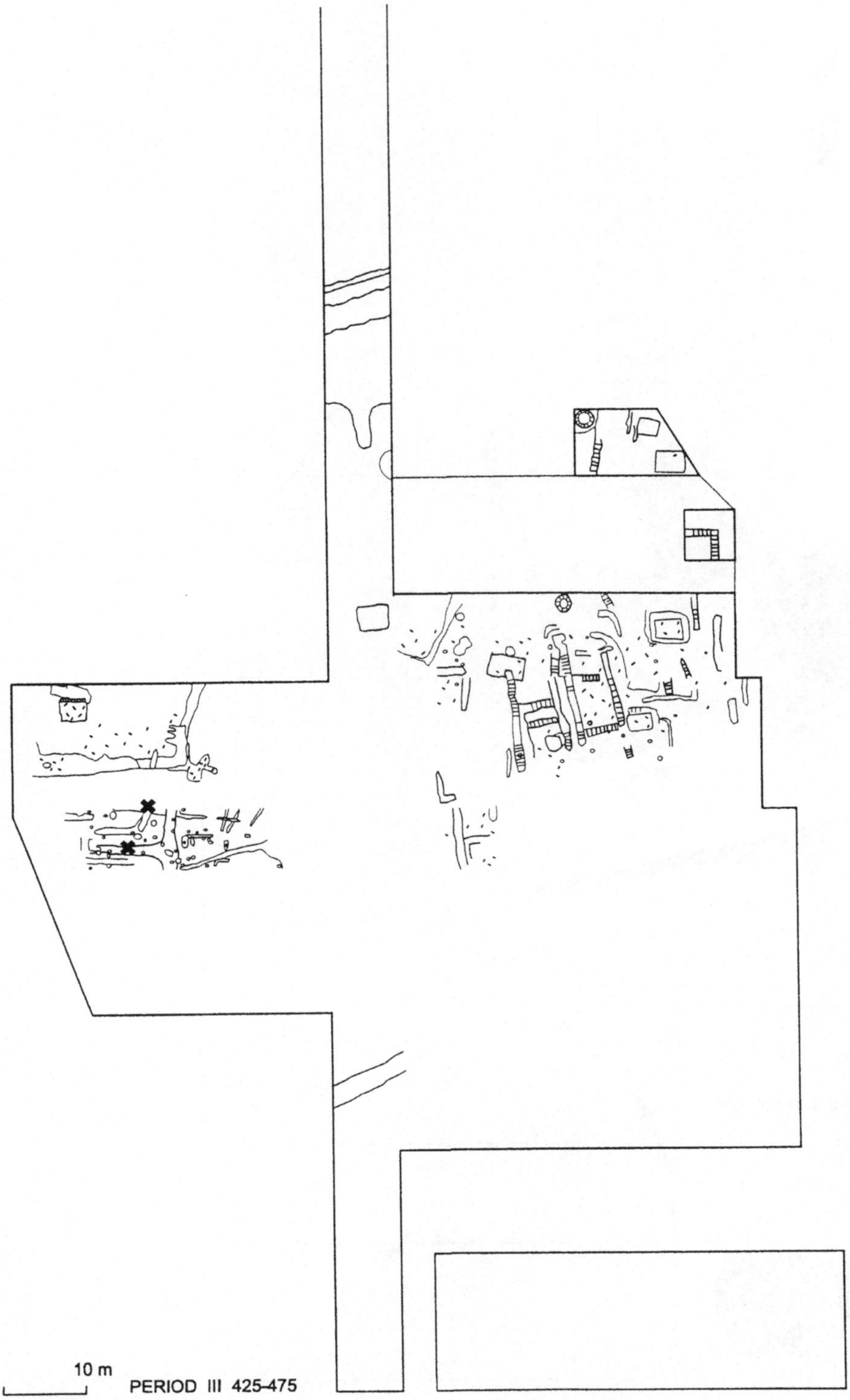

Figure 7. The main features from occupation Phase III A 425-475.

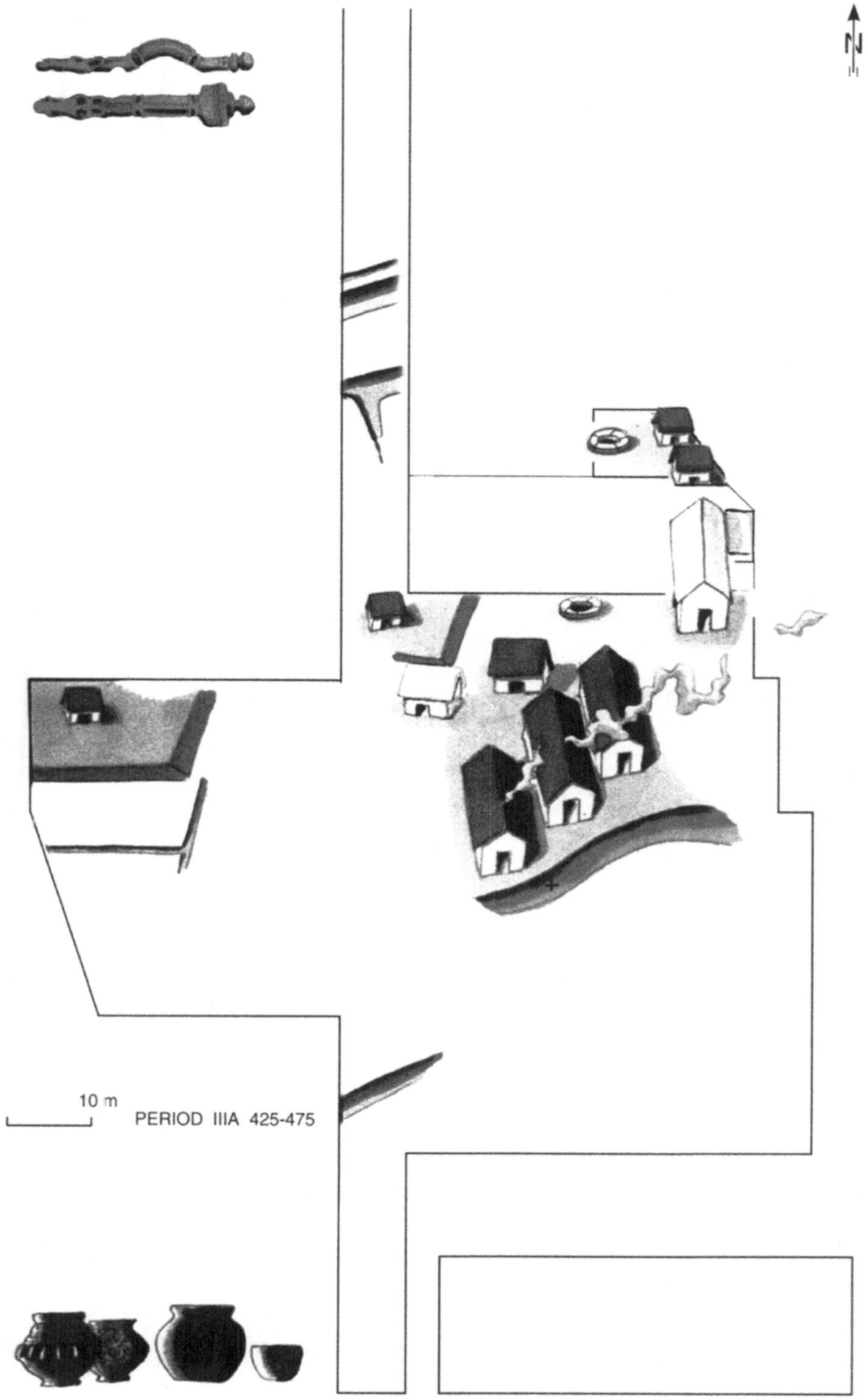

Figure 8. Model-like reconstruction of the settlement between 425 and 475. For explanation of symbols of this figure see Figure 3.

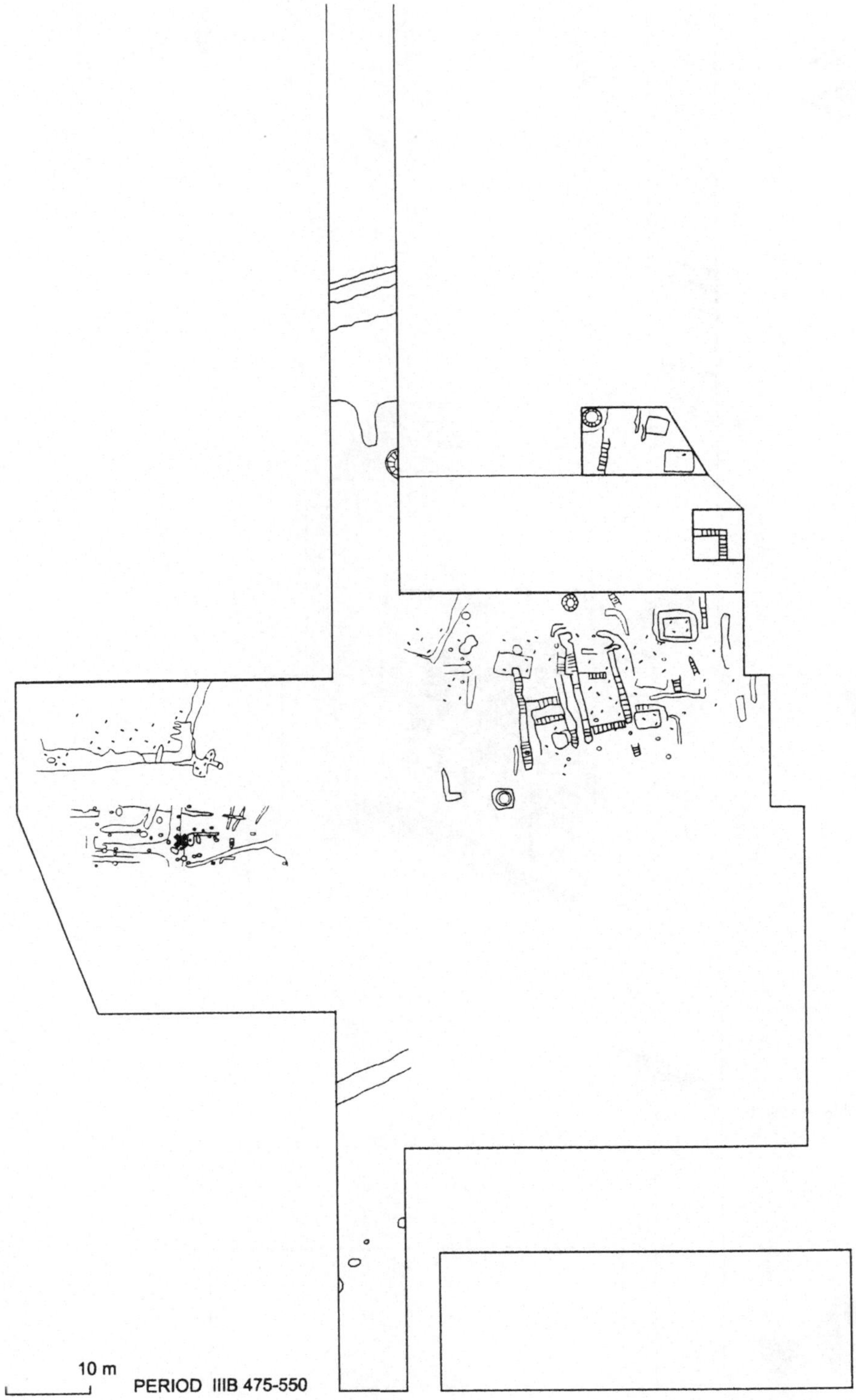

Figure 9. The main features from occupation Phase III B 475-550.

Figure 10. Model-like reconstruction of the settlement between 475 and 550. For explanation of symbols of this figure see Figure 3.

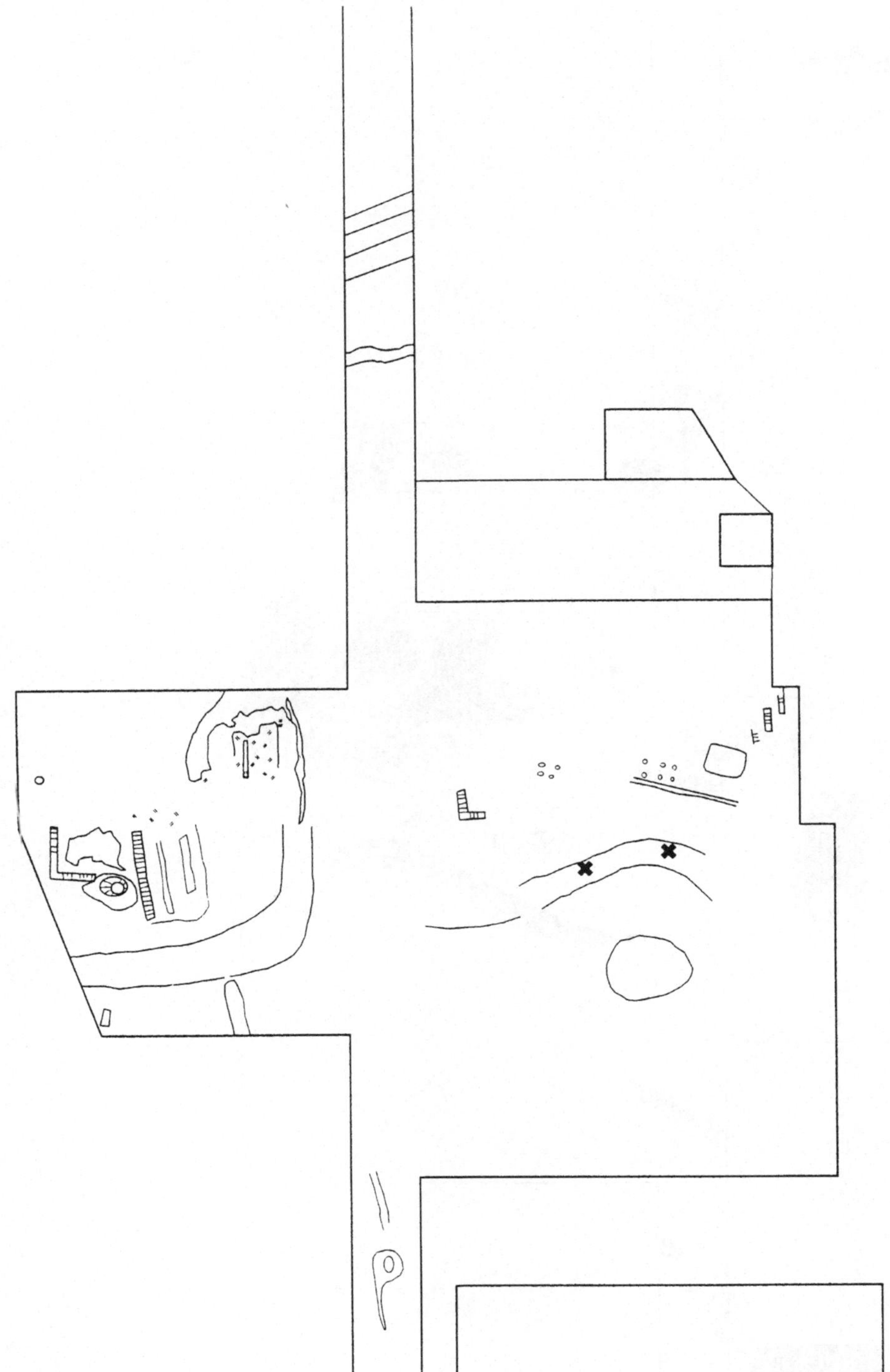

Figure 11. The main features from occupation Phase IV 550-650.

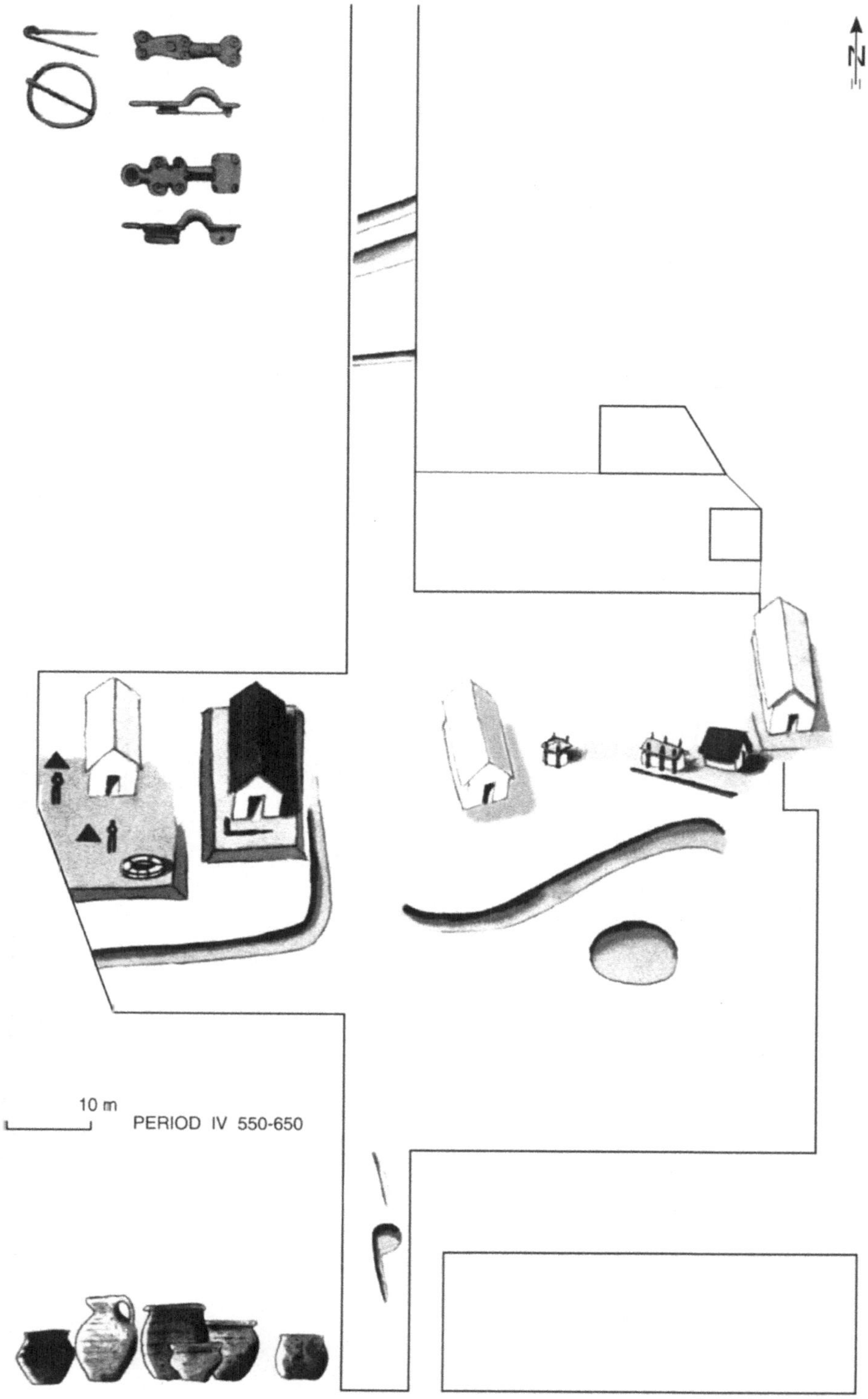

Figure 12. Model-like reconstruction of the settlement between 550 and 650. For explanation of symbols of this figure see Figure 3.

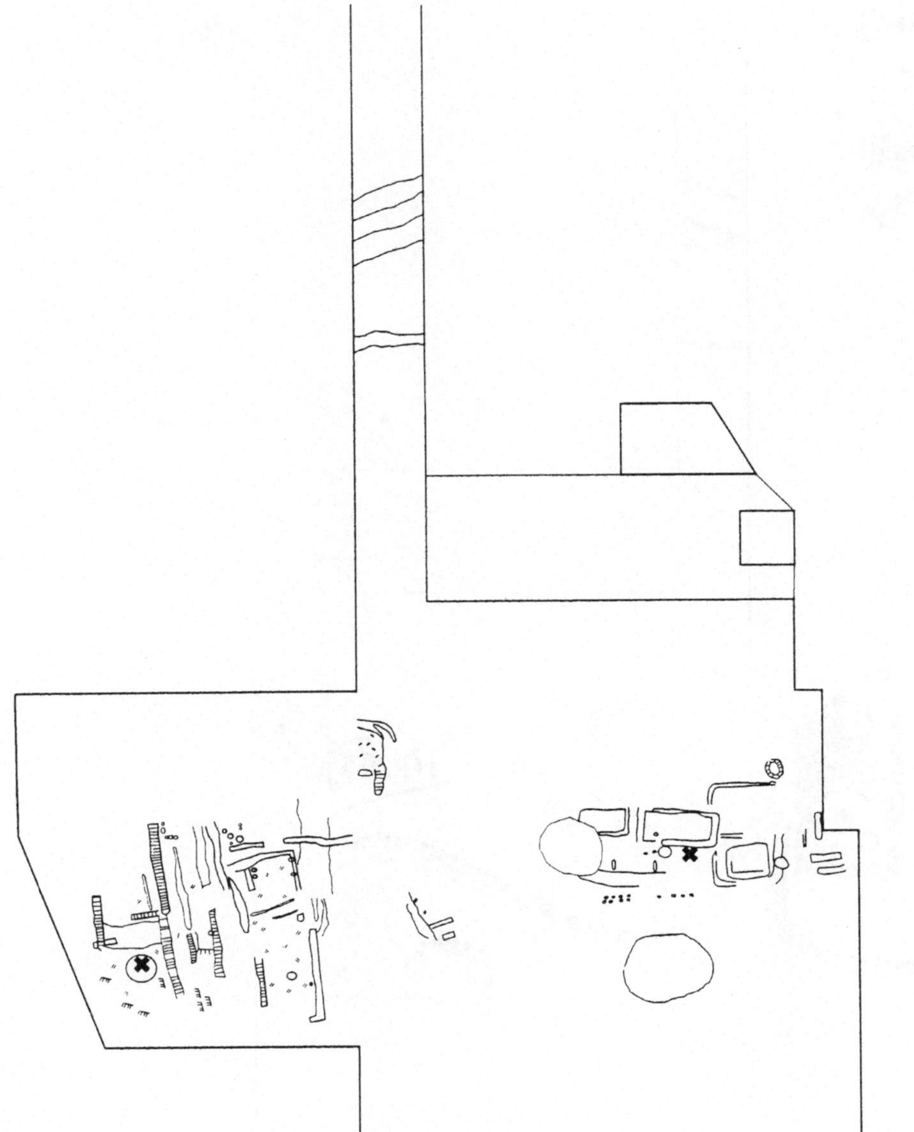

Figure 13. The main features from occupation Phase V 650-750.

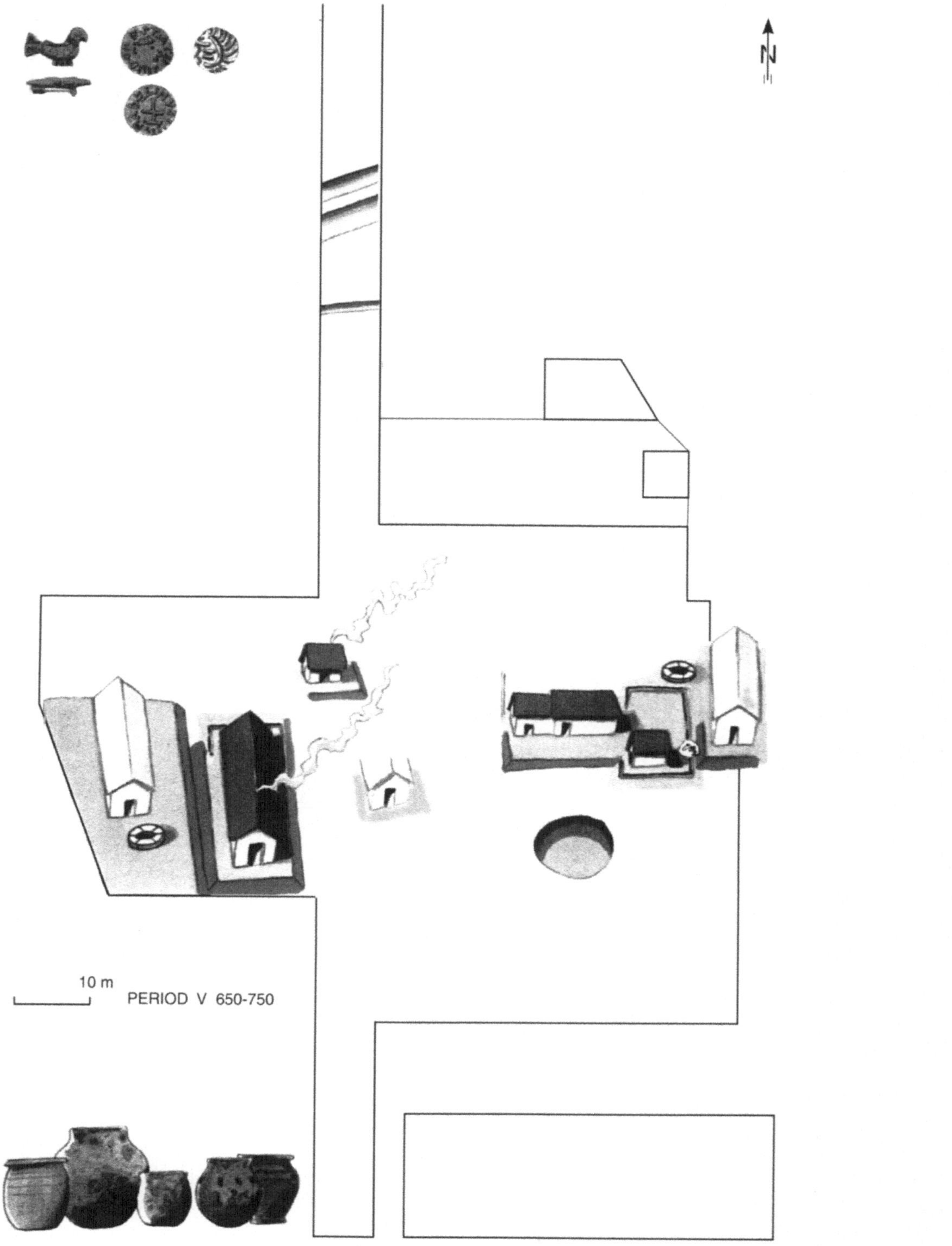

Figure 14. Model-like reconstruction of the settlement between 650 and 750. For explanation of symbols of this figure see Figure 3.

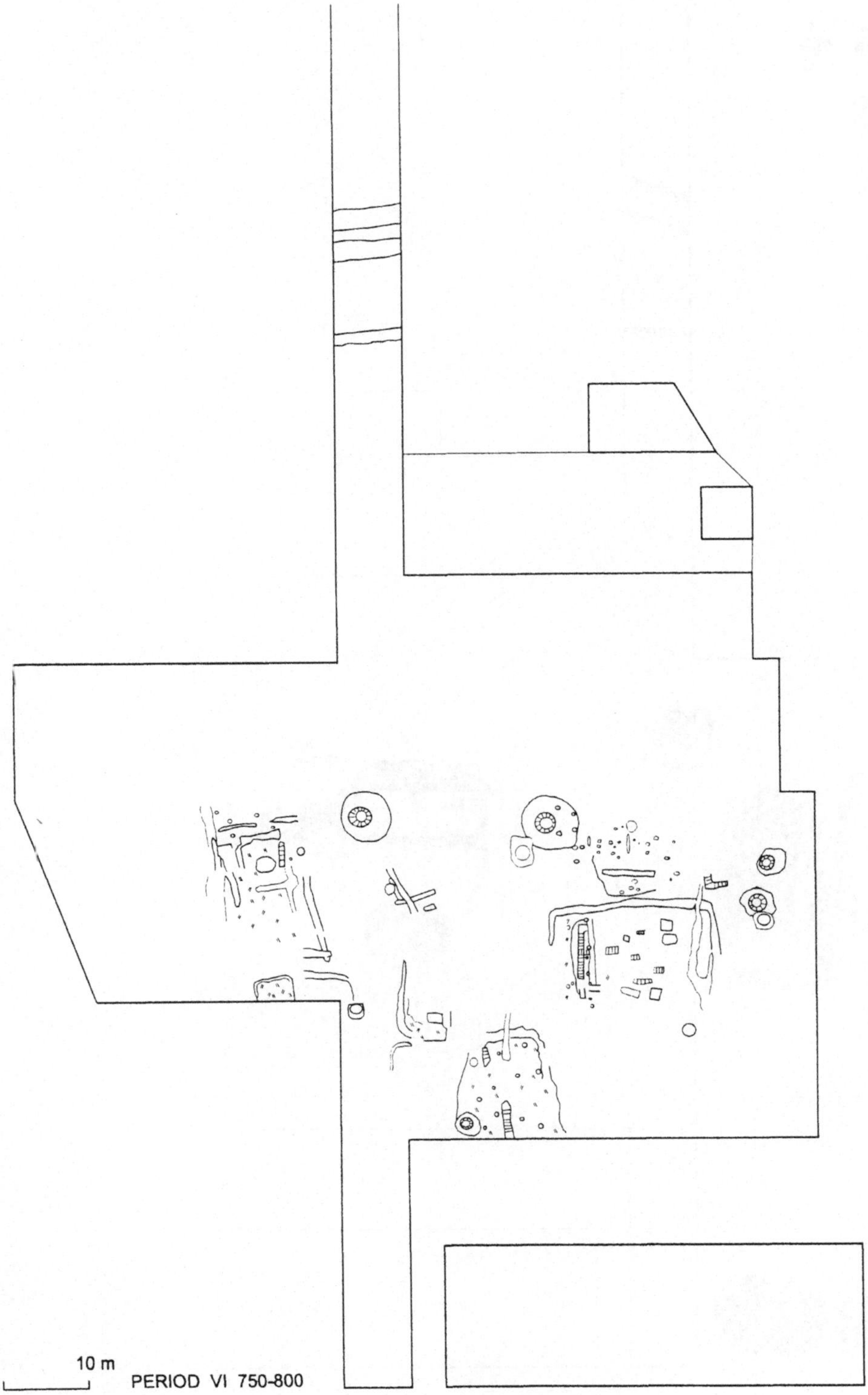

Figure 15. The main features from occupation Phase VI 750-800.

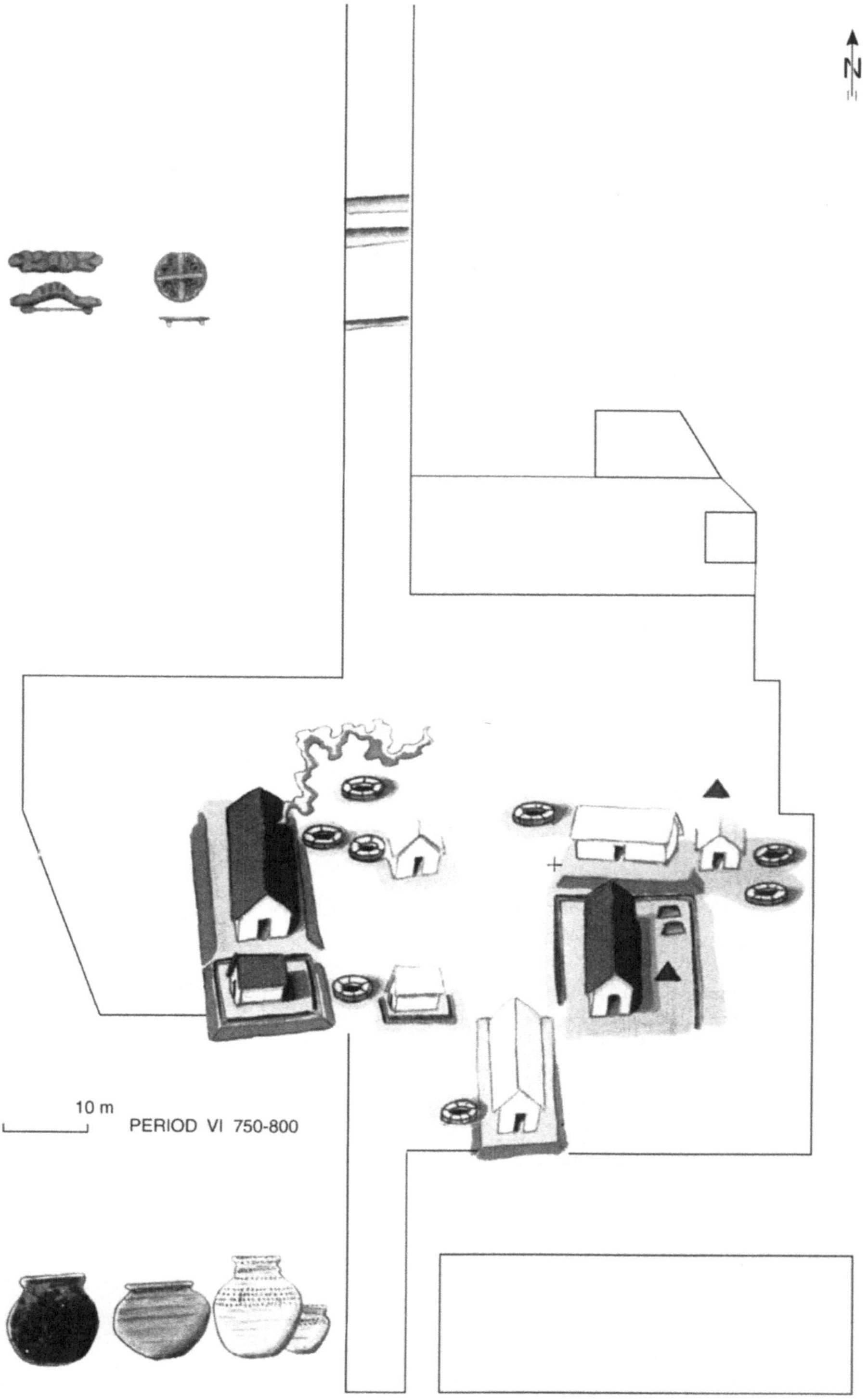

Figure 16. Model-like reconstruction of the settlement between 750 and 800. For explanation of symbols of this figure see Figure 3.

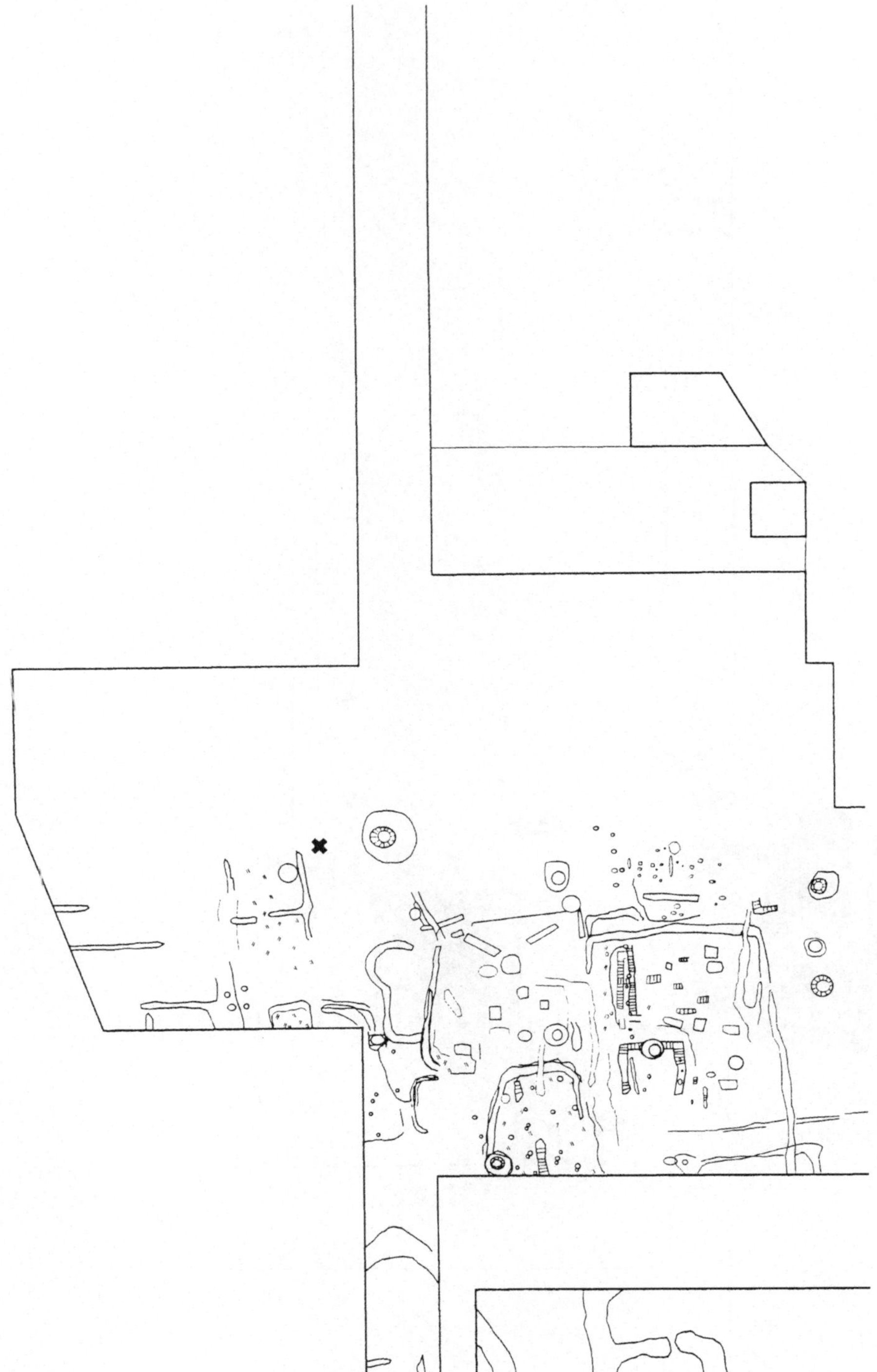

Figure 17. The main features from occupation Phase VII 800-850.

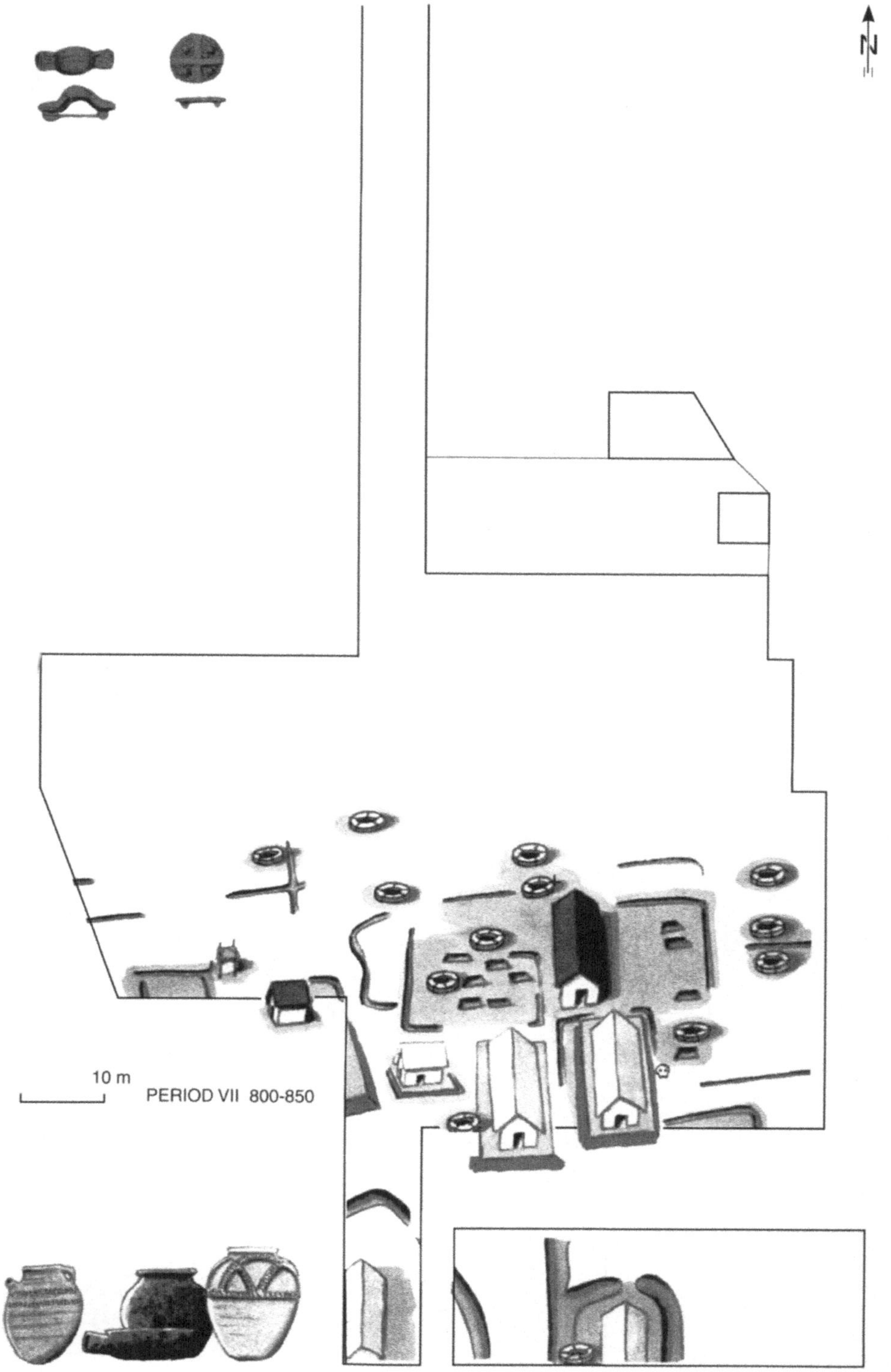

Figure 18. Model-like reconstruction of the settlement between 800 and 850. For explanation of symbols of this figure see Figure 3.

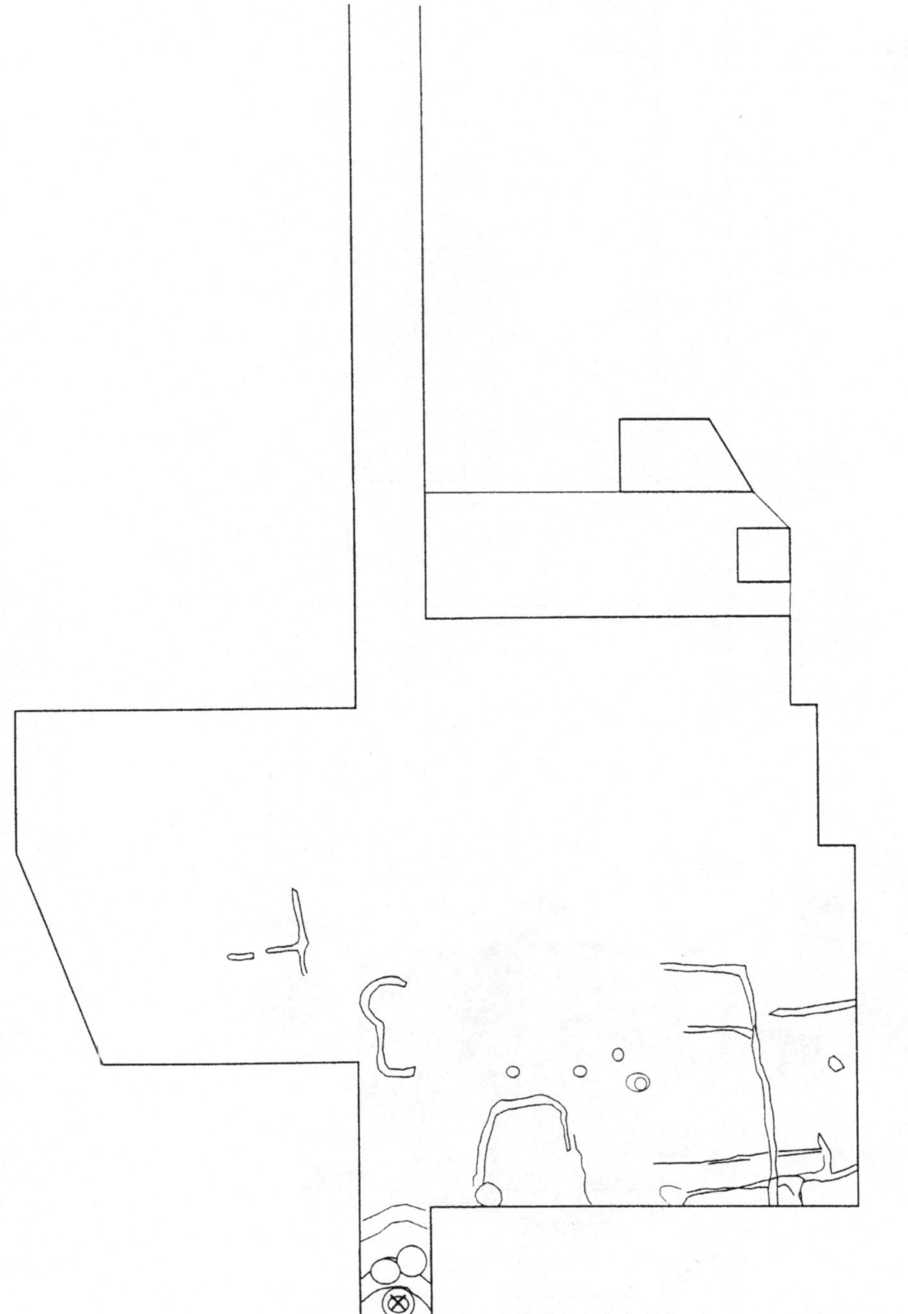

Figure 19. The main features from occupation Phase IX 850-900/950.

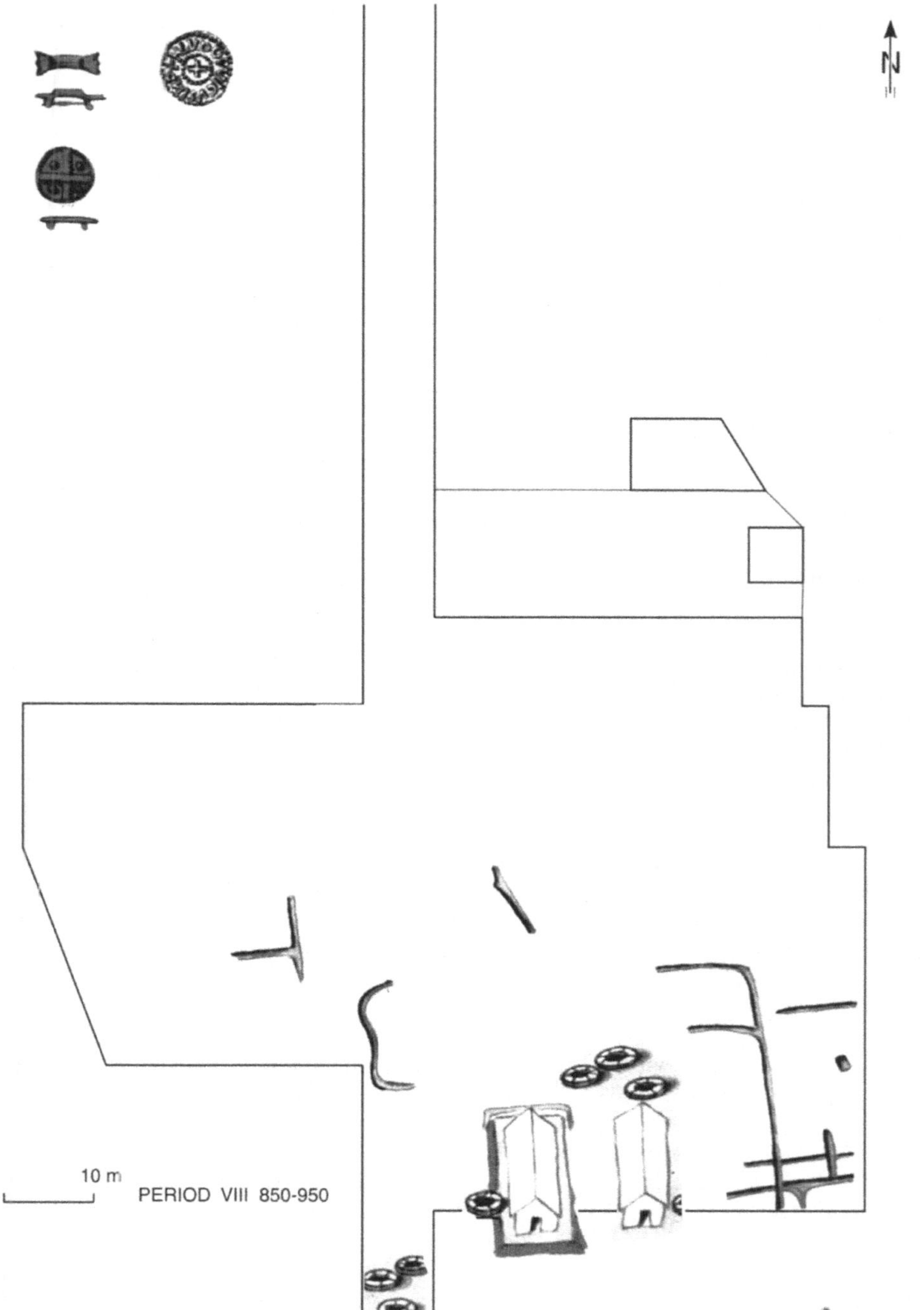

Figure 20. Model-like reconstruction of the settlement between 850 and 900/950. For explanation of symbols of this figure see Figure 3.

tion. The real number of buildings will generally have been larger than what is shown in these figures. Many buildings will have disappeared due to site formation processes like levelling of house remains and erosion. As such, the reconstructions should be regarded as minimum numbers.

Summarizing it can be said that the following occupation phases are based on a complex process of weights and balances, comprising:

- Determining the stratigraphical position of the features,
- Establishing the transitions in pottery assemblages in relation to the stratigraphical position of the features,
- Dating the transitions of the pottery assemblages with the help of well dated pottery in the literature,
- Combining the Harris matrix and the pottery datings,
- The filtering of the influence of transportation of material,
- Checking and cross-checking the resulting datings by comparing them with the more 'absolute' post- or ante quem datings of radiocarbon or dendrochronological datings or by the datings of other well dated finds like brooches, coins etc. as far as these are available.

Period I: ca AD 175-250

This phase is generally defined by the presence of indigenous hand-made pottery from the Roman Period i.e. Taayke types for Oostergo: G6a/b, G7a/b, G8a, V4c, K4b, K4c and K5a; together with very few sherds of terra sigillata.[7] This phase is a relatively short lasting period of 50 or 75 years which will have started somewhere in the last quarter of the 2nd century. This conclusion is supported by the dating of the metal finds from the Roman Period by Erdrich.[8]

For this period a radiocarbon sample of animal bone from an occupation layer (see Fig. 3) produced a dating of AD 82-230 cal (GrN 21393, 1870 ± BP).[9]

Period II: ca AD 250-300/325

This period is characterized by indigenous Roman pottery i.e. Taayke Oostergo-types G6a/b, G7a/b, G8b, G9, K5b and the appearance of terra nigra-like pottery.[10] The framework from a well provided a dendrochronological dating in the summer of 263. As this well was intersected by the eastern main section (section A, see Fig. 5) this post quem dating was of great importance for the absolute dating of this section.

A radiocarbon sample of charcoal from a bronzecaster's workshop (GrN 22068, 1780 ± 50 BP; section A, see Fig. 5) which is also intersected by the same eastern main section provides us with a dating between AD 134 and 386 cal. The modus of the calibration curve makes a dating around 300 most likely. The layers immediately overlying this workshop must be dated in the early 5th century. A second radiocarbon sample from an occupation layer provides a dating between AD 128 and 314 cal (GrN 21394, 1830 ± 30 BP; see Fig. 5). The material content of the feature suggests that it belonged to the latest occupation from the Roman Period. Also here are post-Roman layers immediately overlying the feature. Erdrich suggests on the basis of his study of the metal finds from the Roman Period that occupation at Wijnaldum-Tjitsma will already have diminished to a minimum at the end of the 3rd century.[11]

The content of a very small number of features suggests that occupation did not break off completely after 300. It cannot completely be excluded that some features from the 4th century disappeared because of erosion. At several places in the excavation, however, there is an undisturbed stratigraphical sequence where the 4th/early 5th century ought to be represented, if it were there. As the number of features with material that should be dated in the 4th/early 5th century is very small, it does not seem very likely that occupation of any significance will have lasted much longer than 300/325.[12]

Period III: ca 425-550

During Period III pottery is predominantly handmade, stonegrid tempered 'Anglo-Saxon' pottery from the Migration Period, i.e. Taayke types A1-A4. Grass tempered ware (Taayke type A3) occurs from the 5th century onwards but it is most characteristic for the 6th century.[13] The overwhelming majority of the material is undecorated. Smooth walled biconical wheelthrown pottery, rough walled wheelthrown 'Mayen'-pottery and terra nigra-like pottery occur in small numbers.[14] Cruciform brooches are the fashion during this period. The earliest brooches occurring at Wijnaldum-Tjitsma in Period III are a *Stützarmfibel* of Mahndorf type and four cruciform brooches of Witmarsum-type, all dated by Böhme in the first half of the 5th century.[15]

A sample of cooking remains from sherds used for the construction of a hearth (GrA 1531, 1470 ± 40 BP; see Fig. 9) produced a radiocarbon accelerator dating of 544-656 cal. Regarding its stratigraphical context and the pottery used for its construction, the hearth must be placed at the end of the first half of the 6th century. Samples from two more features, both ditches (see Fig. 7), supplied us with conventional radiocarbon datings of 396-536 cal (GrN 21392, 1625 ± 30 BP) and 408-538 cal (GrN 21397, 1615 ±30 BP). All three samples were collected in the same excavation trench 8 and are stratigraphically related with each other, offering important post and ante quem datings for the long stratigraphical sequence representing Period III in this excavation trench.

Although no chronological subdivision can be made in the material from this period, the large number of intersecting features made it necessary to distinguish two subphases (III A and III B) in the description of settlement development during this period.

Period IV: ca 550-650

This period is characterized by the large number of wheelthrown pottery (63.7% rough-walled pottery; Mayen fabrics but also fine stone grit tempered fabrics from unknown origin), associated with very small numbers of smooth-walled biconical wheelthrown ware. The wheelthrown ware consists almost exclusively of *Wölbwandtöpfe*. Hand-made pottery forms only 36.3% of the total amount of sherds, while the decorated and undecorated Anglo-Saxon wares, initially occurring during this period in large quantities but diminishing in numbers, are replaced by the *Eitopf* and neckless bowls somewhere around 600.[16] The ratio of hand-made to wheelthrown wares is probably somewhat biased by the presence of vertically transported pottery from Period III A/B (predominantly hand-made pottery), which can not typologically be distinguished from hand-made pottery from Period IV, which would mean that the actual amount of wheelthrown pottery will be slightly higher.

From one of the main features from this period, a ditch, the total amount of wheelthrown ware is 88.6% (88.8% in weight). Two radiocarbon samples produced datings for this feature which are quite similar. The first dating is 424-590 cal (Grn 21396, 1570 ± 30 BP), the second dating is 430-594 cal (GrN 21400, 1560 ± 30 BP; see Fig. 11). Regarding its stratigraphical position and its material content, the feature was dated in the second half of the 6th century.

For two other features (a blacksmith/bronzecaster's dump and a well) 61.2% and 66.0% respectively of the total number of pottery (79.8% and 85.1% in weight) is Merovingian wheelthrown ware. Many features from this period are very rich in find material (brooches, glass vessels, e.g. bell beakers, etc.). Four golden tremisses offered important post quem datings for some features of the latter part of the period and for the transition to Period V.

Period V: ca 650-750

During this period important changes take place in the composition of the pottery assemblage. The first change is the decrease in the amount of pottery in general but especially the dramatic decrease of the amount of wheelthrown pottery (from 63.7% in Period IV into 1.2% in Period V), which marks the transition from Period IV to Period V. The dating of this transition is somewhat arbitrary because of the lack of well dated material. A second important transition taking place is the development of the *Eitopf* into the fully developed *Kugeltopf* at the end of the 7th century or the beginning of the 8th century.[17] Some rare fragments of steep-walled pots in 'Badorf'-fabric occur in the first half of the 8th century, together with some very rough-walled predecessors of the Carolingian cooking-pot.[18] Sceattas will have arrived at Wijnaldum-Tjitsma from the beginning of the 8th century onwards.

Two samples offered conventional radiocarbon datings for Period V. The first dating comes from a sample of an occupation layer: 616-688 cal (GrN 21391, 1380 ± 30 BP), which is in accordance with our expectation, regarding the content and the stratigraphical position of the feature. The second dating is from a sample of the fill of a well (Well 24) with a dating of 694-880 cal (GrN 21398, 1245 ± 30 BP; see Fig. 13). Regarding the finds from the fill and its stratigraphical position the well should be dated in the first half of the 8th century.

Period VI: ca 750-800

This period is characterized by the introduction of Carolingian wheelthrown wares like cooking pots and Badorf vessels (13.3%).[19] Hand-made *Kugeltopf*-pottery (86.7%) is initially stonegritted, but from the last quarter of the 8th century onwards growing numbers of *Muschelgruß* and another kind of sand-tempered, thin-walled *Kugeltopf*-pottery are found.[20]

Period VII: ca 800-850
This period is characterized by almost the same kind of pottery as in Period VI, but hand-made pottery (89.7%) now predominantly consists of the more standardized types like the *Muschelgruß*-pottery (42.7% of the total amount of sherds of this period) and the sand-tempered, thin-walled *Kugeltopf* (26.3% of the total amount of sherds of this period).[21]

For one feature which could possibly be a ditch a dating of a radiocarbon sample is available of 648-762 cal (see Fig. 17), which is obviously too old regarding the finds from its content and the stratigraphical position of the feature. Some Louis the Pious denarii offer some important post quem datings for this period.

Period VIII: ca 850-950
This period is hard to define. The beginning of the period is defined by the strong decrease of Carolingian pottery about 850. Chronology for this period is being discussed at the moment.[22] Hand-made *Kugeltopf* pottery (78.7%), at first mainly consisting of *Muschelgruß*, is replaced by stone-grit tempered *Kugeltopf* pottery somewhere around 900. At Wijnaldum-Tjitsma Badorf pottery sometimes coincides with Hunneschans and Pingsdorf pottery, which suggests an earliest date for the Pingsdorf pottery somewhere in the second half of the 9th century. The first introduction of Pingsdorf pottery must have taken place shortly after the arrival of the 8 denarii of Lothar I. Also a dendrochronological dating of 852 ± 8 of the barrel from a well (Well 4; see Fig. 19) offers a post quem for the introduction of Pingsdorf pottery. One single sherd of white glazed pottery was found in one of the youngest preserved features. The youngest finds from a datable context are two sherds of Pingsdorf pottery which should be dated after 900.[23] The amount of Pingsdorf pottery and stonegrit tempered Kugeltopf pottery from post-*Muschelgruß* features is quite small compared with large numbers of handmade pottery of the period before. It does not therefore seem very likely that the occupation will have lasted much longer than the middle of the 10th century. It is, however, hard to determine when exactly the occupation in the excavated area moved more southward, due to erosion. All finds which can be dated with any certainty after the middle of the 10th century come from the topsoil. As all centuries are represented in this topsoil material, these finds seem to have been brought here together with domestic refuse, used as fertilizer.

4 WIJNALDUM-TJITSMA IN THE ROMAN PERIOD – PERIOD I/II: CA AD 175-300/350 (Figs 3-6 and 21)

Hydrological conditions on the high salt-marsh are critical, because with spring-tide and northwesterly storm it will still be inundated. In this kind of wetland we generally see that people seek out the highest places in the landscape like natural levees, dunes etc. to live on. The higher elevation of several decimetres of the linear salt-marsh ridges, compared with the surrounding landscape, meant that in northern Westergo a linear settlement pattern developed.[24]

The first row of terps in northern Westergo was the alignment of Harlingen-Franeker, where occupation started in the 1st century. The border between salt-marsh and tidal flats was several kilometres further north. Several times in a year, especially in winter the salt-marsh will have been inundated by the sea during very high tides. Already for many decennia the high salt-marsh at Wijnaldum-Tjitsma had probably been used as pasture by the same people who started the construction of the settlement in the third quarter of the 2nd century. The pottery of the first people to settle down at Wijnaldum-Tjitsma suggests, that they came from within the region. We know from the study of the indigenous pottery that a strong growth of the population took place in the region during the 2nd century.[25] The rapidly expanding salt-marsh offered new possibilities for the foundation of settlements on the recently developed northern salt-marsh ridge.

The new site-location was carefully selected and prepared as shown by the discovery of a dike-like structure under the house-terp before the construction of the settlement started. It was renewed after being covered by new sediment twice. The small 'dike' was constructed of regular squares of sods with a total width of about 10 m. It reached a final height of 70 cm (0.4 m above NAP (Normaal Amsterdams Peil, Dutch Datum Level)). The purpose of this 'dike' is not yet completely clear, but it could very well be constructed to protect agricultural fields from inundation by the sea.[26] To protect the settlement against the high waves during storms, a complex of platforms of sods was constructed in several phases (see Fig. 24). These phases are separated from each other by findless homogenized sod-layers, showing that the construction was temporarily interrupted. Besides the construction work itself no further indications for the presence of human occupation are found. We assume that the whole complex of platforms was constructed during a number of years. Each year, probably in summer, when there was no threat from the sea and no other work had to be done, people returned to the site to continue their building activities. The sods for the platforms must have come from the immediate environment of the terp, because the only place where the original top of the salt-marsh is still intact is underneath the house-terp itself. To protect the platforms from erosion by the sea during the occasional storm-floods the outside was a systematically constructed sod wall. On the inside of this wall sods were dumped arbitrarily.

The NNE/SSW oriented platforms were built in two phases, on top of and at the south (which is the landside) of the earlier mentioned dike. The highest platform reached a minimum height of 1.2 m (1.40 m above NAP). The final height could not be established due to erosion of the occupation levels which were overlying this platform, but the dug in features still remaining indicate that the first occupation level will not have been much higher. The total length of this highest platform was 23 m. Already before the people finally settled the platform was enlarged in an easterly direction with the construction of a second lower platform (0.65 m above NAP) in at least three phases. This platform was constructed in a very systematic way. One layer of oblique standing sods was followed by the next layer where the sods were standing obliquely in the reverse direction. After AD 250 (Phase II) the complex of platforms was enlarged in a southerly direction several times (see Fig. 25).

Whether the complex of platforms was immediately surrounded by ditches as is often the case at house-terps, is not completely clear. At some distance, north, but especially south of the house-terp several ditches were recorded, which according to orientation and find-material belong to the first two occupation phases. It is clear, however, that they had already lost their function during the earliest occupation phases of the settlement because of sedimentation. Finally some of these ditches were overgrown by peat. The orientation of the ditches suggests, that a regular pattern of parcelling was present at least in the immediate surroundings of the terp. Van Giffen showed that at the terp of Ezinge (Prov of Groningen) parcelling within the settlement by using fences was already present from the Iron Age onward.[27]

A row of postholes on the highest platform shows that a building was constructed with a minimum length of 17.5 m. Only a small part of the building was excavated, which makes it difficult to say anything definite about structure and function. The remains of a hearth in the south part of the building indicates that in this period part of the building probably had a function as a living quarter. A second alignment of postholes with a slightly different orientation shows that the building was rebuilt once. On the lower eastern extension of the house-terp no postholes were observed. Features and finds show that this area probably had no domestic function but was a workshop for metal-working.[28]

A second house-platform was situated about 10 to 15 m southwest of this first house-terp. This platform was also constructed with sod walls on the outside and randomly dumped sods as infill. The large number of postholes suggest that here also a building with a wooden framework had been constructed but no regular patterns could be distinguished. Unfortunately the small area of 10-15 m between the first house-platform and the partly excavated second one could not be investigated. Therefore it is not clear whether we are dealing here with two separate house-terps which grew together at a later stage or with one large complex with several buildings.

5 AN INTERRUPTION IN THE HABITATION FOLLOWED BY THE ARRIVAL OF NEW IMMIGRANTS – CA AD 300/350-425

In his research of hand-made pottery from the Westergo region Taayke concludes that after a continuous 'Frisian' development for centuries a growing influence from the east becomes visible during the second half of the 3rd century.[29] The same seems true of Wijnaldum-Tjitsma, where hand-made pottery, typical of the Frisian coastal region, was replaced by Eddelak-like pottery (Taayke types Oostergo G8 = Driesum-pottery and Groningen G6b), after AD 250. Eddelak pottery was already in use in the coastal region of Groningen from the 1st century onwards. Together with Eddelak pottery stone-grit temper was introduced but only incidentally applied.[30] Roman imports from the south appear to break off after 250, although the presence of some incidental sherds of coarse wheelthrown wares (Mayen fabric) suggest that contacts with the south were maintained on a much smaller scale. At the same time that these

Figure 21. Structure numbers as used in the text for periods I-IV.

Figure 22. Structure numbers as used in the text for periods V-VIII.

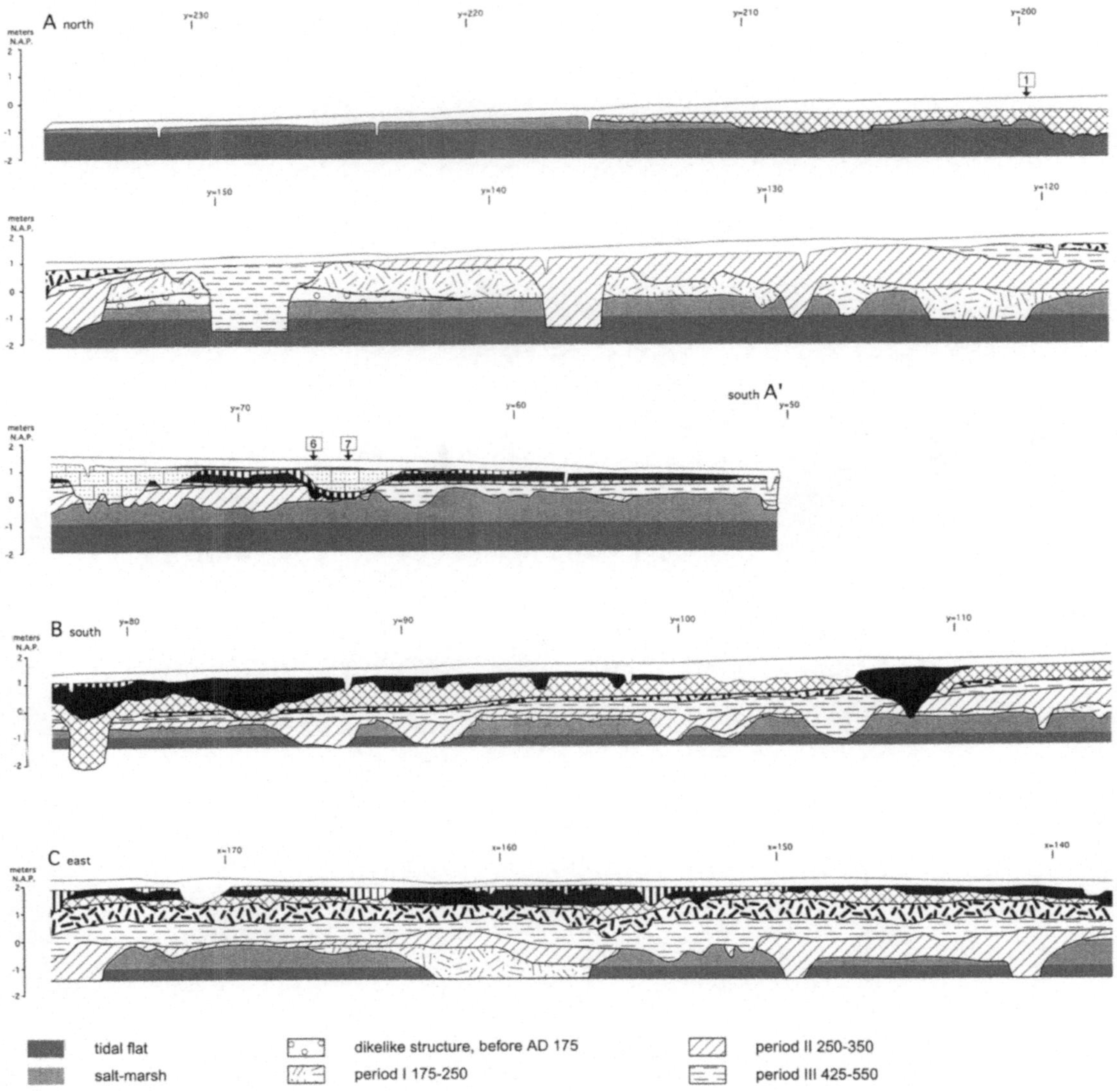

Figure 23. Main sections A-A′, B-B′ and C-C′.

changes took place at Wijnaldum-Tjitsma, settlements on the lower parts of the salt-marsh were abandoned. Wijnaldum-Tjitsma is one of the very few terps in the Westergo region which remained inhabited until the end of the 3rd century. In Groningen a few terps and quite a number of terps in Oostergo (Prov. of Friesland), e.g. Driesumerterp and Hoogebeintum, stayed inhabited in the same period.[31] Finally, about AD 300 Wijnaldum-Tjitsma was also almost completely abandoned. Only few features show possibly the presence of human occupation after 300.

Until recently it was believed that the coastal landscape of the northern Netherlands suffered from the Dunkirk II transgression and that large parts had to be abandoned because of inundation by the sea.[32] Research in northern Westergo shows that the salt-marsh has continuously expanded here at an estimated speed of about 5 m every year.[33] As the low lying hinterland is cut off from the sea by the formation of a large cap ridge, the lower parts of the salt-marsh became less suitable as pasture and as such less favourable for occupation, due to the problematic drainage of the water from the peat

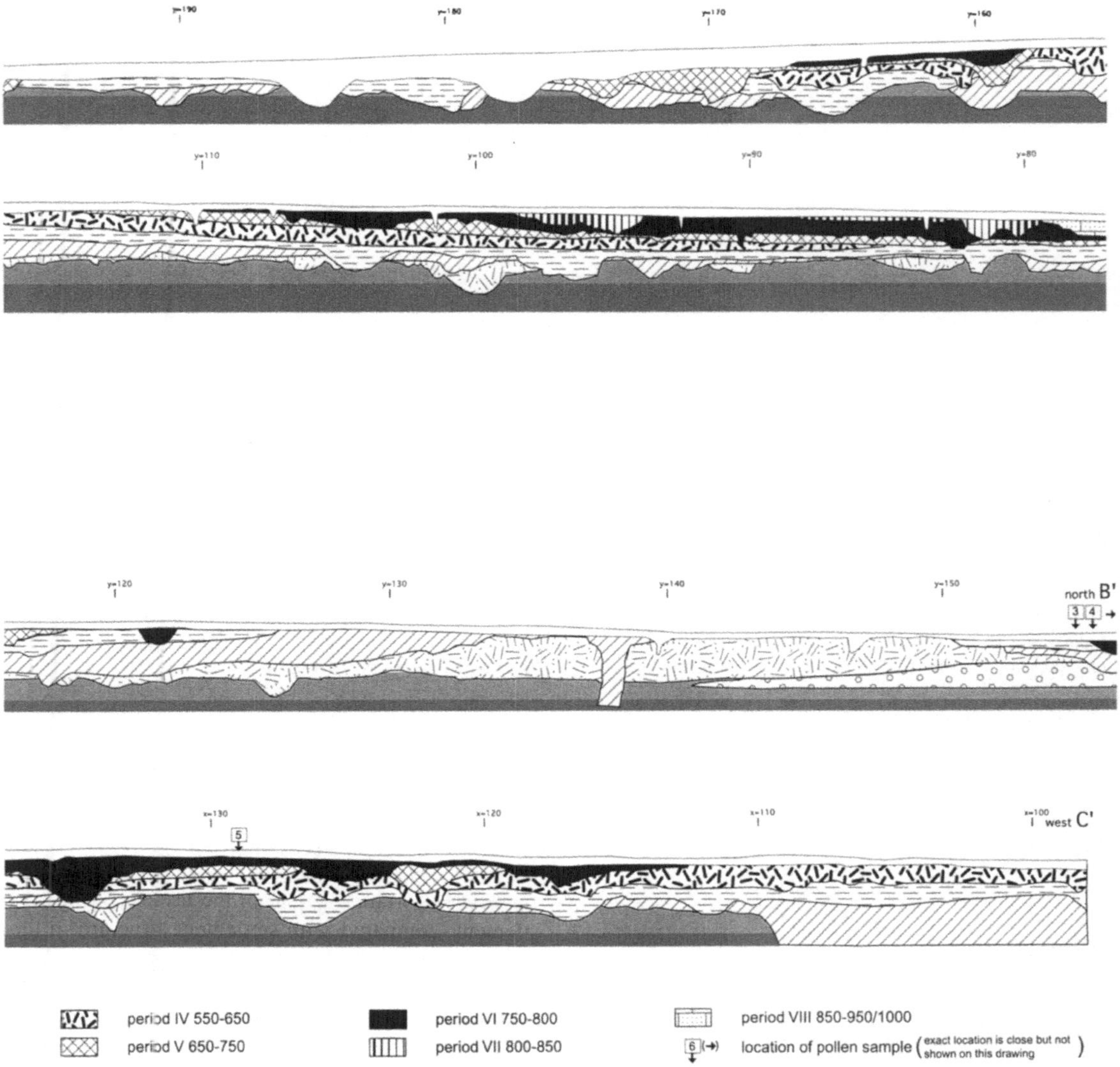

Figure 23. Continued.

area situated further inland. This could be the reason why terps on the low salt-marsh, for example Paddepoel, situated in the northern part of the city of Groningen (Prov. of Groningen), had to be abandoned.[34] Probably this reduction of available pasture also took place in the lower parts of Westergo. The surrounding salt-marsh at Wijnaldum is quite high, compared with southern Westergo and it is not very likely that the inhabitants of the terp Wijnaldum-Tjitsma had to leave the area for environmental reasons. At Wijnaldum-Tjitsma no indication is found for inundation of the terp itself, although sedimentation at the edges of the terp continued during the next centuries until finally the salt-marsh was first dammed in during the 10th century.[35]

Yet the terp was abandoned. Layers which date around the middle of the 5th century immediately overlie layers which can be dated at the end of the 3rd/beginning of the 4th century. Although we selected samples for radiocarbon dating from features with an expected dating in or close to this period, no dating could confirm occupation of any importance in the period between 300/325 and 425 (see Fig. 2).

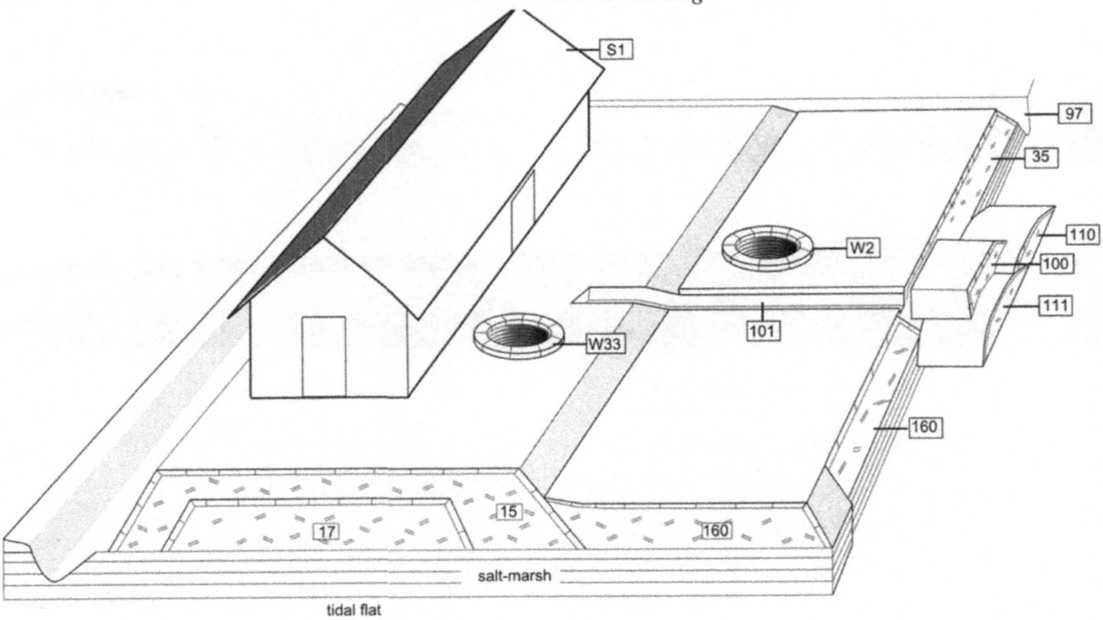

Figure 24. Schematic representation of the different construction phases of Tjitsma in the Roman Period (ca AD 175). The first indications for human presence at the site are shown by the construction of several dike-like structures (111-110-100), followed by the construction of a number of platforms (17-15-160-35) on which a farmhouse was built.

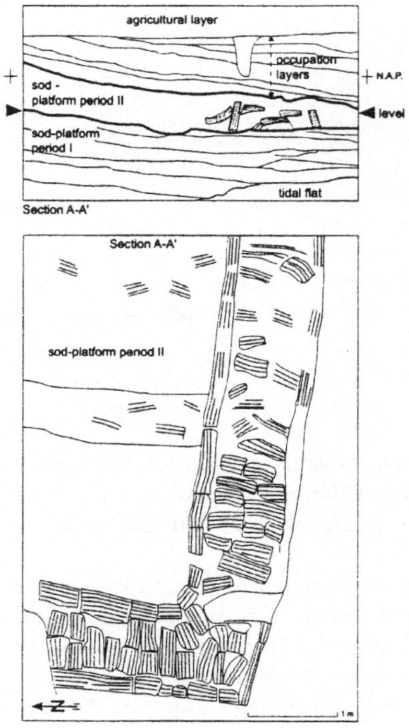

Figure 25. Detail of the enlargement of the complex of sod platforms during Period II (AD 250-300/350).

Most of the archaeologists involved in the excavation project were initially convinced of occupation continuity on the site. It was first during the analysis of the features that we all became convinced of discontinuity. The arrival of immigrants caused important changes in the lay-out of the settlement, compared with settlement structure in the Middle Roman Period. In Period I/II the main orientation of the settlement was NNE-SSW. The main orientation in Period III A has changed to NNW/SSE.

In Period I/II sods were used for the construction of the platforms of the house-terp, but the buildings had a skeleton of roof-supporting posts, as is shown by the numerous postholes, from this period. In Period III A postholes have almost completely disappeared.

A change in the spatial lay-out of the settlement can also be observed at other excavations in the terp region. At the terp of Tritsum in the Westergo region (Prov. of Friesland) a similar change in settlement structure takes place as at Wijnaldum-Tjitsma.[36] In the Middle Roman Period we see a platform, on which a wooden structure is built with a NNE-SSW orientation, which was rebuilt at least once with an orientation slightly more to the north. During the Migration Period sunken huts occur,

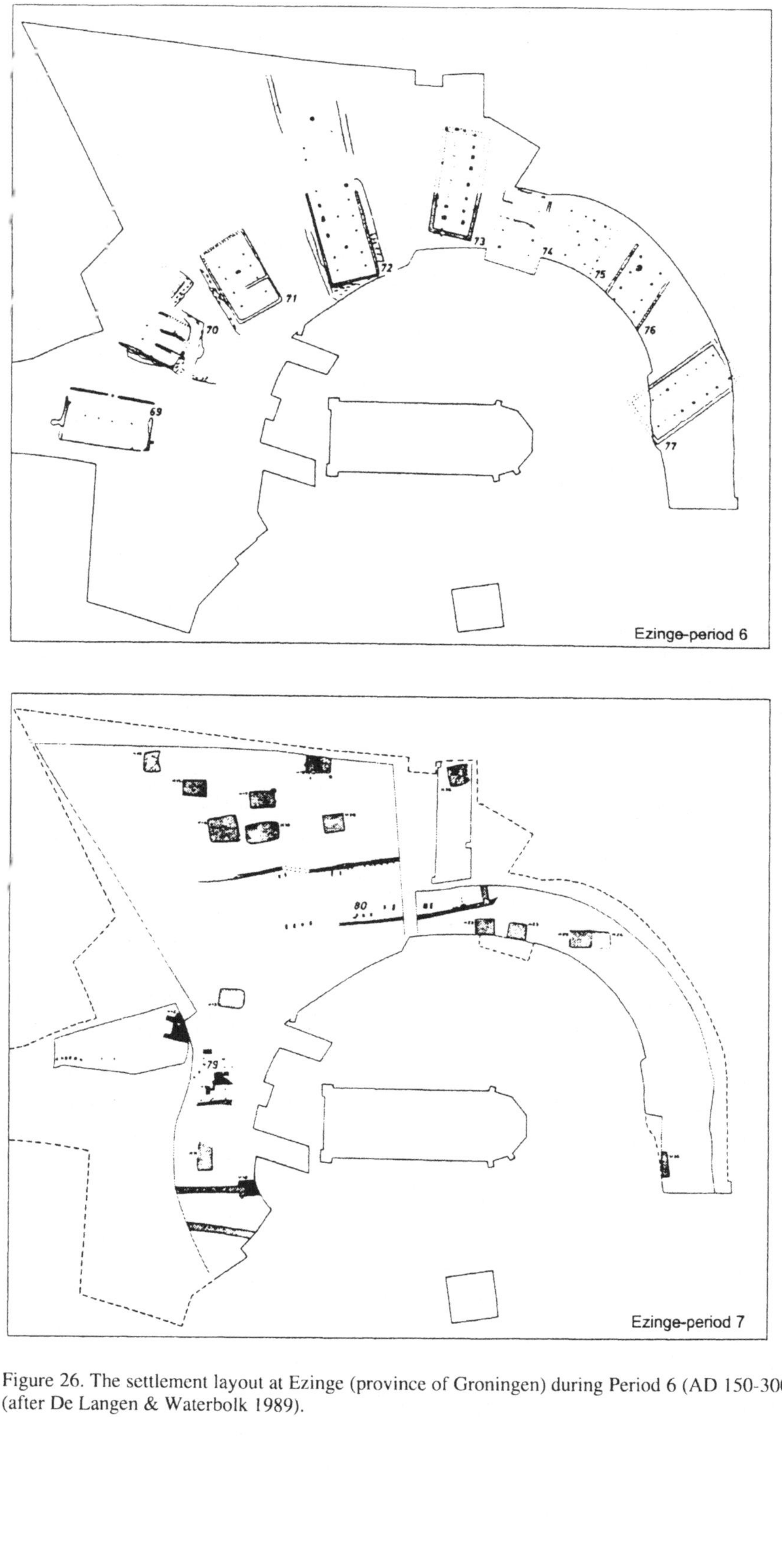

Figure 26. The settlement layout at Ezinge (province of Groningen) during Period 6 (AD 150-300) and Period 7 (AD 300-500) (after De Langen & Waterbolk 1989).

while the orientation of ditches has changed to N-S and E-W. Because of bad preservation circumstances, similar to the Wijnaldum-Tjitsma terp, sod houses were probably not recognized.[37]

A change in settlement lay-out also takes place at Ezinge (Prov. of Groningen) between Period 6 and Period 7 (see Fig. 26).[38] During the Middle Roman Period (Period 6, 150-300) the settlement shows the well-known radial structure. In Period 7 (300-500) the radial structure of the settlement seems to be abandoned and sunken huts are introduced. Also at Ezinge we may assume that besides the sunken huts main buildings were constructed with sods as most important construction material, but that they were not observed by the excavator, due to bad preservation circumstances in the higher occupation levels. Only one possible house plan was discovered by De Langen & Waterbolk, compared with sixty sunken huts.[39]

The material culture of the immigrants is completely different, compared with the material from Period II. The organically tempered, thick-walled pottery typical of the Middle Roman Period is replaced by generally thin-walled, burnished, stonegrit tempered Anglo-Saxon pottery. Cruciform brooches and pottery indicate that the immigrants came from the eastern North Sea region.[40]

6 THE SETTLEMENT IN THE 5TH AND EARLY 6TH CENTURY – PERIOD III A/B: CA 425-550 (Figs 7-10 and 21)

The newly erected buildings at Wijnaldum-Tjitsma are constructed with sod walls, depending on the kind of structure, varying in width between 1.0 and 0.6 m (see Fig. 27). The walls consist of horizontally laid sods, with the former vegetation layer on top. The sods are cut in one piece of 1 m length. The floors of the main buildings show a marbled pattern of sods, lying on their sides, probably to improve their resistance to wear. No complete house plans have been excavated, but the presence of sod walls and sod floors on the inside of these buildings and in one and probably two cases the presence of a hearth make it very likely that we are dealing with houses. One of these hearths was renewed at least once, which suggests that a house had been standing on this location for a considerable time.

The sod buildings on Wijnaldum-Tjitsma are rather small in view of the average width of 4 m, but still fit into the house-building tradition of that time.[41] It is difficult to say anything about the total length of the houses, because only small parts of the houses were excavated. One of the houses (Sod Structure 6, see Fig. 21) had a length of at least 6.5 m, but could have been longer.

Sections in small trial trenches revealed many more sod walls than were visible. After some days of rain sod walls appeared, which had not been recognized before. Site formation processes like the collapsing of walls and levelling of remains of sod walls also made it more difficult to recognize complete house plans.

Postholes, however, were almost completely lacking and no entrances could be observed. A possible explanation for the general lack of postholes in the sod houses at Wijnaldum-Tjitsma where the floor also consists of sods may be that postholes were filled up with sods after the posts were pulled out. This was observed several times in structures from a later period. It does not seem very likely, however, that posts were not observed by the excavators during detailed research of several months, while posts were found elsewhere in the excavation.

Schmid assumes for the small sod houses from the 7th/8th century excavated at Oldorf, where posts are also lacking, that the roof-construction was resting on the sod walls.[42] Voskuil is of the opinion that sod walls could not be higher than their width, because of the limited carrying capacity of these walls.[43] Often in reconstructions of sod buildings the top of the walls is smaller than the bases. At Wijnaldum-Tjitsma, however, sections of sod walls have clearly not a conical shape but are completely vertical inside and outside for all sod buildings, during all occupation phases.

Another possible explanation for the lack of posts could be that they were standing on planks as was the case in the 10th/11th century houses at Assendelft.[44] The posts of the houses in Assendelft were standing on planks, but in postholes, which is a difference with Wijnaldum-Tjitsma where no postholes were observed. As any indication for whatever kind of roof-supporting construction was lacking, we assume that the roof-construction at Wijnaldum-Tjitsma rested on the sod walls.[45]

An indication for the construction of the roof is Uilkema's description of a small early 20th-century sod building at Ruinen (Prov. of Drenthe) where the roof consisted of a small layer of straw, covered with heather and sods.[46] This is strikingly similar to an observation by Kjær of a sod building from the Roman Iron Age at Ginderup (Thy, North Jutland)

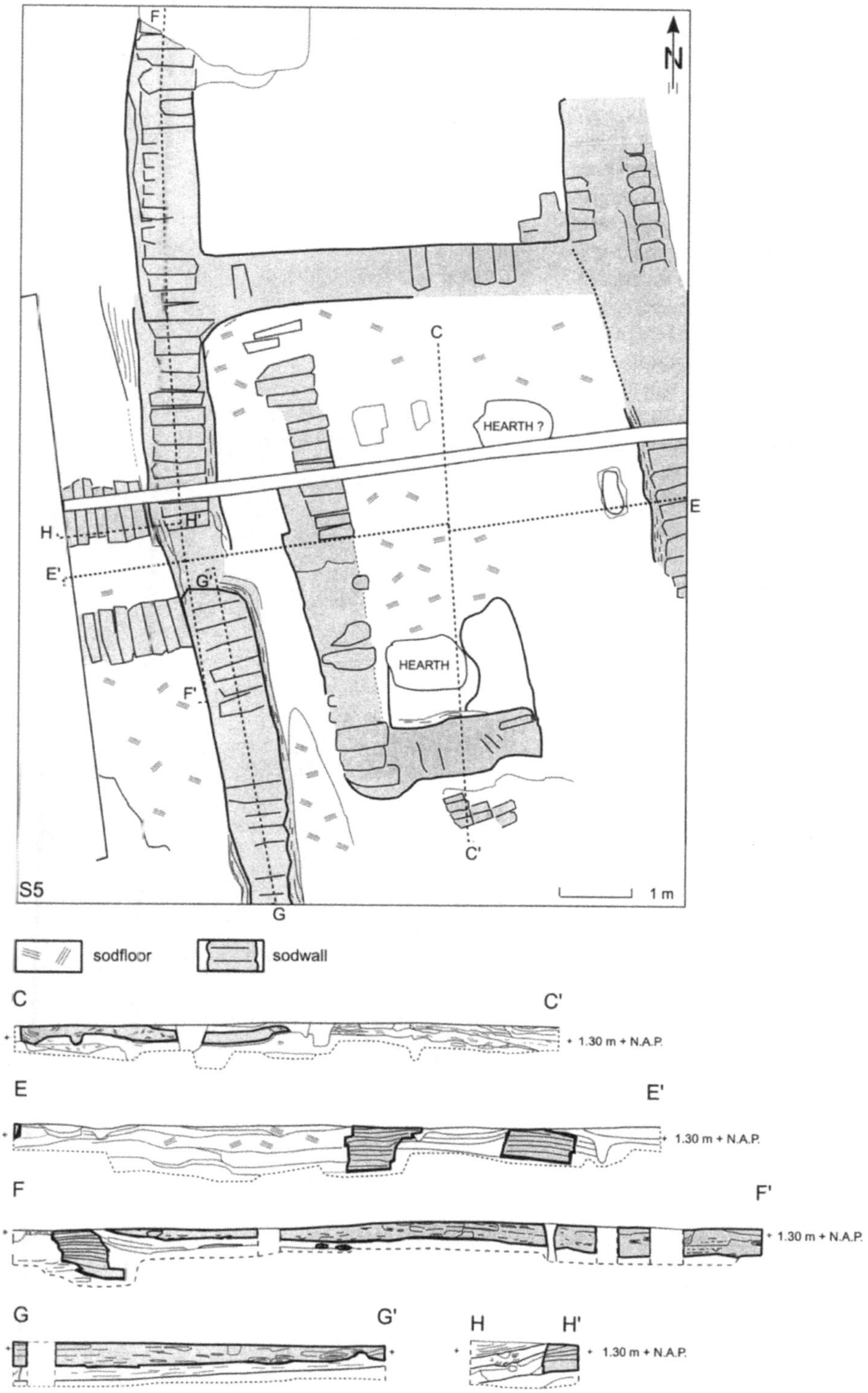

Figure 27. Remains of one of the main sod buildings from Period III A (425-475).

where the roof was also covered first with bundles of straw with heather sods on top of them.[47] Structures with turf walls and turf roofs are also to be found in northern Norway, while structures with walls constructed with a combination of stone and sods and turf roofs are found in the subarctic region from Scotland to northern Norway.[48] At Wijnaldum-Tjitsma several times large numbers of sods were lying on the inside of the building. Generally they were interpreted as part of collapsed sod walls, but regarding the previous observations we have to be aware of the possibility that (part of) these sods were covering the roof.

Terp-like settlements with sod buildings or sunken huts with sod walls dating in the Roman Iron Age/Migration Period in the eastern North Sea region are known from West Jutland in Denmark in the Thy region, the Ringkobing area, the island of Mors, in Southwest Jutland in the Ribe area and from Northern Germany at Archsum-Melenknop on the North-Frisian island of Sylt and at Tofting in the Eiderstedt region of Schleswig-Holstein.[49] These sod buildings have in common that they are generally constructed in areas where there is a lack of wood, like the salt-marsh or where the landscape was heavily deforested as in some areas in North- and West-Jutland.[50] Using sods as construction material for walls means an important reduction of the amount of wood needed to erect buildings. If sod walls were roof-supporting, the thick posts normally needed for a roof-supporting framework could be saved as seems to be the case at Oldorf and Wijnaldum-Tjitsma.

The sod buildings at Wijnaldum-Tjitsma do not necessarily stem from the eastern North Sea region, because sod structures from the Roman Period are also found elsewhere in the coastal area of the Netherlands. But the material culture which accompanies the introduction of the buildings, refers to the eastern North Sea region.[51] The discontinuity in the coastal area of the western Netherlands makes an origin in this part of the country even more unlikely.

As sod walls are sometimes hard to recognize, postholes were lacking, houses were rebuilt several times on the same spot and only a part of the settlement from this period was excavated, it is hard to give a reliable estimate of the total number of houses that have stood on the terp. We assume that in the excavated area at least 3 (SS 5, SS 6 and SS 16) to 5 houses were contemporary, but considering the great number of remains which was excavated, the real number could have been much higher, as is suggested by three excavated hearths which could not immediately be related to a certain structure. The main buildings initially have a NNW-SSE orientation, later changing to N-S and were standing close together. The N-S orientation remains the orientation of the main buildings during the next occupation phases.

Figure 28. Example of sod building from Egilstadir, Iceland, showing many similarities with the sod buildings from Wijnaldum (photographs: D.A. Gerrets).

In the perspective of these new results of the

Wijnaldum excavation, it is very likely that sod buildings were also present at the terps of Ezinge and Tritsum but that due to poor conservation circumstances these buildings were not recognized by the excavators. The occupation levels of Ezinge belonging to the Migration Period, were also the highest documented levels, where features were difficult to recognize.

At Wijnaldum-Tjitsma two well recognizable sunken huts (Sunken Hut 7 and SH 8, see Fig. 29) and one possible sunken hut (SH 3) belonging to Period III A/B were excavated. Three other sunken huts (SH 9, SH 12 and 14) are recognized, but not excavated. As in all later occupation phases (with small aberrations) the sunken huts have an E-W orientation and are often surrounded by a sod wall. The rectangular sunken huts are about 3 m long and 2-2.5 m wide, except for sunken hut 8 which measures 5 m to 3 m. The north and west side of sunken hut 7 was surrounded by a number of oblique standing sods which could have been part of a roof construction or an extra protection against the prevailing northwesterly winds. In sunken hut 9 a wattled timber at the inside of the surrounding sod wall could be recognized. This is the only structure in the excavation where wattle could be observed, but remains of daub tempered with straw and with the imprints of wattle show that wattle and daub

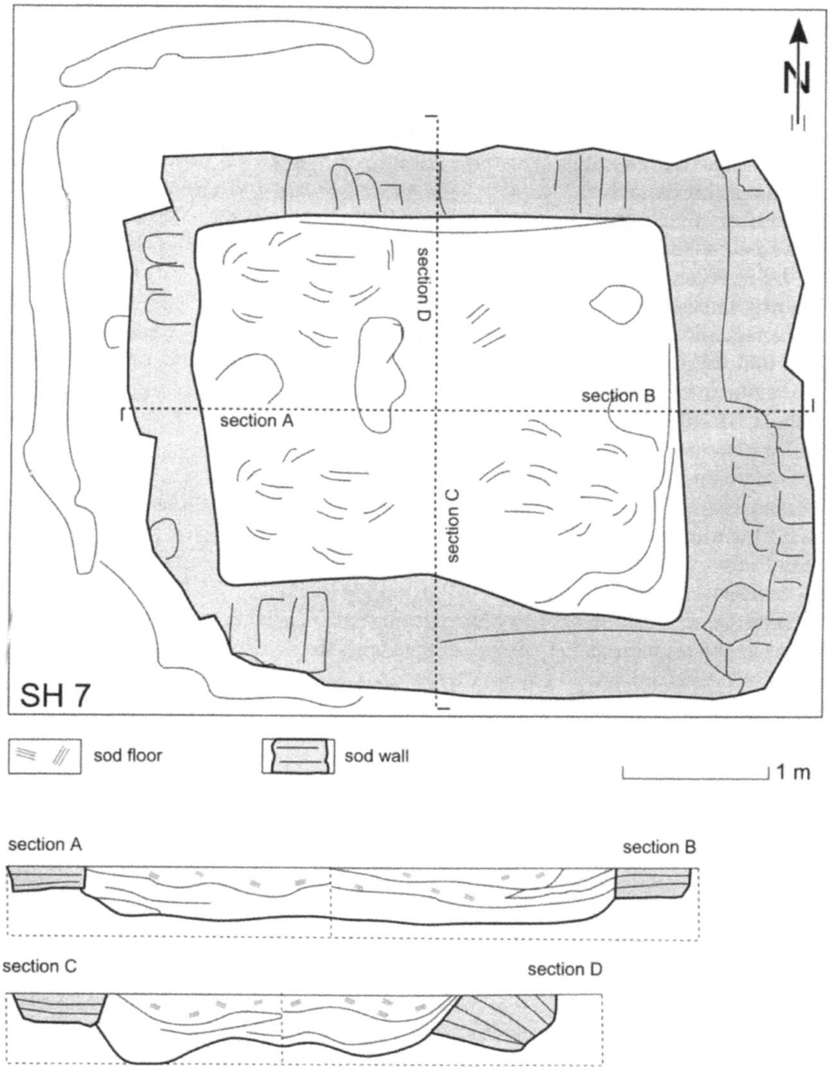

Figure 29. Sunken hut 7 from Period III (425-550).

was more common than is suggested by this single observation. As in the main buildings postholes were missing or at least doubtful. The original floor level in these sunken huts was carefully kept clean and hard to recognize. Some of these sunken huts started a second life as a refuse pit, after they had fallen into disuse and walls had collapsed. This phenomenon could also be observed at later occupation phases.

Most excavated structures were situated at 1.6 m above NAP. Human activities were not restricted to the terp itself but extended over the surrounding salt-marsh. In Period III A/B south of the terp border a hearth (H 17) was excavated on the salt-marsh at a height of 0.1 m above NAP. The presence of eight postholes (which are rare during this period at Wijnaldum-Tjitsma) around the hearth indicate that there was once a structure (SS 33) with a wooden framework here. The postholes had an average width of 25/30 cm, which is quite big for Wijnaldum-Tjitsma. Only part of the structure was excavated, but it is quite possible, that the hearth was inside the building. The main buildings and the sunken huts are still concentrated on the highest part of the terp, but as activities also extend on the salt-marsh and even heavily built structures are erected on the salt-marsh, one gets the impression that the remaining influence of the sea on the environment of the settlement is diminishing. This is underlined by the observation in main section A (see Fig. 23), that no systematical heightening of any significance has taken place after the Roman Period. After Period II we see alternating occupation levels and sod layers. The sod layers seem to be more the arbitrary result of levelling house-remains, than a result of intentional attempts to heighten the site as a protection against the sea. A possible explanation for this diminishing threat of the sea could be the development of the Middelzee, the inlet between Westergo and the Oostergo region. The development of the Middelzee enlarged the storage area, while the storm tidal volume stayed the same, which means that the total amount of water was spread out over a larger area. At the same time the expansion of the salt-marsh reduced the remaining energy of the waves, before they arrived at the terp of Wijnaldum-Tjitsma.

Around 500 the majority of the inhabitants will have moved temporarily to another part of the terp, or perhaps even to another terp. At this time a large part of the former settlement area was covered by an agricultural layer, which was very organic, homogeneous and poor in find material. This agricultural field extended over the border of the dwelling mound on the surrounding salt-marsh. This interruption in the occupation will not have lasted long, because no discontinuity in material culture or settlement structure could be observed. During later periods parts of the Wijnaldum-Tjitsma terp were used for agriculture but these agricultural plots were generally much smaller and did not extend over the surrounding salt-marsh. In the main excavation trench no indications were found for agriculture on the salt-marsh at a larger distance from the settlement before post-medieval times.

Reconstruction of the settlement for this period is, as for most other periods, seriously hampered by erosion. Of sunken hut 3 only the floor level remained. But its presence and the location of well 1 on top of the former house-terp of the Middle Roman Period shows that the settlement in Period III A/B was also situated on top of this house-terp, but had extended in an easterly and southeasterly direction. During the next occupation phases the steep sloping north boundary of the terp stays more or less fixed on the same place. This north boundary is, in contrast with e.g. the south boundary, clearly marked by the presence of large ditches during all periods. Expansion takes place sometimes in more easterly, sometimes in more westerly direction, but always in a southerly direction until the 13th/14th century, when a building was erected on the site of the present farm.

7 WIJNALDUM-TJITSMA DURING THE LATE SIXTH AND EARLY 7TH CENTURY – PERIOD IV: CA 550-650 (Figs 11, 12 and 21)

Features from this period have suffered severely from surface erosion.[52] Especially in the eastern part of the excavation, features over a large area have disappeared due to its high location of 1.6 m or more above NAP, immediately under the ploughing zone. A NW-SE oriented sunken hut (SH 1), two granaries with the same orientation, one constructed with four posts (G 1) and another with six posts (G 2), the remains of two possible N-S oriented sod buildings (SS 16 and SS 19) and some ditches are the only features which indicate that this part of the settlement was occupied during this period. Sunken hut 1 measured 5 to 3 m and was at least partly surrounded by a sod wall. The southern boundary of this part of the settlement is marked by a number of thick layers sloping into a ditch. These layers contained a large amount of material showing that some intense fires must have taken place shortly after each other. Because of the rich content of the layers, burnt loam, charcoal and ashes one gets the impression that some buildings in the neighbourhood burnt down several times in short succession, but the levels where one would expect these buildings were not excavated. South of this boundary of the terp was a depression with a diameter of 8 m and at least 0.9 m deep, probably a drinking-place for cattle, which was completely filled with dung and domestic refuse at a later stage.[53]

In the western part of the excavation features from Period IV are still for the greater part intact, because the contemporary occupation levels are on a much lower altitude of 1 m above NAP and for the greater part are still covered by younger layers. The occupation was concentrated on a number of small elevated platforms, separated from each other by 1.5-2 m wide ditches. The size of the ditches and the fact that they seem to surround the whole platforms suggest that besides a drainage function they also had a function as territorial marker.

We concluded before that the threat of the sea on those rare occasions of spring-tide with northerly storm was decreasing. This is underlined by the observation from this period that buildings are constructed at a level of 1 m above NAP, compared with the 1.6 m in the preceding period.

Sod walls indicate, that at least one (SS 26) and perhaps two N-S oriented buildings were situated here. The width of the sod walls vary between 0.5 m for the smaller outbuildings and 1 m for the main buildings. The width between the sod walls of the main buildings is 4 m.[54] These measurements are more or less representative of most sod buildings on the site, also in other periods. Since one of these sod walls could be followed over a length of ten metres, we may assume that the main buildings were at least 10 m long. Postholes are absent generally, while the presence of a hearth can be regarded as an indication that this structure had a function as a living quarter.

In the most westerly part numerous traces of metalworking (iron, copper alloys, gold and silver and garnet as well) and some remains of bead working were found on a separate platform. The sod buildings were reerected on the same location, but moved in a southerly direction. This is a more common phenomenon at Wijnaldum-Tjitsma, where buildings show a continuity of location sometimes for hundred years or more, but are rebuilt one or more times during that period. Also Van Giffen observed at terps like Ezinge and Leens-Tuinsterwierde (Prov. of Groningen), that houses were lying on top of each other like tiles on a roof.[55] In contrast with the previous period, where expansion of the terp in a southerly direction took place especially in the eastern part of the excavation, now expansion in a southerly direction took place in the western part.

8 THE RE-INTRODUCTION OF A ROOF-SUPPORTING FRAMEWORK – PERIOD V: 650-750 AD (Figs 14,15 and 22)

In the western part of the excavation fragments of a sod house (SS 24) with a hearth (H 16) have been excavated, which were constructed on the southern extension of a platform from Period IV. The building (SS 7) was probably reconstructed at a later stage during the same period, while the platform was extended southward again. On the interior side of the sod walls posts were standing at a distance of 1 m from each other. The interior width of the building was 3.7 m from sod wall to sod wall. The fireplace (H 5) belonging to the house was constructed on a layer of sherds from one and the same pot. This pot represents the transition from *proto-Kugeltopf* to real *Kugeltopf*. The sherds were covered with burnt loam.

Immediately west of this sod house on another house-platform a second building (SS 8) was situated which was reconstructed several times. Buildings were constructed further south again, compared with the preceding period, at 0.8 m above NAP. The artificial heightening of the expanding settlement decreases, while the surrounding ditches from the preceding period are becoming less significant.

A structure (SS 34) situated more to the north is only partly excavated and therefore the orientation is not clear. The surrounding sod wall had a width of 0.7 m. The distance from wall to wall inside the structure is 5 m. The presence of a hearth, close to the east wall, suggests a function as living quarter of the building. The hearth was constructed on a layer of pebbles with a loam cover on top.

In the eastern part of the excavation two sunken huts (SH 13 and SH 5) and a possible third one were situated. These sunken huts measured 5.2 to 2.5/3 m. It is not clear whether both sunken huts were surrounded by a sod wall. The latter sunken hut was replaced at a later stage in Period V by a

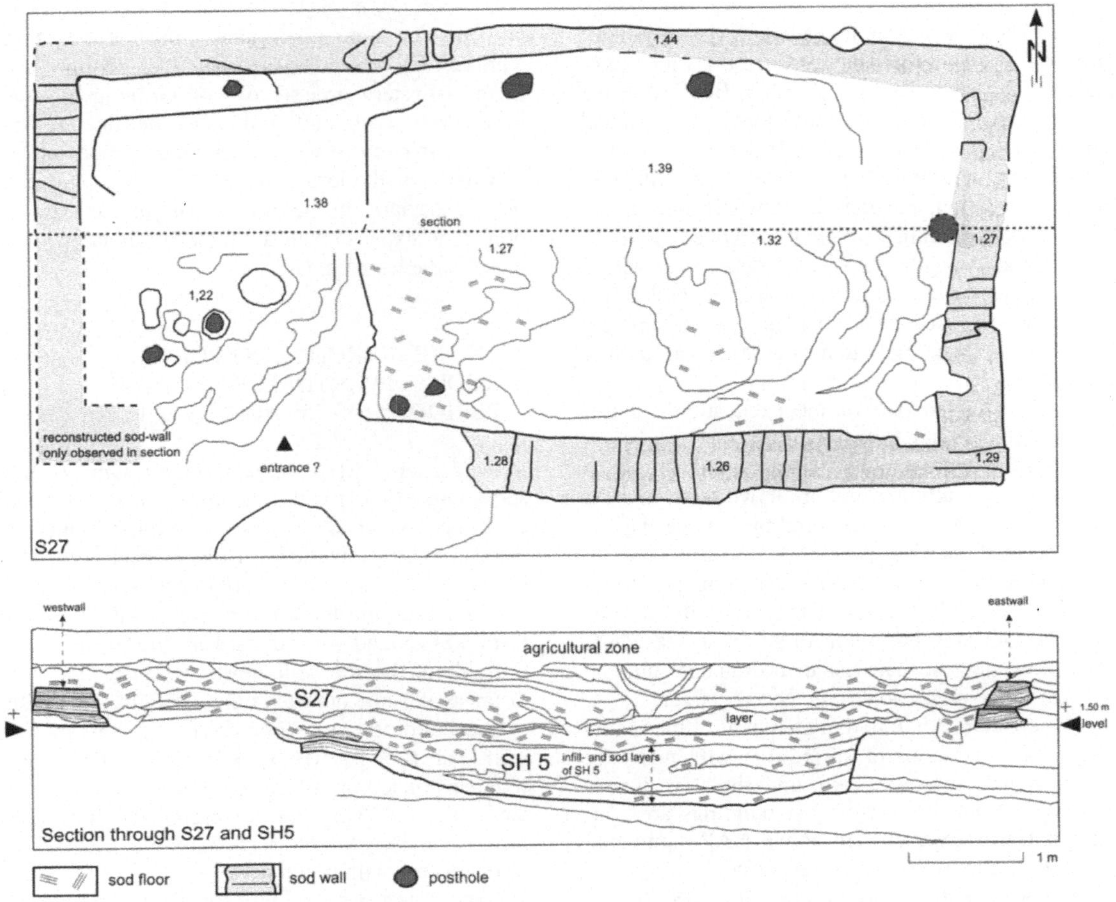

Figure 30. Sunken hut 5 dated in Period V (650-750) replaced by a somewhat larger outbuilding with sod walls (S27) during the same period. This is one of the best preserved structures of the Wijnaldum excavation.

slightly larger sod building (SS 2). Both sunken hut and sod house had a floor made of sods. This E-W oriented sod building was 7.75 m long and 3 m wide. The sod wall was 60 cm thick. Posts were at a distance of 10 to 15 cm from the wall. Besides rows of posts near to the wall there was a row of posts in the middle of the building. This is the first more complete building where postholes could be distinguished since the earlier mentioned buildings of the Mid-Roman Period. At the west side a 1.1 m long annex was constructed. The building was probably rebuilt once. In this building several spindle whorls, loom-weights and bone needles were found. This is the best preserved sod structure of the Wijnaldum-Tjitsma excavation.

A third sunken hut (SH 4, see Fig. 31) was situated immediately southeast of the one last mentioned sunken hut/sod structure. This sod house was surrounded by a 45 cm thick sod wall. Because the sunken hut was constructed on a slope the foundation of the north wall was about 40 cm higher than the foundation of the south wall. In this structure it was clearly visible how the sod walls fell into the structure after the sunken hut fell into disuse and the building collapsed. The building was partially destroyed during Period VI, when a refuse-pit was dug through it. Stratigraphically this sunken hut is somewhat younger than the other two sunken huts belonging to Period V. The sunken hut was surrounded by a ditch, which is different from the previous ditches, because of its much narrower width of 50-75 cm, and its clear rectangular shape.[56] Such ditches become a characteristic feature for the settlement during the next occupation phases and in view of their smaller size they do not seem to play an important role in the drainage sys-

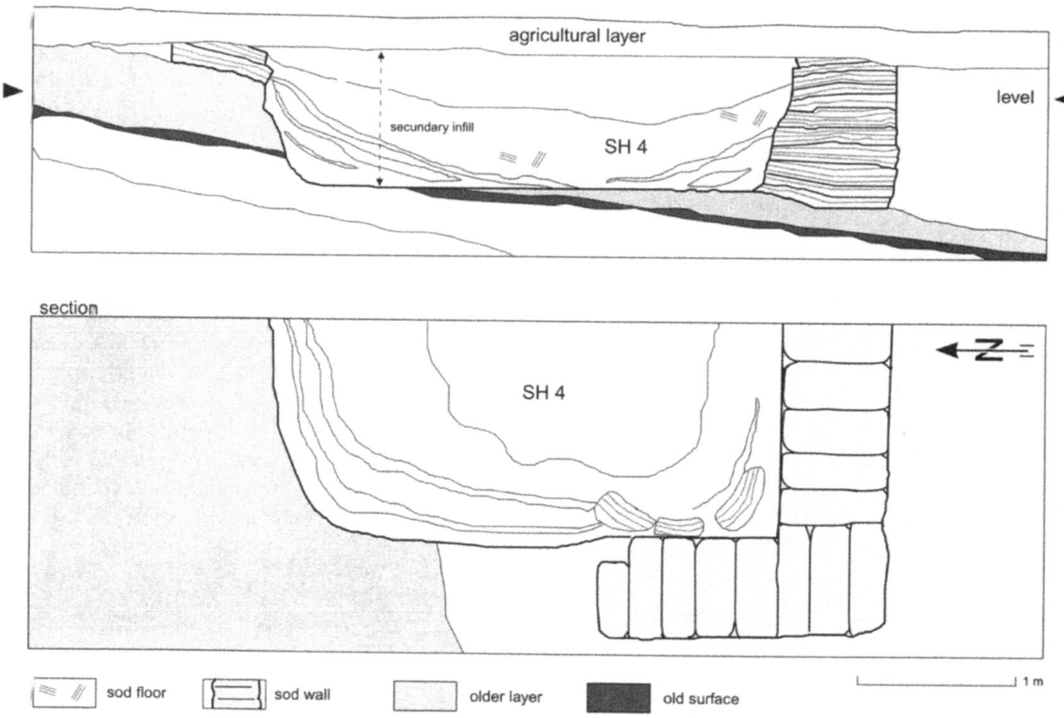

Figure 31. Section and level of sunken hut 4 dated in Period V (650-750). This is a good example why many structures seem to be fragmented in a horizontal level. In this case the fragmentary sod walls in the horizontal level can be explained by the construction of the sunken hut on a steep slope of the terp.

tem. They can be regarded as an indication that territorial boundaries are important.

Two possible buildings were at that time standing in the centre of the excavation, a third possible building was situated at the east edge of the terp. It was not possible to determine the exact orientation and dimensions of these buildings because sod walls were only partially preserved. A growing number of postholes shows the increasing importance of construction-wood but only very seldom is it possible to distinguish any structures. The remaining parts of sod walls show that the actual number of buildings will have been much higher.

9 AN INCREASING POPULATION – PERIOD VI: 750-800 (Figs 15, 16 and 22)

Between 750 and 800 a strong expansion of the settlement in a southerly direction took place. The eight wells and the numerous occupation levels/deposits which can be dated in this period are evidence of intensive occupation. Besides wells a great number of rectangular refuse-pits occur, which is a novelty on the terp. The yard of a main building seems to be surrounded by a rectangular ditch as territorial marker.

Although traces of intensive occupation are abundant, traces of buildings which must have been standing on the terp are rare. From the few constructive elements which remained intact, and the growing number of postholes it seems that we are dealing with main buildings of the three-aisled long-house type, as they are known from the terp Leens-Tuinsterwierde, the terp Het Torp at Den Helder and from the terps at Heveskesklooster and Foudgum, and which seem so characteristic for the terp region.[57]

The best preserved part of such a main building (S22; see Fig. 32) consists of a row of posts immediately at the inside of a 60 cm thick sod wall. Also (part of) a row of roof-supporting posts standing at 30-40 cm from the walls of the building are present.[58] The distance between these posts is about 1.8

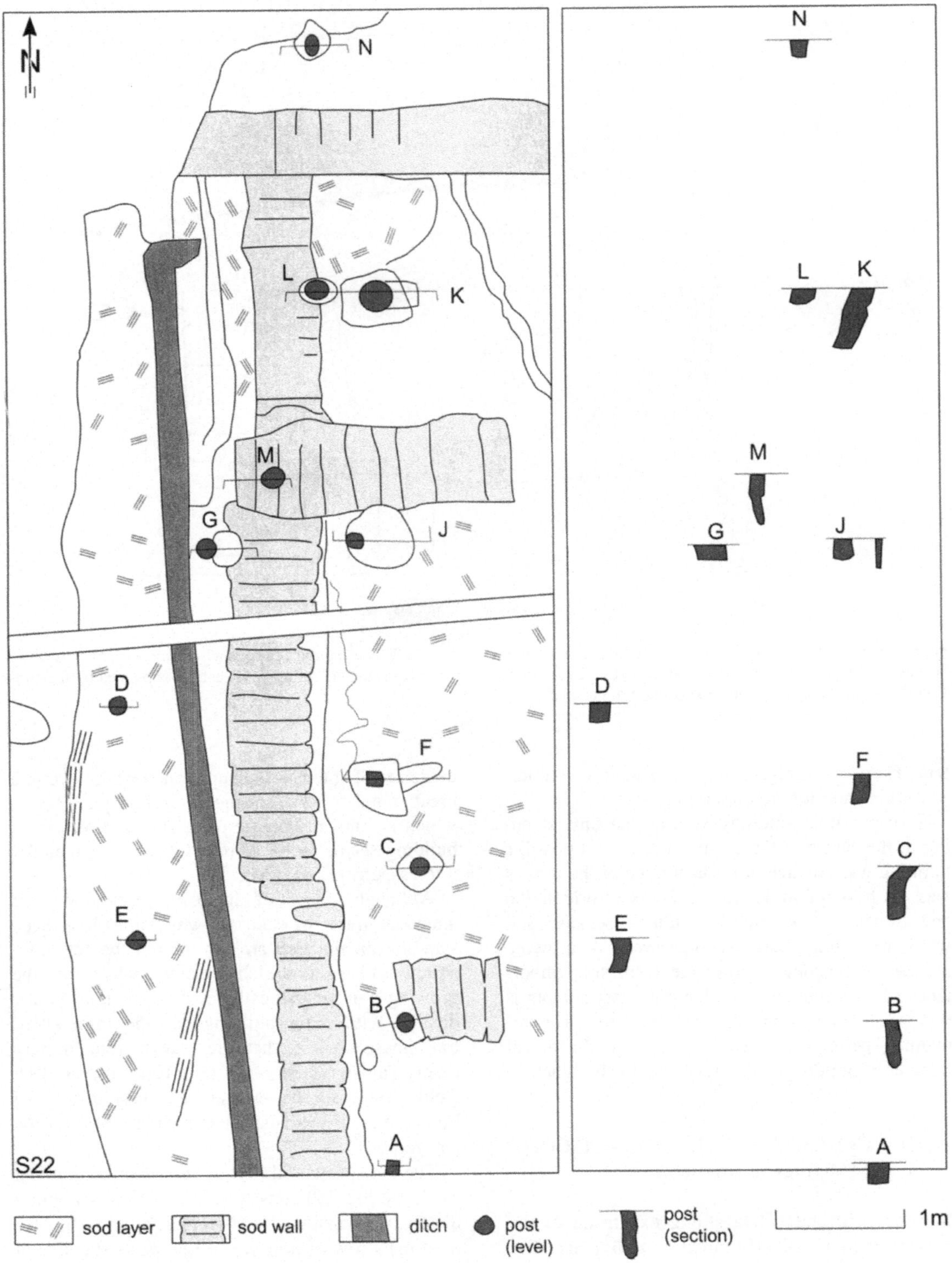

Figure 32. Sod structure 22 dated in Period VI (750-800).

m. Apparently the roof-supporting function of the sod walls, as we assumed for the sod buildings of the earlier periods, is now partly taken over by a wooden framework of two rows of 12-15 cm thick posts and posts situated immediately on the inside of the sod walls. A sod wall possibly divided the interior of the building into two rooms. A 20-cm-wide ditch could be an indication that the roof was hanging over.

A 5.8-m-long outbuilding (PS 31) was surrounded by a 70-cm-thick sod wall. The youngest sunken hut in the excavation (SH 6) was situated south of the house-platform in the western part of the excavation. It measured at least 5.1 m, was surrounded by a 45-cm-thick sod wall and was partially destroyed because a ditch was dug through it during a following occupation phase.

The intensity of the occupation was so high that remains of old buildings were removed by younger occupation. Only under favourable weather conditions were small fragments of former sod walls visible. While the house in the western part of the excavation still seems to be situated on a house-platform already existent during Period V, house platforms disappear during Period VI. The yard is indicated now by small 75-cm-wide rectangular ditches. There are no indications that territorial rights within the settlement were passed down to following generations by e.g. inheritance, because the ditches do not show continuity in space.

We assume that the northern part of the terp was permanently in use as agricultural fields from this period onwards, because no features like wells or ditches from this or later periods were excavated in this area. The ploughing soil belonging to this period will have disappeared because of surface erosion.[59]

10 A HOUSE WITH POSTS ON STONES – PERIOD VII: 800-850 (Figs 17,18 and 22)

During Period VII the settlement moved further south again. In this period the Wijnaldum-Tjitsma terp was most densely populated, but it is also the period which was affected most by the intensive occupation. At least eight and perhaps ten rectangular refuse-pits belong to this period.

There is one building (SS30, see Fig. 33) we know more about. In this sod building the poles were probably placed on large stones, lying on the inside of the 75-cm-wide sod wall. A ditch can be interpreted as sewer. The interior width of this building was 4 m.

Figure 33. Remains of a longhouse (SS30) dated in Period VII (800-850) with sod walls and roof-supporting posts on stones.

It is not clear whether sunken hut 6 was already present in the western part of the excavation during Period VI or not, but this is not impossible. This sunken hut was situated outside the terp on the salt-marsh, in view of its altitude of 20 cm above NAP. Northwest of this sunken hut was a granary with four posts.

11 HEAVILY ERODED FEATURES – PERIOD VIII: 850-900/950 (Figs 19, 20 and 22)

As this period was the last phase of the occupation it probably suffered most from erosion. In the southeast part of the excavation, however, some features were excavated in which Pingsdorf pottery was found. These features were mainly ditches and some postholes perhaps belonging to a building. A well was excavated, much further north than we ever expected, with a lot of Merovingian and Carolingian wheelthrown wares, but also with quite a number of Pingsdorf sherds on the bottom, indicating that the settlement on or near this place was probably still in use. Most of the features belonging to the settlement, however, will have disappeared because of erosion. Also the presence of some denarii of Louis the Pious and Lotharius I in the south part of the excavation show that the settlement was still occupied in the second half of the 9th century and probably the early 10th century.

12 WATER SUPPLY

A total number of 38 wells and possible wells from all periods were (partially) excavated. Roughly speaking three main types of wells were in use at Wijnaldum-Tjitsma. The first type (type A) is the simplest one. It is constructed by digging a straight shaft varying in width between 0.6 and 0.8 m and varying in depth between 1.1 and 1.5 m below NAP. There is no sod lining. It is hard to imagine how exactly it was constructed. This type of well stays in use until the end of the 8th century.

The second type of well (type B) is the funnel-shaped well without sod lining. For this type of well a lever was indispensable. At the bottom of some funnel-shaped wells a wooden barrel prevents the well from gradually filling up with drift sand from the surrounding mud flat sediments. Funnel-shaped wells vary in depth between 0.9 and 2.56 m. below NAP and occur only from 775 onwards.

The third type (type C) is the sod lined well. This type shows great variation in construction. Some are similar to the earlier mentioned wells of type A and B, but with sod lining. Other shapes also occur, however, within this type of wells. Around a core varying in width between 0.55 and 1.6 m a circular coating of sods is constructed, sometimes on a wooden framework. In some of the wells of this type the sod lining immediately covers the wall of the hole but in some other wells there is space between the sod coating and the wall of the hole which was gradually filled up with earth during construction. Some wells with a straight shaft are protected from collapsing by one or more barrels. In at least two cases it was observed that staves were pulled out or had rotted, suggesting that the number of wells with wooden barrels was originally higher.[60] The depth varies between 1.8 and 3 m below NAP, which means that all wells whose depth is known were dug into the tidal flat sediments. Sod lining stays in use with all kinds of variations during the whole period of occupation on Wijnaldum-Tjitsma. Generally, wells were kept clean but in some cases wells were secondarily used as refuse-pit.

Fresh water has a lighter specific density than salt water. Rainwater which penetrates the soil will form a lenticular water supply in the subsoil. The expansion of the terp will also have increased the available water supply. The ground water, however, will have been salt to brackish and an excessive use of the available fresh water supply meant that water became brackish.[61] As animals are more tolerant of salt, this water will have been suitable for cattle. Bantelmann showed that for this reason the water supply for humans and animals on the German Halligen was kept strictly separate.

Period I/II (AD 175-300/350)
During Phase I and II 2 wells (W 31 and W 33) of type A were in use and 1 well (W 2) of type C. A framework of oak was placed at the bottom of well 2 with a dendrochronological felling date in the summer of 263 on which a circular lining of sods was erected. The find of a taxus hoop in one of the possible wells from this period (W37) could be an indication for the use of a wooden barrel.[62]

Period III A/B (425-550)
Three possible wells (W 25, W 32 and W 38) and one certain well (W 1) belong to Period III A/B, of which only the last one was extensively investigated. From this type C well, but without wooden frame at the bottom, water was probably drawn with the help of a lever as is shown by the presence of a contemporaneous posthole just beside the well.

Period IV (550-650)
The lack of wells in this period is remarkable. The only well (W 9) which was in use with certainty during Period IV was excavated in the context of the bronzecaster's workshop and was secondarily filled up with refuse of the metalworking process. This well is type A.

Period V (650-750)
Wells are still scarce in Period V, compared with the following period. Only three wells of type C (W 6, W 20 and W 30) were in use during this period. The latter well (W 6) seems to be constructed in Period V, but probably stayed in use until the end of Period VI. A fourth well (W 24), which was only partially excavated and whose construction is unclear has a terminus post quem of 694 and was

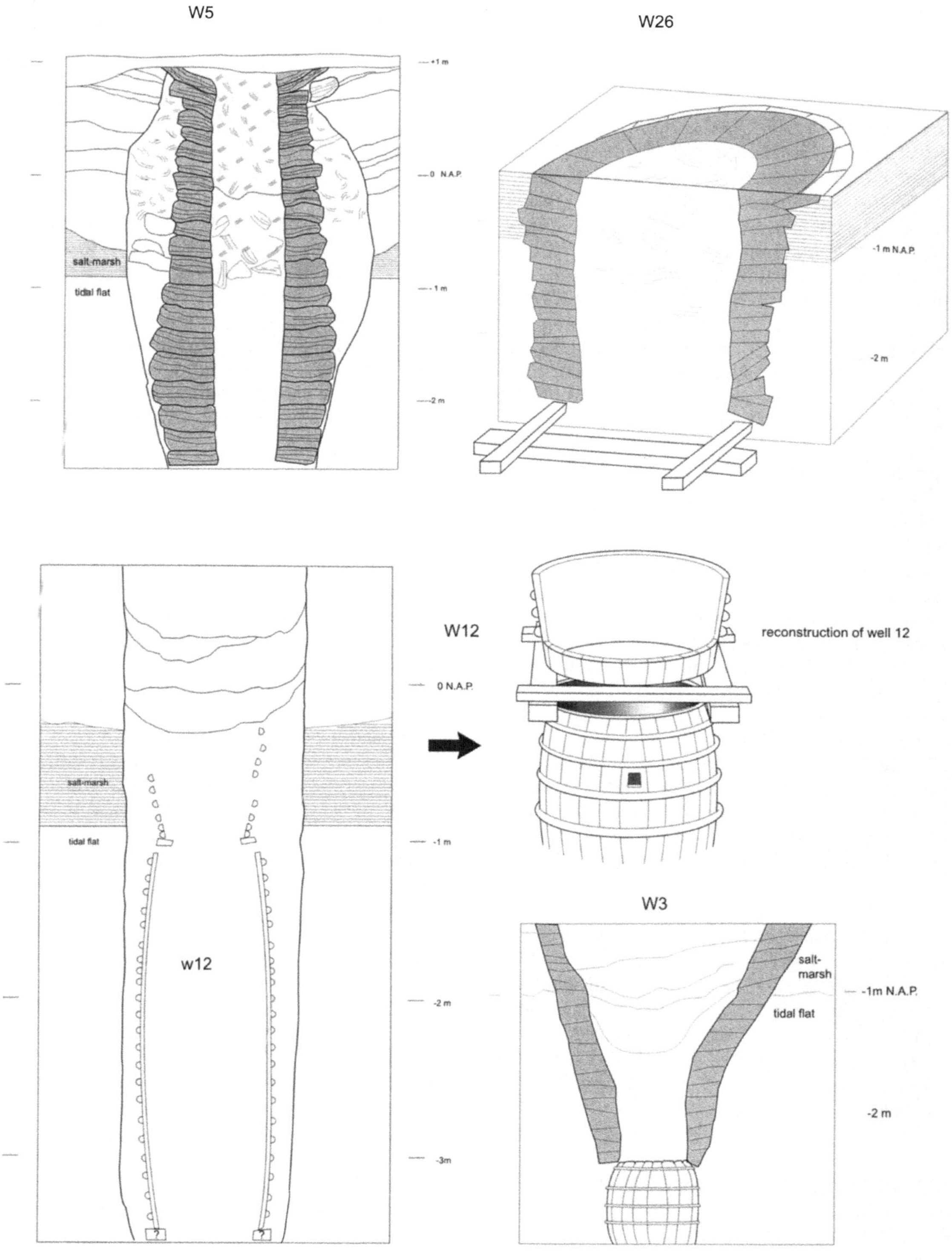

Figure 34. Different types of wells.

also in use during Period V (GrN 21398, 1245 ± 30 BP). A fifth possible well of uncertain construction (W 21) was probably also constructed during Period V and stayed in use until Period VI.

Period VI (750-800)
Twelve wells were constructed during this period: three of type A (W 8, W 28 and W 29), one of type B (W 14), two of type C (W 13 and W 18) and six with an uncertain construction (W 10, W 17, W 22, W 26, W 27 and W 36, see Fig. 34). Well 13 could be a variation on the wells of type C. The top of the hole dug for construction of this well probably had a bottle-neck shape. The core of the well was protected from collapsing by one or more wooden barrels. The staves of the lowest barrel were finally pulled out, while the hoops stayed in place. Fifteen wells were altogether in use during Period VI.

Period VII (800-850)
Ten wells which were already in use during Period VI stayed in use until the end of Period VII. The clearest example of the bottle-neck variation of the type C well (well 5, see Fig. 34) should be dated in Period VII. This well differed from the earlier-mentioned one in the lack of wooden barrels.[63] Only three new wells were constructed: three of type C (W 5, W 7 and W 19) and one of uncertain construction (W 15).

Period VIII (850-900/950)
During Period VIII the total number of wells still in use and newly constructed is rapidly declining. Only two wells (W 7 and W 26, see Fig. 5) from the former period stayed in use. One well of type A (W 12, see Fig. 5), one of type B (W 4) and one of type C (W 3, see Fig. 5) were constructed during this period. The latter two wells were both funnel-shaped. In addition, two wells of uncertain construction (W11 and W16) were in use.

The shaft of well 12 was protected from collapsing by two wooden barrels on top of each other. The staves of the top barrel were pulled out, and only the hoops stayed in place. Between the two barrels were two planks. The lower barrel was 2.4 m high!

13 METALWORKING AND OTHER CRAFTS IN TIME AND SPACE

The crafts executed at Wijnaldum-Tjitsma will be discussed by several specialists. For this reason we will restrict ourselves to discussing the development of craft production and especially metalworking as the most important and long lasting form of handicraft.

Casting of copper alloys started as an incidental activity during Phase I, but after 250 especially copper alloys and to a lesser extent also iron were worked on a relatively large scale.[64] Numerous finds and structures in the southeast corner of the eastern extension of the house-terp indicate that here was a bronzecaster's workshop. At least seven hearths, some of them consisting of sherds with loam on top, were situated beside and on top of each other. Around these hearths fragments of crucibles, a dump with slag material, secondarily collected scrap (fragments of brooches, wire, foil etc.) were found. Some copper alloy ingots were found in contemporaneous layers on the northern side of the house-terp.

The remains of bone and antler working are still being studied at the moment, but during excavation in one sunken hut (SH 7) from Period III A/B several remains of bone and antler working were excavated. The material came from layers related with the second life of the sunken hut as a refuse pit. The amount of remains indicating that bone and antler had been worked for manufacturing e.g combs is very restricted, suggesting that at Wijnaldum-Tjitsma we are dealing with small-scale production for a local group of consumers. The same can be said of the remains of bead making.[65]

In the northeast part of the excavation, the ploughing zone was removed and finds gathered in squares of 4 to 4 m. Because of a lack of money and a shortage of time, only a small part of the total number of features from this area was documented. The finds of an unfinished key and an unfinished brooch with fins, which still had to be filed off and the presence of several hearths and slag-material showed that metalworking must have been an important activity in this area. Finds from the ploughing zone suggest that these features must be dated in Period III A/B (425-550). Part of the metalworking activities, however, will already have moved to the western part of the excavation during this period.

Here, in the most westerly part of the excavation, numerous traces of metalworking (iron and copper alloy as well) were found. In this area it starts in Period III A and increases steadily until it reaches its maximum during Period IV (550-650). Meanwhile iron has replaced copper alloy as the most important metal worked by the local craftsmen. The presence of a large piece of yellow glass paste suggests that the same workshop was used for bead production. One hearth (H 15) probably belonged to this workshop, while a second hearth (H 18) may have been part of a furnace. Only one well (W 9) was in use with certainty during Period IV and used secondarily as a refuse pit for waste material from the metalworking activities in the vicinity.

The presence of gold- and silverworking is hard to prove because waste material from these precious metals will have been collected systematically to be reused in manufacturing new objects. The find of a gold mount of a sword, decorated with filigree and carefully folded together with eight small gold rivets inside apparently to be reused for another purpose forms the first indication that a goldsmith was active at Wijnaldum-Tjitsma from III B onwards. Many other objects like touchstones with gold marks, a crucible fragment with gold drops, gold drops, a small hammer, silver drops and silver bars are evidence that goldworking was not an incidental activity but continued until the end of the settlement. The most striking find was a die for making the cross-hatched pattern in goldfoil, used to improve the sparkling of red garnets. The technical research of the Wijnaldum disc-on-bow brooch has already suggested that this brooch was in the possession of a goldsmith for repair or reuse. The find of the die makes this hypothesis more likely. Because raw materials were probably reused on a large scale and finds are for that reason scarce and often transported unintentionally at a later stage of the occupation, it is almost impossible to localize the gold- and silversmith.

The number of remains of metalworking activities decrease rapidly after 650. A number of iron ingots and semi-manufactured artifacts, however, show that iron is still imported and worked on a smaller scale.

REFERENCES

BAKKER, L. 1981. Spätrömische Befestigungen und frühmittelalterliche Kirche von St. Peter und Paul. In: J. Metzler, J. Zimmer & L. Bakker (eds), *Ausgrabungen in Echternach*, Luxemburg, 269-355.

BANTELMANN, A. 1955. Tofting, eine vorgeschichtliche Warft an der Eidermündung, Neumünster (Offa 12).

BANTELMANN, A. 1975. Die frühgeschichtliche Marschensiedlung beim Elisenhof in Eiderstedt. Landschaftsgeschichte und Baubefunde, Bern etc. 1 (Studien zur Küstenarchäologie Schleswig-Holsteins, Serie A, 1).

BAZELMANS, J., D.A. GERRETS & P.C. VOS 1998. Zoden aan de dijk. Kleinschalige dijkbouw in het Friesland van de Romeinse tijd, *Noorderbreedte* 22(6): 18-22.

BECH, J.-H. 1985. The Iron Age village mound at Heltborg, Thy, *Journal of Danish Archaeology* 4: 129-146.

BERTELSEN, R. & R.G. LAMB 1993. Settlement mounds in the North Atlantic. In: C.E. Batey, J. Jesch & C.D. Morris (eds), *The Viking Age in Caithness, Orkney and the North Atalantic*, Edinburgh, 544-554.

BESTEMAN, J.C. & A.C. GUIRAN 1987. An early peat bog reclamation area in medieval Kennemerland. In: R.W. Brandt, W. Groenman-van Waateringe & S.E. van der Leeuw (eds), *Assendelver Polder Papers 1*, Amsterdam, 297-332 (Cingula 10).

BOERSMA, J.W. 1988. Een voorlopig overzicht van het archeologisch onderzoek van de wierde Heveskesklooster (Gr.). In: M. Bierma, A.T. Clason, E. Kramer et al. (eds), *Terpen en wierden in het Fries-Groningse kustgebied*, Groningen.

BÖHME, H.W. 1986. Das Ende der Römerherrschaft in Britannien und die Angelsächsische Besiedlung Englands im 5. Jahrhunderts, *Jahrbuch des Römisch-Germanischen Zentralmuseums Mainz* 33: 469-574.

BÖHNER, K. 1958. *Die fränkische Altertümer des Trierer Landes*, Berlin (Germanische Denkmäler der Völkerwanderungszeit, Serie B, Die fränkische Altertümer des Rheinlandes 1) Berlijn.

BRANDT, K. 1991. Die mittelalterlichen Wurten Niens und Sievertsborch (Kreis Wesermarsch), *Probleme der Küstenforschung im Südlichen Nordseegebiet* 18: 89-140.

BRØNSTED, J. 1960. *Danmarks Oldtid, Vol. III, Jernalderen*, 2nd ed., Copenhagen.

DIJKSTRA, J. 1998. Archeologisch onderzoek in de binnenstad van Tiel, juni t/m september 1996. Lokaties Koornmarkt en Tol-Zuid, Amersfoort (Rapportage Archeologische Monumentenzorg 57).

ES, W.A. VAN 1970. Paddepoel, excavations of frustrated terps, 200 BC-425 A.D., *Palaeohistoria* 14: 187-352.

ES, W.A. VAN 1973. Terp research; with particular reference to a medieval terp at Den Helder, province of North-Holland, *Berichten van de Rijkdienst voor het Oudheidkundig Bodemonderzoek* 23: 337-345.

ES, W.A. VAN & W.J.H. VERWERS 1980. *Excavations at Dorestad 1. The harbour: Hoogstraat I* (Nederlandse Oudheden 9).

GIFFEN, A.E. VAN 1936. Der Warf in Ezinge, Provinz Groningen, Holland, und seine Westgermanischen Häuser, *Germania* 20(1): 40-47.

GIFFEN, A.E. VAN 1940. Een systematisch onderzoek in een der Tuinster wierden te Leens, *Jaarverslagen van de Vereniging voor Terpenonderzoek* 20-24: 26-117.

HAARNAGEL, H. 1979. Die Grabung Feddersen Wierde, Methode, Hausbau, Siedlungs- und Wirtschaftsformen sowie Sozialstruktur, Wiesbaden (Feddersen Wierde 2).

HARRIS, E.C. 1989. *Principles of archaeological stratigraphy*, 2nd ed., London etc. (Studies in Archaeological Science).

HARRIS, E.C., M.R. BROWN & G.J. BROWN (eds) 1993. *Practices of archaeological stratigraphy*, London.

HEERINGEN, R.M. VAN, P.A. HENDERIKX & A. MARS (eds) 1995. *Vroeg-Middeleeuwse ringwalburgen in Zeeland*, Goes etc.

HEERINGEN, R.M. VAN & F. VERHAEGHE 1995. Het aardewerk. In: R.M. van Heeringen, P.A. Henderikx & A. Mars (eds), *Vroeg-Middeleeuwse ringwalburgen in Zeeland*, Goes etc.

HUIJTS, C.S.T.J. 1992. De voor-historische boerderijbouw in Drenthe. Reconstructiemodellen van 1300 vóór tot 1300 na Chr., Arnhem.

KJÆR, H. 1928. Oldtidshuse ved Ginderup i Thy. *Fra Nationalmuseets Arbejdsmark*, 7 ff.

KLUGE-PINSKER, A. 1988. Die Entwicklung der Keramikproduktion in Duisburg im 9. und 10. Jahrhundert, Frankfurt am Main (thesis).

KNOL, E. 1993. De Noordnederlandse kustlanden in de Vroege Middeleeuwen, Groningen (thesis Vrije Universiteit, Amsterdam).

KOOI, P.B., G. DELGER & K. KLAASSENS 1987. A chieftain's residence at Peelo: a preliminary report on the 1987 excavations, *Palaeohistoria* 29: 133-144.

KOSSACK, G., O. HARCK & J. REICHSTEIN 1975. Zehn Jahre Siedlungsforschung in Archsum auf Sylt, *Berichte der Römisch-Germanischen Kommision* 55: 261-429.

KOSSACK, G., J. REICHSTEIN & O. HARCK 1980. Archäologische Geländeforschung 1963-1978. In: Kossack, G. et al. (eds), *Archsum auf Sylt. Teil I. Einführung in Forschungsverlauf und Landschaftsgeschichte*, Mainz am Rhein, 173-193 (Studien zur Küstenarchäologie Schleswig-Holsteins Serie B, Archsum 1 = Römisch Germanische Forschungen 39).

LANGEN, G.J. DE 1992. Middeleeuws Friesland. De economische ontwikkeling van het gewest Oostergo in de Vroege en Volle Middeleeuwen, Groningen (thesis).

LANGEN, G.J. DE & H.T. WATERBOLK 1989. De archeologie van Ezinge, nederzettings- en onderzoeksgeschiedenis van een Gronings terpdorp, *Jaarverslagen van de Vereniging voor Terpenonderzoek* 66-72: 78-111.

OLST, E.L. VAN 1991. Uilkema, een historisch boerderijonderzoek. Boerderij-onderzoek in Nederland 1914-1934, II, Arnhem.

PIRLING, R. 1966. *Das römisch-fränkische Gräberfeld von Krefeld-Gellep*, Berlin (Germanische Denkmäler der Völkerwanderungszeit, Serie B, 2).

RIENKS, K. & G.L. WALTHER 1954. *Binendiken en slieperdiken yn Fryslân*, Leeuwarden.

SCHMID, P. 1994. Oldorf – eine frühmittelalterliche friesische Wurtsiedlung, *Germania* 72: 231-267.

STUIVER, M. & G.W. PEARSON 1993. High precision bidecadal calibration of the radiocarbon time scale, AD 1950-500 BC and 2500-6000 BC, *Radiocarbon* 35: 1-23.

TAAYKE, E. 1990/1996a/1996b. Die einheimische Keramiek der nördlichen Niederlande, 600 v Chr. bis 300 n. Chr., Teil I: Westergo, *Berichten van de Rijkdienst voor het Oudheidkundig Bodemonderzoek* 40: 101-222; Teil IV: Oostergo, 42: 87-161; Teil V: Übersicht und Schlußfolgerungen, 42: 163-208.

TAAYKE, E. & E. KNOL 1992. Het vroeg-middeleeuwse aardewerk van Tritsum, Gem. Franekeradeel (Fr.), *Paleoaktueel* 3: 84-88.

THERKORN, L.L. 1987. The structures, mechanics and some aspects of inhabitant behaviour. In: R.W. Brandt, W. Groenman-van Waateringe & S.E. van der Leeuw (eds), *Assendelver Polder Papers 1*, Amsterdam, 177-225 (Cingula 10).

TRIER, B. 1969. Das Haus im Nordwesten der Germania Libera, Münster.

UNVERZAGT, W. 1968. *Die Keramik des Kastells Alzey*, Bonn (Materialien zur römisch-germanischen Keramik 2).

VERHOEVEN, A.A.A. 1998. *Middeleeuws gebruiksaardewerk in Nederland (8ste-13de eeuw)*, Amsterdam (Amsterdam Archaeological Studies 3).

VOSKUIL, J.J. 1979. Van vlechtwerk tot baksteen: Geschiedenis van de wanden van het boerenhuis in Nederland, Arnhem.

WATERBOLK, H.T. 1961. Beschouwingen naar aanleiding van de opgravingen te Tritsum, gem. Franekeradeel, *It Beaken* 23/24: 216-226.

WATERBOLK, H.T. 1979. Siedlungskontinuität im Küstengebiet der Nordsee zwischen Rhein und Elbe, *Probleme der Küstenforschung in südlichen Nordseegebiet* 13: 1-21.

WATERBOLK, H.T. 1991. Ezinge. In: H. Beck et al. (eds), *Hoops Reallexicon der Germanischen Altertumskunde* 8, Berlin etc., 60-76.

WATERBOLK, H.T. 1992. Das mittelalterliche Siedlungswesen in Drenthe. Versuch einer Synthese aus archäologischer Sicht. In: Böhme, H.W. (ed.), *Die Salier. Siedlungen und Landesausbau zur Salierzeit, Vol. 1: In den nörlichen Landschaften des Reiches*, Sigmaringen, 47-109.

WILLEMS, W. 1981. Romans and Batavians: a regional study in the Dutch Eastern River Area, I, *Berichten van de Rijksdienst voor Oudheidkundig Bodemonderzoek* 31: 7-217.

WOLTERING, P.J. 1975. Occupation history of Texel, I: The excavations at Den Burg: Preliminary report, *Berichten van de Rijksdienst voor het Oudheidkundig Bodemonderzoek* 25: 7-37.

NOTES

1 See also Bos & Gerrets (this volume).

2 Harris (1989), Harris (1993).

3 Böhner (1958), Pirling (1966).

4 See also the contribution of Gerrets (this volume).

5 Stuijts' study (next volume) shows that wood was reused on a large scale.

6 See also the contributions of Erdrich, Galestin, Volkers and Van der Vin (all in this volume).

7 An extensive typochronological study of the indigenous handmade pottery was recently published by Ernst Taayke (see Taayke 1990, 1996a and 1996b). Because the classification of the Oostergo material is more specific, the Oostergo typology was applied by Ernst Taayke to the Wijnaldum material. The types G6, G7, V4 and K4 are dated by Ernst Taayke between AD 100 and 250 (Taayke 1996b, p. 140). The Oostergo types G8, V5 and K5 are dated between 250 and 350. At Wijnaldum-Tjitsma the transition seems to be a gradual one (Taayke, next volume). Of all datable sherds of terra sigillata only 9 sherds belong perhaps to Period I (see Volkers, this volume).

8 Except for some first-century metal finds from a later context, the bulk of the metal finds has to be dated after AD 175 (Erdrich, this volume).

9 For most radiocarbon datings large pieces of animal bone were selected in order to get the smallest margins in the datings. Generally animal bone gives more reliable datings than e.g. charcoal. All radiocarbon datings presented here are based on two standard deviations. Calibration according to Stuiver & Pearson (1993).

10 For the indigenous handmade pottery: Taayke (1996a, p. 140). Oostergo type G 9 is partly contemporaneous with G8 but is generally later (Taayke, next volume). The wheelthrown terra nigra-like pottery and its dating is still a subject of debate (see Galestin, this volume). The latter pottery shows in shape similarities with the Oostergo Type K5b (Taayke 1996a, pp 111 and 124 and note 75; Galestin this volume).

11 For instance the early-fourth century types of the Almgren VII brooches are completely absent (Erdrich, this volume).

12 Also well dated finds from the fourth century are rare. It is remarkable that rough-walled pots of the so-called Alzey 27 type (Bakker 1981, p. 335 and Fig. 245, pp 9-11) which are characteristic of the fourth century are completely lacking. It must immediately be added that 'early' rough-walled Roman pottery in general is lacking.

13 For the chronology and typological classification Taayke (next volume).

14 The smooth-walled biconical wheelthrown ware consists mainly of small beakers, which are hard to classify, because only small rim fragments are preserved. The rough-walled material consists mainly of Alzey 32/33 type cooking-pots (Unverzagt 1968, p. 35 and Fig. 22, pp 13-15; Bakker 1981, pp 338-340 and Fig. 246, pp 36-41; Willems 1981, pp 168-170 and Fig. 42) and three rim sherds of the Alzey 27 type, two with the long tender sickle shaped rim and one with the reduced sickle shaped rim (Unverzagt 1968, p. 34 and Figs 21 and 22, p. 10; Bakker 1981, pp 336-338 and Fig. 245, pp 24-26), which are all generally dated in the fifth century. Black glittering (volcanic?) inclusions in almost all sherds suggest an origin in the Mayen region.

15 Böhme (1986, pp 527-538).

16 Types H III and H IV in the Dorestad typology (Van Es & Verwers 1980, pp 119-121). Some late undecorated Anglo-Saxon wares from Wijnaldum-Tjitsma show many similarities with the earliest H III and H IV type pottery. One of the main differences is that burnishing or polishing of the the surface on the inside or outside of the pottery is now omitted. It seems likely that one kind of pottery developed into the other.

17 The difference between the *Eitopf* and the earliest 'small' *Kugeltöpfe* with rims of type H IA is sometimes hard to distinguish (Van Es & Verwers 1980, p. 122). The transition of the *Eitopf* to the fully developed *Kugeltopf* must have taken place long before the arrival of sceattas at Wijnaldum-Tjitsma at the beginning of the 8th century. Van Es and Verwers (1980, pp 122 and 152) assume that the *Kugeltopf* was introduced at Dorestad in the middle of the 8th century. The Wijnaldum material suggests that this happened here about half a century earlier. Also Verhoeven (1998, p. 251) assumes that the *Kugeltopf* was introduced in the beginning of the eighth century in the western and central parts of the Netherlands and that the *Eitopf* disappeared soon after.

18 Type W IX and the W III types in the Dorestad typology (Van Es & Verwers 1980, p. 99 ff).

19 The Carolingian cooking-pots and the Badorf vessels (W III and W II in the Dorestad typology are dated by Van Es and Verwers after 725 (Van Es & Verwers 1980, 81 ff.). As their first appearance at Wijnaldum is clearly after the occurance of the sceattas, we assume a dating for the introduction of the Wijnaldum material after 750 AD, but clearly before 775 when the first shellgrit-tempered *Kugeltöpfe* are introduced.

20 Stratigraphical analysis at Ribe also suggests a date for the introduction of shell-grit tempered pottery around 770 (Claus Faveile, pers. comm.).

21 Especially characteristic of the beginning of this period are the very standardized H IB rims of the shell-tempered, wide mouthed *Kugeltöpfe* (Van Es & Verwers 1980, p. 156).

22 During the analysis we accepted the final date for Carolingian imports from the Middle Rhine region suggested by Van Es and Verwers (1980, pp 86 and 152) of 850. and the appearance of Pingsdorf at the beginning of the 10th century. This last date is recently discussed by Van Heeringen & Verhaeghe (1995, pp 145-148) and Kluge-Pinsker (1988, p. 317) who suggest that Badorf pottery was still produced after 850 and co-existed with Pingsdorf pottery. See also Dijkstra (1998, pp 27-43).

23 Juke Dijkstra (Archaeological Service Centre, Bunschoten, pers. comm), Dijkstra (1998, pp 28-35 and pp 40-43).

24 See Vos (this volume).

25 Taayke (1996b, p. 200).

26 Bazelmans et al. (1998, p. 18 ff).

27 Van Giffen (1936, Figs 2 and 5). Parcelling by using ditches and fences is also observed in other settlements of the Roman Period, e.g. Peeloo in the Province of Drenthe (Kooi et al. 1987, p. 137, Waterbolk 1992, p. 83).

28 In this workshop eight hearths were excavated. Some of these hearths were constructed on a layer of sherds, which were covered with a layer of loam. Bantelmann (1955, pp 66-67) remarks that hearths constructed on a layer of sherds are common in Lower Saxony. He sees a development towards hearths contructed on pebbles. At Wijnaldum-Tjitsma hearths constructed on sherds are represented from Period II until Period IV. During Period V (650-750) a hearth (H 12) is constructed on a layer with pebbles, covered with loam. After Period V hearths are constructed with loam only.

29 Taayke (1996b, p. 194).

30 See for the dating of Roman imports the contributions of Erdrich, Galestin and Volkers (this volume).

31 Taayke (1996b, pp 179-181).

32 Boersma (1988, p. 77), Waterbolk (1979, p. 17). Knol (1993, pp 19-24) gives an overview of datings of the Dunquerque II transgression. It shows that dating of the transgression and the impact of the sea on the landscape differed very much from one region to the other.

33 See Vos (this volume).

34 Van Es (1970, pp 199-20).
35 Rienks & Walter (1954, pp 20-21).
36 Taayke & Knol (1992, p. 87).
37 Waterbolk (1961, pp 224-225).
38 De Langen & Waterbolk (1989, 92 ff., Figs 12 and 13), Waterbolk (1991, 64 ff., Figs 11 and 12).
39 De Langen & Waterbolk (1989, pp 88/89 and 104), Waterbolk (1991, 70/71). At Ezinge Period 7 was the uppermost documented level which was documented by Van Giffen. Preservation circumstances at this level were much more difficult than at the lower levels, where the well preserved house plans were excavated.
40 Taayke (next volume).
41 At Leens sod buildings with a similar interior width were excavated (Van Giffen 1940, p. 38). Also at Foudgum the width from sod wall to sod wall measured about 4 m (De Langen 1992, p. 178). A width of the interior of main buildings of 5 m or more, however, is more common.
42 Schmid (1994, p. 246). Similar small sod buildings were also excavated by Bantelmann at Elisenhof (Bantelmann 1975, p. 126), by Brandt at Niens (Butjadingen, Lower Saxony; Brandt 1991, 105 ff.) and by Van Giffen at Leens (Van Giffen 1940, p. 26). Brandt (1991, p. 105) assumes also for the three-aisled house 2 at Niens, that an important part of the weight of the roof construction was resting on the sod walls regarding the width of the roof-supporting posts 20 × 10 cm. Van Giffen supposed for Leens that the roof was constructed on the more than 1-m-thick sod walls. A similar small kind of sod buildings (Type Ia/b) were excavated in the ringfort of Oost-Souburg (Van Heeringen et al. 1995, pp 132-133).
43 Voskuil (1979, p. 16). Trier (1969, p. 73) also holds the opinion that sod walls will have been too unstable to support a roof construction. Trier believes that the stability of sod walls rapidly decreases when the roots of the plants start to rot. This also explains why sod-structures are restricted to the zone with a moderate climate and a precipitation spread over the whole year. Uilkema (In: Van Olst 1991, pp 45, 366-367 and 368-369) observed before World War II in Ruinen (Prov. of Drenthe) and in Houtigehage (Prov. of Friesland) small sod buildings without a roof-supporting framework. Another sod structure at Houtigegehage showed walls being pushed outward because of the pressure of the roof construction. Uilkema (In: Van Olst 1991, p. 367) observed in one sod building that the roof-construction formed a very stiff entity. He also assumes that the outward pressure was partly neutralized by increasing the thickness of the sod walls. Also Therkorn (1987, pp 206-207) contradicts Voskuil's assumption on the basis of ethnographical evidence from the 19th-century United States.
44 Besteman & Guiran (1987, p. 320).
45 The carrying capacity of sod walls will increase when sod walls are made of sods of one piece. At Wijnaldum-Tjitsma sod walls are made almost exclusively of single sods. At other terp excavations this is not always the case. At Leens for example sod walls one sod thick and sod walls of more sods beside each other are represented (Van Giffen 1940, Figs 13 and 16).
46 In: Van Olst (1991, p. 45).
47 Kjær (1928, p. 16).
48 Bertelsen & Lamb (1993, 544 ff).
49 For references about excavated sod structures in Thy and the other regions in West-Jutland see Bech (1985), Brønsted (1960, 129 ff.) and Bantelmann (1955), the Ribe area: at Herredsbjerget a sunken hut with sod walls very similar to the Tjitsma sunken hut was excavated with a radiocarbon date 400-580 ± 110 and finds from the 4th/5th century: U. Näsman (pers. comm.), Archsum-Melenknop: Kossack et al. (1975, 1980), Tofting: Bantelmann (1955).
50 Bech (pers. comm). Voskuil also sees sod wall-constructions as a solution for a general lack of construction wood (1979, p. 16).
51 At Assendelft sod buildings from the 1st century AD were excavated (Therkorn 1987). These sod buildings, however, differ considerably from the Tjitsma-buildings. On the island of Texel (Prov. of North-Holland) sod buildings from the Roman Period were excavated which show, according to the preliminary publication, striking similarities in size (Woltering 1975, pp 26-27). One house measures 4 m × ca 20 m and a second one less than 3.75 to ca 15 m. At Site VIII two more sod buildings were excated, which are possibly three-aisled, measuring 6.0 × 17.5 m and 5.5 × 10 m. At Ezinge three-aisled sod buildings were excavated, which belong to Period 6 (de Langen & Waterbolk 1989, Fig 12).
52 See Bos & Gerrets (this volume).
53 See also Haarnagel (1979, pp 170-171 and Tables 154 and 155/1).
54 See also note 26. These measures are not exceptional compared with other sod buildings excavated elsewhere. Sod walls of main buildings are usually between 1.0 m and 1.2 m thick. The thickness of the sodwalls of secondary buildings varies between 0.45 m and 0.75 m and occasionally up to 1 m.
55 Such a continuity of buildings is not uncommon on terps. Van Giffen (1940, 26 ff.), see also Knol (1993, pp 124-128 and Figs 26-27) observed at Leens-Tuinsterwierde, that a house was rebuilt five times on the same place. The same is true of many other terps e.g. Ezinge (Van Giffen 1940, p. 33) and Elisenhof (Bantelmann 1975, pp 65-67).
56 Also at Foudgum similar rectangular ditches are observed (De Langen 1992, 173 ff.).
57 Leens: Van Giffen (1940), Den Helder: Van Es (1973), Heveskesklooster: Boersma (1988), Foudgum: De Langen (1992, pp 173-186).
58 Posts standing at the inside of the sodwall are also observed at Terp Het Torp by Van Es (1973, p. 343). Van Es concludes that these posts were the main constructive element of these buildings, because they were supporting the roof. It is not clear yet whether this is a more common development, more or less similar with the high sandy soils where Waterbolk (e.g. 1992, pp 68-73) made clear that the main constructive development is from three-aisled buidings with rows of roof-supporting poles in the middle of the building towards one-aisled boatshaped buildings where roof-supporting posts became part of the wall (see also Huijts 1992, 117 ff.). The boatshaped longhouse, however, is completely absent in the terp region.

59 For the erosion: see also Bos & Gerrets (this volume).

60 This observation underlines Stuijts' conclusion (next volume), that wood was scarce and used secondarily on a large scale.

61 During an open day for the public a farmer from the neighbourhood said, that he knew from his childhood that water from the well used for human consumption became too brackish. Another farmer from the village of Ezinge said that they had a sod-lined well, which was too brackish for human consumption and was therefore used for the cattle.

62 At Feddersen Wierde most wells, including the sod-lined wells, were rectangular in contrast with Wijnaldum-Tjitsma. Sometimes the walls were covered with planks (Haarnagel 1979, 168 ff. and Figs 149-153).

63 It is now hard to imagine how this type of well was constructed. A similar kind of well is described for the German Halligen by Bantelmann (Bantelmann 1975, 20 ff and Figs 15 and 18). The meaning of the bottle-neck shape was that these wells could easily be closed at very high storm floods.

64 Metalworking at Wijnaldum-Tjitsma will be extensively discussed by Tulp in the next volume.

65 See Sablerolles (this volume).

The vegetation on the terp

W. GROENMAN-VAN WAATERINGE

1 INTRODUCTION

Samples for pollen analysis were taken over the entire terp area to obtain information on the natural vegetation and the changes brought along by agricultural activities, both crop cultivation and animal husbandry. Another question was whether pollen analysis might shed some light on continuity in occupation or not.

Pollen analysis of a terp is rather problematical for several reasons. There is in the first place the preservation of the pollen, rather bad in clay samples with a high pH and a poor humic content, a bit better in ditches, dug by men, with a filling of humic layers. Secondly there is the question of the origin of the pollen in the samples. Clay, specifically marine clay, has a high amount of secondary pollen. Pollen caught in the marine sediments is either windblown, or transported by rivers towards the sea. Thus a clayey subsoil contains a lot of pollen with an origin far beyond the near surroundings. To gain insight in the vegetation on such soils means to try to discern between this allochthonous pollen and the pollen produced by the vegetation growing on the clay. In the case of clay sods used for the building of the terp the same holds true. Samples from man-made ditches provide data highly influenced by man's activities. Dung, an ideal medium for pollen preservation, is by its nature, also a product of man's agricultural practices.

From a pollen preservation viewpoint wells are ideal for sampling, the problem here, however, being the interpretation of the data because of formation processes in the wells.

2 MATERIAL

Samples were taken from sections in different parts of the terp: two sections mainly through natural sedimentation layers (Wijnaldum 1-2) and from ditches, to be dated from the Roman to the Carolingian period (Wijnaldum 3-8). In the case of ditch sampling one or two samples from beneath the ditch were also taken if possible. This could be the natural subsoil or terp material, depending on the depth of the ditch. Samples from wells taken for the analysis of botanical macroremains[1] will also be used for pollen analysis, but this work has not yet begun. By the nature of the samples the question of continuity in habitation is difficult to answer as all the samples were either pre-terp period or they were taken in immediate connection with habitation phenomena, ditches etc., thus only if one of the ditches showed a natural silting following layers with culture indicators could one argue for discontinuity.

Samples were prepared in the usual way by hydrofluoric acid treatment followed by acetolysis. When necessary the residues were sieved through a sieve with mesh size 10/14 μ. Included in the pollen sum is pollen of trees and shrubs, upland herbs, culture indicator plants and halophytes.

Data are presented as separate spectra, as the formation processes in a ditch do not necessarily lead to a natural succession of layers. For example, the deposition processes may have been disrupted by clearing-out, or deliberately filling-in with all kinds of material amongst which clay from the subsoil, or material from the old surface may have slumped-in the ditch through trampling by man and animals along the ditch banks.

3 DESCRIPTION OF THE PROFILES

For the location of most of the samples in the main sections of the excavation[2].

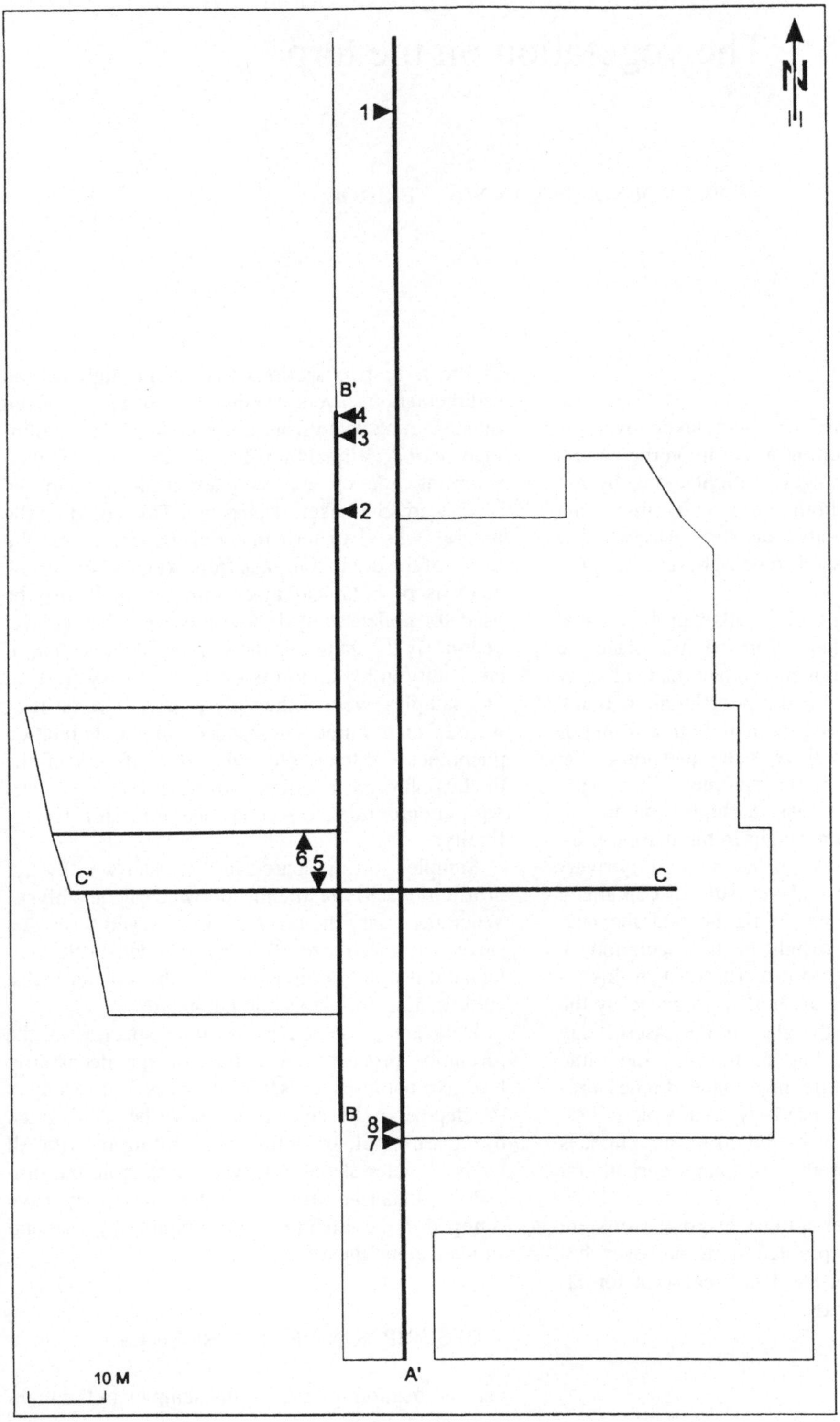

Figure 1. Location of pollen diagrams 1-8.

Profile 1 (Figs 1 and 2)
Trench IV, east section
Length of section 18 cm [0.46 to 0.64 m below NAP (Nieuw Amsterdams Peil, Dutch datum level)]
Samples 1-2: salt marsh
Sample 3: transition tidal flat/salt marsh
Sample 4: tidal flat

Profile 2 (Figs 1 and 3)
Trench I, west section, No. 4. Samples taken by F. Vos[3]
Length of section 1.35 m (0.32 to –1.03 m NAP)
Samples 1-3 (0.30, 0.0 and –0.31 m NAP): terp
Sample 4 (–0.62 m NAP): salt marsh
Sample 5 (–0.97 m NAP): tidal flat
Samples taken in between –0.18, –0.44 and –0.72 m NAP did not contain enough pollen

Profile 3 (Figs 1 and 4)
Trench I, west section, ditch
Length of section 1.08 m (0.06 to –1.02 m NAP)
Sample 1: 425-750 [4][4]
Sample 2: 425-550 [4]
Sample 3: 250-350 [4]
Samples 4-5: 250-350 [1]

Profile 4 (Figs 1 and 5)
Trench I, west section, ditch
Length of section 0.76 m (0.36 to –0.30 m NAP)
Samples 1-12: 425-640

Profile 5 (Figs 1 and 6)
Trench XXXVII, south section, from bottom to top: ditch, agricultural layers (Samples 10-7) and occupation layers
Length of section 1.51 m (1.37 to –0.24 m NAP)
Sample 1: 550-650 [1]
Samples 2-3: 550-650 [4]
Samples 4-6: 550-650 [1]
Samples 7-10: 450-525 [3]
Samples 11-14: 250-350 [1]
Sample 15: 250-350 [4]
Samples 16-19: 175-250 [4]

Profile 6 (Figs 1 and 7)
Trench VIII, north section, ditch
Length of section 0.85 m (1.45 to –0.60 m NAP)
Samples 1-2: topsoil
Samples 3-6: 770-850 [2]
Sample 7: 600-650 [1]
Samples 8-13: 575-625 [4]
Below Sample 13 the pollen content was nil

Profile 7 (Figs 1 and 8)
Trench V, east section, ditch
Length of section 0.50 m (0.44 to 0.94 m below NAP)
Samples 1-5: 850-900 [1]
Samples 6-7: salt marsh
Sample 8: tidal flat

Profile 8 (Figs 1 and 9)
Trench V, east section, ditch
Length of section 0.50 m (0.04 to 0.54 m below NAP)
Samples 1-3: 850-900
Sample 4: salt marsh

4 RESULTS

4.1 *The tidal flat and salt marsh samples* (Figs 2, 3, 8 and 9)

Tidal flat samples are represented in Profile 1 (Samples 4, partly 3), Profile 2 (Sample 5) and Profile 7 (Sample 8). These samples are characterised by rather high values for plants belonging to a vegetation type that had certainly not been present locally: tree pollen (between 40 and 60%), Ericaceae (heather family) and bog plants, for example *Sphagnum* (bog moss) and *Lycopodium* (club moss). Analysis of macro-remains of the terp Heveskesklooster revealed that only *Quercus* (oak), *Salix viminalis* (common osier), *Betula pendula* (silver birch), *Betula pubescens* (brown birch) and *Sambucus nigra* (elder) may have grown on the terp.[5] Extremely scarce are halophilous plants and culture indicators. It is obvious that not much vegetation occurred locally.

There are a few samples of the salt marsh which do seem not to have been under direct human influence. These are Samples 1-2 in Profile 1, Sample 4 in Profile 2, Samples 6-7 in Profile 7 and Sample 4 in Profile 8. There is not much difference with the samples from the tidal flat. They are again characterised by allochthonous pollen. As it is not feasible that the tidal marsh did not have a lush vegetation, the near absence of pollen from the local vegetation has to be explained. Two reasons can be brought forward. Firstly, the ample pollen production and thus presence of the allochthonous pollen suppresses the local vegetation in the pollen spectrum, secondly the local pollen is rather badly preserved.

Wijnaldum, profile 1

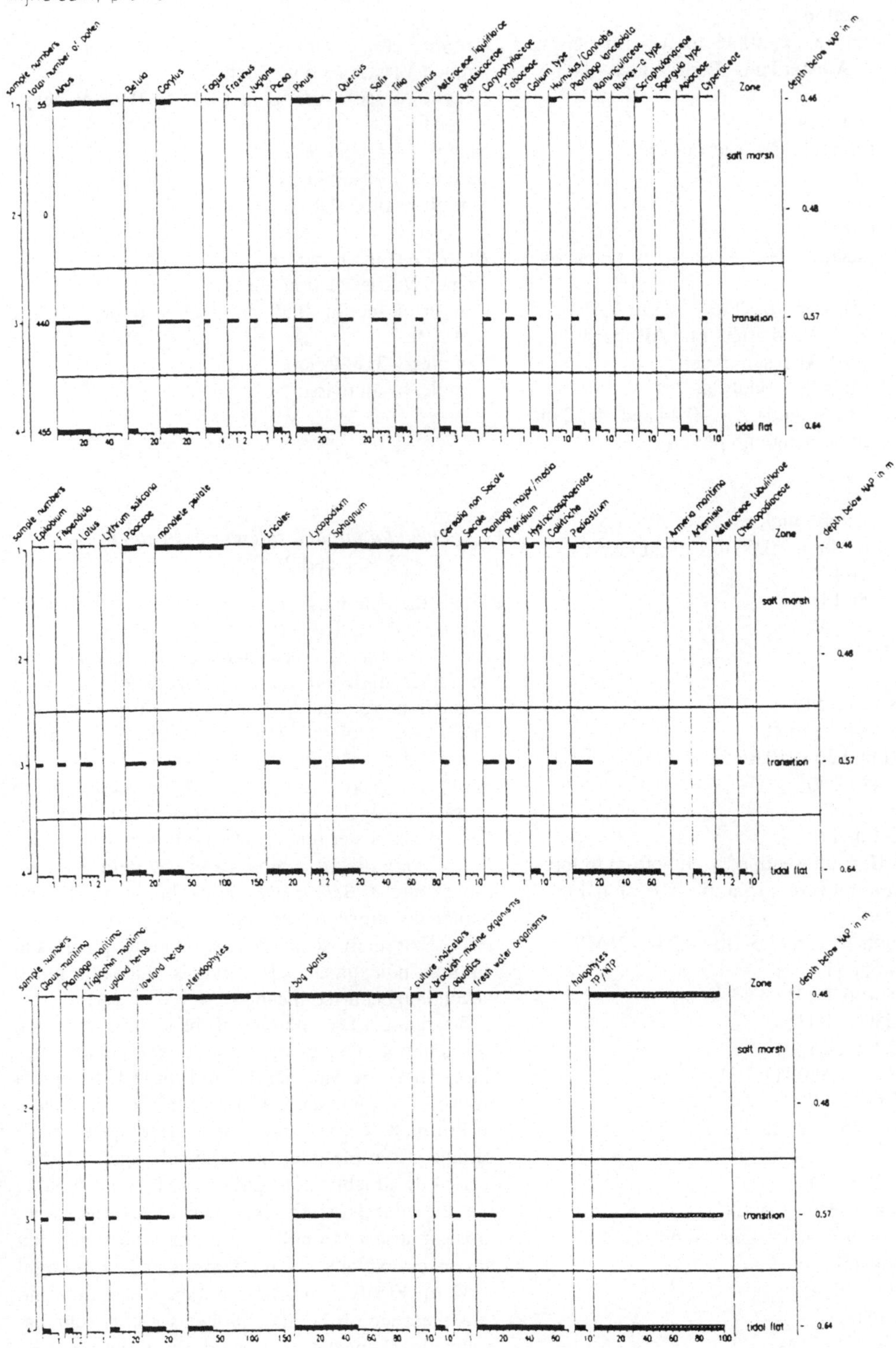

Figure 2. Pollen diagram 1.

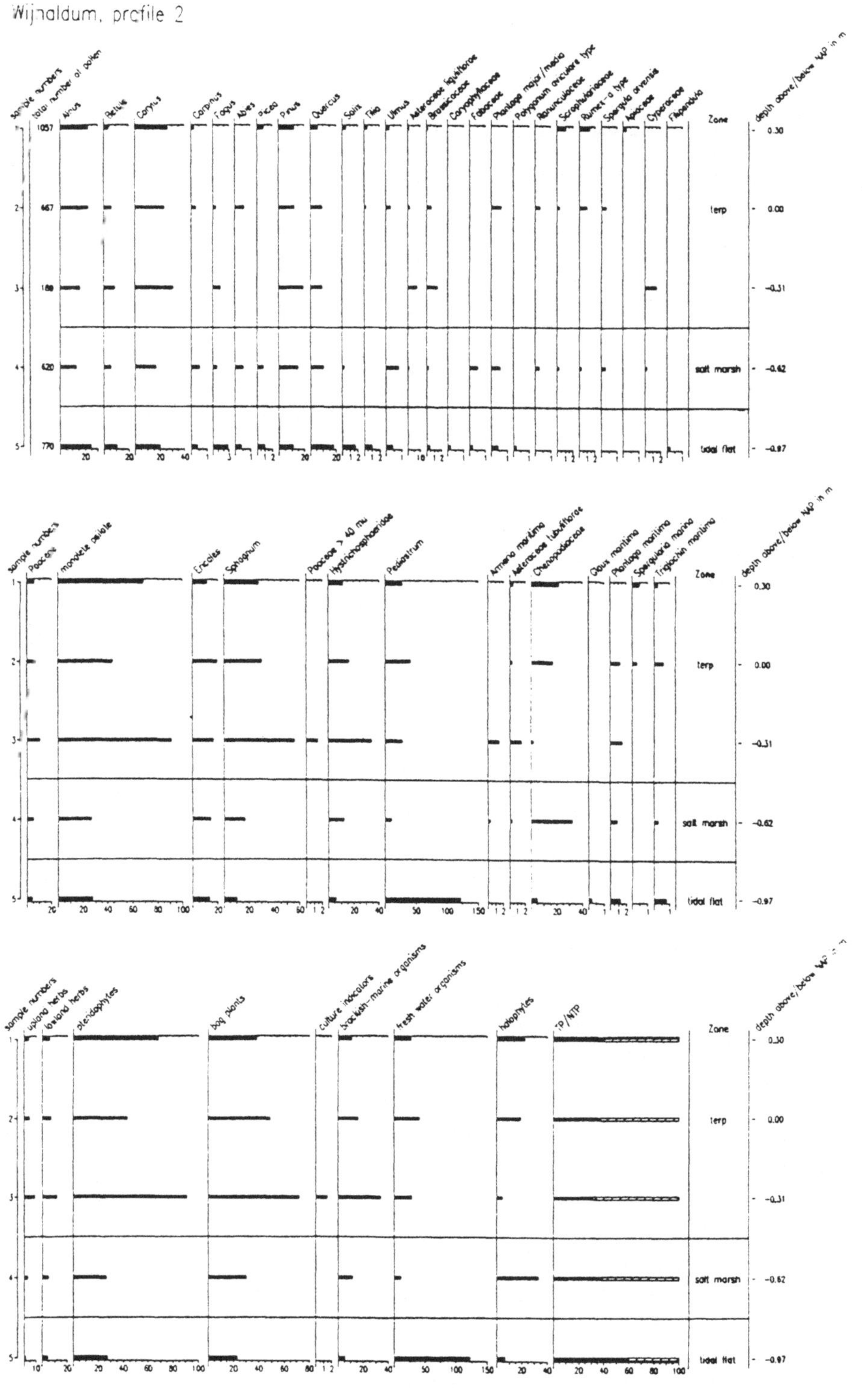

Figure 3. Pollen diagram 2.

Wijnaldum, profile 3

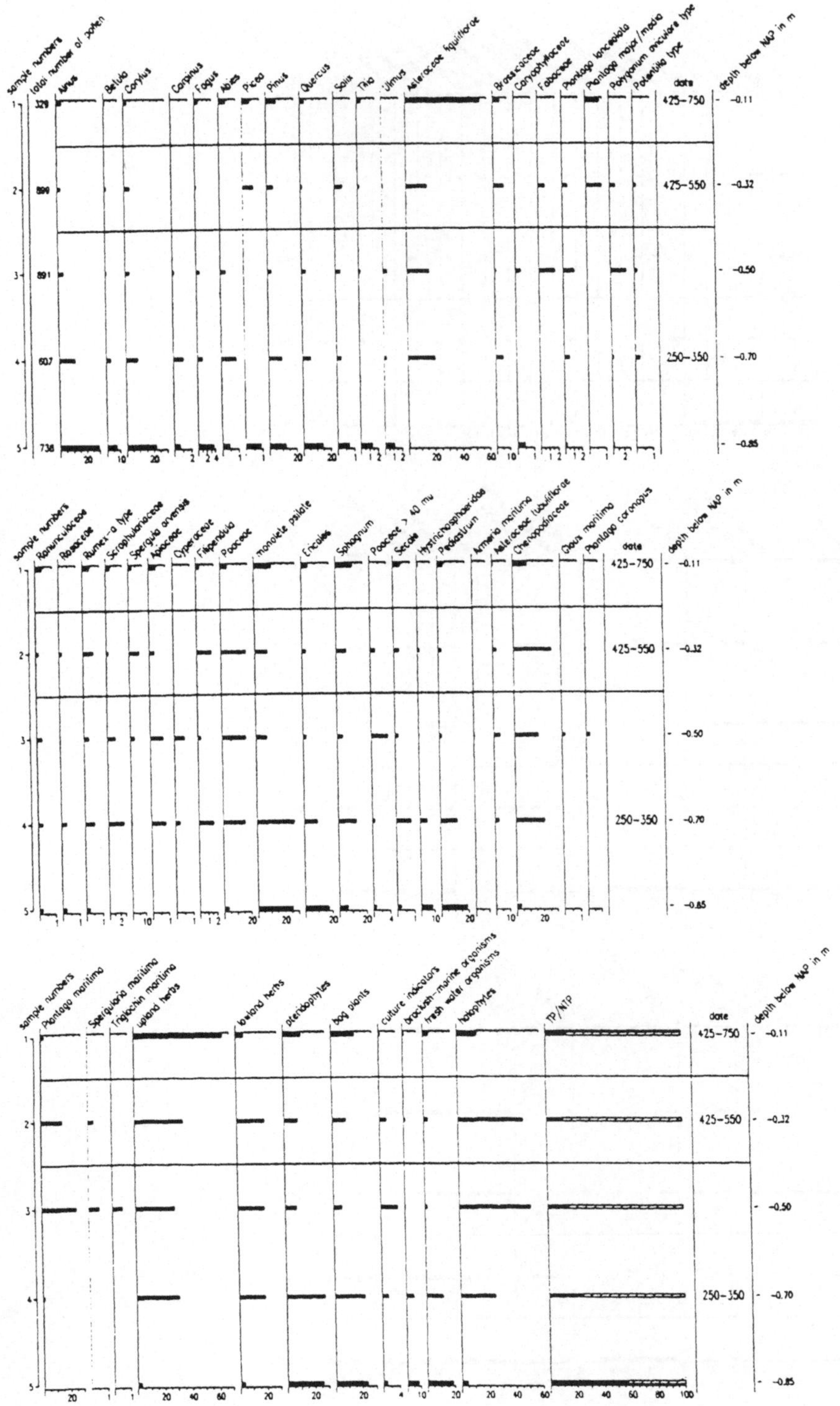

Figure 4. Pollen diagram 3.

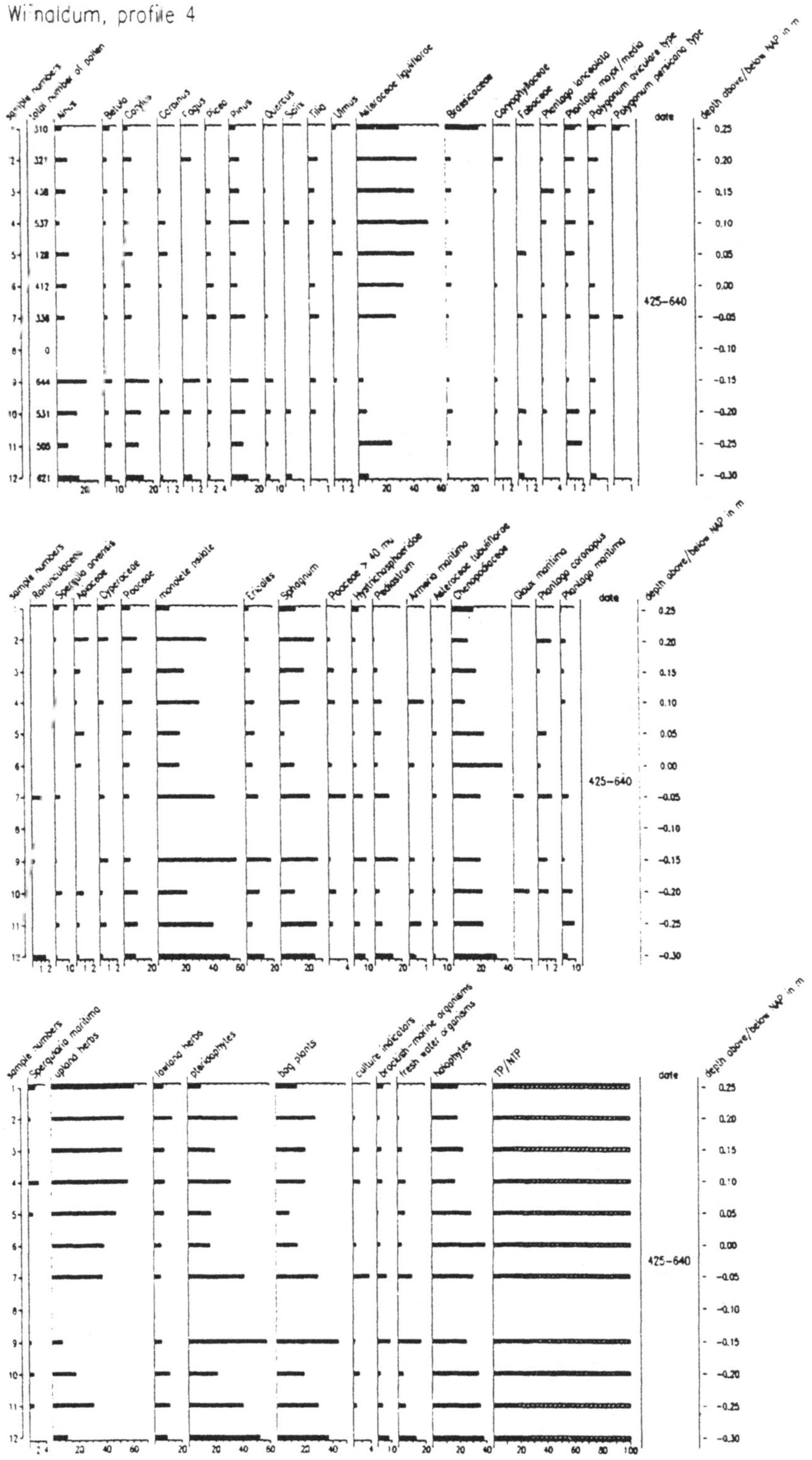

Figure 5. Pollen diagram 4.

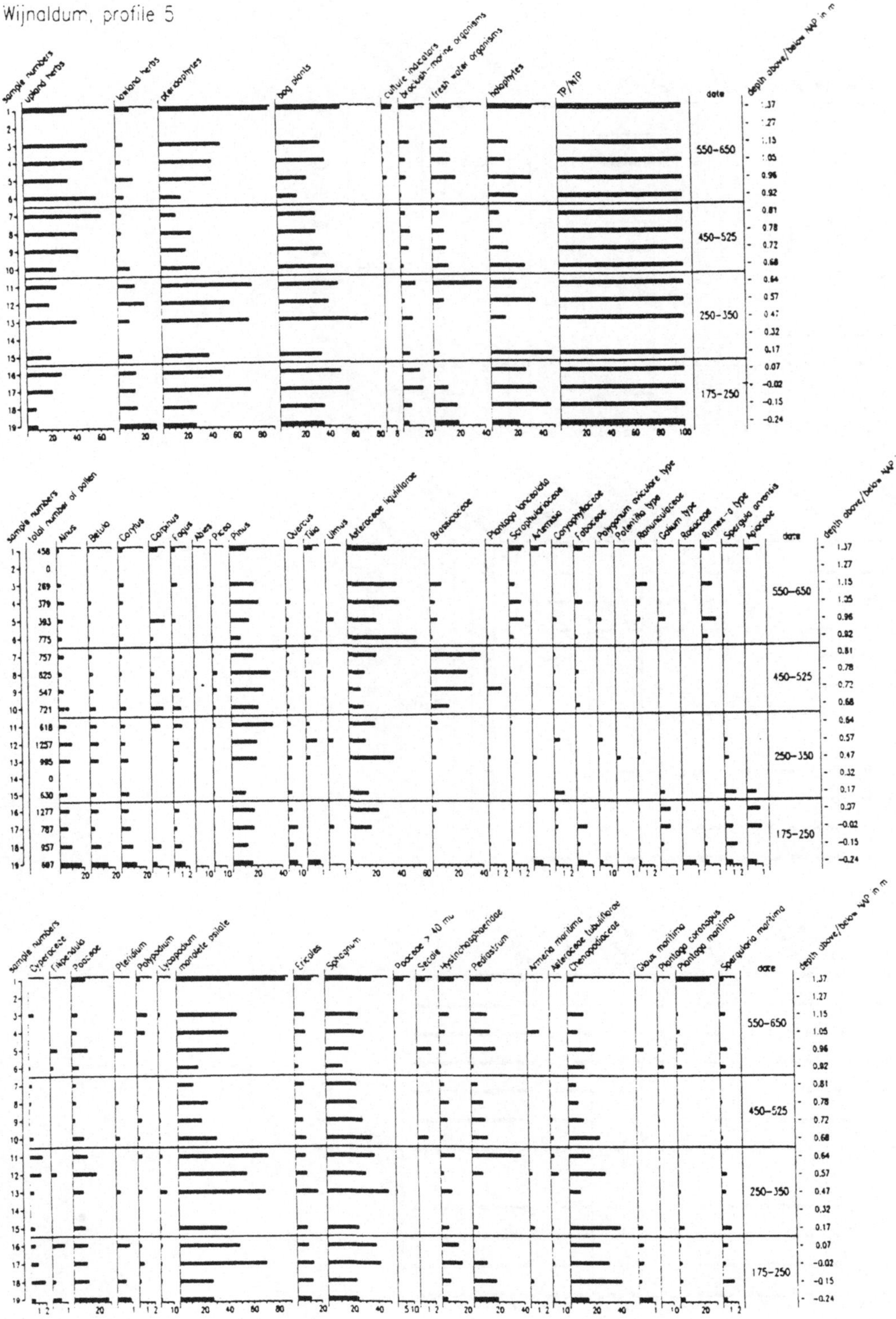

Figure 6. Pollen diagram 5.

Wijnaldum, profile 6

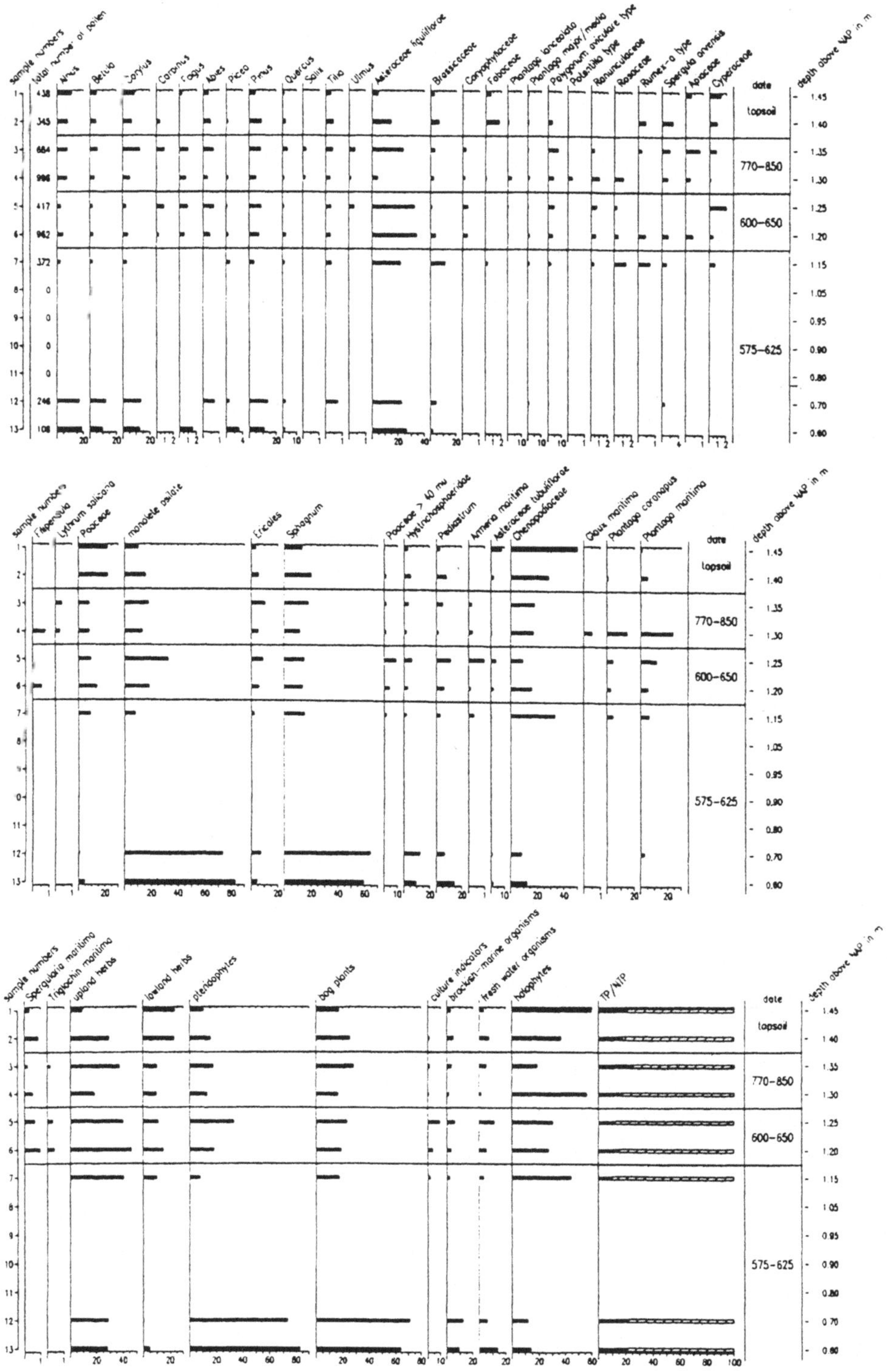

Figure 7. Pollen diagram 6.

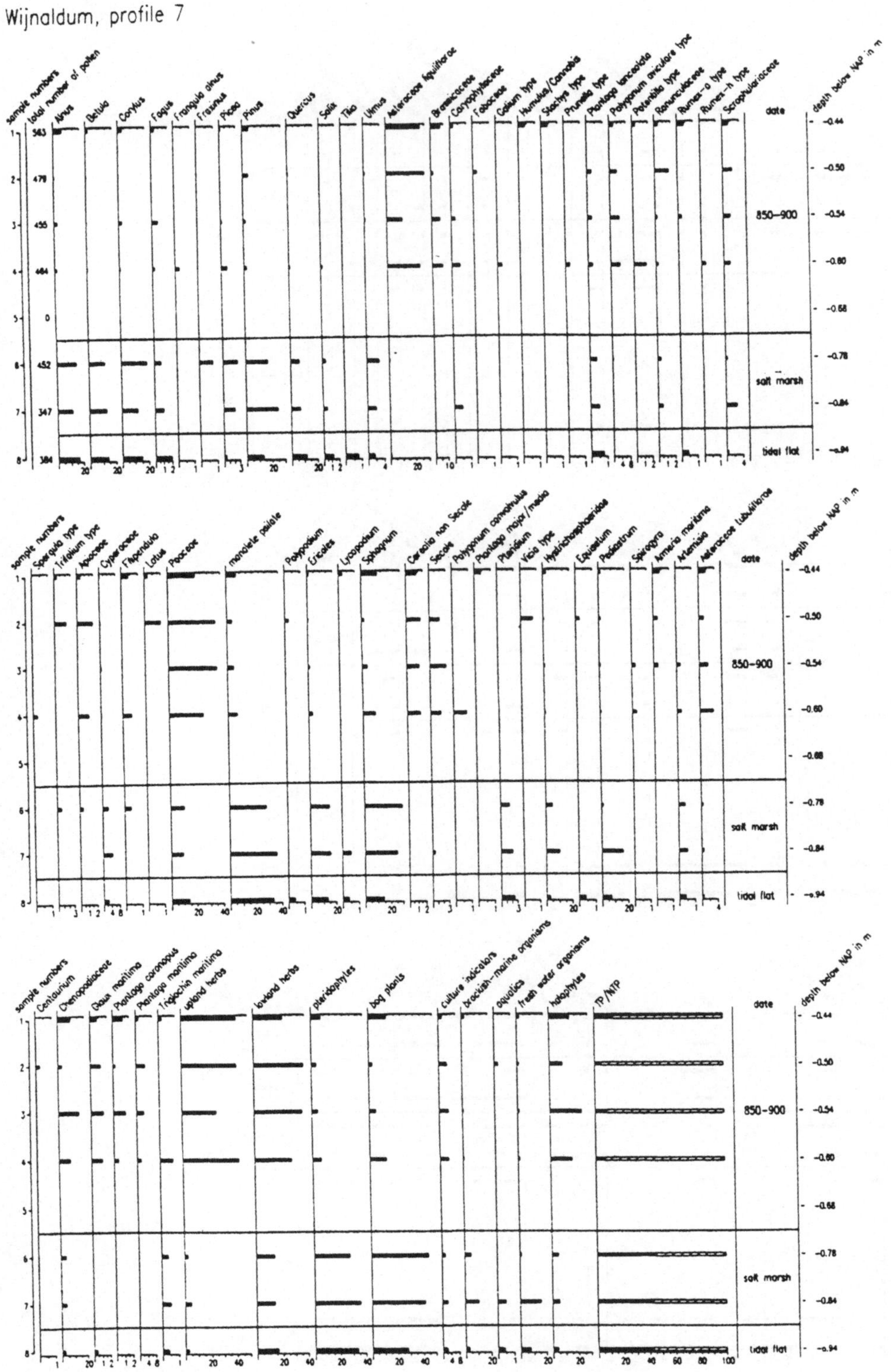

Figure 8. Pollen diagram 7.

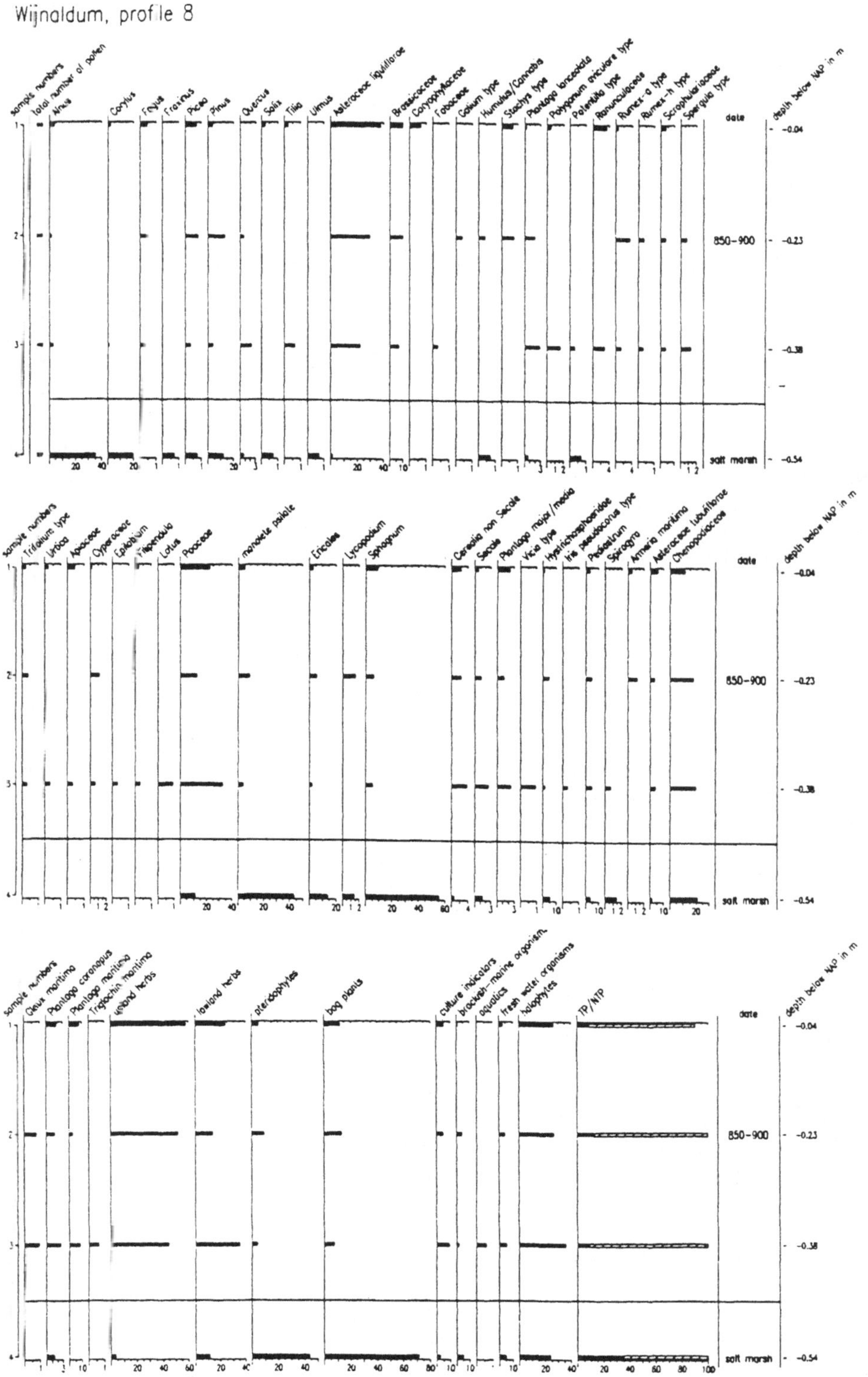

Figure 9. Pollen diagram 8.

Both agents must have played a role, the first is rather obvious as under the allochthonous pollen high pollen producers are represented, the second becomes clear when we look at samples from ditches, wells etc, with a higher humic content and better conditions for pollen preservation because of protection against oxydation.

Anyway, a few changes are visible. The percentage of tree pollen is slightly less, *Pediastrum* (a freshwater green alga) is much lower, the halophytes seem to increase a bit and, most important, the amount of pollen taxa increases.

4.2 *The terp samples*

Period AD 175-350 (Figs 4 and 6)
Samples belonging to this period can be found in Profile 5, Samples 16-19, dating 175-250 and Samples 11-15, dating 250-350 and in Profile 3, Samples 3-5, dating 250-350.

What are the main characteristics of this period and how do they do compare with the tidal flat and salt marsh samples?

There is in the first place a drastic decline in the percentage of allochthonous pollen; secondly there is an increase in the amount of pollen taxa and thirdly, specifically in the first part of the time slice (175-250) there is an increase in halophytes, such as Chenopodiaceae, *Plantago maritima* (sea plantain), *Glaux maritima* (saltwort) and *Spergularia*, most probably *marina* (sea spurry). Culture indicators are low. *Secale* (rye) first occurs towards the end of the 3rd century. However, the percentages are so low that local cultivation on any scale of the wind-pollinated rye is out of the question and so is deliberate import from the diluvial soils.

The high tree pollen percentage in the lowermost sample in both profiles points to a derivation of the samples from the natural salt marsh. It is probably subsoil material slumped-in or washed-in from the ditch sides shortly after the ditch had been dug. The halophytes seem to decline somewhat in the second time slice. The vegetation on the terp gradually becomes less halophilous.

Period 425-770 (Figs 4-7)
Samples belonging to this period can be found in Profile 3, Samples 1-2, dated between 425-750, Profile 4, dated between 425-640, Profile 5, Samples 7-10 dated 450-525 and Samples 1-6 dated between 550-650, Profile 6, Samples 8-13 dated 575-625 and Sample 7 dated 600-650.

What are the main characteristics of this period and how do they compare with the previous time slice?

The halophytes do not show a very distinct pattern. Their decrease does not seem to continue. This may have three reasons, firstly the influence of the late-Roman to early-medieval Dunkerque II transgression phase, secondly the origin of the material from which samples were taken (probably subsoil material, see above) and thirdly the family of the Chenopodiaceae, forming the bulk of the halophytes, consists of halophilous plants but also contains quite a few arable weeds. Thus with an increase of agriculture, in one way or another, the Chenopodiaceae will increase too. In Profile 5 there is an increase in the Brassicaceae (family consisting of halophilous plants and arable weeds) in the period 450-525, declining between 550-650. For the Asteraceae liguliflorae it is the other way round. The increase of the latter is also to be seen in Profile 3 towards the end. Profile 6, covering the period 575-770 shows a good comparability with the second time slice of Profile 5 and so does Profile 4, with the exception of the percentages for the Brassicaceae, which are fairly low in this Profile. Salt marsh plants such as *Plantago maritima*, *Plantago coronopus*, (buck's-horn plantain), *Glaux maritima*, *Armeria/Limonium* (thrift/sea lavender) are well represented and provide evidence for the presence of a salt marsh vegetation prior to and during the human occupation. Culture indicators are increasing somewhat, although the percentages are still very low. Evidence for grazing is low, but must have been excellent on the salt marsh. *Plantago lanceolata* (ribwort plantain), the classical indicator in pollen diagrams for grazing, is extremely rare. Cappers found the same for the terp Heveskesklooster.[5] I do not agree, however, with his supposition that this species is underrepresented in the palynological record in general. There is evidence enough for high values for this species in palynological analyses from underneath burial mounds.[6] Blom was able to show that in compacted soils *Plantago* germinates poorly and with difficulty.[7] The clayey subsoils of terps in general must have been extremely vulnerable to compaction, so it is no wonder that Plantago will not be found profusely in terp samples. The amount of cereal pollen is also extremely low. This can be understood for the self-pollinating genera of *Triticum* (wheat), *Hordeum* (barley) and *Avena* (oats), but not for the wind-pollinating *Secale*. The extremely low percentages for this species is evidence that this cereal was not cultivated locally. For the other cereals it cannot be said with certainty if local cultivation took place or not. The answer to this question has to come from analysis of macroremains.

Period 770-950 (Figs 7-9)
Samples belonging to this period are to be found in Profile 6, Samples 3-6, dated 770-850, Profile 7, Samples 1-5, dated 850-900 and Profile 8, Samples 1-3, also dated 850-900.

The questions are again, what are the characteristics of this time slice and how do the spectra compare with the previous period?

Typical salt marsh plants, such as *Plantago maritima* and *Plantago coronopus* have become rare. In contrast there is an increase in the Chenopodiaceae. We can safely assume that this concerns the weed plant and ruderal section of this family. There is also an increase in the culture indicators (higher values than ever in Profiles 7 and 8 between 850 and 900),[8] although in Profile 6 there seems to be a decline between 770-850 in comparison to the previous period. Cereal pollen remains low, *Secale* was certainly not cultivated on the terp. Some pollen of fresh water plants in the ditch of Profile 7 may point to increasing desalination of the terp milieu.

5 COMPARISON WITH OTHER SITES

As pollen analysis of terps is not very rewarding, thanks to the preservation condition and the formation processes of the pollen content of the samples, comparison with other sites is restricted to only a few.

Van Zeist analysed samples from the Frisian coastal area, but these concerned peat layers and not terp samples.[9] From the Groningen marine clay district vegetation horizons were analysed.[10] Cappers analysed, besides the basal peat layer, Dunkerque layers and two intercalated vegetation horizons beneath the terp Heveskesklooster, the lower part of the terp belonging to the second occupation phase, ca 800-1300.[11]

Remarkable is the percentage of tree pollen, around 20%, in the second vegetation horizon and the immediately following lower part of the terp, the decrease in Chenopodiaceae and *Dryopteris* type (our monolete psilate spores), the increase in Gramineae (Poaceae in the Wijnaldum diagrams), Compositae liguliflorae and culture indicators. The background vegetation of the salt marsh is comparable with the data from Wijnaldum.

Körber-Grohne analysed samples from the terp Feddersen Wierde, specifically samples taken from layers with plant remains.[12] One sample was taken from sedimentation layers in a creek below the terp and one sample from a ditch inside the habitation area. Only the last two samples can be compared with the sampling from Wijnaldum: The sample from the creek with samples in Profiles 1 and 2, the ditch sample with the similar samples from Wijnaldum.

The higher tree pollen percentages in the creek samples compared to that in the ditch is comparable with the situation in Wijnaldum between the tidal flat, c.q. salt marsh samples and the ditch samples. The halophytes are, however, much better represented in the creek samples from Feddersen Wierde than in the salt marsh samples from Wijnaldum. This is probably caused by differences in the preservation conditions. Poaceae (Gramineae), Chenopodiaceae and culture indicators are better represented in the ditch sample. Cereal pollen or pollen from other cultivated crops, for which there is ample evidence in the macroremains, were not found in the ditch sample. *Plantago* pollen is extremely low, although animal husbandry must have been important there as well as in Wijnaldum.

6 SYNOPSIS

Human habitation took place once the flora had the character of a well-developed salt marsh vegetation. Evidence for animal husbandry is difficult to conclude from pollen analysis, as the classical indicators could either belong to the natural vegetation anyway, or were missing because of less favourable soil conditions.

Arable cultivation, certainly of barley, must have been possible. Rye cultivation never took place. The somewhat higher pollen of Secale in the Carolingian samples may point to import of rye in the form of grains.

Towards the Carolingian period the environmental conditions definitely seem to have changed towards a more freshwater environment.

REFERENCES

Blom, C.W.P.M. 1974. De invloed van bodemverdichting en betreding op het voorkomen van enkele plantesoorten, *Contactblad voor oecologen* 10: 17-21.

Bohncke, S.J.P. 1984. Palaeobotanical study on vegetation horizons in the young holocene coastal plain of Groningen (northern Netherlands), *Dissertationes Botanicae* 81: 221-256.

CAPPERS, R.T.J. 1994. An ecological characterization of plant macro-remains of Heveskesklooster (the Netherlands), a methodological approach, Groningen (Ph.D. thesis Groningen University).

CASPARIE, W.A. & W. GROENMAN-VAN WAATERINGE 1980. Palynological analysis of Dutch barrows, *Palaeohistoria* 22: 7-65.

KÖRBER-GROHNE, U. 1967. *Geobotanische Untersuchungen auf der Feddersen Wierde*, Wiesbaden (Feddersen Wierde, die Ergebnisse der Ausgrabung der vorgeschichtlichen Wurt Feddersen Wierde bei Bremerhaven in den Jahren 1955 bis 1963).

ZEIST, W. VAN 1955. *Pollen analytical investigations in the northern Netherlands, with special reference to archaeology*, Amsterdam.

NOTES

1. See Pals (this volume).
2. See Gerrets & De Koning (this volume, Fig.23).
3. For a detailed description of this section see Vos (this volume).
4. [4] indication for the reliability of the dating: [1] finds and context/stratigraphy consistent = very reliable; [2] finds and context/stratigraphy non-consistent = unreliable; [3] finds and context/stratigraphy partly consistent = under certain conditions reliable; [4] no finds, date based on context/stratigraphy = reliable (see Gerrets & De Koning, this volume).
5. Cappers (1994, p. 150).
6. Cappers (1994, p. 150).
7. See Casparie & Groenman-van Waateringe (1980).
8. See Blom (1974), Casparie & Groenman-van Waateringe (1980, p. 59).
9. This could, however, also be caused by better preservation conditions in the two ditches of Profiles 7 and 8.
10. Van Zeist (1955, chapter VII).
11. See Bohncke (1984).
12. Cappers (1994, pp 147-151).
13. Körber-Grohne (1967, pp 86-97, Table 14).

Preliminary notes on crop plants and the natural and anthropogeneous vegetation

J.P. PALS

1 INTRODUCTION

The aim of the archaeobotanical investigations at Wijnaldum-Tjitsma is to elucidate man-plant relationships, in order to contribute to the reconstruction of human behaviour at the level of both the site and the region, and possibly to establish trends and changes in time. Specific questions relating to this aim are:

– Which crop plants were used?

– Were these crops grown locally, or where they imported from elsewhere?

– What was the distribution of crops in space and time?

– Is it possible to establish crop processing activities, and, if so, can these be related to specific contexts?

– How were the environmental conditions around the settlement especially in terms of salinity and moisture?

Although crop plants will be discussed in this paper, we will concentrate on the reconstruction of the vegetation cover on and around the terp. The other aspects will be dealt with in a future publication.

2 FIELD AND LABORATORY PROCEDURES

Samples for investigation of macroscopic plant remains were taken from as many different contexts as possible. Over two hundred samples have been taken, of which about 30% have been investigated.

Conditions for preservation of organic remains were relatively poor, or at least not as favourable as was expected, considering the fact that we are dealing with material from a terp excavation.[1]

Depending on preservation conditions, broadly two types of samples may be distinguished:

1. From occupation layers, pits, ditches, postholes, etc.. From these contexts mainly charred plant remains were extracted. The origin of this material is secondary or even tertiary. It represents 'settlement noise', consisting of remains of plants that have a chance to become carbonized, which is mainly the harvested crop and its associated weeds, but possibly also remains of dung used as fuel.

These samples had a volume of 10-12 litres, and were processed by flotation during the excavation, the floating remains being sieved over meshes of 2 mm and 0,5 mm respectively.

2. From wells and ditches, preserved below the water table. In this type of context, the bulk of the plant remains is waterlogged. They have been deposited by natural agencies directly from the environment, and are expected to represent the nearby vegegation, the regional vegetation, or possibly, refuse from the settlement. These samples had a volume of ca 1 litre, and have been wet-sieved in the laboratory, using meshes down to 0.25 mm.

In terms of vegetation cover, the charred material would, broadly speaking, reflect the environmental conditions in the fields, the waterlogged plant remains would represent the 'natural' vegetation.

The results of the analysis are presented in Table 2 (charred plant remains) and Table 3 (waterlogged plant remains).

3 AGRICULTURE IN THE SALT MARSH

Experiments have shown that it is possible to practise agriculture in the unprotected salt marsh,[2] provided that several limiting factors are taken into consideration. Thus, the growing of crops can only be practiced at places with the highest elevation, because that is where the risk of inundation by salt water is relatively small, and the salt is washed out quickly by precipitation. Furthermore, because of the risk of flooding during autumn and winter, only spring-sown crops could have been used.

A not altogether surprising conclusion from these experiments is that it is possible to cultivate the four species most frequently found in the botanical material from the terps, viz. barley (*Hordeum vulgare*), gold-of-pleasure (*Camelina sativa*), celtic bean (*Vicia faba var. minor*) and flax (*Linum usitatissimum*), more or less successfully in the unprotected salt marsh. Oats (*Avena sativa*) could also be grown with a reasonable result, but was less successful than barley. Turnip (*Brassica rapa*) proved to be a satisfactory crop, whereas cultivation of broomcorn millet (*Panicum miliaceum*), spelt wheat (*Triticum spelta*) and bread wheat (*T. aestivum*) failed completely. The results of emmer wheat (*T. dicoccum*) were too poor to consider cultivation of this crop in the unprotected salt marsh a feasible option. Rye (*Secale cereale*) and peas (*Pisum sativum*) were not used in the experiments.

Although the results of these experiments give a reasonable idea of what was feasible and what was not, they should be interpreted carefully: the plants used were present day cultivars not adapted to brackish conditions, and apart from the fact that the terp dwellers possibly had a different (i.e. less demanding) opinion of what was 'successful' than the present day investigators, it is conceivable that their crops had developed a relatively high salt tolerance in the course of time by natural selection. During the experiments it was observed that a number of salt marsh plants occurred as weeds in the fields, although a consistent association of weeds had not developed after five years.

4 PLANTS OF ECONOMIC IMPORTANCE

A synopsis of plants of economic importance is presented in Table 1. In view of the limited number of samples analysed, no attempt has been made to make a quantification per period.

It is not possible to establish the relative importance of agriculture and stock-raising on the basis of palaeobotanical investigations alone. The finds of crop plants are compared to those from other medieval terps.[3] One of these, Leeuwarden,[4] developed into a relatively important centre (trading settlement), the others are single farmsteads or rural villages. Heveskesklooster[5] was an agrarian settlement in its early stages. During late medieval times (Period 3, from 1300 onwards) it was a monastery.

Cereals

Barley, the main crop in the Dutch terp area, was by far the most numerous in the botanical material from Wijnaldum-Tjitsma.

The fact that not only grains of barley, but also chaff remains (rachis internodes, culm nodes and occasional awn fragments) were found, is an indication for local cultivation and processing of this crop, which would have been sown in spring.

The find of grains and spikelet forks of emmer wheat is remarkable, because this cereal, as was mentioned earlier, is not especially suitable for cultivation in the salt marsh. One other record of *Triticum dicoccum* is from Iron Age Paddepoel,[6] but no comment on this find is given.

A single grain of bread wheat (*T. aestivum*) was found in a well from Period 8 (AD 800-850). The only other record of bread wheat is from Leeuwarden. A single grain of club wheat (*T. aestivum var. compactum*) was found in late-medieval Heveskesklooster. In view of the impossibility to grow wheat in the salt marsh, it is likely that the *Triticum* grains had been imported, or possibly were just admixtures in the barley crop.

Table 1. Macroscopic remains of plants of economic importance.

Period	1	2	4	5	6	7	8/9	Total	Pres.	
Barley	x	x	x	x	x	x	x	1889	c	*Hordeum vulgare*
Emmer wheat	.	x	.	x	x	x	.	9	c	*Triticum dicoccum*
Bread wheat	.	.	.	.	.	.	x	1	c*	*Triticum aestivum*
Oats	.	.	x	.	x	x	x	43	c	*Avena* sp.
Celtic bean	.	.	x	.	.	x	x	37	c	*Vicia faba* var. *minor*
Pea	.	.	x	.	.	x	.	3	c	*Pisum sativum*
Flax	.	.	.	.	.	.	x	45	w	*Linum usitatissimum*
Gold-of-pleasure	.	.	.	x	.	.	.	1	c	*Camelina sativa*
Turnip	.	.	x	x	x	x	x	103	c	*Brassica rapa*
Black mustard	.	.	x	x	.	.	.	2	c	*Brassica nigra*
Apple?	.	.	.	.	x	.	.	1	c	cf. *Malus sylvstris*

c = carbonized, w = waterlogged. *Found in waterlogged sample.

Table 2. Synopsis of carbonized plant remains. Context: dt = ditch, hea = hearth, met = refuse context of metal processing, ph = posthole, pit = large pit, ref = general refuse context, srf = occupation surface, till = layer with traces of tillage, well = well. Quantification Stem fragments: x = 10-50, and xx = >50.

Period	1 (AD 175-250)	Total	2 (AD 250-350)					Total	4a	4b	4b	4b	4/5	4a/5	Total
Sample no.	2377	= 1	1629	2346	2358	2382	2461	= 5	2278	2874	2879	3954	3923	4568	= 6
Context	srf		well	srf	srf	srf	srf		well	till	till	srf	srf	dt	
Hordeum v. vulgare	3	3	2	10	9	2	7	30	3	2	4	33	5	73	120
Hordeum, internodes	.	.	.	1?	.	.	1	1	.	1	.	8	4	8	21
Gramineae culm nodes	.	.	.	.	.	.	.	.	.	.	.	1	1	5	7
Gramineae stem frags.	.	.	.	.	.	5	x	x	.	.	.	.	4	x	x
Triticum dicoccum	.	.	.	.	1	.	.	1	.	.	.	.	.	.	.
Avena sp.	.	.	.	.	.	.	.	.	.	.	.	.	.	20	20
Cerealia indet.	.	.	.	.	.	.	.	.	1	.	.	.	.	.	1
Brassica nigra	.	.	.	.	.	.	.	.	.	.	.	.	1	.	1
Brassica rapa	.	.	.	.	.	.	.	.	.	.	.	.	.	10	10
Pisum sativum	.	.	.	.	.	.	.	.	.	.	.	2	.	.	2
Vicia faba	.	.	.	.	.	.	.	.	.	.	.	.	.	12	12
Weeds															
Atriplex sp.	.	.	.	2	1	1	3	7	.	.	.	1	7	7	15
Bromus secalinus/mollis	2	2	.	1	.	1	.	2	?1	.	.	.	.	.	?1
Galium aparine	.	.	.	.	.	.	.	.	.	.	.	.	.	7	7
Gramineae	11	11	.	.	1	.	.	1	.	.	.	.	.	2	3
Potentilla anserina	.	.	.	.	.	.	.	.	.	.	.	.	.	2	2
Rumex sp.	1	1	.	.	.	.	.	.	.	.	.	.	.	.	.
Trifolium sp.	.	.	1	.	.	.	5	6	1	.	.	.	.	.	1
Vicia sativa ssp. nigra	.	.	.	.	.	.	.	.	.	.	.	1	1?	.	1
Salt marsh															
Juncus gerardii (fruits)	.	.	.	.	.	.	.	.	.	.	.	.	.	2	2
Plantago maritima	.	.	.	.	.	.	1	1	.	.	.	.	1	.	1
Scirpus maritimus	.	.	.	.	.	.	.	.	.	.	.	1	3	1	5
Other															
Indet.	.	.	.	.	.	.	8	8	.	.	.	1	.	4	5

Table 2. Continued.

Period	5	5	5	5	5	5	5	5	5	5	5	5	5	5	5	5	5	5	5	5	5	Total
Sample no.	1380	2293	2513	2517	2541	2546	2552	2555	2648	2651	2658	2662	2814	2815	2816	2891	2944	3019	3789	3790	3907	= 21
Context	srf	pit	ref	srf	srf	dt	dt	pit	srf	srf	srf	?	met	met	met	sod?	met	well	well	srf	srf	
Crops																						
Hordeum v. vulgare	81	91	.	74	20	8	.	42	18	26	2	2	7	5	47	.	.	88	28	.	58	597
Hordeum, internodes	.	10	.	10	12	.	.	9	4	3	.	.	1	.	5	.	.	15	2	.	20	91
Gramineae culm nodes	4	.	.	.	1	.	.	5	.	.	.	.	.	.	.	.	.	.	.	.	10	20
Gramineae stem frags.	.	.	.	.	.	.	.	5	.	.	.	.	.	.	.	.	.	x	1	.	x	6
Avena sp.	.	.	.	.	.	.	.	.	.	.	.	.	.	.	.	.	.	.	.	.	.	.
Triticum dicoccum	.	.	.	.	.	.	.	.	.	.	.	.	.	.	.	.	.	1?	.	.	.	1?
Triticum, internodes	.	.	.	.	.	.	.	.	.	.	.	.	.	.	3	.	.	11?	.	.	.	3
Cerealia indet.	.	.	.	.	.	.	.	.	1	.	.	.	.	.	.	.	2	.	.	.	.	3
Brassica nigra	.	.	.	.	.	.	.	.	.	.	.	.	1	.	.	.	.	.	.	.	.	1
Brassica rapa	1	1	.	6	.	.	.	2	.	.	.	.	.	.	.	.	.	.	.	.	2	12
Camelina sativa	.	.	.	.	.	.	.	.	.	.	.	.	.	.	1	.	.	.	.	.	.	1
Weeds																						
Agrostemma githago	.	.	.	1	.	.	.	.	.	.	.	.	.	.	.	.	.	.	.	.	.	1
Atriplex sp.	47	11	.	38	4	.	.	7	28	1	.	.	.	1	.	.	.	.	6	5	42	190
Bromus secal./mollis	.	.	.	.	.	.	.	7	.	.	.	.	.	.	.	.	.	.	2	.	.	9
Chenopodium album	.	.	.	.	.	.	.	.	.	.	.	.	.	.	4	.	.	25	.	.	.	29
Gramineae	8	4	.	6	1	1	.	2	1	.	.	.	.	.	5	.	.	4	3	.	12	57
cf. Medicago lupulina	.	.	.	.	.	.	1	.	.	.	.	2	.	.	.	.	.	.	.	.	.	3
Polygonum aviculare	2	.	.	.	.	.	.	1	1	.	.	.	.	.	1	.	.	2	1	.	.	8
Potentilla anserina	.	.	.	.	.	.	.	6	.	.	.	.	.	.	.	.	.	.	.	.	1	7
Raphanus raphanistrum	.	.	.	.	.	.	.	.	.	.	.	.	.	.	2	.	.	.	.	.	.	2
Rumex acetosella	.	.	.	.	.	.	.	.	1	.	.	.	.	.	.	.	.	.	.	.	.	1
Rumex sp.	.	1	.	.	.	.	.	.	1	.	.	.	.	.	.	.	.	.	.	.	4	6
Trifolium sp.	.	.	.	.	.	.	.	1	.	.	.	.	.	.	.	.	.	.	.	.	.	1
Vicia sat.ssp. nigra	1	.	.	.	.	.	.	3	.	.	.	.	.	.	.	.	.	.	.	.	.	4
Salt marsh																						
Armeria maritima	.	.	.	.	.	.	.	3	.	.	.	.	.	.	.	.	.	.	.	.	3	6
Carex cf. extensa	.	.	.	.	.	.	.	.	.	.	.	.	.	.	.	.	.	.	.	.	10	10
Juncus sp. (fuits)	.	1	.	1	.	.	.	.	.	.	.	.	.	.	.	.	.	.	.	.	.	2
Plantago maritima	.	.	.	4	.	.	.	1	5	.	.	.	.	.	.	.	.	.	1	.	18	29
Ranunculus sardous	.	.	.	.	.	.	.	.	1	.	.	.	.	.	.	.	.	.	1	.	3	5
Scirpus maritimus	3	2	.	1	.	3	.	.	31	.	.	.	.	.	1	.	.	.	.	.	37	78
Scirpus rufus	.	.	.	.	.	.	.	.	.	.	.	.	.	.	.	.	.	.	.	.	1	1
Triglochin maritima	.	.	.	.	.	.	.	.	.	.	.	.	.	.	.	.	.	.	.	.	7	7
Other																						
Carex sp.	.	1	.	.	.	.	.	.	.	.	.	.	.	.	.	.	.	.	.	.	.	1
Carex cuprina	.	.	.	.	1	.	.	12	4	.	.	.	.	.	.	.	.	1	3	.	21	42
Carex paniculata	1	.	.	.	.	.	.	.	.	.	.	.	.	.	.	.	.	.	.	.	.	1
Carex riparia	1	.	.	.	.	.	.	.	.	.	.	.	.	.	.	.	.	.	.	.	.	1
Carex rost/vesic.	.	.	.	.	.	.	.	.	1	.	.	.	.	.	.	.	.	.	.	.	.	1
Eupatorium cannabinum	.	.	.	.	.	.	.	1	1	.	.	.	.	.	.	.	.	.	.	.	.	2
Rhinanthus sp.	.	.	.	.	.	.	.	1	.	.	.	.	.	.	.	.	.	.	.	.	.	1
Sambucus nigra	.	.	.	.	.	.	.	.	.	.	.	.	.	.	2	.	.	.	.	.	.	2
Indet.	4	1	.	3	.	.	.	4	9	.	.	.	.	.	.	1	.	8	.	.	4	34
Compositae	.	.	.	.	.	.	.	.	.	.	.	.	.	.	.	.	.	.	1	.	.	1
Leguminosae	.	.	.	.	.	.	.	.	.	.	.	.	.	.	.	.	1	.	.	.	.	1

Table 2. Continued.

Period	6	6	6	6	6	6	6	6	6	Total	7	7	7	7	7	7	7	7	7	7	7	7	Total
Sample no.	925	930	2210	2211	2682	2806	3914	4312	4055	= 9	924	928	1079	1405	1406	2522	3504	3662	3668	3670	3692	3693	= 12
Context	pit	srf	dt	dt	dt	srf	pit	ref	srf		srf	pit	srf	pit	pit	ph	pit	srf	hea	hea	hea	srf	
Crops																							
Hordeum v. vulgare	5	3	23	3	120	5	86	2	4	251	5	16	30	1	14	7	3	103	250	.	15	7	456
Hordeum, internodes	.	.	.	1	12	.	4	.	.	17	.	.	1	1	1	.	.	.	1	.	.	.	4
Gramineae culm nodes	1	.	.	.	2	.	5	.	1	9	.	.	2	.	.	.	.	.	.	.	.	.	2
Gramineae stem frags.	.	.	.	.	.	.	xx	x	.	xx	.	.	.	.	x	.	1	.	x	.	.	.	x
Avena sp.	.	.	.	?1	.	.	.	.	.	?1	.	.	.	.	1	.	.	16	.	.	?1	.	17
Triticum dicoccum	.	.	.	.	.	.	.	.	.	.	.	.	.	.	?1	.	.	.	.	.	.	.	?1
Triticum, internodes	.	.	.	1	.	.	.	.	.	1	.	.	.	.	.	1	.	.	.	.	.	.	1
Cerealia indet.	.	.	.	.	2	.	.	.	.	2	.	.	2	.	.	2	.	.	.	.	.	.	4
Brassica rapa	.	.	.	.	2	.	1	.	.	3	1	.	.	.	.	.	.	.	1	.	.	.	2
Malus sylvestris	.	.	.	.	.	.	.	?1	.	?1	.	.	.	.	.	.	.	.	.	.	.	.	.
Pisum sativum	.	.	.	.	.	.	.	.	.	.	.	.	.	.	.	.	.	1	.	.	.	.	1
Vicia faba	.	.	.	.	.	.	.	.	.	.	1	.	2	.	2	.	?2	10	.	1	.	.	17
Weeds																							
Atriplex sp.	.	.	.	1	18	2	12	4	.	37	4	1	1	4	.	3	.	.	7	.	1	.	21
Bromus secalinus/mollis	.	.	.	.	.	.	1	.	.	1	1	.	.	.	.	.	.	.	.	.	.	.	1
Galium aparine	.	.	.	.	.	.	.	.	1	1	.	.	.	.	.	.	.	.	.	.	.	.	.
Gramineae	.	.	.	.	2	.	.	.	.	2	.	.	.	.	.	.	.	.	2	.	.	.	2
Polygonum aviculare	.	.	.	.	.	.	.	.	.	.	.	2	.	.	.	.	.	.	.	.	.	.	2
Potentilla anserina	.	.	.	1	.	.	1	.	.	2	.	.	.	.	.	.	.	.	.	.	.	.	.
Rumex sp.	.	.	.	.	.	.	1	.	.	1	.	.	.	.	.	.	.	.	.	.	.	.	.
Spergula arvensis	.	.	.	.	.	.	.	.	.	.	.	.	.	.	1	.	.	.	.	.	.	.	1
Vicia sativa ssp. nigra	.	.	.	.	.	.	1	.	.	1	1	.	.	.	.	.	.	.	.	.	.	.	1
Salt marsh																							
Armeria maritima	.	.	.	.	.	.	2	.	.	2	.	.	.	.	.	.	.	.	.	.	.	.	.
Glaux maritima	.	.	.	.	.	.	.	.	.	.	.	.	1	.	.	.	.	.	.	.	.	.	1
Juncus sp. (fruits)	.	.	.	.	.	.	1	1	.	2	.	.	.	.	.	.	.	.	1	.	.	.	1
Juncus gerardii (fruits)	.	.	.	.	.	.	.	.	.	.	.	.	.	.	.	.	.	.	1	.	.	.	1
Plantago maritima	.	.	.	.	.	.	1	.	.	1	.	.	.	.	.	.	.	.	.	.	.	.	.
Ranunculus sardous	.	.	.	.	1	.	.	.	.	1	.	.	.	.	.	.	.	.	.	.	1	.	1
Scirpus maritimus	.	.	.	.	1	.	2	.	.	3	.	.	1	1	.	.	.	.	.	.	2	.	4
Zostera marina	.	.	.	.	.	.	30	.	.	30	.	.	.	.	.	.	.	.	.	.	.	.	.
Other																							
Carex sp.	.	.	.	.	.	.	.	.	.	.	.	1	.	3	.	.	.	.	.	.	.	.	4
Carex cuprina	.	.	.	.	.	.	12	.	.	12	.	.	.	.	.	.	.	.	.	.	.	.	.
Carex riparia	.	.	.	.	?1	.	.	.	.	?1	.	.	.	.	.	.	.	.	.	.	.	.	.
Leguminosae	.	.	.	.	.	.	.	.	.	.	.	3	.	.	.	.	.	.	.	.	.	.	3
Rhinanthus sp.	.	.	.	.	.	.	3	.	.	3	.	.	.	.	.	.	.	.	.	.	.	.	.
Compositae	.	.	.	.	.	.	3	.	.	3	.	.	.	.	.	.	.	.	.	.	.	.	.
Indet.	.	.	.	.	.	1	.	.	2	3	.	.	.	5	.	.	.	.	.	.	1	.	6

Table 2. Continued.

Period	8	?8	8	8	8	8	8	8	8	9	9	Total
Sample no.	923	3494	3638	3642	3657	3703	3722	3724	4532	3315	4569	= 11
Context	well	?	srf	srf	ph	?	?	?	ref	srf	dt	
Crops												
Hordeum v. vulgare	15	2	14	133	21	2	9	2	73	10	.	281
Hordeum, internodes	.	2	.	9	2	2	2	1	18	6	5	47
Gramineae stem frags.	x	.	.	.	1	.	.	x	.	.	.	1
Ave na sp.	.	.	.	3	.	.	.	3	.	.	.	6
Cerealia indet.	.	.	.	.	.	.	.	.	2	.	3	5
Brassica rapa	.	.	.	75	.	1	.	.	.	.	.	76
Vicia faba	3	.	.	.	4	.	.	.	.	.	.	7
Weeds												
Atriplex sp.	5	.	.	12	1	1	1	2	20	1	1	44
Chenopodium album	.	.	.	.	.	.	.	.	.	3	.	3
Chenopodium ficifolium	.	.	.	1	.	.	.	.	.	.	.	1
Gramineae	.	.	.	16	.	.	.	2	5	.	1	24
Matricaria inodora	.	.	.	.	.	.	.	.	1	.	.	1
Polygonum aviculare	.	.	1	8	.	.	.	.	1	.	.	10
Polygonum lap./pers.	.	1	.	.	.	.	.	.	.	1	.	2
Potentilla anserina	.	.	.	.	.	.	.	1	.	.	1	2
Raphanus raphanistrum	.	.	.	1	.	.	.	.	.	.	.	1
Rumex sp.	.	.	.	2	.	.	.	1	.	.	.	3
Vicia sativa ssp. nigra	.	1	.	.	.	.	.	.	.	.	1	2
Salt marsh												
Armeria maritima	.	.	.	1	.	.	.	.	.	.	.	1
Juncus gerardii (fruits)	.	.	.	3	.	.	.	.	.	.	.	3
Juncus sp. (fruits)	.	.	.	.	.	.	.	.	1	.	2	3
Plantago maritima	.	.	.	.	.	.	.	.	1	.	.	1
Scirpus maritimus	.	.	.	1	.	1	.	2	.	.	13	17
Other												
Carex sp.	.	.	.	3	.	.	.	.	.	.	.	3
Cruciferae (fruits)	.	.	.	5	.	.	.	.	.	.	.	5
Umbelliferae	.	.	.	.	.	.	.	.	.	.	1	1
Puparia	.	.	.	.	.	.	.	.	.	.	1	1
Alnus (charcoal)	.	.	.	.	.	1	.	.	.	.	.	1
Indet.	.	1	1	7	1	.	.	.	.	1	.	11

Of oats (*Avena* sp.) only grains were found, without the characteristic lemma bases on the basis of which it is possible to distinguish the wild plant (*Avena fatua*) from the cultivated one (*A. sativa*). Although experiments have shown that it is possible to cultivate *A. sativa* in the salt marsh, it is still a matter of doubt whether or not oats have ever been grown in the terp area, because of the limited possibilities for identification of the archaeobotanical material.

Pulses

Charred celtic beans (*Vicia faba var. minor*) and field peas (*Pisum sativum*) have not only been found during the analysis of the archaeobotanical samples, but also during the examination of the archaeological sieving residues. A synopsis of the latter category of finds is not yet available, but it concerns several hundreds of specimens at least. Celtic beans are commonly found in the terp area, and there is no doubt that they were grown at Wijnaldum-Tjitsma. The only other terp besides Wijnaldum where field peas have been found is Leeuwarden. They are not especially adapted to saline conditions, but they 'could have been cultivated in gardens on the mounds'.[7]

Oil seeds

Turnip (*Brassica rapa*) is regularly found in Wijnaldum-Tjitsma from the 5th century onwards. This species is present in the archaeobotanical record in the Netherlands from the Bronze Age onwards. It is present in low numbers in the botanical material from the terps. Although turnip is an element of the wild flora, Van Zeist assumes that it was cultivated in the terp area.[8] Its status as a crop is questioned by Körber-Grohne, who regards the turnip seeds from the Iron Age terp Feddersen Wierde (Northern Germany) as being derived from wild plants.[9]

A similar problem is met in the interpretation of black mustard (*B. nigra*) seeds. It remains unclear whether *Brassica rapa* and *B. nigra* were grown, or that these plants occurred as ruderals or weeds in the synanthropic vegetation on the terp.

The only find of flax (*Linum usitatissimum*) at Wijnaldum-Tjitsma is among the waterlogged remains in a well from Period 8 (AD 800-850). The presence of capsule remains indicates that flax was cultivated on or near the terp. Flax was also grown for the fibres, but archaeobotanical evidence[10] that this was the case at Wijnaldum-Tjitsma is lacking as yet.

5 NATURAL AND SYNANTHROPIC VEGETATION

Cappers'conclusion of a comparative analysis of the botanical records of the Dutch terps is that they are largely characterized by both salt marsh and fresh water plants, wet and moist soils rich in nutrients, and plants of pioneer vegetations (including ruderals) and grasslands.[11]

The reconstruction of the vegetation at Wijnaldum-Tjitsma fits into this rather broad characterization. The amount of samples investigated is as yet insufficient to draw any detailed conclusions in terms of vegetation changes in time.

The species in Table 3 are classified according to the ecological characterization of vegetation units described by Westhoff & Den Held.[12] The main ecological groups characteristic for the aspect of the landscape will be briefly discussed.

Salt marsh

All the different stages of the 'halosere', i.e. the spatial succession of vegetation types from tidal flat to high salt marsh, are represented. In the lowest parts of the littoral zone, on the flats that are inundated during every high tide, the vegetation consists almost exclusively of *Salicornia*[13] (glasswort). Like other salt marsh plants such as *Aster tripolium* (sea aster), this plant is very suitable for consumption as a vegetable. In former times, it may have formed an important source of vitamin C for the people living in the coastal area, although there is no direct evidence that it was used as such. At a somewhat higher elevation, inundated regularly for a short time, a grassy type of vegetation is found (the Puccinellietum maritimae), dominated by *Puccinellia maritima* (common salt marsh grass). This species, as well as other taxa occurring in the present day Puccinellietum (*Salicornia, Limonium vulgare*) may also occur in other halophytic plant communities.

The high salt marsh, the zone which is flooded only during spring tide and storm surges, is also covered by grasslands (vegetation types of the Armerion maritimae), often dominated by *Juncus gerardii* (salt mud rush) (Fig. 1b) and grasses such as *Agrostis stolonifera* (creeping bent) and *Festuca rubra* (red fescue). Other species characteristic for this type of grassland and present in the botanical record are *Aster tripolium* (sea aster), *Glaux maritima* (sea milkwort), *Armeria maritima* (thrift) (Fig. 1c), *Limonium vulgare* (sea lavender), *Triglochin*

Table 3. Waterlogged plant remains, related to types of vegetation. 1. Salicornietum = tidal flats, 2. Puccinellietum maritimae = low salt marsh, 3. Armerion = high salt marsh, 4. Puccinellietum distantis = high salt marsh (unstable), 5. Thero-Suaedetalia = drift deposits, 6. Scirpetum maritimi = brackish rush vegetations, 7. Scirpo-Phragmitetum = reed land, 8. Secalietea = weeds of autumn sown crops, 9. Polygono-Chenopodietalia = weeds of spring sown crops, 10. Sisymbrietalia = ruderals, 11. Bidention = pioniers from damp, nitrate-rich environment, and 12. Artemisietalia = perennials from dry, nitrate rich environment.

Period:	4	5	5	8	Vegetation type											
Sample nr:	7408	4834	4873	4920												
Context:	ditch	ditch	ditch	well	1	2	3	4	5	6	7	8	9	10	11	12
Salicornia europaea	–	5	–	21	*	.	.	.	.	.	.	.	.	.	.	.
Puccinellia maritima	–	–	1	3	.	*	.	.	.	.	.	.	.	.	.	.
Aster tripolium	2	17	3	16	.	*	*	*	.	*	.	.	.	.	.	.
Glaux maritima	–	2	–	5	.	*	*	.	.	.	.	.	.	.	.	.
Juncus gerardii	17	830	85	280	.	*	*	*	.	.	.	.	.	.	.	.
Limonium vulgare	–	–	–	7	.	*	*	.	.	.	.	.	.	.	.	.
Plantago maritima	–	6	–	12	.	*	*	.	.	.	.	.	.	.	.	.
Spergularia media	–	112	–	67	.	*	*	.	.	.	.	.	.	.	.	.
Triglochin maritima	–	18	8	32	.	*	*	*	.	*	.	.	.	.	.	.
Agrostis sp.	–	–	–	17	.	.	*	*	.	.	.	.	.	.	.	.
Armeria maritima	–	–	–	5	.	.	*	.	.	.	.	.	.	.	.	.
Carex cf. extensa	5	–	–	2	.	.	*	.	.	.	.	.	.	.	.	.
Festuca rubra	–	–	–	1	.	.	*	.	.	.	.	.	.	.	.	.
Leontodon autumnalis	4	2	–	5	.	.	*	.	.	.	.	.	.	.	.	.
Medicago lupulina	2	–	–	2	.	.	*	.	.	.	.	.	.	.	.	.
Poa pratensis/trivialis	–	–	–	8	.	.	*	.	.	.	.	.	.	.	.	.
Ranunculus sardous	18	–	7	31	.	.	*	.	.	.	.	.	.	.	.	.
Sagina sp.	–	–	–	55	.	.	*	.	.	.	.	.	.	.	.	.
Trifolium sp.	6	–	–	3	.	.	*	.	.	.	.	.	.	.	.	.
Scirpus rufus	–	–	–	1	.	.	*	.	.	.	.	.	.	.	.	.
Alopecurus geniculatus	–	–	–	11	.	.	*	.	.	.	*	.	.	.	.	.
Potentilla anserina	19	9	6	23	.	.	*	.	.	.	.	.	.	.	.	.
Atriplex hastata	5	9	3	45	.	.	*	*	.	.	.	*	*	.	.	.
Spergularia marina	–	–	–	12	.	.	.	*	.	.	.	.	.	.	.	.
Atriplex littoralis	–	–	–	1	.	.	.	.	*	.	.	.	.	.	.	.
Beta maritima	–	–	–	2	.	.	.	.	*	.	.	.	.	.	.	.
Matricaria inodora	–	–	–	2	.	.	.	.	*	.	.	.	.	.	.	.
Scirpus maritimus	37	–	–	23	.	.	.	.	.	*	.	.	.	.	.	.
Carex cuprina	–	–	–	2	.	.	.	.	.	*	*	.	.	.	.	.
Carex paniculata	–	–	–	1	.	.	.	.	.	.	*	.	.	.	.	.
Carex riparia	–	–	–	1	.	.	.	.	.	.	*	.	.	.	.	.
Carex rostrata/vesicaria	–	–	–	1	.	.	.	.	.	.	*	.	.	.	.	.
Eupatorium cannabinum	–	–	–	2	.	.	.	.	.	.	*	.	.	.	.	.
Juncus effusus	–	–	–	16	.	.	.	.	.	.	*	.	.	.	.	.
Mentha aquatica	–	11	–	–	.	.	.	.	.	.	*	.	.	.	.	.
Stachys palustris	–	–	–	1	.	.	.	.	.	.	*	.	.	.	.	.
Rhinanthus sp.	–	–	–	3	.	.	.	.	.	.	*	.	.	.	.	.
Typha sp.	–	–	5	–	.	.	.	.	.	.	*	.	.	.	.	.
Sinapis arvensis	10	–	–	17	.	.	.	.	.	.	.	*	.	.	.	.
Agrostemma githago	–	–	–	5	.	.	.	.	.	.	.	*	.	.	.	.
Bromus secalinus	–	–	–	1	.	.	.	.	.	.	.	*	.	.	.	.
Galium aparine	–	–	–	7	.	.	.	.	.	.	.	*	.	.	.	*
Raphanus raphanistrum	–	–	–	2	.	.	.	.	.	.	.	*	.	.	.	.
Vicia sat. ssp. nigra (carb.)	1	–	–	–	.	.	.	.	.	.	.	*	.	*	.	.
Rumex acetosella	–	1	–	12	.	.	.	.	.	.	.	*	*	.	.	.
Spergula arvensis	–	–	–	4	.	.	.	.	.	.	.	*	*	.	.	.
Atriplex prostrata	–	–	–	9	.	.	.	.	.	.	.	*	*	.	.	.
Polygonum aviculare	2	2	1	1	.	.	.	*	.	.	.	.	*	*	.	.
Stellaria media	6	2	4	8	.	.	.	.	.	.	.	.	*	*	.	.
Brassica rapa	–	3	1	12	.	.	.	.	.	.	.	.	*	.	.	.
Chenopodium album	1	8	4	17	.	.	.	.	.	.	.	.	*	*	*	.
Hyoscyamus niger	27	–	–	–	.	.	.	.	.	.	.	.	*	.	.	.
Polygonum lapathifolium	1	–	–	5	.	.	.	.	.	.	.	.	*	*	*	.
Polygonum persicaria	–	–	–	17	.	.	.	.	.	.	.	.	*	.	.	.
Sonchus arvensis	4	–	–	4	.	.	.	.	.	.	.	.	*	.	.	.
Chenopodium ficifolium	7	1	–	6	.	.	.	.	.	.	.	.	.	*	.	.
Urtica urens	–	6	–	2	.	.	.	.	.	.	.	.	.	*	.	.
Cirsium sp.	–	1	–	1	.	.	.	.	.	.	.	.	.	.	.	*
Characeae	–	–	6	–	.	.	.	.	.	.	.	.	.	.	.	.

maritima (sea arrow-grass), *Plantago maritima* (sea plantain) and *Spergularia* spp. (sea spurrey).

A number of species with a less halophytic character (characteristic for the Molinio-Arrhenateretea, grasslands in a fresh water environment, and for the disturbance communities of the Lolio-Potentillion), such as *Leontodon autumnalis* (smooth hawkbit), *Medicago lupulina* (black medick), *Trifolium* sp. (clover), *Ranunculus sardous* (hairy buttercup), *Potentilla anserina* (silverweed) and *Poa pratensis/trivialis* (meadow grass), may be found in the salt marsh, and are indicative for slight desalination and/or grazing.

The general aspect of the salt marsh is one of extended natural grassland ('Atlantic prairies'), providing almost unlimited possibilities for grazing.

Brackish and freshwater marsh vegetation

Freshwater reed lands (Scirpo-Phragmitetum) are represented by some obligate freshwater plants (*Carex* spp, *Rhinanthus, Eupatorium*), but many species occurring in this type of vegetation are more or less salt tolerant. The habitats of the species mentioned would have been restricted to the highest (desalinated) parts of the salt marsh and the terp. Reed vegetations would have covered the banks of creeks and ditches, gradually passing into the brackish marsh vegetations of the Scirpetum maritimi.

Pioneers, ruderals and field weeds

The interpretation of the salt marsh plants found in the samples with carbonized plant remains is problematic. The charred remains of 'wild' plants (i.e. plants other than those mentioned in Table 3), are considered to be derived fom plants that in one way or another have been involved in agrarian activities. This implies that they are interpreted either as impurities of the fields or, in case of ingredients of animal dung used as fuel, as reflecting grasslands where the livestock had been grazing. Thus, in Table 1, the separation of 'weeds' and 'salt marsh plants' is only made to indicate which species are obligate halophytes, and has nothing to do with the archaeological interpretation.

Remarkably, some weeds characteristic for winter cereal fields on sandy soils (Secalieta) appear sporadically: corn cockle (*Agrostemma githago*), wild radish (*Raphanus raphanistrum*) and sheep's sorrel (*Rumex acetosella*) in Period 5 (AD 550-650), and corn spurrey (*Spergula arvensis*) in Period 7 (750-800). In the waterlogged well Sample 4920 from Period 8 (800-850), Secalietea species are well represented. Since the presence of both winter cereals and fields on sandy soils seem to be quite unlikely in the vicinity of Wijnaldum, these weed species might indicate that grain was imported from the Pleistocene sandy soils in Drenthe and southern Friesland. A similar conclusion was drawn by Van Zeist in his study of the plant remains from Leeuwarden.[14] Van Zeist, however, emphasizes that the evidence from Leeuwarden for the import of grain from the sand district should not be over-estimated in general terms of contacts between the salt marsh and the Pleistocene area, the more so since no sites from the sand district are known where contacts with the salt marsh area have been established.[15] Because rye pollen is found from Period 5 (550-650) onwards,[16] rye may have been imported to Wijnaldum-Tjitsma since Merovingian times. These possible contacts between the clay (salt marsh) district and the high sandy soils more to the south are remarkable, since the terps in northern Germany seem to have been economically isolated from the Pleistocene hinterland.[17]

Two plant species are discussed in more detail, because they are of botanical and archaeological interest:

Common grass-wrack (Fig. 1a)

In Sample 3914, a pit from Period 6 (7th-8th C), 30 charred seeds of common grass-wrack (*Zostera marina*) were found. Since *Zostera* is a water plant, interpretation as a field weed can be ruled out.

Before *Zostera* became practically extinct in the Netherlands due to an epidemic in 1932, it was gathered extensively and used for a broad spectrum of purposes,[18] e.g. as stuffing for mattresses and pillows, for cane chairs, thatch, as isolation for silage, to prevent sand drifts, as building material for dikes, as stable litter (which was subsequently put on the fields as fertilizer), as fuel and even as fodder. Linnaeus gives an extensive description of *Zostera Oceanica*, including a chapter 'The Use of the Sea Weed', where he states: 'The ashes of the Sea Weed provide, by leaching, a salt, that not only serves to preserve the Meat from decay, but is also often used in the Glassworks in England, as well as for the manufacture of Alum and Soap, instead of Potash, under the name of Kelp'. In an 'Addendum' is mentioned that 'the famous Mr Gunnerus tells that the Jutlanders make a salt of it, which they use to salt down fish for preservation, which, however, turns blackish by this treatment'.[19] This prac-

Figure 1. a) *Zostera marina*, b) *Juncus gerardii fruit*, and c) *Armeria maritima*.

tice of salt making is confirmed in a recent publication by Brøndegaard, who reports that 'from times immemorial' until the 18th century, along the east coast of Jutland, large amounts of *Zostera marina* were burnt in the open field to extract salt.[20] This practice was apparently known in the Netherlands much earlier as well: recently, at Kolhorn (prov. of Noord-Holland), medieval deposits of ashes have been found, rich in well recognizable remains of *Zostera*.[21]

The fact that the seeds are charred would restrict the possibilities for interpretation to salt burning or use as fuel, either as dung, after consumption of the plants by animals, or directly.

Henbane

Sample 7804, a ditch from Period 4 (5th-6th century), yielded 24 seeds of henbane (*Hyoscyamus niger*). This plant has a certain fame as a medicinal herb, especially as a sedative in rheumatic pains and toothache. The hallucinogenic effect of henbane is also well known.

In the Netherlands, its natural habitat is formed by dry, warm places with a friable soil which is rich in nitrates. The range of this originally Mediterranean plant extends nowadays to Scandinavia, up to a latitude of 65° North. In northern areas, the occurrence of henbane is often associated with former occupation, e.g. gardens of old mansions and monasteries, ruins and ruderal places in old town centres,[22] which suggests dispersal into these areas by man.

Other finds of henbane in the Dutch terps are restricted to Leeuwarden[23] and Heveskesklooster.[24] It is lacking in the rural settlements.

So, it is tempting to characterize *Hyoscyamus* as a kind of medicinal index fossil for 'high status' settlements. If this holds true for Wijnaldum, this settlement would have had a status of similar importance as Leeuwarden and late medieval Heveskesklooster as early as the 5th/6th century.

However, there are several records of *Hyoscyamus* from the North Sea coast from contexts that indicate that it was a wild plant. At Feddersen

Wierde, it is sporadically found in dung layers, almost without exception associated with threshing remains.[25] In Iron Age Assendelft (site F), charred henbane seeds were found in a ditch, together with cereals and weeds.[26]

5 CONCLUSIONS

The landscape around the terp was dominated by salt marsh vegetations. Barley was the main crop during all periods of occupation. Flax, Celtic beans, peas and possibly oats and turnip were also cultivated on or near the terp. Sporadical finds of emmer wheat, bread wheat and gold-of-pleasure do not give clear proof of local cultivation. There are indications for contacts with the high sandy soils (the import of rye) from the 6th century onwards.

REFERENCES

BEHRE, K.-E. 1986. Ackerbau, Vegetation und Umwelt im Bereich früh- und hochmittelalterlicher Siedlungen in Flussmarschgebiet der unteren Ems, *Probleme der Küstenforschung im südlichen Nordseegebiet* 16: 99-125.

BOTTEMA, S., T.C. VAN HOORN, H. WOLDRING et al. 1980. An agricultural experiment in the unprotected salt marsh, part II, *Palaeohistoria* 22: 128-140.

BRØNDEGAARD, V.J. 1987. Folk og flora 1. *Danske etnobotanik*, Rosenkilde og Bagger.

CAPPERS, R.T.J. 1994. An ecological characterization of plant macro-remains of Heveskesklooster (the Netherlands). A methodological approach, Groningen (thesis University of Groningen).

GROENMAN-VAN WAATERINGE, W. & J.P. PALS 1987. The Assendelver Polders Project: Integrated ecological research. In: M.K. Jones (ed.), *Integrating the subsistence economy*, Oxford, (BAR International Series 181) 135-161.

HEIMANS, E., H.W. HEINSIUS & J.P. THIJSSE 1935. *Geïllustreerde flora van Nederland*, Amsterdam.

LINNAEUS, C. 1783. Natuurlijke historie of Uitvoerige beschryving der dieren, planten en mineraalen, volgens het samenstel van de Heer Linnaeus, II.

KÖRBER-GROHNE, U. 1967. *Geobotanische Untersuchungen auf der Feddersen Wierde*, Wiesbaden.

LEMPIÄINEN, T. 1992. Macrofossil finds of henbane (*Hyoscyamus niger*) in the old settlement layers in southern Finland. In: J.P. Pals, J. Buurman & M. van der Veen (eds), Festschrift W. van Zeist, *Review of Palaeobotany and Palynology* 73: 227-239.

ODUM, S. 1965. Germination of ancient seeds, *Dansk Botanisk Arkiv* 24: 1-70.

OOSTSTROOM, S.J. VAN & TH.J. REICHGELT 1964. Zosteraceae, *Flora Neerlandica*, Amsterdam I(6): 88-92.

PALS, J.P. & M.C. VAN DIERENDONCK 1988. Between flax and fabric: cultivation and processing of flax in a Medieval peat reclamation settlement near Midwoud (prov. of Noord-Holland), *Journal of Archaeological Science* 15: 237-251.

WESTHOFF, V. & A.J. DEN HELD 1975. *Plantengemeenschappen in Nederland*, Zutphen.

ZEIST, W. VAN 1974. Palaeobotanical studies of settlement sites in the coastal area of the Netherlands, *Palaeohistoria* 16: 223-371.

ZEIST, W. VAN 1988. Botanical evidence of relations between the sand and clay districts of the north of the Netherlands in medieval times. In: H. Küster (ed.), Festschrift U. Körber-Grohne, *Forschungen und Berichte zur Vor- und Frühgeschichte in Baden-Württemberg* 31: 381-387.

ZEIST, W. VAN, T.C. VAN HOORN, S. BOTTEMA et al. 1976. An agricultural experiment in the unprotected salt marsh, *Palaeohistoria* 18: 111-153.

ZEIST, W. VAN, R. CAPPERS, R. NEEF et al. 1987. A palaeobotanical investigation of medieval occupation deposits in Leeuwarden, the Netherlands, *Proceedings of the Koninklijke Nederlandse Academie van Wetenschappen B90*: 371-426.

NOTES

1. Cf. Bos & Gerrets (this volume).
2. Körber-Grohne (1967), Van Zeist et al. (1976) and Bottema et al. (1980).
3. See synopsis by Cappers (1994).
4. Prov. of Friesland, AD 800-1525 (Van Zeist et al. 1987).
5. Prov. of Groningen, Period 1: AD 50 BC-400, Period 2: 800-1300, Period 3: 1300-1610, Period 4: 1610-1975 (Cappers 1994).
6. Van Zeist (1974).
7. Van Zeist et al. (1987, p. 417).
8. Van Zeist (1974).
9. Körber-Grohne (1967).
10. Cf. Körber-Grohne (1967) and Pals & Van Dierendonck (1988).
11. Cappers (1994).
12. Westhoff & Den Held (1975).
13. It is impossible to distinguish the different species of *Salicornia* occurring in the Netherlands *(S. europaea* s.s., *S. procumbens* and *S. disarticulata*) on the basis of the seeds. The latter species has seeds somewhat less hairy than the other two, but this is not a reliable character in subfossil material.
14. Cf. Van Zeist et al. (1987) and Van Zeist (1988).
15. Van Zeist (1988).
16. Groenman-van Waateringe (this volume).
17. Körber-Grohne (1967) and Behre (1986).
18. Cf. Heimans et al. (1935) and Van Ooststroom & Reichgelt (1961).
19. Linnaeus (1783, p. 238 – 'bijvoegsel na p. 698').
20. Brøndegaard (1987, Vol. I, p. 18).
21. Dr B. van Geel (pers. comm. 1996).
22. Cf. Odum (1965) and Lempiäinen (1992).
23. Van Zeist et al. (1987).
24. Cappers (1994).
25. Körber-Grohne (1967).
26. Groenman-van Waateringe & Pals (1987).

The terra sigillata from Wijnaldum-Tjitsma in regional perspective

T.B. VOLKERS[1]

1 TERRA SIGILLATA FINDS FROM FRIESLAND

The distribution of Roman pottery north of the limes has traditionally drawn much attention from archaeologists. In this respect the Frisian terp region is outstanding for a remarkable quantity of terra sigillata finds. Therefore the presence of 63 fragments of terra sigillata in the finds of the excavation of terp Tjitsma at Wijnaldum was not surprising. Most of the Frisian terra sigillata finds, ca 2200 fragments, are kept in the collection of the Fries Museum, Leeuwarden. These were mainly collected in the 19th and first half of the 20th centuries, during large-scale soil extractions, as a result of which many terps were levelled.

The many finds of terra sigillata show a wide distribution and induced the Frisian archaeologist Boeles to conclude that this pottery was present in all the terps from the Roman period.[2] Nearly all the objects found during these soil extractions in the terps were discovered by chance, although the Fries Genootschap voor Geschied-, Oudheid en Taalkunde (Frisian Society of History, Archaeology and Philology) encouraged the workers to collect archaeological finds by offering an award. In relation to terra sigillata 5 cents were paid for an undecorated fragment and 10 cents for a decorated one. It has been claimed that the collection of the Fries Genootschap, which is the basis of the present collection of the Fries Museum, is unjustifiably large, because the workers broke the fragments in half in order to receive double payment. However the terra sigillata fragments show hardly any recent breakages and matching fragments hardly ever occur.

Apparently the policy of the Fries Genootschap had good results, considering the large number of very small fragments of terra sigillata. The distinctive red colour of the pottery obviously also contributed to the success. Because the collection had been accidentally assembled and find records are almost lacking the scientific value is mainly determined by the large number of finds and of sites. The fragments in the Fries Museum have been collected from more than 90 different terps, which show remarkable differences in frequency. There are 12 terps which contained more than 50 fragments, the following 5 of which had more than 100: Hatsum I (452), Hatsum II (256), Burmania (241), Ritsumaburen (148) and Jousumburen (124).

The different frequencies of terra sigillata in Frisian terps cannot simply be explained by a more thorough search by the labourers of those terps that provided many and valuable finds and, therefore too, payment. Concerning the quantity of terra sigillata the terp of Hatzum even is comparable to a settlement south of the limes, such as Rijswijk (province of South Holland) (ca 500 fragments).[3] However, the vast majority of the terps in Friesland have yielded no more than a few, or at the most, a few dozen fragments, together accounting for less than 20% of the collection in the Fries Museum.

2 THE TERRA SIGILLATA FROM THE EXCAVATION OF WIJNALDUM-TJITSMA

In contrast to the almost recordless Museum collection, the accurate find record of the Wijnaldum excavation offers a rare opportunity for the research of terra sigillata in Friesland. In comparing the typochronology and frequency of the Wijnaldum terra sigillata with the Museum collection there appears a remarkable resemblance between the two groups. Of the terra sigillata in the Fries Museum, by far the largest part is dated to the second half of the 2nd and the first half of the 3rd centuries. Apart from this, there is a small number (10 fragments) from the first half of the 1st century, and a few fragments from the 4th and 5th centuries. Of the Wijnaldum excavation, all terra sigillata belongs to the first large group dated to the 2nd and

Table 1. Frequency of terra sigillata types from the Wijnaldum excavation and in the Fries Museum.

	Wijnaldum		Fries Museum	
Type	Number	%	Number	%
D37	21	60%	850	64%
D31 (18/31)	5	14%	177	14%
D33	5	14%	102	8%
Mortars	2	5.6%	85	7%
D36	2	5.6%	16	1%
D41	1	2.8%	Exact quantity unknown	
D32	–	–	76	6%
Dech.72	–	–	32	2%

Table 2. Frequency of terra sigillata finds in relation to the periodisation of the archaeological context.

	Certain	Possible	Total
AD 150-250	–	–	–
AD 250-350	8	1	9
AD 408-550	3	6	9
AD 650-700	1	1	2
AD 700-900	4	5	9

3rd centuries except for two 4th to 5th-century fragments. The typology of the Wijnaldum terra sigillata seems to correspond completely to the Museum collection either, apart from the absence of bowl D32, which is considered to succeed plate D18/31, 31 and which is therefore dated somewhat later.

The quantity of terra sigillata finds from the Wijnaldum excavation, 63 fragments, is not exceptional, especially considering the accurate method of collection in an excavation, in which the soil content of almost all archaeological features had been sieved. This can easily be demonstrated by the 80 stray finds of terra sigillata collected from the terp to the west of the excavated terp Tjitsma, belonging to the same row of terps, which constitutes even more sigillata than from the excavation, while at the more easterly situated terp Voorrijp, 58 fragments were found. In view of the relatively small quantity of terra sigillata and the similarity of the assemblage, the terp Tjitsma at Wijnaldum seems to be representative of the majority of Frisian terps with sigillata finds, which may help us to answer general questions about Roman import finds in Friesland.

3 THE DATING OF THE WIJNALDUM TERRA SIGILLATA AND OF ITS ARCHAEOLOGICAL CONTEXT

The accurate recording of the find circumstances of the terra sigillata of the Wijnaldum excavation allows us to draw interesting conclusions which may explain the high frequency of this Roman pottery in central Frisia, a region far north of the limes. From the excavation record it appears that 29 of the 63 terra sigillata fragments were found in soil features which were dated by archaeological evidence.[4] It is remarkable that 20 of these sherds were found in an early-medieval context. Even the nine remaining fragments originating from features dated to AD 250-350 are actually in a context that is later than the dating of the terra sigillata types concerned. This might be concluded from the remarkable absence in Wijnaldum of the younger type D32 which is characteristic for the middle of the 3rd century. Moreover the dating to AD 190-225 of the only decorated fragment (No. 19) does not really correspond to the reliable dating to AD 250-350 of the find context, unless a very long period of use is assumed. However for the vast majority, the 20 fragments from an early-medieval context, this explanation is excluded. The 29 fragments from a dated find context appear to be evenly distributed over the different periods. There are only two fragments from the period 650-700, which covers only 50 years.

The intriguing questions in relation to Table 2 are how to explain the difference between the typological dating of the terra sigillata finds and the much younger date of the archaeological context, and what are the consequences concerning the remaining finds with an undatable context. For the first question there are two possibilities: either the terra sigillata fragments originate from older (AD 150-250) habitation layers of the terp itself or they were transferred from elsewhere. In the first case the pottery was used in the same period and ended up as discarded fragments. Then accidental finds of discarded fragments in a younger period by inhabitants of the terp might explain the younger context. If this should be correct one should expect a considerable quantity of finds of terra sigillata from the many habitation layers of AD 150-250 which could explain the stray finds from a younger context. However, not a single fragment of terra sigillata has been found in that context. So we are left with the possibility that the terra sigillata was transferred from elsewhere. Considering the limited quantity of terra sigillata found elsewhere in Friesland and the general lack of find context, it is unlikely that the Wijnaldum finds were transferred from other Frisian terps. For these terps as for Wijnaldum we assume that this pottery was directly imported from outside Frisia. From research into the production centres of decorated terra sigillata of the 2nd and 3rd centuries found in Friesland, it appeared that the origin of the Frisian finds show a remarkable difference compared to the finds elsewhere in the Netherlands and a striking resemblance to those in the German Rhineland for instance from Neusz, Holzhausen, Butzbach, Butzbach-Degerveld and Inheiden.[5] In Friesland as well as in the Rhineland most of the terra sigillata originates from Trier, followed by Rheinzabern, whereas a small amount was produced at Argonne. At Dutch archaeological sites outside Friesland much more Argonne ware is found. We suppose that it is most likely that the Frisian finds originated from the German Rhineland and have been brought from there in a later period.

The above conclusion may perhaps throw some light on some remarkable Roman finds in Friesland, whose function in Frisia could not properly be explained and whose character as a Roman import was doubted. That is the case, for instance with the Hludana-votive stone from the terp of Beetgum whose function is difficult to explain there. The same doubt applies to the wax tablet of Tolsum with its Latin text on the sale of a cow. Such typical Roman, but hardly prestigious finds are difficult to understand far north of the limes and their true character will stay uncertain.[6]

Apart from Friesland, a few sites are known with Roman import finds in a younger archaeological context in the adjacent province of North Holland. Vons and Bosman mention fragments from the same object found in both a native settlement at Assendelft and the castellum at Velsen.[7] This, however, is easy to explain because the then deserted fortification was in the vicinity of the native settlement. This does not apply to a medieval site at Schagen where, besides early-medieval pottery, 5 fragments of terra sigillata were found but not a single find of native pottery from the Roman period.[8]

The presence of the large number of 2nd and 3rd century fragments of Roman pottery in Friesland is usually explained against the background of existing trade relations and in the socio-political sphere of gift exchange in the Roman period.[9] The find circumstances of the Wijnaldum terra sigillata show that after the Roman period and perhaps also from the second half of the 3rd century older terra sigillata was transferred to Wijnaldum. If the Wijnaldum excavation should indeed be representative of Friesland, this has consequences for the interpretation of other terra sigillata finds without a proper Roman context in Friesland and perhaps north of the limes in general.

Finally this raises the question why terra sigillata was brought to Wijnaldum in a later period and whether it concerned complete pottery objects or fragments. Of the 63 fragments from Wijnaldum not a single one fits another or matches any other fragment in form, colour. A clear function for these isolated sherds is not visible. They might just have been attractive for their red colour and shine. For 52 fragments no other reason can be given. To the eleven remaining sherds a possible function can be attributed. There are seven playing chips, one loomweight and a decorated fragment with a perforation might have been used as a pendant. There are also two fragments that have been rounded off and one fragment shows damage on the inside as though it served as a protection against a pointed object like, for example, a needle.[10] The function of some of the pottery fragments does not add new information on the question of the later transport to Friesland.

4 CONCLUSION

The hypothesis of later transport of Roman terra sigillata to Friesland is a remarkable conclusion of the Wijnaldum research and does not stand alone. It is supported by the research on the other Roman finds from the excavation such as a fragment of a Roman column and a few dozen fragments of Roman roof-tiles and tegulae. The research into this has not been completed, however all these finds appear to be connected with younger archaeological features. The same conclusion is drawn from the Roman metal finds, the glass finds and the Roman coins from Wijnaldum.[11] This later, mostly early medieval dating of the archaeological context of the major part of the Roman finds which had a dated context has consequences for the interpretation of Roman finds elsewhere in Friesland. An interesting possibility to confirm the above conclusion is offered by the chronological differences in the habitability of the various salt marsh ridges in Friesland as explained by Vos.[12] The ridge of Wijnaldum-Ried became habitable in the Roman period and then was the most northerly ridge. The few (10) terra sigillata fragments from the first half of the 1st century, when Frisia was under Roman control, were all found on this ridge, whereas the numerous 2nd-3rd-centuries fragments have a much wider distribution. However the youngest salt marsh ridge of Pieterbierum-Minnertsga which was settled only in the Carolingian period, seems to have no terra sigillata find until now, except from some dubious finds by metal detector amateurs. Although this youngest settlement area is by far the lest investigated, one should also expect terra sigillata finds from there, if Roman finds reached Friesland after the Roman period. This may show that the important conclusion of the Wijnaldum excavation, that many of the Roman imports seem to have reached Frisia only later, even after the Roman period, needs further investigation. Nevertheless the results of the Wijnaldum excavation proof that the traditional explanation of the Roman finds in Friesland in relation to contemporary trade relations needs to be extended to later contacts also. Further research into the Roman finds in Friesland may put a light on Frisian relations not only in the Roman period but perhaps in the Early Middle Ages as well.

5 CATALOGUE OF THE WIJNALDUM TERRA SIGILLATA FRAGMENTS

The catalogue of the terra sigillata finds follows the usual Dragendorff or Chenet typology.[13] [14]The fragments have been numbered per type. Then rim, wall and basefragments and sometimes a function of the fragment have been distinguished as far as possible, followed by the width, height of the fragment and diameter (d) of the rim or footstand in cm. The record number and feature are given, followed by the dating of the archaeological context of the find and the reliability of this dating as concluded by the excavators.[15] The reliability of the archaeological datings are indicated by the following numbers: 1. Finds and context/stratigraphy are consistent = very reliable; 2. Finds and context/stratigraphy non-consistent = unreliable; 3. Finds and context/stratigraphy partly consistent = under certain conditions reliable; 4. No finds, date based on context/stratigraphy = reliable. Finally, if possible a more precise date of the pottery is given with reference to the relevant literature.

No.	Description/size	Record no./feature	Context date/ reliability	Remark/reference
Dragendorff 31 (18/31)				
1	Rim, 3/3/d 22	W91/4634,		
2	Rim,4/2.5/d 22	W93/1 excavated soil		
3	Rim,4.5/3.5/d 20	W93/9583		
4	Rim, 3/3/d ca 22	W93/9726	250-300/1	
Probably Dragendorff 31 (18/31)				
5	Wall, 2.5/2	W91/4240, occupation level	450-525/3	
Dragendorff 33				
6	Rim, 2/2/d ca 12	W92/8212, well	from 425/3	
7	Rim, 2.5/2/d 12	W93/1, excavated soil		
8	Base, 4.5/2.5	W93/9451, sod structure	250-350/1	
9	Rim, 2/2,/d 10	W93/11139, top soil		
10	Wall, 2.5/1.5	W93/11205, occupation level	250-350/1	
Probably Dragendorff 35/36				
11	Wall, 2/1.5	W91/1, excavated soil		
12	Wall, 4.5/3	W92/7282, ditch	408-538/3	
Dragendorff 37				
13	Wall, 6/4	W91/8, top soil		'Springender Hirsch' Ricken & Fischer (1963) T90a, Rheinzabern
14	Base, 2.5/3	W92/1, excavated soil		
15	Wall, 2/1.5	W92/2130		
16	Wall, 3/4	W92/5903, arable layer	550-650/1	Garland
17	Wall, d 2.5	W92/6495, refuse pit/occupation level	770-900/1	Playing chip
18	Wall, 4/3	W92/6984		
19	Wall, 6/6	W92/7414, ditch	250-350/1	Ovolo Gard (1937) R14, Trier, Amator-Attilus/Pusso, d. 190-225

20	Base, 4/2/d 7	W93/1, excavated soil		
21	Wall, 1.5/1	W93/9025		
22	Wall, 2.5/2	W93/9489		
23	Base, 6/6/d 11	W93/11142, top soil		
24	Rim, 6/5/d 30	W93/11152, top soil		
25	Wall, 3/2.5	W93/11705		Smoothened, perforated
Probably Dragendorff 37				
26	Rim, 5/4/d 22	W91/1333		
27	Rim, 3/1.5/d 21	W92/5887		
28	Rim, 3/3/d 32	W92/5918, sod layer	425-500/3	
29	Wall, 3.5/3	W92/8203, layer		
30	Rim, 5/4/d 24	W92/8253		Innerside damaged by pitmarks
31	Base, 3/1	W93/9911		
32	Wall, 3/2	W93/11060	450-500/1	
33	Rim, 2/11/d ca 20	W93/11731 well	800-850/3	
Dragendorff 41				
34	Wall, 4/3.5	W92/1 excavated soil		
Mortar				
35	Wall, 6.5/5	W91/1 excavated soil		
36	Wall, 6/5.5	W92/7934	425-550/1	
Chenet 320				
37	Base, 7/6/d 8	W91/559, pit	775-850/3	Possibly of bowl with chequered decoration, 425-475[16]
Chenet 324				
38	Rim, 7/6/d 22	W93/9778, sunken hut	450-500/1	Trace of barbotine decoration, 375-475[17]
Indeterminable				
39	Wall, 3.5/3.5	W91/332		Smoothened on two sides
40	Base, 4/3/d 10	W91/1335		Smoothened
41	'Playing chip', d 1	W91/1526, sod layer	700-750/1	
42	Wall or base	W91/2010, occupation layer	750-850/3	
43	Wall or base	W91/2458, occupation layer	300-350/3	
44	Rim, 2/3	W91/2876, ditch	750-770/3	
45	Wall, 2.5/2	W91/2902, sod layer	650-750/3	
46	Wall or base, 1.5/1	W91/3520, well	775-850/3	
47	Playing chip d 1.5	W91/4422		
48	Wall, d 6	W91/4686, occupation layer	300-350/1	Quarter of a disc-shaped loom-weight
49	Rim, 3.5/2,/d 22	W91/4808		
50	Wall or base	W92/5296		
51	Wall or base, 2.5/2	W92/6314, ditch	408-538/3	
52	Wall/base, 1.2/0.8	W92/6389, big pit	650-700/3	Burned
53	Wall/base, 6/5.5	W92/6418		Innerside burned
54	Playing chip, d 1	W92/6545		
55	Playing chip, d 2	W92/7933		
56	Base, 6/3.5,/d 7	W92/7962		
57	Wall or base, 1/1	W92/7292, occupation level	250-350/1	
58	Base, 3.5/3	W92/7402, occupation level	250-350/1	
59	Wall, 3/3	W93/9468		
60	Wall or base, 2/2	W93/9608, ditch	425-550/3	
61	Playing chip, d 2.8	W93/9969, refuse pit/occupation level	450-500/1	
62	Playing chip, d 1.5	W93/10754		
63	Wall or base, 0.5/1	W93/11153		

REFERENCES

BAKKER, L. 1981. *Spätrömische Befestigungen und frühmittelalterliche Kirche van St. Peter und Paul*, Luxembourg.

BLOEMERS, J.H.F. 1978. *'De Bult' eine Siedlung der Cananefaten*, Amersfoort (Nederlandse Oudheden 8).

BOELES, P.C.J.A. 1951. *Friesland tot de elfde eeuw*, 2nd edition, 's Gravenhage.

BOS, J.M. 1996. Beetgum en Tolsum: heilige huisjes in de Friese archeologie, *It Baeken* 58: 42-46.

CHENET, G. 1941. *La céramique gallo-romaine du IVe siècle et la terre sigillée decorée a la molette*, Macon.

DIEDERIK, F. 1982. Molenweg, Waldervaart. Een vroegmiddeleeuwse nederzetting te Schagen, *Westerheem* 31: 53-68.

DIJKMAN, W. 1992. La terre sigillée décorée à la molette à motif chrétiens dans la stratigraphie maastrichtoise (Pays-Bas) et dans le nord-ouest de l'Europe, *Gallia* 49: 129-172.

DRAGENDORFF, H. 1895. Terra Sigillata; ein Beitrag zur Geschichte der griechischen und römischen Keramik, *Bonner Jahrbücher* 96(7): 18-155.

ES, W.A. VAN 1966. Friesland in Roman times, *Berichten van de Rijksdienst voor het Oudheidkundig Bodemonderzoek* 15(6): 37-68.

GARD, L. 1937. Reliefsigillata des III. und IV. Jahrh. aus den Werkstätten von Trier, Tübingen (unpubl. Ph D thesis University of Tübingen).

RICKEN, H. & C. FISCHER 1963. *Die Bilderschlüsseln der römischen Töpfer von Rheinzabern*, Bonn.

VOLKERS, T.B. 1991. Terra Sigillata aus friesischen Wurten, *Germania* 69: 176-186.

VONS, P. & A.V.A.J. BOSMAN 1988. Inheemse boeren bezochten de verlaten Romeinse versterkingen Velsen I en II, *Westerheem* 37: 1-16.

NOTES

1. The Dutch text was translated by Jane Green, Harlingen.
2. Boeles (1951, p. 152).
3. Bloemers (1978, p. 241).
4. See Gerrets & De Koning (this volume).
5. Volkers (1991).
6. Bos (1996).
7. Vons & Bosman (1988).
8. Diederik (1982).
9. Boeles (1951, pp 134-177) and Van Es (1966).
10. This is also observed in terra sigillata fragments found elsewhere in Friesland.
11. Erdrich (this volume), Sablerolles (this volume) and Van der Vin (this volume).
12. Vos (this volume).
13. Dragendorff (1895).
14. Chenet (1941).
15. Gerrets & De Koning (this volume).
16. Description and date W. Dijkman; Dijkman (1992, p. 151).
17. Description and date M. Erdrich; Bakker (1981, p. 328).

Roman wheelthrown pottery, terra nigra-like bowls and tiles

M.C. GALESTIN

1 INTRODUCTION

The excavation in the Tjitsma terp, near Wijnaldum, in the years 1991-1993 was the first large-scale excavation of a terp for many years. A new element in this excavation was the method of excavation with contexts, by which a more precise dating could be obtained by careful stratigraphical analysis. As a result eight occupation phases, dating from the 2nd century AD into the 10th century could be discerned. It also turned out that Roman objects were not only discovered in contexts dating to the Roman Period but also in contexts of much later periods.

Quite a number of different types of Roman artefacts were discovered in the terp which was situated far beyond the border of the Roman Empire. In this contribution the Roman wheelthrown pottery (except for the terra sigillata), the terra nigra-like bowls and the tiles will be presented. The objects will first be described and dated and their provenance will be discussed. Furthermore the presence of Roman objects in Roman as well as in much later contexts will be discussed.

2 DESCRIPTION OF THE MATERIAL

The Roman material presented here consists of six fragments of three different amphoras, five fragments of fine wares, thirty fragments of tiles and objects made of tiles and 21 fragments of Roman wheelthrown terra nigra-like bowls.

Amphoras (Fig. 1)
Two fragments, perfectly fitting, form the handle of an amphora. The short flat handle with a central groove and smooth surface indicates a flat-based amphora, datable from the 2nd to mid 3rd century. This type of amphora with a flat base has been called 'jar-amphora'.[1]

Three amphora fragments belong to a type of two-handled containers that is known to have been manufactured in the Lower Scheldt Valley. This type of pottery is found in the civilian sites of Valkenburg (Marktveld) and Nijmegen (Ulpia Noviomagus). A fragment from London suggests a wide distribution as is also indicated by the fragments from Wijnaldum. The Scheldt Valley amphoras may be dated according to their rim type into three groups. The first group dates to the end of the 1st century while groups 2 and 3 date from the middle of the 2nd century. The fragments from Wijnaldum do not have a rim but in the context of the Wijnaldum terp a date from the middle of the 2nd century would be in accordance with the date of the terp in general.[2]

The last fragment is difficult to determine. It is a wall fragment that may be part of a Roman amphora.

Fine wares (Fig. 2)
A rim fragment of a flagon (jar) of Roman fine ware, datable to the 2nd century, may be attributed to one of the factories south of the Rhine. This type is frequently encountered in Nijmegen.[3]

The rest of the fine wares are all fragments of beakers, two fragments of black-slipped ware and two of colour-coated ware. The black-slipped ware comes principally from workshops at Trier, the production ranges from the end of the 2nd century to the mid 3rd century.[4] The colour-coated wares cannot be attributed to a specific workshop and have the same date as the black-slipped wares except for the small fragment of a beaker with a yellowish brown coating which dates in the 2nd century.[5]

Wheelthrown terra nigra-like bowls (Fig. 2)
Some twenty fragments of terra nigra-like bowls were excavated. The ware is fired in a reduced atmosphere. The colour of the fabric varies from very

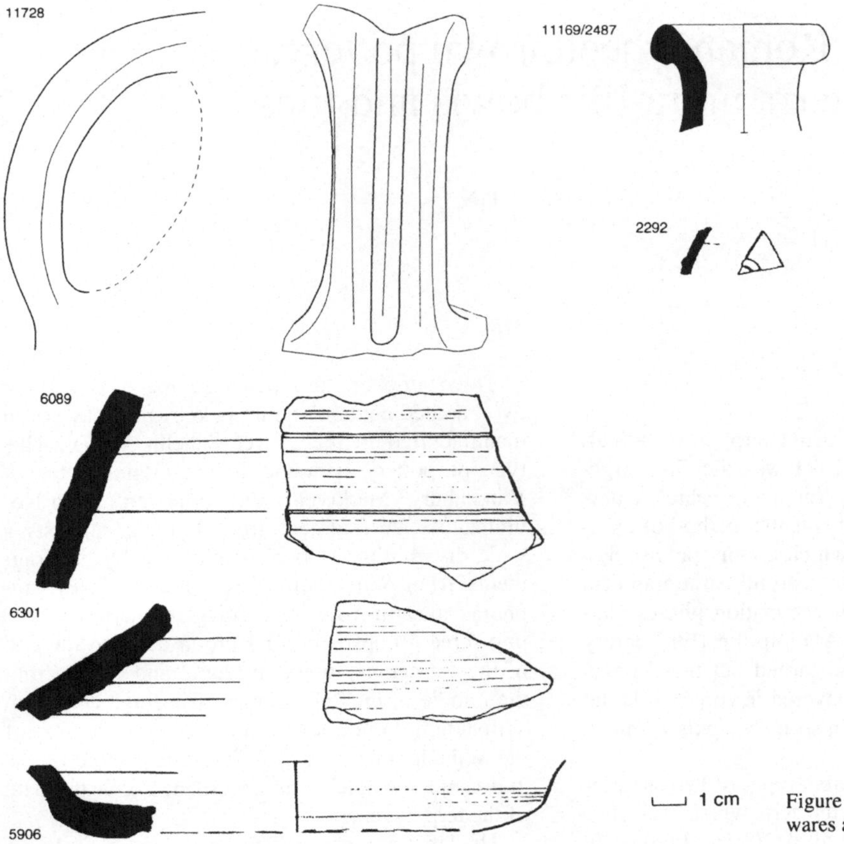

Figure 1. Fragments of Roman fine wares and amphoras. Scale 1:2.

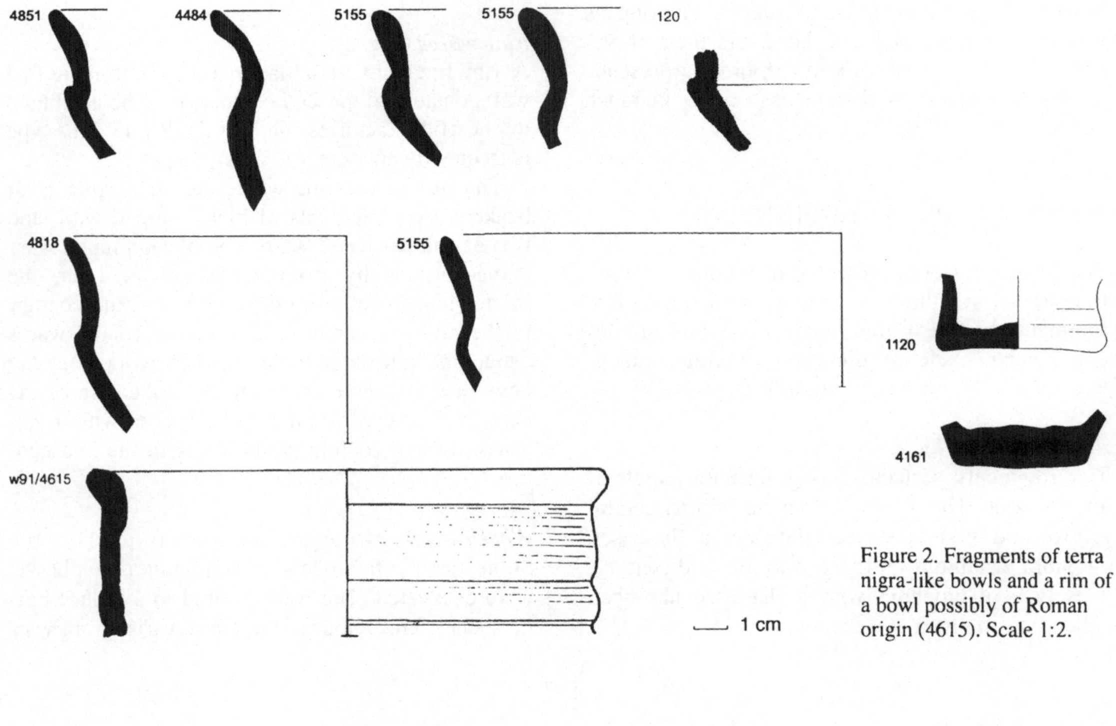

Figure 2. Fragments of terra nigra-like bowls and a rim of a bowl possibly of Roman origin (4615). Scale 1:2.

dark grey to light grey and the surface often has the same colour as the fabric but it may also be darker in colour, due to polishing.

This pottery is not only found at Wijnaldum but also on other terp sites in Friesland and Groningen. It was first identified by Van Es, who named it 'Black wheelthrown pottery from Northern Holland'.[6] The examples cited by him were all funnel-shaped bowls, closely resembling the local hand-made forms.

Similar pottery was discovered at Rijswijk, De Bult, and it was classified by Bloemers among the terra nigra and terra nigra-like pottery.[7] A comparison was made with the pottery from North Holland, published by Van Es and from that time on the term terra nigra or terra nigra-like was also used to describe the wheel-made and funnel-shaped bowls from the northern part of the Netherlands.

The grey wheelthrown pottery from Wijnaldum displays only one form with some variants: the funnel-shaped bowl. The comparable hand-made form is called *Trichterschale* or *Situla*.[8] Characteristic is the small diameter of the flat foot compared to the width of the rim. The ratio is about 1:4. The diameter of the rim varies from circa 17 to 22 cm. There are, however, only a very small number of rims to measure this. From the foot upward, the wall widens into a concave or convex belly.

The date attributed to this type of bowl is roughly 2nd-4th century. Van Es proposed to date the pottery by comparing it with the hand-made local bowls. He noticed that they were in fact strikingly similar to the native bowls which he dated to the Period 150-300 and maybe into the 4th century.[9] The recent study of hand-made pottery from the northern provinces by Ernst Taayke has shown that the hand-made so-called *trichterformige Keramik* appears in the province of Friesland from the 2nd century onward but becomes popular in the course of the 3rd century.[10] Among the forms discussed by Taayke we notice forms that are very similar to the types discussed here. Taayke also mentions the fact that at the Wijnaldum excavation different forms of regional hand-made pottery are found together with terra nigra-like pottery with the same form. According to him the types of hand-made pottery date to the 3rd and beginning of the 4th century.[11] The fragments excavated at Wijnaldum also date to this period. Some fragments appear in contexts dated to the first half of the 4th century for instance the rim fragments 5155 and 4851. A wall fragment, 2495, was discovered in a context that was dated by ^{14}C between 128 and 314 and without any material of post-Roman date. The other fragments from dated contexts make it clear that the majority of datable fragments attributable to the category of grey wheelthrown terra nigra-like bowls discovered at Wijnaldum date to the 4th century.

The last question to be answered is the provenance of these bowls. The term often used to indicate the ware of the pottery is terra nigra or terra nigra-like. With this term the pottery is classified among the provincial Roman pottery, made within the Roman Empire and occasionally even outside.[12] Terra nigra pottery was also exported to regions outside the Empire, compare for instance the cups found in Wijster. The grey wheelthrown bowls discussed here are, however, different from these terra nigra cups which are smaller in size (diameter between 11 and 18 cm). They often have a rim with a beaded lip and some have a rolled decoration on the shoulder or belly.[13] The striking similarity in form with the native pottery from the northern provinces and the difficulty of distinguishing hand-made from wheelthrown bowls as indicated by Van Es may even lead to the conclusion that wheelthrown grey ware was made in the same region as the hand-made pottery and that among the local potters (or by immigrant potters) the potter's wheel had been introduced. Compare for instance the probably local production of wheelthrown terra nigra bowls at Heeten, also situated outside the Roman Empire, in the 4th century.[14] This, however, is not very likely because of the small number of finds from the province of Friesland. This small number is not in accordance with the conditions for wheelthrown production of pottery, namely: a certain degree of specialization, a high degree of standardization of the forms and a market, i.e. trade over a large area. None of this applies to the Frisian terp area. Further investigation is strongly needed to provide more details on typology, date and distribution of this very interesting class of grey wheelthrown pottery.

Tiles (Figs 3-4)

Tiles may be identified as Roman by their form: a combination of the flat part of the tile with a flange is characteristic for the *tegula* while the curved surface of a tile is characteristic for the *imbrex*. Fragments lacking a flange may be identified as Roman tiles by their thickness. The thickness varies from 19 to 36 mm. Tiles from the Medieval period are much thinner.[15]

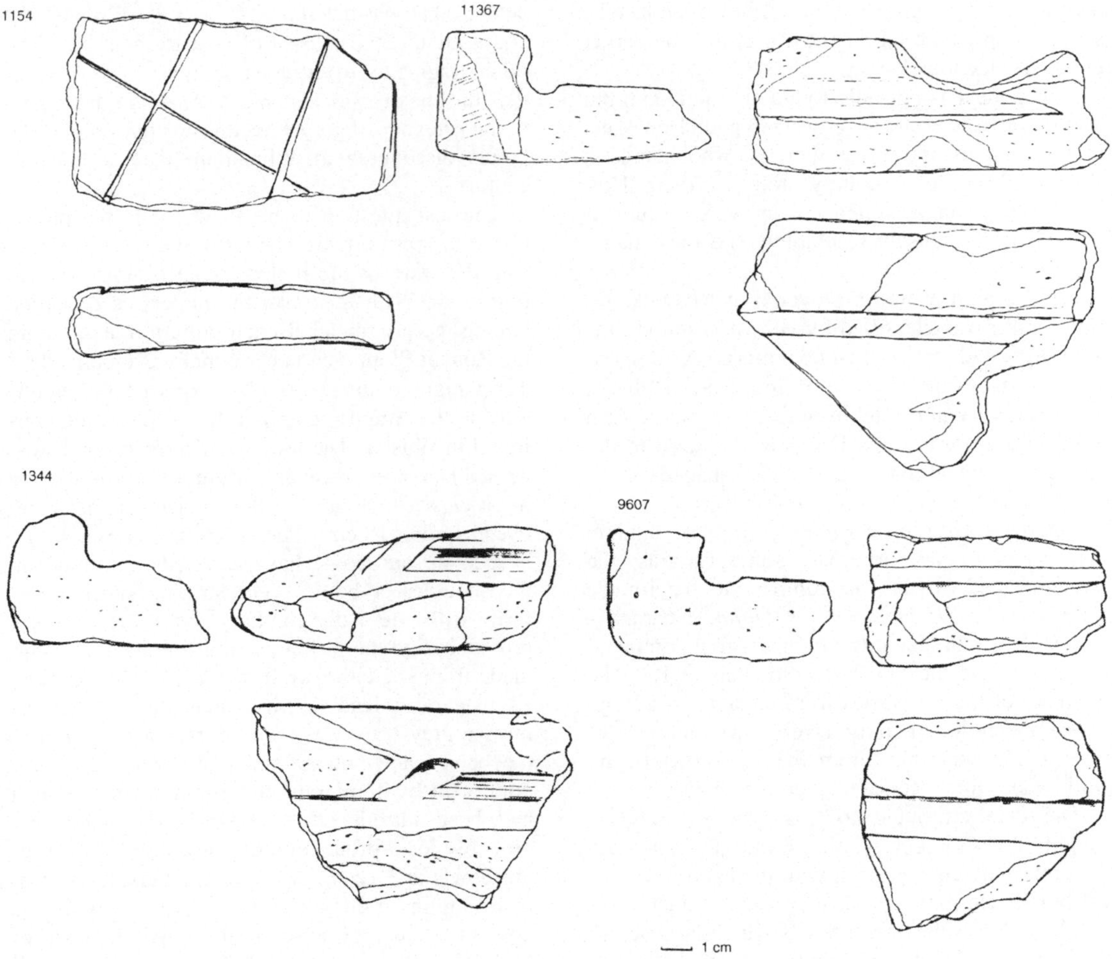

Figure 3. Fragments of Roman tiles and of a box flue tile (1154). Scale 1:2.

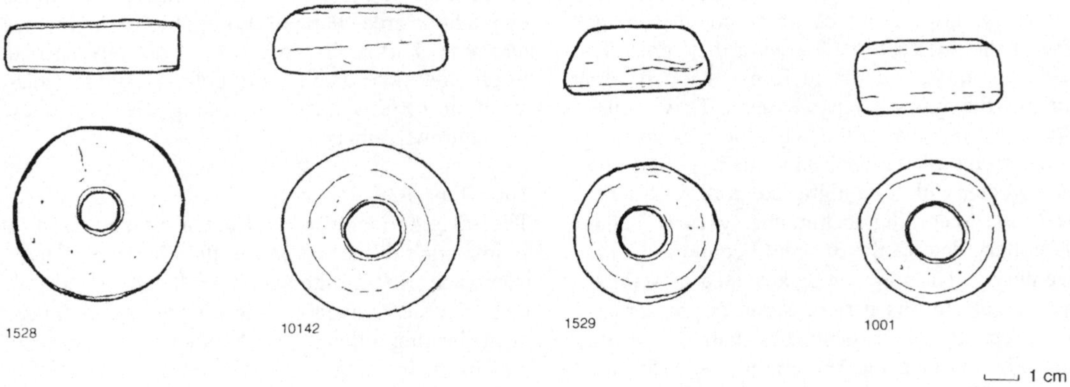

Figure 4. Spindle whorls made out of Roman tiles. Scale 1:2.

The Roman tiles from the Netherlands can be dated to the 2nd and 3rd centuries. The tiles found at Wijnaldum-Tjitsma do not have any legionary stamps and so their provenance cannot be ascertained. Tiles from the Netherlands have not yet been analysed chemically and cannot be divided into different categories as was done in England. Roman tiles from England have been distinguished by their fabric and form.[16] The fabrics differ in colour and composition and for the most part the colour varies from red and orange to brown while the temper is generally quartz. Also the flange of the *tegulae* differ. They may have a finger groove on the inside edge of the flange, while the profile of the flange can vary from rectangular to a gently sloping outward edge.

The tiles from Wijnaldum show differences in shape, colour and fabric. The lack of data from other sites in the Netherlands makes it impossible to know whether the presence of tiles with different shape, colour and composition is a normal feature in sites from the Roman Period in the Netherlands. The presence of many different wares and colours in the London wharf excavations have led the excavators to the idea that the tiles may have arrived from many different places and that the tiles were used as ballast in ships and were not imported as building material.[17] The tiles from Tjitsma may also have come from different places and may have had a similar function, as ballast. Their presence in layers of post-Roman date indicates that the tiles were collected and brought to the terp in order to serve other purposes, for instance the fabrication of spindle whorls. Two of these were discovered in a sunken hut and a third (no. 1001) in a sunken hut together with other spindle whorls, needles and loomweights dating to the 7th and 8th century. Apart from the roof tiles, there is one fragment of a different building material, a box flue tile, originally meant to be used as part of a wall in a Roman hypocaust system. This fragment of a box flue tile from Wijnaldum (Tegula fragment no. 26) has a so-called key for mortar, which is a groove made by a sharp utensil (a knife or pointed stick) in a diamond pattern on the outer surface. Similar tiles have been discovered in the Roman Period farmhouse at Rijswijk.

Before we consider the presence of the Roman objects in the terp we first have to discuss the date of the context of the finds and the location of the finds in the terp.

3 ROMAN OR MEDIEVAL: THE DATE OF THE FINDS IN RELATION TO THEIR CONTEXT

3.1 *Introduction to the problem*

It appears that in the excavation at Tjitsma, many Roman finds or finds from the Roman Period have been found in much later contexts.[18] This poses the problem whether these Roman items arrived in the Roman Period and were re-used in later times or whether they arrived at the Tjitsma terp in much later times, and consequently are of no value for the interpretation of the terp in the Roman Period.[19]

The answer to this question must be sought in the first place in the results of the excavation itself. Their assemblage together with the stratigraphical context were used to date the features. Depending on the measure of homogenety of the finds within their stratigraphical context, four values for the reliability of a dating were distinguished. The 4 degrees of reliability are:

1 Most reliable (finds and stratigraphy consistent),
2. Unreliable, generally because downward transport of younger material has taken place,
3. Under certain conditions reliable, and
4. Quite reliable, no other datable finds present, fairly reliable dating based on stratigraphical context.

As a consequence finds from contexts with values 1 or 4 offer a quite reliable dating, while contexts with a value 2 or 3 do not provide a reliable dating. It is therefore very important to take into consideration the measure of reliability attributed to the date of the context in the discussion of the date of the context in which the Roman objects are found.

3.2 *Discussion*

In Tables 1-3 the material is presented with its own date, with the date of the context, and with the evaluation of the date of the archaeological context indicated by a number:

1. Finds and context/stratigraphy consistent = very reliable,
2. Finds and context/stratigraphy non-consistent = unreliable,
3. Finds and context/stratigraphy partly consistent = under certain conditions reliable,
4. No finds, date based on context/stratigraphy = reliable.[20]

The value 'none' in the column Reliability is attributed to those fragments that are not found in a feature but are discovered while excavating in other ways, e.g with a digging machine. The material is divided in three groups: Roman pottery, tiles and wheelthrown terra nigra-like bowls.

For the three groups of material mentioned in Tables 1-3, there are different relations between the original date (of manufacture) and the date of the contexts in which they were excavated. The first relation is that the date of the context corresponds to the date of the fragment. The second possibility is that the date of the context is later than the date of the fragment. A third possibility, fragments discovered in an earlier context, is not present among the finds discussed here.

Table 1. Roman wheelthrown pottery, except terra nigra-like pottery.

Find number	Date	Context	Reliability	Find number	Date	Context	Reliability
2292 Beaker wall frgm.	150-250	550-650	3	6301 Scheldt amphora	150-260	425-550	None
3798 Amphora handle (fits to frgm 11728)	100-300	500-550	3	7650 Beaker wall frgm.	150-300	175-230	1
				9246 Beaker wall frgm.	100-300	425-700	3
5906 Scheldt amphora	150-260	425-550	None	10287 Beaker wall	100-200	770-850	1
5907 Scheldt amphora	150-260	425-550	None	11169 Rim of flagon	100-200	750-900	None
6089 Wall frgm. Amphora?	150-250	800 till present	None	11728 Amphora handle	100-300	175-550	None

Table 2. Terra nigra-like bowls.

Findnumber	Date	Context	Reliability	Findnumber	Date	Context	Reliability
1120 Base	200-400	750-770	3	4923 Wall frgm.	200-400	425-550	None
2495 Wall frgm.	200-400	300-350	1	5155 Rim/shoulder, 3×	200-400	300-350	1
3628 Wall frgm.	200-400	800-850	1	5299 Wall frgm.	200-400	450-500	1
3813 Wall frgm.	200-400	425-500	1	5693 Wall frgm.	200-400	300-350	1
3902 Wall frgm.	200-400	575-625	1	6428 Wall frgm.	200-400	425-550	3
4161 Base	200-400	800-900	None	7395 Wall frgm.	200-400	250-350	1
4615 Rim frgm.	200-400	425-650	None	11060 Wall frgm.	200-400	450-500	3
4484 Rim/shoulder	200-400	175-350	None	11151 Wall frgm.	200-400	Recent (topsoil)	None
4818 Rim/shoulder	200-400	>1500	None				
4820 Wall frgm.	200-400	425-550	None	11728 Rim frgm.	200-400	175-550	None
4851 Rim/shoulder	200-400	300-350	1	11876 Wall/shoulder	200-400	425-500	None

Table 3. Tile fragments.

Findnumber	Date	Context	Reliability	Findnumber	Date	Context	Reliability
394	Roman	750-850	None	1875	Roman	800-900	None
587	Roman	800-900	None	2953	Roman	Uncertain	None
721	Roman	720-750	1	3689	Roman	775-850	1
751	Roman	750-850	None	4242	Roman	750-900	None
1001 Spindle whorl	Roman	650-750 sunken hut	1	4274	Roman	800-900	None
				4881	Roman	775-850	1
				6026	Roman	770-800	None
1078	Roman	750-850	3	6027	Roman	770-800	None
1154	Roman	650-770	1	6039	Roman	770-900	None
1344	Roman	Uncertain	None	6780	Roman	750-900	None
1528 Spindle whorl	Roman	700-750 sunken hut	1	7541	Roman	750-900	None
				9607	Roman	775-900	3
				10142 Spindle whorl	Roman	650-750	4
1529 Spindle whorl	Roman	650-750 sunken hut	1	11173	Roman	700-900	None
				11184	Roman	650-750	3
				11367	Roman	700-900	None
1779	Roman	800-900	None	11400	Roman	700-900	None
1863	Roman	800-900	None				

Of the Roman pottery (terra nigra-like not included), amounting to 11 pieces, only one fragment (7650) was found in a Roman context with a value 1. A second fragment was found in a context with a value 1 dated to the 8th to 9th century. Three fragments were discovered in much later contexts than the date of the fragments but their reliability value is 3 which means that the date of the finds in the context is not consistent and a combination of finds from different dates is found, probably due to soil disturbance due digging up (*opspit*).

This implies that only two fragments have a certain date. One fragment belongs to the Roman Period of the terp and arrived at the terp in that period. The other fragment belongs to a much later period of the terp. Here I see two possible answers: the fragment was brought to the terp after being picked up at a Roman site, outside the terpen area or after being picked up in the neighbourhood, e.g in the terpen area of Friesland. It is impossible to decide which of the two occurred. It is known from other terpen in Friesland that Roman material may turn up in much later contexts. This is demonstrated by the Roman sigillata fragments found in a medieval grave (burial) on the Oldehove terp . The decision whether the fragments were brought to the terp in medieval times or were found on the terp itself cannot be made with certainty. However, the fact that in the same terp a Roman habitation phase was discovered, together with Roman sigillata may point to a locally found Roman object.[21] Since many terpen were inhabited from the Roman Period, or even earlier, and well into the medieval Period it is not astonishing that Roman material turned up in later times and found its way as an isolated (single) Roman find in much later contexts. In later times items from the Roman Period were regarded as very precious, as may be seen from the Roman sigillata fragment found in an 8th-century grave in Nijmegen (exposition, Museum Kam). Items from older periods may have been of great value in later periods. In England for instance objects from earlier periods were used in the Anglo-Saxon Period as Amulets or Curing Stones.[22]

Among the Roman tiles and spindle whorls made out of tiles, discovered in the terp, there is no fragment that was discovered in a Roman context. The majority of the Roman tiles are from contexts dating to the 7th, 8th or 9th century. The evaluation of these finds is with a value 'none'. These dates, however, are relatively reliable because of the relation with other finds in higher and lower levels. Only two of the contexts have a reliability 3, some have no context value and only one example has a context value 1, dating to 775-850. It is a spindle-whorl made out of a tile. We cannot decide whether this spindle whorl is a Roman or a much later product, since the form of the spindle whorl cannot be dated very precisely. This type of spindle whorl may be dated on the base of similar forms of locally made spindle whorls in the first millennium.[23] The context of the finds, in a sunken hut together with other implements of textile working, indicates a date for the spindle whorls in the 8th or 9th century.

There are indications of Roman tiles in the Roman Period of the terpen area, compare the fragment of a Roman tile discovered in an Iron Age farm building in the wattle of a longhouse.[24] In this example the tile may be interpreted as a foundation deposit. The tile fragments from Wijnaldum all seem to have arrived in a later period, as raw material for the manufacture of different items.

The third category of finds are the grey wheelthrown terra nigra-like bowls. In most cases it is very difficult to decide from wall fragments whether these fragments fired in oxidizing conditions to a grey or dark grey, are Roman or medieval. Rim fragments and base fragments are often less difficult to date. The datable rim and base fragments and wall fragments are presented here.

From the total of 21 fragments, eleven have a well-dated context. For five of these the date of the context, first half of the 4th century, corresponds to the date of the fragment. This date corresponds well with the date of comparable hand-made pottery and with the date of similar fragments from Castricum. For both a date in the second half of the 3rd century is also possible (as we discussed above) while the Wijnaldum fragments seem firmly dated to the first half of the 4th century.

The other six fragments also have a well-dated context but appear to date much later although the small (wall)fragments do not appear very different from the 4th century fragments.

4 ROMAN ARTEFACTS IN THE TERPEN AREA IN GENERAL, AND IN THE TJITSMA TERP IN PARTICULAR

In the foregoing we have been able to distinguish between different periods in which the Roman material has been used in the excavated parts of the terp. As a consequence we will have to give an explanation for three phenomena: the presence of

Roman material in an area far beyond the Roman Empire in three different periods: first the Roman Period (contemporary with the date of the object itself) and second Roman artefacts in post-Roman contexts which are sometimes a combination of finds from different periods (as a result of soil disturbance by digging) and which sometimes, the third option, have a purely post-Roman dating among material from a certain period, so that the Roman object is an exception.

The presence of Roman artefacts in the terp region situated at that time outside the Roman Empire, has been given much attention in the last few years. The explanation for their presence varies from trade, homecoming soldiers to a regulated and controlled supply of objects. The phenomenon of soil disturbance causing objects from different periods to be found in a single context is typical of a spot which was occupied for a long time, in this case the terp. The presence of Roman material in (early-)medieval contexts on the other hand is a new phenomenon which has been attributed to the plundering of Roman castella in the medieval period and to the excavation of Roman objects on the site[25]. The first and last phenomena can also be observed in areas south of the Roman limes where Roman objects have been found in (graves from) a later period. It is difficult to establish whether the Roman artefacts came from the surroundings or whether they were brought from abandoned castella. It would appear that there is no convincing explanation for the presence of Roman objects in later contexts. It will have to be sought in each group of materials and within these, according to type. This is evident also from the data given by Van der Vin in his contribution on the Roman coins. In his opinion, there is a different explanation for the presence of silver coins and for that of copper money. The first category appears to have slowly shifted northwards, while the second category, which does not occur in the castella along the limes, probably reached the terpen area in the Roman Period only to turn up again there later, thus finding its way into later contexts.

For the finds discussed in this article, the various kinds of pottery and the roof tiles, there are also various explanations, in my view. The Roman pottery from Tjitsma discussed in this article was only found once in an exclusively Roman context. One fragment in an 8th-9th-century context whereas most fragments were found in layers dating in all probability from after the Roman Period, although they do not have a reliable dating. The handle of an amphora found in two pieces in different find conditions (one context dating from 175 and the other from 500-550) shows that there may well be Roman material present in the 6th-century layers. Apparently just a few sherds from a layer below somehow got into them. For this reason I assume that the amphora sherds from the 425-550 context (just as half the handle of the amphora) belong to the Roman Period of the terp. I also assume that this type of pottery (fragments) did not receive the attention in the post-Roman Period that the Roman sigillata sherds did. After all, the latter are found in post-Roman graves. Amphoras and beakers most probably came to the terp in the Roman Period. This is certainly the case with the terra nigra-like pottery. The dates of a number of well-dated contexts in the first half of the 4th century correspond to the dating of the material itself. Nevertheless, fragments are also found here in a later, well-dated context, varying from the 5th to the 9th century. On the other hand, there is no evidence that roof tiles were imported in the Roman Period. A number of fragments which were found in a well-datable, post-Roman context (including spindle whorls made from a Roman tile) show that Roman roof tiles and tile fragments were used as material for other products until into the 8th or 9th century. Take, for instance, their use as a material for spindle whorls.

As a result of the excavations in the terp Tjitsma we now have more information on the presence of Roman artefacts in an area beyond the limits of the Roman Empire in different periods of time. Future excavations may help to elucidate the many new questions that have been raised by the excavation at Wijnaldum.

ACKNOWLEDGEMENTS

My thanks are due to the following persons: R. van Dierendonck, M. Erdrich, J.-K. Hagers, W.J.H. Verwers and E. Taayke, who provided very useful information on the identification of the material.

CATALOGUE

The date of the archaeological context is followed by a number in square brackets indicating the reliability of the date. Finds with inventory numbers marked with * are shown in the figures.

Black-slipped and colour coated finewares

1. Wall fragment of a beaker, fine black-slipped ware, soft orange red fabric with black coating and white barbotine. Probably Oelmann (1914), type 33a, Technique d.
Measurements: 1.2 × 1.3 × 0.2 cm
Date: IIB-III
Context date: 550-650 [3]
Inv. no. W91/2292*

2. Wall fragment of a beaker, fine black-slipped ware, soft orange red clay with black coating. Technique d. Cf. no. 1.
Measurements: 1.7 × 1.2 × 0.2 cm
Date: IIB-III
Context date: 175-230 [1]
Inv. no. W92/7650

3. Wall fragment, probably of a beaker, fine colour-coated ware. White clay with black coating and in one spot light brown. Fabric Oelmann Technique b.
Measurements: 3 × 3.5 × 0.5 cm
Date:II-III
Context date: 425-700 [3]
Inv. no. W93/9246

4. Wall fragment of a beaker ? White clay with yellowish brown coating. The wall shows a very slight rib on the outside. For the fabric compare Stuart (1977), Technique b.
Measurements: 2 × 1.3 × 0.3 cm
Date: II
Context date: 770-850 [1]
Inv. no. W93/10287

Smooth-walled pottery (Fig. 1)

1. Rim fragment of flagon (jug). Hard fabric, white with orangy red to grey smoothed surface (slip?). The form of the rim comparable to Stuart type 110.
Measurements: 3.4 × 5.2 × 0.6 cm, diameter lip 5 cm
Date: II
Context date: 750-900, no value
Inv. no. W93/11169*

Flat-based Amphoras

1. Flat handle with a central groove. The form of the handle may be of a flat-based amphora, compare Stuart, types 131 and 132. Hard, fine textured fabric, light orange-brown. Two fitting fragments from different contexts.
Measurements: 10.5 × 4.3 cm
Date: II-III
Context date: 175-550 (11728), no value and 500-550 (3798) [3]
Inv. nos. W91/3798 and W93/11728*

2. Three fragments of an amphora with flat base and with traces of white slip on the outside: a base fragment with slight base-ring, a wall fragment and a shoulder fragment. Fabric, reddish yellow (5 YR 6/6) and on the interior pinkish grey to light brown (7.5 YR 6/2 to 6/4). The core is light grey (2.5 Y 7/1).
The amphora belongs to the Scheldt Valley amphoras, originating from the Lower Scheldt Valley (Van der Werff et al. 1997).
Measurements: base 4 × 7.8 × 0.7 cm, diameter base 12 cm
Wall fragment: 5.8 × 6.8 × 0.9 cm
Wall fragment 6.5 × 5.2 × 0.7 cm
Date: ca 90-260
Context date : 425-550, no value
Inv. nos. W92/5906*, 5907, 6301

3. Wall fragment of an amphora (?) The wall is ribbed horizontally. Soft fabric, orange with traces of white slip on the exterior. The fabric is similar to the foregoing.
Measurements: 6.2 × 8.4 × 1 cm, diameter 44 cm?
Date: 120-260
Context date: unknown
Inv. no. W92/6089*

Terra nigra-like bowls (Fig. 2)

Rim fragments

1. Rim fragment, broken in two. Pronounced shoulder with straight slightly everted lip. Grey fabric. Traces of burnishing on both sides.
Measurements: 4.9 × 10.3 × 0.4 cm, diameter 20 cm
Date: II-IV
Context date: 300-350 [1]
Inv. no. W92/5155(a)*

2. Rim fragment, pronounced shoulder with straight slightly everted lip. Dark grey fabric. Surface on the outside worn away. The clay shows small particles of a brownish red colour.
Measurements: 5.4 × 6.2 × 0.5 cm, diameter 17 cm
Date: III-IV
Context date 300-350 [1]
Inv. no. W92/5155(b)*

3. Rim fragment. Angular shoulder with everted lip. Dark grey fabric, the outer surface is worn away. The clay shows small particles of brownish red colour.
Measurements: 4 × 3.1 × 0.5 cm
Date: III-IV
Context date: 300-350 [1]
Inv. no. W92/5155(c)*

4. Rim fragment, angular shoulder with everted lip. Hard fine textured, dark grey fabric burnished and partly worn away on both surfaces.
Measurements: 4.8 × 3.3 × 0.5 cm
Date: III-IV
Context date: 300-350 [1]
Inv. no. W91/4851*

5. Rim fragment (and small shoulder fragment), angular shoulder and slightly everted lip. Hard fine textured light grey fabric.
Measurements: 6 × 6 × 0.5 cm, diameter ca 20 cm
Date: III-IV
Context date: after 1500, no value
Inv. no. W91/4818*

6. Rim fragment and two wall fragments. Everted lip and slightly angular shoulder. One of the wall fragments has a horizontal groove on the outside at the transition from body to shoulder. Fine textured grey fabric with black slip shining on the outside.
Measurements: 3 × 3 × 0.5 – 0.4 cm, 2 × 2.2 × 0.3 cm, 3.5 × 0.8 × 0.3 cm
Date: III-IV
Context date: 300-350 [1]
Inv. no. W91/4851*

7. Rim fragment with everted lip, rounded shoulder. Grey fabric.
Measurements: 6 × 6 × 0.5 (lip) – 0.7 (wall), diameter 22 cm
Date: III-IV
Context date: 175-350, reliability: none
Inv. no. W91/4484*

8. Rim fragment with everted lip and horizontal grooves on the lower part of the lip. Compare with Taayke's (1996, pp 34 and 55, Fig.28) type 7.
Grey fabric burnished on both sides and with a dark grey core. Small red-brown particles (iron ores) in the clay.
Measurements: 3 × 5.1 × 0.5 cm, diameter 20 cm
Date: III-IV
Context date: 175-550, reliability: none
Inv. no. W93/11728

Wall fragments

1. Wall fragment with angular transition to the shoulder on the interior. Grey fabric burnished on both sides.
Measurements: 2.7 × 3.3 × 0.4 cm
Date III-IV
Context date: 250-350 [1]
Inv. no. W../7395

2. Wall fragment. Grey fabric with dark grey core.
Measurements: 2.9 × 3.1 × 0.4 cm
Date: III-IV
Context date: 300-350 [1]
Inv. no. W92/5155

3. Wall fragment with horizontal groove on the outside and an angular transition to the shoulder on the inside. Grey fabric.
Measurements: 4.2 × 4.8 × 0.4 cm
Date: III-IV
Context date: 300-350 [1]
Inv. no. W91/2495

4. Wall fragment. Grey fabric, burnished to a dark grey on the exterior.
Measurements: 2.5 × 3.6 × 0.4 cm
Date: III-IV
Context date: 300-350 [1]
Inv. no. W92/5693

5. Wall fragment with horizontal groove on the exterior and angular transition to the shoulder. Grey fabric. The form may be compared with Taayke's (1995, pp 28 and 60, Fig. 27 'Situlen' from North Drenthe) type K3c, and from the coastal area of northern Germany, in central Groningen the type K7 dated to the 4th century is similar.[26]
Measurements: 4.5 × 4 × 0.4 cm
Date: III-IV
Context date: unknown
Inv. no. W93/11151

6. Wall fragment. Hard, fine textured grey fabric, surface burnished on the exterior. The fragment shows an inside curving wall and probably belongs to the foregoing fragment.
Measurements: 6.5 × 4 × 0.5 cm
Date: III-IV
Context date: unknown
Inv. no. W93/11151

7. Two wall fragments with rounded shoulder flanked by a groove on either side. Grey fabric. Compare Taayke's (1996, pp 33-55, Fig. 28.4) *Standfussschale* type K7 from central Groningen and wheelthrown terra nigra-like cups Group III from Rijswijk Bloemers 1970, 260).
Measurements: 2.2 × 3.5 × 0.4 cm, 2.5 × 3 × 0.4 cm
Date: III-IV
Context date: unknown
Inv. no. W93/11876

8. Wall fragment with angular shoulder. Compare the rim fragment Inv. 4851 above. Hard fine textured, dark grey fabric burnished on both sides.
Measurements: 5.3 × 3.9 × 0.6 cm
Date: III-IV
Context date: 575-625 [1]
Inv. no. W91/3902

9. Wall fragment. Hard, fine textured grey fabric.
Measurements: 4.4 × 4.6 × 0.4 cm
Date: III-IV
Context date: 425-550, reliability none
Inv. no. W91/4820

10. Wall fragment. Hard, fine textured grey fabric.
Measurements: 5.8 × 3.8 × 0.5 cm
Date: III-IV
Context date: 425-550 [1]
Inv. no. W91/4923

11. Wall fragment. Hard, fine textured grey fabric.
Measurements: 2.5 × 3.4 × 0.5 cm
Date: III-IV
Context date: 800-850 [1]
Inv. no. W91/3628

12. Three wall fragments. Hard fine textured grey fabric, surface burnished. One of the fragments has a slightly angular shoulder and an everted rim.
Measurements: 3 × 3 × 0.5 cm, 2 × 1.9 × 0.3 cm, 1.4 × 3.5 × 0.4 cm
Date: III-IV
Context date: 300-350 [1]
Inv. no. W91/4851

13. Wall fragment. Hard, fine textured grey fabric, surface burnished.
Measurements: 3 × 2 × 0.4 cm
Date: III-IV
Context date: 425-500 [1]
Inv. no. W91/3813

14. Wall fragment. Hard, fine textured grey fabric with burnished dark grey surface.
Measurements: 2.8 × 2.6 × 0.5 cm
Date: III-IV
Context date: 450-500 [1]
Inv. no. W92/5299

15. Wall fragment. Hard, fine textured grey fabric with burnished grey surface.
Measurements: 2.2 × 1.3 × 0.5 cm
Date: III-IV
Context date: 450-500 [3]
Inv. no. W93/11060

Base fragments

1. Foot fragment. Grey fabric. The foot is high and slightly concave.
Measurements: diameter foot 5.5 cm, height 2 cm
Date III-IV
Context date: 750-770 [3]
Inv. no. W91/1120*

2. Foot fragment. The foot is flat and low, the wall curves out obliquely.
Grey fabric.
Measurements: diameter foot 4.5 cm, height 1.5 cm
Date: III-IV
Context date: 800-900 [none
Inv. no. W91/4161*

Possibly Roman pottery (Fig. 2)

Roman Bowl?

1. Rim fragment of bowl. Grey fabric with dark shining slip. The lip is straight and the rim is rounded outwards. A horizontal band bordered with a groove on each side is the transition to the body of the bowl. The form of the bowl with the rounded rim is Roman. No parallel came to my notice.
Measurements: 7.7 × 4 cm
Date: III-IV?
Context date: 425-650, reliability: none
Inv. no. W 91/4615*

Tiles (Fig. 3)

1. Tegula fragment with rectangular flange profile. Finger groove on the inside edge of the flange. Orange in colour, fine fabric with occasional quartz.
Measurements: 13 × 9.5 × 2.5 cm, height flange 5 cm
Date: Roman Period
Context date: unknown
Inv. no. W93/11367*

2. Tegula fragment with rectangular flange profile. Finger groove on the inside edge of the flange. Orange in colour, fabric with quartz.
Measurements: 9 × 8 × 3 cm, height flange 4.5 cm
Date: Roman Period
Context date: 775-900 [3]
Inv. no. W93/9607*

3. Tegula fragment with rounded flange profile, sloping outwards at the outside. Finger groove on the inside edge of the flange. Colour brownish orange. Fabric with quartz.
Measurements: 13 × 7 × 2.5 cm, height flange 5 cm
Date: Roman Period
Context date: uncertain
Inv. no. W91/1344*

4. Tegula fragment with rounded flange profile. Finger groove on the inside edge of the flange. Colour pink, well-fired, small quartz grains.
Measurements: 6 × 4.2 cm, height flange 4.7 cm
Date: Roman Period
Context date: 770-800, reliability none
Inv. no. W92/6026

5. Tegula fragment. Colour orange-red.
Measurements: 6.4 × 5.5 × 2.4 cm
Date: Roman Period
Context date: 770-800, reliability none
Inv. no. W92/6026

6. Tegula fragment. Flange flat on top. Colour light brown. Fabric with many different inclusions.
Measurements: 7.3 × 5.2 cm, height flange 4.7 cm
Date: Roman Period
Context date: 750-850 [3]
Inv. no. W91/1078

7. Tegula fragment. Colour brownish-orange.
Measurements: 10 × 8.8 cm
Date: Roman Period
Context date: 750-850, reliability none
Inv. no. W91/1078

8. Tegula fragment. Colour pink. Fabric with quartz up to 0.6 cm.
Measurements: 9.2 × 5 cm
Date: Roman Period
Context date: 750-850, reliability none
Inv. no. W91/1078

9. Tegula fragment. The outer straight side of the flange still

remains and a small part of the finger groove on the inside edge of the flange. Colour orange. Sandy fabric with quartz inclusions.
Measurements: 8.1 × 3.6 cm, height flange 5.1 cm
Date: Roman Period
Context date: 775-850 [1]
Inv. no. W91/3689

10. Tegula fragment. Colour orange and various shades of brown.
Measurements: 3.5 × 4.3 cm, height flange ca 4.9 cm
Date: Roman Period
Context date: unknown
Inv. no. W92/7541

11. Tegula fragment. Colour orange.
Measurements: 4.5 × 4 cm
Date: Roman Period
Context date: 800-900, reliability none
Inv. no. W91/4274

12. Tegula fragment. Finger groove on the inside edge of the flange. Colour greyish brown. Fabric with inclusions of different sizes.
Measurements: 5.8 × 5.5 cm, height flange ca 5.3 cm
Date: Roman Period
Context date: unknown,
Inv. no. W93/11400

13. Tegula fragment. Colour light brown and orange.
Measurements: 11.3 × 8.5 cm, thickness 3.5 cm
Date: Roman Period
Context date: 775-850 [1]
Inv. no. W91/4881

14. Tegula fragment. Finger groove on the inside edge of the flange and a rounded profile. Orange-red colour. Fabric with fairly common quartz of different sizes.
Measurements: 8.4 × 6.5 cm
Date: Roman Period
Context date: 650-750 [3]
Inv. no. W93/11184

15. Tegula fragment. Colour yellowish-brown. Fabric with small air-holes and imprints of organic material.
Measurements: 7.5 × 5 cm
Date: Roman Period
Context date: 650-750 [3]
Inv. no. W93/11184

16. Tegula fragment. Colour orange brown. Fabric with different sizes of inclusions.
Measurements: 4.5 × 4.7 cm
Date: Roman period
Context date: 800-900, reliability none
Inv. no. W91/1875

17. Tegula fragment. Colour orange. Fairly fine sandy fabric.
Measurements: 6 × 6.2 × 2.5 cm
Date: Roman Period
Context date: unknown
Inv. no. W93/11173

18. Tegula fragment. Colour red with pink surface. Hard well-fired tile.
Measurements: 5 × 4.5 × 2.6 cm
Date: Roman Period
Context date: 750-900, reliability none
Inv. no. W92/6780

19. Tegula fragment. Colour orange. Sandy fabric.
Measurements: 5.4 × 3.8 cm
Date: Roman Period
Context date: 770-800, reliability none
Inv. no. W92/6027

20. Tegula fragment? Colour yellow and orange. Fabric with marbled composition in orange and yellow bands
Measurements: 2.8 × 3.1 cm
Date: Roman Period
Context date: 770-800, reliability none
Inv. no. W92/6027

21. Tegula fragment. Colour yellowish-brown. Sandy fabric.
Measurements: 5.8 × 3.2 cm
Date: Roman Period
Context date: 750-850, reliability none
Inv. no. W91/751

22. Tegula fragment. Colour light brown. Sandy fabric with different sizes of grains.
Measurements: 9 × 8.6 × 2.5 cm
Date: Roman Period
Context date: unknown
Inv. no. W91/578

23. Tegula fragment. Colour orange. Fairly fine, sandy fabric.
Measurements: 3.5 × 3.2 × 2.6 cm
Date: Roman Period
Context date: 770-900, reliability none
Inv. no. W92/6039

24. Tegula fragment. Colour brownish-orange. Fabric with inclusions of different sizes.
Measurements: 8 × 6.3 × 2.6 cm
Date: Roman Period
Context date: 750-850, reliability none
Inv. no. W91/394

25. Imbrex fragment. Shallow parallel grooves at right angles to the convex surface. Colour red. Fine, sandy fabric.
Measurements: 8.5 × 6 × 2.2 cm
Date: Roman Period
Context date: 750-850, reliability none
Inv. no. W91/394

26. Fragment of box flue tile with a diamond pattern impressed by a knife to provide a key for mortar. Compare a fragment from Rijswijk (Bloemers 1979, p. 313, no. 714) and from London (Betts 1986, p. 250, no.16. 58).
Measurements: 12.5 × 7 × 2.3 cm
Date: Roman Period
Context date: 650-770 [1]
Inv. no. W91/1154*

Spindle-whorls made out of Roman tiles (Fig. 4)

1. Globular spindle-whorl with vertical hole and flattened at both sides
Measurements: diameter 3.5, height 2 cm
Date: 0-1000
Context date: 650-750 [1]
Inv. no. W91/1001*

2. Disc-shaped spindle-whorl with vertical hole
Measurements: diameter 4.5, height 1.5 cm
Date: 0-1000
Context date: 700-750 [1]
Inv. no. W91/1528*

3. Globular spindle-whorl with vertical hole and flattened at both sides
Measurements: diameter 4 cm, height 2 cm
Date: 0-1000
Context date: 650-750 [1]
Inv. no. W91/1529*

4. Globular spindle whorl with vertical hole and flattened at both sides
Measurements: diameter 5 cm, height 2 cm
Date: 0-1000
Contex date: 650-750 [4]
Inv. no. W93/10142*

REFERENCES

BETTS, I.M. 1986. Brick and tile. In: T. Dyson, *The Roman quay at St Magnus House, London. Excavations at New Fresh Wharf*, London (London and Middlesex Archaeological Society, special paper no. 8), 246-252.

BLOEMERS, J.H.F. 1978. *Rijswijk (Z.H.), 'De Bult' – Eine Siedlung der Cananefaten*, Amersfoort (Nederlandse Oudheden 8).

ERDRICH, M. (in press). Terra Nigra Fussschalen wie Chenet 342 oder Gellep 273: eine salisch-frankische Keramikgattung.

ES, W.A. VAN 1967. *Wijster. A native village beyond the imperial frontier*, Groningen.

ES, W.A. VAN & M. MIEDEMA (eds) 1970-1971. Leeuwarden: small terp under the Oldehove cemetery, *Berichten van de Rijksdienst voor het Oudheidkundig Bodemonderzoek* 20-21: 89-117.

GALESTIN, M.C. 1991. De opgravingen door van Giffen in Hatsum, bij Dronrijp (1920-'22), *Jaarverslagen van de Vereniging voor Terpenonderzoek* 75: 86-93.

GIFFEN, A.E. VAN 1924. Voortgezet onderzoek in de terp te Hatsum bij Dronrijp, *Jaarverslagen van de Vereniging voor Terpenonderzoek* 7-8: 9-47.

HAGERS, J.-K.A. & M.M. SIER (eds) (in press). *Castricum-Oosterbuurt, bewoningssporen uit de Romeinse tijd en middeleeuwen*, Amersfoort (Rapporten Archeologische Monumentenzorg, 53).

HALPAAP, R. 1983. Verzierte Terra Nigra-Fussschalen vom Typ 'Gellep 273'. Anhaltspunkte zur zeitlichen Einordnung und Provenienz, *Boreas* 6: 291-303.

LAMMERS, M. 1994. Dakpannen op De Horden. In: W.A. van Es & W.A.M. Hessing (eds), *Romeinen, Friezen en Franken in het hart van Nederland. Van Traiectum tot Dorestad 50 v.C.-900 n.C.*, Utrecht etc. 161-168.

MEANEY, A.L. 1981. *Anglo-saxon amulets and curing stones*, Oxford (British Archaeological Reports, British series 96).

MIEDEMA, M. 1983. Vijfentwintig eeuwen bewoning in het terpenland ten noordwesten van Groningen, Amsterdam (thesis Vrije Universiteit).

OELMANN, F. 1914. *Die Keramik des Kastells Niederbieber. Materialen zur römisch-germanischen Keramik I*, Frankfurt a. M.

STUART, P. 1977. *Gewoon aardewerk uit de Romeinse legerplaats en de bijbehorende grafvelden te Nijmegen*. Beschrijving van de verzamelingen in het Rijksmuseum G.M. Kam te Nijmegen, 2nd edition, Leiden.

TAAYKE, E. 1990/1995/1996. Die einheimische Keramik der nördlichen Niederlände, 600 v.Chr. bis 300 n.Chr., Teil I: Westergo (Friesland), Teil II: Nord-Drente, Teil III: Mittel Groningen, Oostergo (Friesland), *Berichten van de Rijksdienst voor het Oudheidkundig Bodemonderzoek* 40: 101-222, 41: 9-102, 42: 9-161.

WERFF, J. VAN DER, H. THOEN & R. VAN DIERENDONCK 1997. Amphora production in the lower Scheldt Valley (Belgium)? The Valkenburg-Marktveld evidence, *Rei Cretariae Romanae Fautorum Acta* 35: 63-70.

NOTES

1. Compare Stuart (1977, p. 55, type 131).
2. Van der Werff et al. (1997).
3. Stuart (1977, pp 44-45, type 110).
4. Oelmann (1914, type 33a, technique d).
5. Compare Stuart (1977, p. 20, technique b).
6. Van Es (1967, pp 166-168).
7. Bloemers (1978, pp 260-261, type III).
8. Taayke (1996, p. 110, type K5).
9. Van Es (1967, p. 324, Fig. 175).
10. Taayke (1996, p. 179, Fig. 9).
11. Taayke (1966, p. 127 note 75).
12. Halpaap (1983) and Erdrich (in press).
13. Van Es (1967, pp 158-160).
14. Erdrich (in press).
15. Lammers (1994, pp 161-166).
16. Betts (1986).
17. Betts (1986, p. 252).
18. See the contributions of Van der Vin, Volkers and Erdrich (this volume).
19. Volkers (this volume) and Van der Vin (this volume).
20. See Gerrets & de Koning (this volume).
21. Van Es & Miedema (1971).
22. Meaney (1981).
23. Miedema (1983, p. 131, type XXB, Fig. 85).
24. Van Giffen (1924, p. 35, Fig. 16) and Galestin (1991).
25. See the explanations given by Van der Vin (this volume).
26. Taayke (1995, Fig. 27, pp 28 and 60; 1996, p. 55, Fig. 28).

Continuity or discontinuity: Native and Roman metal finds

M. ERDRICH

1 INTRODUCTION

During the large-scale excavations that took place at the Tjitsma terp, near Wijnaldum in addition to finds and soil traces from the Early Middle Ages, clear evidence was also found of Roman habitation.

The results of the archaeological investigation at the Tjitsma terp gave rise to the formulation of two problems: first the question of when settlement of the Tjitsma terp began, and second, and even more important, the question of settlement continuity during the transition period between the Roman period and the Early Middle Ages, from the late 3rd and 4th to the 6th century. The answer to these questions must be sought in an analysis of the native Germanic and, though to a lesser extent, the imported Roman finds.

In this article, attention is focused on bronze and silver finds from the Wijnaldum excavation. A number of severely cut bronze objects of unknown origin, mainly bronze sheet or small bronze bars, have been excluded. These intentionally cut bronze fragments were intended for bronze scrap recycling. Finds from private collections, mostly metal-detector finds, will not be discussed in this article, even if they were discovered on the site of the later excavation. Already during the excavation it was observed that some of the Roman finds were discovered in a clearly early-medieval context. This observation will be discussed on the basis of a comparison of the context dating of Germanic and Roman finds.

2 GERMANIC FINDS (Figs 1 and 2)

The recognisable native-Germanic metal finds from the excavations at Wijnaldum are confined to bronze and silver brooches. Ornaments such as pendants or pins, bracelets, neck-rings, belts or other clothing accessories familiar to us from Germanic cemeteries and settlements in other parts of the Western and Central European barbaricum,[2] were not found at Wijnaldum. This picture is confirmed by an inventory of several private collections with hundreds of metal finds from the province of Friesland.[3]

None of the Germanic brooches show any trace of cutting or hacking, all recognized damage is due to functional use. Several fibulae lack their pin or spring, the parts which are most susceptible to wear. Even the silver brooches are complete and show no signs of any secondary damage which might be associated with scrap recycling.

Almgren 19 (Nos 1-2)
Among the oldest finds is an Almgren 19 fibula (9644*[4], Fig. 1), which was worn in the first half of the 1st century until the Flavian period (ca AD 0-70). The fragment of an undeterminable, rather large iron brooch with a round perforation in the catchplate probably belongs to the same period (3203*, Fig. 1).[5]

Almgren group VII (Nos 3-22)
The majority of all Germanic finds from Wijnaldum belong to the well-known Almgren VII brooches with high catchplate. Based on the form of the bow and the foot, W. Matthes distinguishes four different 'series' within the Elbian circle. His typological description is generally accepted.[6]

- Ser. 1 (S-shaped bow): 9827,
- Ser. 2 (foot cut off straight): 5141*,
- Ser. 3 (knee-shaped bow, foot cut off straight): 3081*, 3851* (lead), 3867*,
- Ser. 4 (bow and catchplate merge): 2430, 6133, 6266, 6624*, 8053, 8197*, 9164, 9312*, 9325*, 10758*, 11093*, 11875.

In addition to these generally known variants we find some small disc-shaped brooches closely related to the Almgren VII brooches with high catch-

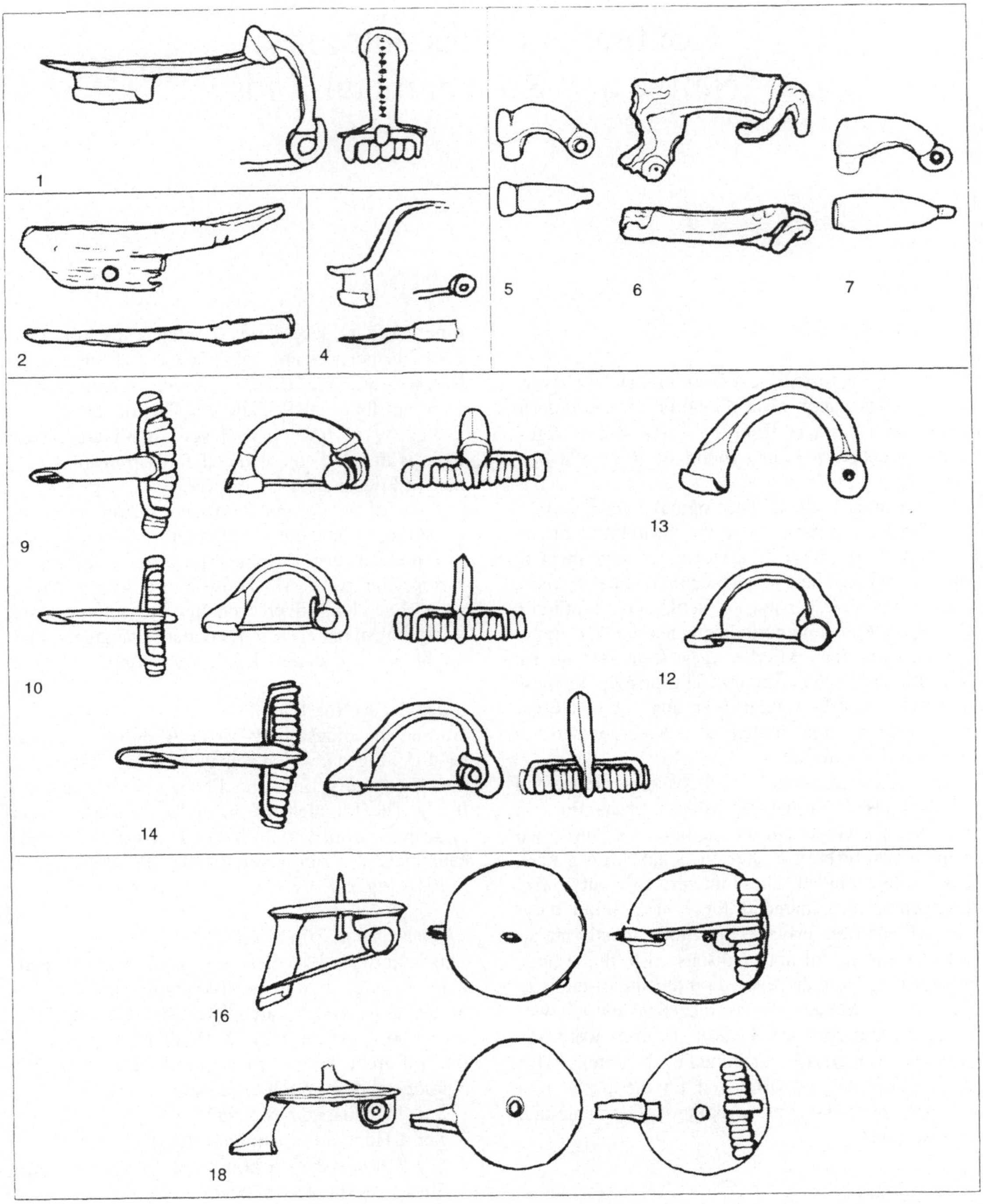

Figure 1. Germanic brooches from Wijnaldum (1:1).

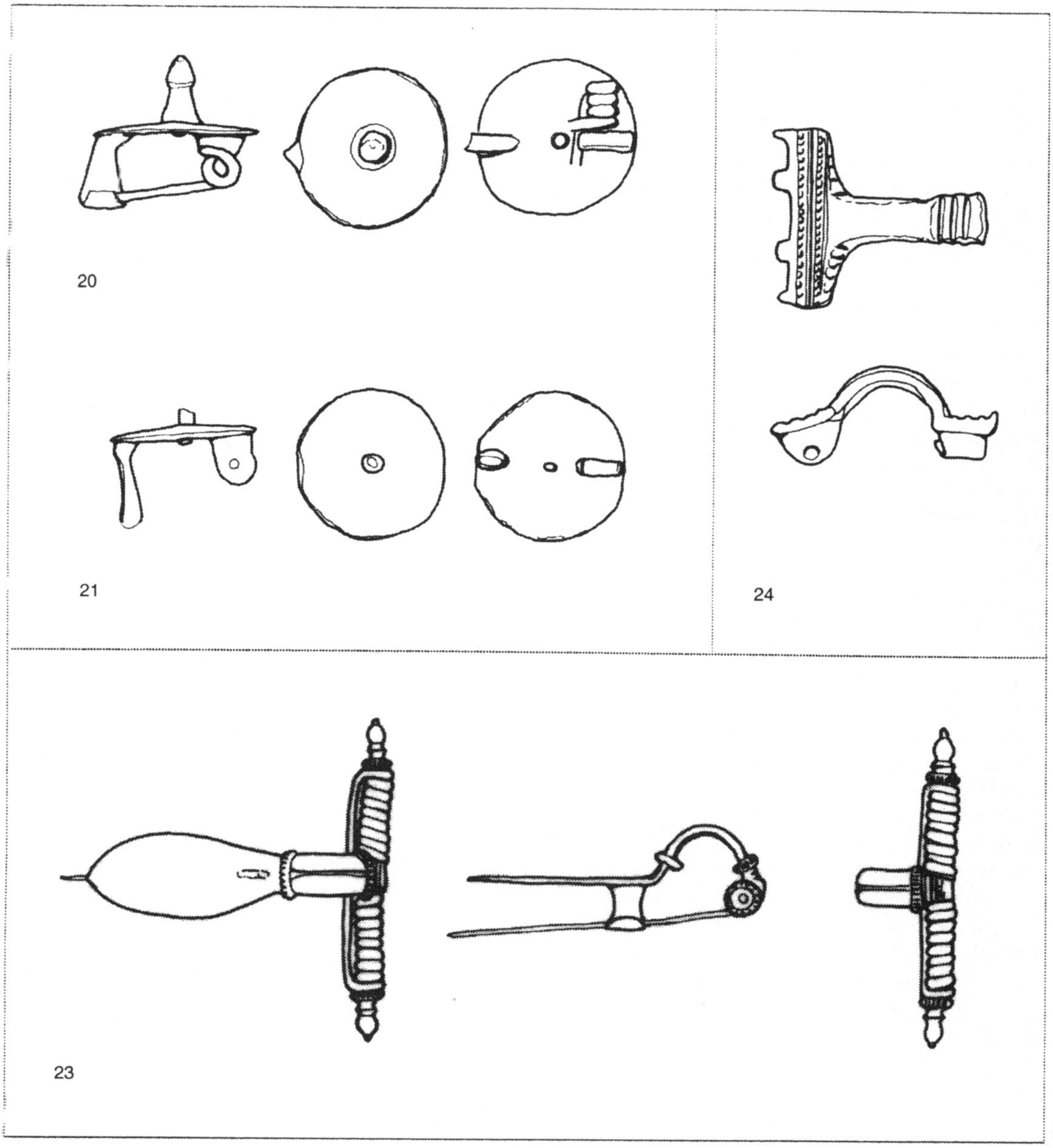

Figure 2. Germanic brooches from Wijnaldum (1:1).

plate. The construction of this type of disc fibula shows great similarities to that of the two-piece bow brooch with high catchplate Almgren VII, Ser. 3 and 4. Instead of a bent bow, it was hammered out into a disc by the craftsman. A comparison between the disc fibulae (11093* or 11443*, Figs 1 and 2) and the bow brooch Almgren VII, Ser. 4 (6133 of 8053, both Fig. 1) makes the similarities clear. Characteristic of the disc-shaped variant of the two-piece bow brooch is that catchplate, disc and the plate with the hole for the spring consist of a single piece. A bronze or iron pin was subsequently hammered through the disc to which a probably rather thin ornamental plate was attached, and which is usually missing. As in the case of the other two-piece bow brooches Almgren VII, Ser. 3

or 4, spring was set in the springholder. Almgren VII, Ser. 1-4 brooches date from the late 2nd to the end of the 3rd century. In view of the great morphological and technical similarities, it seems likely that the disc-shaped variant of the Almgren VII group may also be dated to this period.

The two-piece bow brooch with high catchplate Almgren VII, Ser. 1-4 and the disc-shaped variant may be considered the dominant fibula form in the coastal area of the North Netherlands during the late 2nd and 3rd centuries. Several hundred brooches of this type are known from excavations, but particularly from private collections. Although the majority are made of bronze, silver bow brooches are not unusual. The regional distribution of the two-piece bow brooch Almgren VII, Ser. 3 and 4 was most recently investigated by A. Böhme and J.K. Haalebos. Apart from an obvious concentration north of the River Elbe in Brandenburg and Mecklenburg, their presence in several auxiliary forts in the Roman provinces of Upper and Lower Germany is striking. Several fibulae of this type are also found in the hinterland of the immediate border zone.[7] The scarcity of the disc-shaped variant of the Almgren VII, Ser. 3 and 4 among the finds of the Saalburg and Zugmantel castella is striking.[8]

The discovery of a lead fibula of the Almgren VII, Ser. 3 type (3851*, Fig. 1) should not be regarded as evidence of any production on the Tjitsma terp. Serious damage has made a detailed analysis of the construction of this fibula virtually impossible. Strong evidence against the assumption that this item may have been used as a mould for Almgren VII, Ser. 3 brooch, is the fact that some coils of the spring have remained. As this type of brooches consists of two pieces, namely the bow with the catchplate and a separately manufactured spring and pin, which is placed into a hole in the upper part of the bow, this object cannot be interpreted as a model for a two-piece brooch.

Two-piece bow brooch with oval foot plate (No. 23)
The discovery of a silver two-piece bow brooch with oval foot plate (11315*, Fig. 2) is remarkable. Almost identical brooches are known from Maurik (Prov. of Gelderland) and Haren (Prov. of Groningen). Fibulae of this type, three of which are now known of in the Netherlands, may be related to a variant of the two-piece brooch with trapezium-shaped foot described by H. W. Böhme. Fibulae of this type are characterised, just as our specimens, by a broad spring with terminal knobs on its axis, a short and sharply bended bow and a comparatively high catchplate, which is set into an incision in the footplate just behind the bow. The footplates are usually covered with an ornamental (gilded) foil. Most of these fibulae are made of silver. No remains of these covering foils were left on any of the Dutch specimens, although their presence is highly likely. These brooches can be dated into the second half of the 4th and the early 5th century.[9]

Stützarm-brooches (No. 24)
Only a single 'Stützarm'-brooch was found during excavations at the Tjitsma terp. This specimen belongs to a group with a rectangular-shaped foot dating in the late 4th and first half of the 5th century.[10]

3 ROMAN FINDS (Fig. 3)

Compared to the Germanic, the spectrum of Roman finds is fairly heterogeneous. Apart from bronze vessels, brooches and other ornaments, fragments of military equipment were also found. A bronze spatula and perhaps also a bone pyxis may be included among the medical or cosmetic equipment.

In contrast to the Germanic copper alloy finds, a great number of Roman finds show damage which cannot be ascribed to functional use. Cutting or hammering traces are repeatedly observed. Several objects are badly cut up. We may conclude that at least some of the Roman copper alloy finds were scrap material prepared for recycling.

3.1 *Bronze vessels* (Nos 25-26)

Not more than two of all the copper alloy pieces found during the excavation could be determined as being part of Roman bronze vessels. One of them is a solid cast escutcheon belonging to a bucket with a small base and slightly outward sloping wall like the Eggers type 36. On the reverse of the escutcheon traces of solder, probably a tin-lead alloy remained (9618*). The production of these buckets dates to the Hadrianic and Antonine Period (ca AD 120-180). Buckets of this type are rather exceptional finds in native settlements or cemeteries beyond the Imperial frontiers between the Rhine and the Elbe.[11]

The other piece is a rim fragment of a sieve or strainer like Eggers type 161, their production took place during the second half of the second and the first half of the 3rd century. The rim fragment clearly shows traces of deformation by cutting and

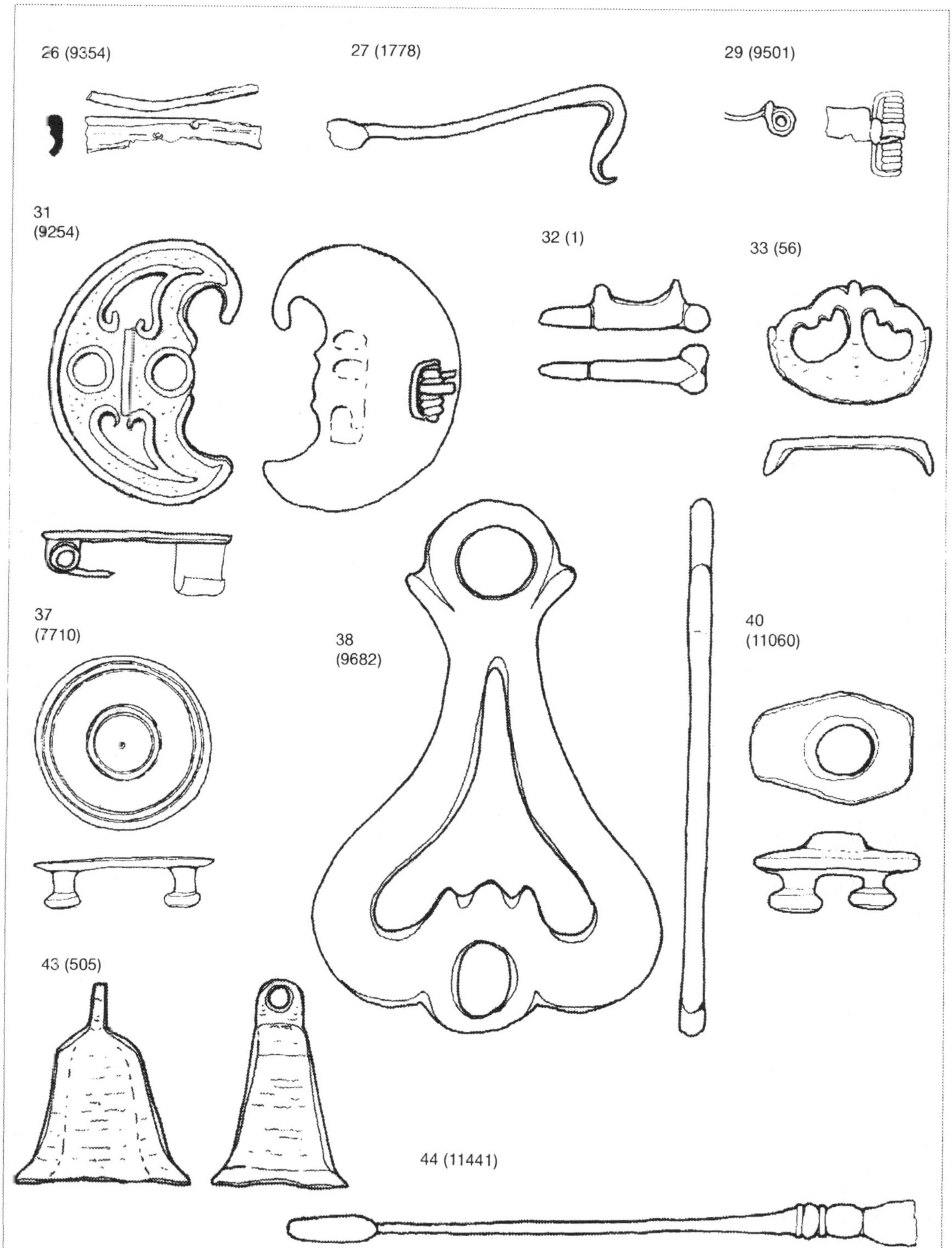

Figure 3. Roman finds from Wijnaldum (1:1).

hammering, which usually indicates metal recycling (9354).

3.2 *Brooches* (Nos 27-31)

A total of 5 Roman brooches or fragments of Roman brooches were found during the excavations.

Almgren 15 (No. 27)
Fragments of four Roman one-piece brooches formed from a round-sectioned wire could be determined. One of them is a fragment of a bronze a so-called 'soldier's brooch'. Its spring has been cut off (No. 1778).

The distribution of these fibulae shows a clear concentration in the Rhine area, mainly in the military forts. Numerous specimens were also found to the north of the Rhine border. Although some brooches of this type may have been worn until the end of the 2nd century, this type seems typical for the Flavian Period and the early 2nd century (ca AD 70-125).[12]

Almgren 16 (Nos 28-30)
Three fragments belong to the one-piece brooches formed from a flat hammered wire. None of the three has been preserved intact. Even though there is no clear evidence of cutting, the pattern of fracture of all three fibulae does not appear to have been caused by daily wear: all three fragments have been broken right across the bow.

Haalebos regards this type as the successor to the *Bogenfibel* Almgren 22 and, based on evidence from Nijmegen, dates it mainly in the second half of the 2nd century. Van der Roest compares them with the Almgren 15 and assumes them to be more or less contemporaneous, i.e. dating to the Flavian Period and early Antonine Period (ca AD 70-140/150).[13]

Enamelled brooch Exner type III, 24 (No. 31)
Flat pelta-shaped brooch with spring. Enamelled brooches are widespread within the provinces of the Empire and date to the late 2nd and early 3rd century. Beyond the frontiers, round brooches with brightly coloured millefiori-coverings are known in some quantity.[14] Except of the one from Wijnaldum, pelta-shaped enamelled brooches are virtually unknown beyond the imperial frontiers[15].

3.3 *Roman military equipment* (Nos 32-41)

Altogether 10 pieces or fragments of Roman military equipment were recovered during the excavation.

The Roman military equipment found during the excavations of the Tjitsma terp generally dates to the late 2nd and 3rd century. Similar finds are known from various native settlements and cemeteries beyond the frontiers of the Roman Empire between the Rhine and the Elbe.[16] It is striking that most of the Roman military equipment from Wijnaldum is heavily damaged. In this respect, they differ from comparative finds from other sites in the Northern Netherlands or North Germany where Roman military objects found were almost always intact, apart from incidential traces of wear. The pattern of fracture on the Wijnaldum specimens can in no way be explained as resulting from normal use. For example, the rather massive loop of the phallus pendant (1*) is broken on both sides at the same level. The reverse of the chape (56*) was cut off exactly in the middle. On the other hand, several small, but cast pieces such as the hexagonal (11060*, 11153) and the flat-headed studs (71, 1211) are undamaged.[17]

3.4 *Miscellaneous* (Nos 42-46)

Roman finger rings of the 2nd and 3rd century are known in some quantity beyond the imperial frontier,[18] so the fragment of a single bronze ring from Wijnaldum is not surprising. The small ring is broken in half and shows the fitting of a lost gem (9531). The Roman origin of another, heavily corroded iron ring cannot be proved satisfactorily and has consequently been omitted.

Bronze bells like the pyramid-shaped bell with a rectangular base from Wijnaldum (505*) belong to the rather frequent Roman imports in barbarian Central Europe.[19] The chronology of bronze as well as iron bells is far from sound, only a general dating in the 2nd to 4th century appears possible.

4 CONTEXT AND DATING OF GERMANIC FINDS

For this survey, there has been a deliberate option for broad dating of the individual Germanic fibulae (Table 1). It is striking that both early-Roman fibulae were found in a context which can certainly be dated in the Early Middle Ages. The same applies to the lead (?) brooch 3851 which is difficult to determine on all counts. It was discovered about 50 m from the traces of habitation dating from the

Table 1. Find and context dating of Germanic fibulae at Wijnaldum.

Find No.	Find-dating	Context-dating	Context-reliability
9644*	1-70	550-650	3
3203	1-70	550-600	1
9827*	175-300	250-350	1
5141	175-300	175-250	1
3081*	175-300	(650-?)	–
3851*	175-300	800-850	3
3867*	175-300	150-250	1
2430	175-300	(800-?)	–
6266*	175-300	(425-?)	3
6624*	175-300	(175-?)	–
9164	175-300	(250-?)	–
9312*	175-300	(250-?)	–
9325*	175-300	300-350	1
10758*	175-300	(800-?)	–
9021	175-300	300-350	–
8197*	175-300	650-650	–
9319	175-300	(175-?)	–
11093*	175-300	(175-?)	–
11151	175-300	(175-?)	–
11154*	175-300	(175-?)	–
11443*	175-300	200-300	1
5135	175-300	300-350	1
11215*	350-400	250-350	1
4697*	350-425	425-550	3

Roman Period. All features close to the area where the lead (?) fibula was found, date to the Early Middle Ages. As far as the context datings are known for brooches with high catchplate Almgren VII, Ser. 1-4 and their disc-shaped variant, they correspond with the dating of the brooches themselves. The same applies to the two late-Roman fibulae.

In view of the absence of Germanic brooches from the 1st and the first half of the 2nd century,the Almgren VII, Ser. 1-4 fibulae and their disc-shaped variant found in a contemporaneous context date from the beginning of the occupation of the terp settlement in the course of the second half of the 2nd century. There are no Germanic metal finds dating from the period following that in which the Almgren VII brooches fell into disuse, i.e. the late 3rd and the first half of the 4th century. Two fibulae, the silver two-piece foot-bow brooch and the *Stützarmfibel*, both dating to the second half of the 4th and perhaps even the first half of the 5th century, appear to be isolated finds. Not until after the middle of the 5th century do Germanic clothing accessories reappear, together with the cruciform fibulae.

Summing up, it may be stated that settlement of the Tjitsma terp at Wijnaldum began during the second half of the 2nd century and probably sharply decreased, if not ended, before 300. The idea of a serious decline of the population of late-Roman terp settlements in Friesland is supported by a preliminary survey of several private collections[20]. Against a large quantity of Almgren VII, Ser. 1-4 brooches and their disc-shaped variants is the incidental occurrence of several Germanic metal finds dating certainly to the 4th and first half of the 5th century. From this it may be concluded that outside Wijnaldum there was a considerable drop in population even before 300.

5 CONTEXT AND DATING OF THE ROMAN FINDS

In the case of the Roman small finds too, a broad dating of the individual find was opted for in order to gain insight into the relation between the dating of the individual find and that of the find context (Table 2). Roughly half of the objects is generally, though not exclusively, associated with Roman military equipment. On the other hand, especially the brooches with flattened bow Almgren 16 are generally expected in more civilian surroundings.

Table 2. Find- and context dating of Roman finds from Wijnaldum.

Find No.	Find-dating	Context-dating	Context-reliability
9618	130-180	(250-?)	–
9354*	150-250	250-350	4
1778*	70-150	(800-?)	–
3624	70-150	800-850	3
9501*	70-150	450-500	3
9564	70-150	500-550	3
9254*	170-230	250-350	1
1*	150-300	Stray find	–
56*	150-300	(175-?)	–
34	150-300	(250-?)	–
71	150-300	(800-?)	–
1211	150-300	750-800	3
7574	150-300	(175-?)	–
7710*	150-300	(425-?)	–
9682*	150-300	450-525	3
10839	150-300	(800-?)	-
11060*	150-300	450-550	1
11153	150-300	(175-?)	–
9531	100-300	250-350	1
505*	100-300	1900-	3
11441*	100-300	(175-?)	1
9253	100-300	(250-?)	–
9470	100-300	(250-?)	–

Figure 4. Distribution of Germanic and Roman finds.

Figure 5. Distribution of Germanic and Roman finds.

Thus, contrary to Germanic finds from Wijnaldum, the Roman objects display a rather heterogeneous structure. They date from the late 1st to the 3rd century.

Just a few of these Roman bronze objects were found in 2nd to 4th-century contexts. None of the four early brooches Almgren 15 and 16 were among these finds, which adds to the idea, that settlement on the terp Wijnaldum did not begin before the late 2nd century. The remaining Roman objects, found in a more or less contemporaneous context fit into the range of Roman finds known from other sites – settlements and grave finds – beyond the imperial frontiers.[21]

The majority of Roman finds from Wijnaldum, however, were found in features dating in post-Roman Periods. Apart from the Roman bronze find, virtually no, or far too few other native Roman Period Germanic finds such as pottery were recovered from these features. This on the first sight rather awkward picture cannot be explained by accidentially digging up older material. Thus a large number of the metal finds, originally coming from the Roman provinces and dated to the late 1st until the 3rd century, did not come to Wijnaldum until after the end of the Roman empire. From the context datings of the Roman copper alloy finds it may be cautiously concluded that Roman bronze objects from the late 1st to the 3rd century were brought to Wijnaldum in the 5th century and again in the 8th century. The cutting and hammering traces which were regularly observed indicate preparation for metal recycling.

6 WIJNALDUM IN A WIDER FRISIAN PERSPECTIVE

The Germanic brooches found during excavations at the Tjitsma terp indicate a beginning of the settlement which was not before the late 2nd century. Both older brooches were found in an early Migration Period context, cutting marks indicate their function as metal scrap prepared for recycling. Surprisingly, typically 4th-century fibulae like the various types of *Stützarmfibel* or other contemporaneous brooches are rare. Not until the second half of the 4th or first half of the 5th century do we see the appearance of a considerable number of early Migration Period brooches.

The dating of the Germanic fibulae leads us to suspect that settlement on the Tjitsma terp came more or less to an end even before the end of the 3rd century, or at least dropped to a level which is hardly evidenced archaeologically. The same conclusions emerge from a – preliminary – inventory of the Germanic fibulae from the present provinces of Friesland and North Holland. Various variants of the Almgren VII types predominate, and less than 1% belong to clearly 4th-century types.

Only a small proportion of the Roman bronze finds was found in a context dating, on the basis of native pottery and stratigraphical considerations, to the Roman Period. The majority of Roman bronze finds, militaria, fibulae and bronze vessels were found in a definitely post-Roman context. These objects must be regarded as scrap metal for recycling, in view of the almost ever-present cutting marks.

CATALOGUE

The catalogue includes all the Roman Period silver, bronze objects and several recognizable iron objects from the excavations at the Wijnaldum-Tjitsma, municipality of Harlingen (Prov. of Friesland). The catalogue is divided into two sections, Germanic finds and Roman finds and appear in order of find numbers. Each object is described and dated separately. Finds which are no longer typologically identifiable due to corrosion or fragmentation have not been included. The same applies to production refuse. Unless stated, all finds are of copper alloy (bronze). * indicates that the object is illustrated.

GERMANIC FINDS

Almgren 19

No. 1 9644*
One-piece brooch with a prominent knob on the bow.
Date range: mainly Ia

No. 2 3203*
Fragment of an undetermined iron brooch with a round hole in the catchplate.
Date range: IA

Almgren Gruppe VII

No. 3 9827
Two-piece bow brooch with high catchplate Almgren VII, Ser. 1. S-shaped bow. Heavily corroded silver, probably gilded.
Date range: IIc-IIIc

No. 4 5141*
Two-piece bow brooch with high catchplate Almgren VII, Ser. 2a. Flattened bow. Heavily corroded.

Date range: IIc-IIIc

No. 5 3081*
Two-piece bow brooch with high catchplate Almgren VII, Ser. 3. Spring and pin missing.
Date range: IIc-IIIc

No. 6 3851*
Two-piece bow brooch with high catchplate Almgren VII, Ser. 3. Heavily corroded lead(?). Part of spring and pin missing.
Date range: IIc-IIIc

No. 7 3867*
Two-piece bow brooch with high catchplate Almgren VII, Ser. 3. Spring and pin missing.
Date range: IIc-IIIc

No. 8 2430
Iron two-piece bow brooch with high catchplate Almgren VII, Ser. 4 Rectangular-sectioned bow. Spring and pin missing.
Date range: IIc-IIIc

No. 9 6266*
Two-piece bow brooch with high catchplate Almgren VII, Ser. 4. Diamond-sectioned bow. 8-turn spring.
Date range: IIc-IIIc

No. 10 6624*
Two-piece bow brooch with high catchplate Almgren VII, Ser. 4. Diamond-sectioned bow. 12-turn spring.
Date range: IIc-IIIc

No. 11 9164
Two-piece bow brooch with high catchplate Almgren VII, Ser. 4. Square-sectioned bow. Spring and pin missing.
Date range: IIc-IIIc

No. 12 9312*
Two-piece bow brooch with high catchplate Almgren VII, Ser. 4. Rectangular-sectioned bow. Spring and pin missing.
Date range: IIc-IIIc

No. 13 9325*
Two-piece bow brooch with high catchplate Almgren VII, Ser. 4. Diamond-sectioned bow. Spring and pin missing.
Date range: IIc-IIIc

No. 14 10758*
Two-piece bow brooch with high catchplate Almgren VII, Ser. 4. Diamond-sectioned bow. 14-turn spring, pin missing.
Date range: IIc-IIIc

No. 15 9021
Two-piece bow brooch with high catchplate Almgren VII. Completely broken.
Date range: IIc-IIIc

Disc-shaped variant Almgren VII, Ser. 3-4

No. 16 8197*
Two-piece silver disc brooch with high catchplate Var. Almgren VII, Ser. 4. Central silver rivet, decorative sheet missing.
Date range: IIc-IIIc

No. 17 9319
Two-piece disc brooch with high catchplate Var. Almgren VII, Ser. 4. Heavily damaged.
Date range: IIc-IIIc

No. 18 11093*
Two-piece disc brooch with high catchplate Var. Almgren VII, Ser. 4. Central rivet, decorative sheet missing.
Date range: IIc-IIIc

No. 19 11151
Two-piece disc brooch with high catchplate Var. Almgren VII, Ser. 4. Central rivet, decorative sheet missing. Heavily corroded.
Date range: IIc-IIIc

No. 20 11154*
Two-piece disc brooch with high catchplate Var. Almgren VII, Ser. 4. Central rivet, decorative sheet missing. Heavily corroded.
Date range: IIc-IIIc

No. 21 11443*
Two-piece disc brooch with high catchplate Var. Almgren VII, Ser. 4. Central rivet, decorative sheet missing.
Date range: IIc-IIIc

No. 22 5135
Spring of a two-piece bow brooch with high catchplate Almgren VII.
Date range: II-III

No. 23 11315*
Silver two-piece foot-bow-brooch. Short sharply bended bow, broad spring with knobs on the endings, thin oval foot and a rather small high catchplate.
Date: ca IVc-VA

Stützarmfibel

No. 24 4697*
Stützarmfibel with rectangular sectioned bow and rectangular foot.
Date: ca IVc-VA

ROMAN FINDS

Bronze vessels

No. 25 9618
Almond-shaped escutcheon of a bucket similar to Eggers type 36. Handle-loop slightly bent inwards. On the reverse remains of solder, tin-lead alloy.
Date: ca 120/130-170/180

No. 26 9354*
Rim-fragment from an unidentified type of bronze vessel, probably sieve or strainer. Secondary cutting marks and hammered.
Date: unknown

Brooch Almgren 15

No. 27 1778*
Bow of a one-piece bow brooch *(Soldatenfibel)* with round-sectioned bow. Secondary cutting marks.
Date range: ca IA-IIA

Brooches Almgren 16

No. 28 3624
Fragment of an iron one-piece bow brooch. On the slightly bent flattened bow two parallel lines of incised dots.
Date range: ca IA-IIA

No. 29 9501*
Fragment of an one-piece bow brooch. On the slightly bent flattened bow two parallel lines of incised dots.
Date range: ca IA-IIA

No. 30 9564
Bow of an unidentified brooch. On the flattened bow two rows of C-shaped incisions, separated by a faint curled double line (probably of one-piece bow brooch Almgren 16).
Date range: ca IA-IIA

Exner type III, 24

No. 31 9254*
Flat, pelta-shaped enamelled disc brooch, slightly damaged, pin missing. Two heart-shaped leaves and two circles in yellow field. Enamel discoloured.
Date range: II-IIIa

Roman military equipment

No. 32 1*
Fragment of a phallic pendant. Loop broken off (Oldenstein type 403-405; Jones & Miket 3.587).
Date range: IB-IIIA(?)

No. 33 56*
Fragment of a scabbard chape with openwork pelta decoration. Reverse cut off (Oldenstein 112; Allason-Jones & Miket 3.401).

No. 34 71
Stud with circular sectioned shank and two flat square heads (Oldenstein 494-503).

No. 35 1211
Stud with circular sectioned shank and two flat circular heads (Oldenstein 494-503; Allason-Jones & Miket 3.869).
Date range: IIB and III

No. 36 7574
Fragment of a scabbard chape (?) with openwork pelta decoration. Rather thin bronze sheet. Secondary cutting marks (cf. Oldenstein 112-113; Allason-Jones & Miket 3.401).
Date range: IIB and IIIc

No. 37 7710*
Large flat circular stud with two circular sectioned disc-headed shanks on the reverse. On the face side two pairs of incised concentric circles. Slightly damaged (Oldenstein type 540).
Date range: IIB and IIIc

No. 38 9682*
Flat Baltheus-pendant. Roughly heart-shaped openwork pendant with two openwork loop holes on either side. In the centre openwork heart- or leaf-shaped opening (cf. Feugère, cat. no. 199, Fig. 21).
Date range: IId-IIIa (?)

No. 39 10839
Fragment of a bronze sheet with pelta-shaped openwork decoration, prob. scabbard chape. Secondary cutting marks (cf. Oldenstein 112-113).
Date range: IIB and III

No. 40 11060*
Long hexagonal stud with raised oval boss, which has a deep median groove (Oldenstein 267-269; Allason-Jones & Miket 3.870).
Date range: IIB and IIIc

No. 41 11153
Heavily corroded long hexagonal stud with raised oval boss, which has a deep median groove (Oldenstein 267-269; Allason-Jones & Miket 3.870).
Date range: IIB and IIIc

Various objects

No. 42 9531
Fragment of a finger-ring, D-shaped sectioned. Oval stone or glass setting is missing.
Date range: II-III

No. 43 505
Pyramid-shaped bell with rectangular base.
Date range:II-III(?)

No. 44 11441*
Fragment of a spoon-probe. Broken, one end of the instrument is missing (cf. Künzl, Figs 14, 15 and 78).
Date range: II-III

No. 45 9253
Fragment of a rightangled decorated plate, prob. of a *Schminkkästchen*. Faint traces of a Perlstab. Corroded. Secondary cutting marks (cf. Korvig Taf. 37).
Date range: II-III

No. 46 9470
Segment of an undecorated cylindrical pyxis, bone.
H = 39 mm, D = rim 37 mm, and D = base 40 mm.
Date range: II-III

REFERENCES

ALLASON-JONES, L. & MIKET, R. 1984. *The catalogue of small finds from South Shields Roman fort*, Newcastle upon Tyne (Society of Antiquaries of Newcastle upon Tyne, Monogr. Ser. 2).

ALMGREN, O. 1923. *Studien über nordeuropäische Fibelformen der ersten nachchristlichen Jahrhunderte mit Berücksichtigung der provinzialrömischen und südrussischen Formen*, 2nd edition, Leipzig (Mannus-Bibliothek 32).

BECKMANN, C. 1969. Metallfingerringe der römischen Kaiserzeit im Freien Germanien, *Saalburg Jahrbuch* 26: 5-106.

BÖHME, A. 1972. Die Fibeln der Kastelle Saalburg und Zugmantel, *Saalburg Jahrbuch* 29: 5-112.

BÖHME, H.W. 1974. *Germanische Grabfunde des 4. bis 5. Jahrhunderts zwischen unterer Elbe und Loire*, München (Münchener Beiträge zur Vor- u. Frühgeschichte 19).

BÖHME, H.W. 1986. Das Ende der Römerherrschaft in Britannien und die angelsächsische Besiedlung Englands im 5. Jahrhundert, *Jahrbuch Römisch-Germanisches Zentralmuseum* 33(II): 469-574.

COSACK, E. 1979. *Die Fibeln der älteren römischen Kaiserzeit in den Germania Libera (Dänemark, DDR, BRD, Niederlande, CSSR)*, Neumünster (Göttinger Schriften zur Vor- und Frühgeschichte 19).

CRUMMY, N. 1983. *The roman small finds from excavations in Colchester, 1971-1979*, Colchester (Colchester Archaeological Reports, 2).

EGGERS, H.-J. 1951. *Der römischen Import im Freien Germanien*, Glückstadt (Atlas der Vorgeschichte 1).

ERDRICH, M. 1994. Waffen im mitteleuropäischen Barbaricum: Handel oder Politik, *Journal of Roman Military Equipment Studies* 5: 199-209.

ERDRICH, M. & M.C. GALESTIN (in prep.). Die Kaiserzeitliche Fibeln aus Friesland (provisional title).

FEUGÈRE, M. 1981. Objets de parure et mobilier metallique de Frejus, *Documents d' Archéologie Méridionale* 4: 130-167.

GECHTER, M. 1980. Die Fibeln des Kastells Niederbieber, *Bonner Jahrbücher* 180: 589-610.

GODŁOWSKI, K. 1970. *The chronology of the Late Roman and Early Migration Periods in Central Europe*, Kraków.

KORVIG, E. 1937. Die Haupttypen des kaiserzeitlichen Emaillarbeiten in Pannonien, Budapest (Dissertationes Pannoniae II, 4.)

KÜNZL, E. 1982. Medizinische Instrumente aus Sepulkralfunden der römischen Kaiserzeit, *Bonner Jahrbücher* 182: 227-241.

MACZYNSKA, M. 1989. *Das Frauentrachtzubehör des mittel- und osteuropäischen Barbaricums*, Bonn.

MATTHES, W. 1931. *Die nördlichen Elbgermanen in spätrömischer Zeit*, Leipzig (Mannus-Bibliothek 48).

NOWAKOWSKI, W. 1988. Metallglocken aus der römischen Kaiserzeit im europäischen Barbaricum, *Archeologia Polona* 27: 69-146.

OLDENSTEIN, J. 1976. Zur Ausrüstung römischer Auxiliareinheiten. Studien zu Beschlägen und Zierat an der Ausrüstung der römischen Auxiliareinheiten des obergermanschen-raetischen Limesgebiets aus dem zweiten und dritten Jahrhundert, *Berichte der Römisch-Germanischen Kommission* 57: 49-284.

RIHA, E. 1979. *Die Fibeln aus Augst und Kaiseraugst*, Augst (Forschungen in Augst 3).

ROEST, J. VAN DER 1988. Die römischen Fibeln von 'De Horden', *Berichten van de Rijksdienst voor het Oudheidkundig Bodemonderzoek* 38: 141-202.

SCHNURBEIN, S. VON & ERDRICH, M. 1992. Das Projekt. Römische Funde im mitteleuropäischen Barbaricum, dargestellt am Beispiel Niedersachsen, *Berichte der Römisch-Germanischen Kommission* 73: 5-29.

SCHULZE, M. 1977. *Die spätkaiserzeitlichen Armbrustfibeln mit festem Nadelhalter*, Bonn (Antiquitas 3).

THOMAS, S. 1960. Die provinzial-römischen Scheibenfibeln der römischen Kaiserzeit im Freien Germanien, *Berliner Jahrbücher der Vor- u. Frühgeschichte* 6: 119-148.

VÖLLING, T. 1994. Fibelformen der vorrömischen Eisen- und römischen Kaiserzeit, *Berichte der Römisch-Germanischen Kommission* 75: 147-282.

ZIJLSTRA, J. 1993/1995. Friese bodemvondsten, archeologische, historische en naamkundige aspecten, Leeuwarden (unpublished typescript).

NOTES

1. See Gerrets & De Koning (this volume).
2. Cf. Godłowski (1970) and Maczynska (1989).
3. Zijlstra (1993, 1994). Inventory and publication of all Roman period brooches – both of Roman and Germanic origin – from the Dutch provinces of North Holland, Friesland and Groningen is in progess (Erdrich & Galestin, in prep.).
4 Illustrated finds have their find numbers marked with *.
5. Haalebos (1986, pp 30-35), Van der Roest (1988, pp 150-151) and Völling (1994, pp 207-216).
6. Matthes (1931, pp 19-26).
7. Böhme (1972, Figs 3 and 4), Gechter (1980, pp 604-607, Fig. 9) and Haalebos (1984/1985, Fig. 26).
8. Böhme (1972, pp 41-43; Nos 1126 and 1130).
9. Haalebos (1986, pp 66 and 100, Figs 26, 49 and 182), Böhme (1974, pp 8-10, Fig. 1, plates 144, 1-3, 75, 17) and Schulze (1977, p. 80 – Gruppe 135).
10. Böhme (1974, pp 13-14; 1986, pp 529-530 and Figs 50-52).
11. Eggers (1951) and Erdrich & Galestin (in prep).
12. Cosack (1979, pp 22-23), Riha Typ 1.6: Haalebos (1986, pp 51-52) and Van der Roest (1988, pp 153-155).
13. Haalebos (1986, p. 52) and Van der Roest (1988, pp 155-156).
14. Thomas (1966).
15. Exner type III, 24: Böhme (1972, pp 37-38, cat. no. 981-983) and Riha (1979, cat. no. 307).
16. Erdrich (1994.)
17. Oldenstein (1976).
18. Beckmann (1969).
19. Nowakowski (1988).
20. Erdrich & Galestin (in prep).
21. Von Schnurbein & Erdrich (1992) and Erdrich & Galestin (in prep).

Roman coins from Wijnaldum

J.P.A. VAN DER VIN

1 INTRODUCTION

Friesland is rich in Roman coins. During the last as well as the present century, large and small hoards of coins were found in the soil at dozens of sites in this province. At present, Wijnaldum with its total of 87 coins out of a registered 1100 Roman and Byzantine finds, is one of the most important findspots. Approximately half these coins, 41 specimens, were discovered at Wijnaldum over many years and reported by various finders. The majority of the finds, however, is of recent date; only 24 specimens were mentioned in *Fundmünzen der Römischen Zeit in den Niederlanden, Provinz Friesland*, whose catalogue was completed on 1 January 1991.[1] Of these 41 coins, no further specification of the findspot is known, apart from the place of origin – Wijnaldum. The other half of the coins found, 46 specimens altogether, were discovered during excavations of the terp of Wijnaldum-Tjitsma carried out by the Biological-Archaeological Institute of the Rijksuniversiteit Groningen (the present Groningen Institute of Archaeology) and the Institute of Pre- and Protohistorical Archaeology of the University of Amsterdam (the present Amsterdam Archaeological Centre) from 1991 to 1993. This group of coins may be regarded as a single find complex, and in the case of 19 specimens an archaeological context is known thanks to the precise registration of the finds. This archaeological context, determined by archaeological finds other than coins, reveals, for the first time some information about the duration of use of Roman coinage in the Frisian territory north of the *limes*. So far nothing was known due to insufficient excavation data.

2 THE SERIES OF COINS

Since the pattern of coins found during excavation does not basically differ from that of the coins previously found at Wijnaldum, it is wise to make use of the complete find material.

The data from Wijnaldum are recorded in a table which includes all the known coin finds.[2] The specimens from the excavation are indicated by an asterisk (*) and also those specimens for which an archaeological context could be determined (**).

Apart from a somewhat deviant halved as from the first half of the 1st century AD, the consecutive series of coins begins with a denarius of Domitian from 81. A first peak is found in the middle of the 2nd century, the Antonine period (138-192), which is amply represented with 20 specimens. The number of coins from the Severan period drops sharply to 7 specimens. There is another striking peak of 20 items from the period of Gallienus up to and including Tetricus II (253-273) together with their imitations made in the two following decennia. The period of the Tetrarchy (294-313) is totally lacking, but again there are 10 specimens from the later part of the reign of Constantine (306-337) and his sons (337-361). The last phase of the Roman money consists of the small copper coins of Valentinian I and his successors (364-395), four of which were found. The series is finally completed by two Byzantine coins from the first half of the 6th century and a Visigoth imitation of a Byzantine tremissis from the same period.

The majority of coins is copper; only 2 gold tremisses and 18 silver denarii were discovered, and the only antoninian which should have been silver (under Gordian III, circa 238) appears to be plated. The best represented coin is the brass sestertius (23 specimens), followed by the antoninian which was strongly devalued after circa 260 and therefore made of copper (20 specimens). Of the 4th-century Roman copper follis, 9 items were found, and of the late-4th-century aes-III and aes-IV, another 5. Other denominations, such as the as, dupondius and Byzantine follis are only represented by a few specimens.

Among the coin finds from Wijnaldum, there were a relatively large number of fragments (5 specimens), halved coins (3 specimens) and even a quarter of a denarius (1 specimen).

Since Friesland was situated outside the Roman Empire and, apart from a brief period in the early first century (until AD 47) never even belonged to it, it is not wise to interpret this series of coins from Wijnaldum in the same way as one would a series from a settlement in the Roman Empire. There is not any indication that the coins found actually circulated in a market situation and it is far from certain whether they travelled so far north shortly after issue or whether this took place during a very gradual process of decades or even centuries. On the other hand, compared with other Frisian places[3] large numbers of Roman coins were discovered at Wijnaldum, which leads to the inevitable conclusion that the coins there were in some way of special importance to the local Frisian population. It is the link between a number of coins to their archaeological context which now offers a possibility of discovering more about the function and use of Roman coinage in *Germania Magna*.

3 INTERPRETATION OF THE FINDS

The series of coins from Wijnaldum appears to fit in with what is known about Roman coins in the provinces of Friesland[4] and Groningen[5]. In both provinces the silver denarius and the brass sestertius, both mainly from the Antonine period, predominate. Material from the Severan Period is scarce throughout the North Netherlands. Several parallels can be found in Germanic territory for the substantial increase in material from the short period in which the southern Netherlands were part of the Gallic empire under the usurpers Postumus and the two Tetrici (259-273). Coins dating from the Tetrarchy are virtually lacking throughout the Netherlands, whereas material since Constantine (from circa 320/330) was discovered at several sites in the North Netherlands. At most places in the Netherlands with 4th-century coin finds, the series ends about 400.[6] This is the moment when the import of new material to this country came to an end, due to the concentration of the Roman authorities on Italy. However, in Groningen and Friesland the coin finds continue in places until into the 7th century.[7] The three 6th-century coins from Wijnaldum therefore fit into a wider regional development.

4 THE STRATIGRAPHICAL INFORMATION

A closer examination of the 19 coins with an archaeological and stratigraphical context yields surprising results. First of all there are several coins whose dating and that of the stratigraphical context correspond perfectly. This makes it likely that the coin landed in Frisian soil not very long after the date of production. However, several coins whose dating does not altogether or does not at all fit that of the context are much more interesting. The difference between the date of production and the date of context sometimes amounts to several decennia, occasionally even more than a century. In extreme cases the difference may even span more than five centuries! So far, no evidence for this phenomenon has been found in the Netherlands apart from Wijnaldum.

A similar development – but exclusively with Roman silver coinage – has been known in Germanic territory for much longer. In South Sweden and on the Baltic island of Gotland thousands of Roman denarii from the late 1st and mainly 2nd century were found in a 4th or even 5th-century context.[8] The denarii from South Sweden mostly ended up there via a westerly route along the Rhine and through North Germany (*Germania Magna*), and those from Gotland along an easterly route from Italy via the eastern Alps, Bohemia and Poland. At present it is assumed that these denarii gradually moved northwards in a slow process. Within the Germanic societies of Germany, Bohemia and Poland, they must have had their own function as an object of value, as a result of which they were preserved for a long time.

The denarius of Antoninus Pius (147-161), found in a 4th-century context at Wijnaldum, seems to be a similar case. A Domitian denarius from 81 may also belong to the same category, although the 425-550 context in this case is less reliable, according to the excavators.

The copper coins appear to follow the same development in Wijnaldum as the silver coins. There are great differences in the case of some sestertii and asses from the 1st-2nd century: the dates of coin and context differ by 2 to 4 centuries. According to the excavators, it would appear that some of these coins only ended up in the soil at Wijnaldum at a very late stage; they are found in a dated, undisturbed layer. In other cases there is no indication that the coins entered Wijnaldum at a late date, and it remains uncertain how and where these coins braved the centuries.

Apart from the 1st and 2nd-century sestertii and asses, several copper antoniniani from the 3rd century and folles from the 4th century were found at Wijnaldum in a much later context. Here too, the dates differ, sometimes as much as 500 years! Material from the two Tetrici is sometimes found in a 6th or 7th-century layer; for one antoninianus of Tetricus II a reasonably reliable context was established between 770 and 900. Small folles of Constantine and his sons and coins of Valentinian I also occur in this 6th and 7th-century context, 2 to 3 centuries after their date of production.

5 CONCLUSION

The link between the date of the coins and the stratigraphical date produces quite remarkable results in Wijnaldum, both from an archaeological and a numismatical point of view. It was already clear that, due to the small number of denominations in Friesland, there can never have been a full coin circulation, enabling all payments both large and small to be made with coins. This means that the occurrence of hundreds of Roman coins in the North Netherlands must have had a different cause. Their primary function was not payment, but probably of a more psychological nature, for example, as objects of value which were preserved within Germanic societies for a long time. One may refer to concepts such as 'gift exchange' and 'prestige' within a Germanic elite class.[9] In the case of the silver coins this is quite feasible; and, moreover, it is completely in keeping with the Germanic preference for Roman silver coinage observed by Tacitus in his Germania.[10] Such a preference is also imaginable for the heavy brass sestertii. In ancient times, brass was an attractive and valuable metal.

However, the antoniniani and folles finds, two lightweight coin types of poor-quality copper are more problematical. What the Frisians saw in these remains a mystery. The suggestion of the excavators who discovered several of these coins together with some other bronze fragments in the vicinity of a bronzeworking workshop, may perhaps provide some indication: these coins and fragments were only looked on as raw material and were intended for smelting. The relatively large number of fragments and halved specimens may also point in this direction.

Where the inhabitants of Wijnaldum got their Roman coins from is uncertain. One might consider the suggestion that during the 6th and 7th centuries digging activities took place in the former castella along the limes, and the material found there was taken north and sold as a commodity.[11] The problem is, however, that sestertii and folles are not the types of coin found in large numbers in the castella. One would expect to find asses. It would therefore be more advisable to start from an earlier transport date – still in the Roman Period – to the Frisian territory, where the coins, now functional in another way, were preserved for centuries.

The surprising dates from Wijnaldum have yet another consequence. For the first time in archaeological investigation in the North Netherlands, the stratigraphical information from the Wijnaldum excavation is available on a large scale. This adds a new dimension to the investigation. The coins from Wijnaldum, however, completely fit the pattern of Roman coin finds in the rest of Friesland and Groningen. This means that the same probably applies to many of the dates there: they are not what they appear!

CATALOGUE OF ROMAN COINS FROM WIJNALDUM

*Wijnaldum excavation find. **With an archeological context.

No.	Type	Mint	Emperor	Date	Archaeological date	Inventory/ find no.	Remarks
1	as	Rome	IA<	54		FML A 10421	halved
2 **	denarius	Rome	Domitian	81	425-550	W 141	
3 *	sestertius	Rome	Nerva	96-98		W 1185	
4	denarius	Rome	Trajan	98-99			
5	denarius	Rome	Trajan	98-117			fragment
6	denarius	Rome	Trajan	114-117			
7 **	sestertius	Rome	Hadrian	119-121	300-350	W 203	

*Wijnaldum excavation find. **With an archeological context.

No.	Type	Mint	Emperor	Date	Archaeological date	Inventory/ find no.	Remarks
8 *	dupondius	Rome	Hadrian	117-122		W 7477	
9 *	denarius	Rome	Trajan/Hadrian	98-138		W 11101	
10	sestertius	Rome	Hadrian?	117-138			
11 *	denarius	Rome	Antoninus Pius	148-149		W 4540	fragment
12 **	denarius	Rome	Antoninus Pius	147-161	300-350	W 11838	
13	sestertius	Rome	Antoninus Pius	138-161			
14	sestertius	Rome	Antoninus Pius	138-161		FML A 10422	halved
15 *	sestertius	Rome	Antoninus Pius (?)	138-161		W 3751	
16	sestertius	Rome	Faustina II	145-161			
17 *	sestertius	Rome	Faustina II	145-161		W 4538	
18 **	sestertius	Rome	Faustina II	145-161	425<	W 6266	
19 *	denarius	Rome	Marcus Aurelius	166-166		W 11152	
20	denarius	Rome	Marcus Aurelius	170-171			
21	denarius	Rome	Marcus Aurelius	161-180			
22	sestertius	Rome	Marcus Aurelius	161-180			fragment
23	dupondius/as	Rome	Faustina II	145-175			
24 **	denarius	Rome	Faustina II	161-175	200-300	W 130	
25 *	sestertius	Rome	Commodus	184-185		W 4537	
26 *	denarius	Rome	Commodus	184-185		W 11152	
27 *	sestertius	Rome	Commodus	180-192		W 1184	halved
28	sestertius	Rome	M. Aurelius/Commodus	161-192			
29	denarius	Rome	Antonine period	138-192			fragment
30 *	sestertius	Rome	Antonine period	138-192		W 4535	
31 *	sestertius	Rome	Antonine period	138-192		W 4536	
32 **	as ?	Rome ?	I-II	1-192	550-650	W 4604	
33	denarius	Rome	IB-II	69-192			quarter
34 *	sestertius	Rome	II	98-192		W 527	halved
35 **	as ?	Rome	II	98-192	300-350	W 9939	
36	denarius	Rome	Antonine-Severian period	161-211			
37	sestertius	Rome	Septimius Severus	193-211			
38	sestertius	Rome	Severian ?	193-217			
39	sestertius	Rome	Julia Domna	196-211			
40 *	denarius	Rome	Elagabalus	218-222		W 9138	
41	denarius	Rome	Severus Alexander	222-235			
42	antoninian	Rome	Gordian III	238-239		FML N 314	fragment
43 **	as	Rome	I-III	1-250	263-350	W 5143	quarter
44	denarius	Rome	I-III	69-253			
45	sestertius	Rome	I-III	69-270			fragment
46	sestertius	Rome	II-III	98-270			
47	sestertius	Rome	II-III	98-270			
48	sestertius	Rome	II-III	98-270			halved
49 **	as	Rome	II-III	98-250	775-850	W 9226	
50 *	antoninian	Rome	Saloninus	259		W no no.	
51	antoninian	Rome	Gallienus	260-268			
52	antoninian	Rome	Gallienus	260-268			
53 *	antoninian	Rome	Gallienus	260-268		W 7619	
54	antoninian		Postumus	259-268			
55	antoninian		Postumus	259-268			
56 *	antoninian	Gallia	Postumus	259-268		W 7730	
57 **	antoninian	barbarian	Postumus	259-268	450-525	W 9156	
58 *	antoninian	Rome	Claudius II	268-270		W 9490	
59	antoninian	Gallia	Divus Claudius II	270-275			
60	antoninian	Gallia	Tetricus I	270-273			
61	antoninian	Gallia	Tetricus I	270-273			
62 **	antoninian	Gallia	Tetricus I	270-273	575-625	W 108	
63 *	antoninian	Gallia	Tetricus I(?)	270-273		W 8201	
64 **	antoninian	Gallia	Tetricus II Caesar	270-273	425-550	W 140	
65 *	antoninian	Gallia	Tetricus II Caesar	270-273		W 4543	

*Wijnaldum excavation find. **With an archeological context.

No.	Type	Mint	Emperor	Date	Archaeological date	Inventory/ find no.	Remarks
66 **	antoninian	Gallia	Tetricus II Caesar	270-273	770-900	W 5979	
67 **	aes	?	IIIB	260-295	500-550	W 9084	quarter fragment
68	antoninian	?	IIId	270-295			
69	antoninian	Gallia ?	IIId	270-290			
70 **	antoninian	barbarian	IIId	280-295	175-250	W 297	
71	follis	Treves	Constantine I	321-323			
72 **	follis	?	Constantine I	330-335	650<	W 17	
73 *	follis	Arles	Constantine I	336		W 8054	
74 **	follis	Treves	Constantine II Caesar	330-335	450–	W 4394	
75	follis		Constantine II Caesar	330-335			
76 *	follis		Constantine I or sons	330-335		W 7575	
77 *	follis	?	Helena	337-340		W 2860	
78	follis	Treves	Constantius II/Constans	340			
79 **	follis	barbarian	(sons of) Constantine I	335-350	650-750	W 4389	
80 *	aes-III (?)		IIIB?	351–		W 11150	
81 *	aes-III?		Valentinian I/Valens	364-378		W 243	
82 **	aes-III		Valentinian I/Valens	364-378	550-650	W 6117	
83 *	aes-IV		Theodosius I and family	378-402		W 9005	
84 *	aes-IV		Theodosius I and family	378-402		W 8029	
85	follis	Constantinople	Justinus I	518-527			
86	tremissis	Visigothic	type of Justinian I	540-550		FML A 2250	
87	tremissis	Constantinople	Justinian I	527-565			

REFERENCES

ALFÖLDI, M.-R. & J.P.A. VAN DER VIN (eds) 1992. *Die Fundmünzen der römischen Zeit in den Niederlanden(FMRN), Band I, Provinz Friesland,* Berlin.

ALFÖLDI, M.-R. & J.P.A. VAN DER VIN (eds) 1996. *Die Fundmünzen der römischen Zeit in den Niederlanden (FMRN), Band II, Die Provinze Groningen, Drenthe, Overijssel, Flevoland,* Berlin.

ANDERSON, J.G.C. (ed.) 1961. *Cornelii Taciti, De origine et situ Germanorum,* Oxford.

BESTEMAN, J.C., J.M. BOS & H.A. HEIDINGA 1992. *Graven naar Friese koningen, de opgravingen in Wijnaldum,* Franeker.

LIND, L. 1975. Roman denarii found in Sweden. In: *Studien zu Die Fundmünzen der römischen Zeit in den Niederlanden I,* Berlin, 137-142.

VIN, J.P.A. VAN DER (1996). Roman coins in the province of Friesland. In: *Proceedings of the 13th Oxford Symposion on Coinage and Monetary History, Coin Finds and Coin Use in the Roman World, 25-27 March 1993, Berlin.*

NOTES

1. Alföldi & Van der Vin (1992).
2. Find archive, Royal Coin Cabinet, Leiden.
3. Alföldi & Van der Vin (1992, passim).
4. Van der Vin (1996, 357-371).
5. Alföldi & Van der Vin (1996).
6. Find archive Royal Coin Cabinet, Leiden.
7. Alföldi & Van der Vin (1992, passim).
8. Lind (1975, pp 137-142).
9. Besteman et al. (1992, especially Chapter I).
10. Anderson (1961, Tacitus V, pp 4-5).
11. Result of a discussion with Dr M. Erdrich about the problem of context of the Roman material from the Wijnaldum-Tjitsma terp.

The Wijnaldum brooch

J. SCHONEVELD & J. ZIJLSTRA

1 INTRODUCTION

The principal motive of the research into the terp of Wijnaldum-Tjitsma was the Wijnaldum brooch. At that time this gem was still hanging between other ornaments in a side display case in the Fries Museum at Leeuwarden. It mainly drew the attention of connoisseurs, and a lot of visitors of the archaeological department of the Fries Museum may even have passed the ornament without noticing it. Yet it does not often occur that one comes face-to-face with such an ornament from the early Middle Ages. In the northern part of the Netherlands, art of that period is seldom found. In this part of the country the early Middle Ages do not excel in producing monumental works of art in the field of architecture, sculpture or painting. Characteristic of the art of the early Middle Ages is the making of small objects, particularly using precious metals. Also striking is the fact that, relatively speaking, the art of the early Middle Ages lacks naturalistic, figurative elements. The human figure in particular is hardly used in the form of language. In the expression of art there are mainly cases of decorative art, with a sacred if not magical-religious purpose.[1] Especially the latter category is very infrequently present in our area. And so the Fries Museum in Leeuwarden may consider itself lucky that its collection contains one of the highlights of medieval art from the Netherlands: 'the Wijnaldum brooch'. Unfortunately this important piece of art is not wholly complete. But thanks to chance, diligent survey and industrious digging we were still able to get an idea of the original shape and lustre. Until recently the brooch was divided among different collections. But now the piece can be admired, almost complete, in a display in the Fries Museum. This chapter will describe the ornamental motifs of the Wijnaldum brooch. In this respect, the work of R. Bruce-Mitford and B. Arrhenius was essential. Furthermore this article will attempt to discuss its international context and give some more information about when the brooch was made and by whom it was made. Finally an attempt will be made to ascribe the brooch to a possible aristocrat or owner.

2 THE FIND CIRCUMSTANCES

In the beginning of the 1950s drainage operations were carried out in Wijnaldum. For this purpose a series of ditches was dug by hand in the arable land behind Jan Bergman Sr's farm with the topographical name Tjitsma terp. From this dwelling mound a farm-hand happened to dig up the most important piece of the Wijnaldum brooch: the foot-plate. According to the farmer, the foot-plate, together with two human skulls, that in all probability have nothing to do with the brooch, was stored in the barn that belonged to the farm for a couple of years. The skulls were used to frighten the milk-hand and the brooch seems at one point to have moved to Bergman Sr's sister's window-sill. Through the mediation of Mr Van der Kooi, who was a general practitioner in Dronrijp, the piece of art, by means of exchange, ended up in the Fries Museum in Leeuwarden. Mr Wassenbergh, the then director of the museum, went to London on business and had the opportunity to have the piece assessed by the British Museum. The comment of Mr Bruce-Mitford of the British Museum on the piece was: 'It has been a long time since I have seen something so exciting...'.[2] He considered it to be a part of a buckle. He probably arrived to this suggestion because of the similarity in form and decoration with a buckle from the Sutton Hoo ship burial.

A survey that was done several years after the find on the field in question at Wijnaldum remained without result. The exact find-spot could not be traced any more. The memories seemed to have faded to such an extent that the precise place of the drainage ditch on the large and extensive field could no longer be pointed out. Attempts to examine the findspot any further were abandoned. However, the mound was investigated again in the 1980s by amateur archaeologists with a metal detector. They discovered that the alignment of terps at Wijnaldum was a rich findspot. Through the efforts of one of the authors, J. Zijlstra, the objects that by this time had been found, did not end up scattered. He made contact with different amateur archaeologists who were working with a metal detector in the area and collected their finds. Thus he was able to contribute to the knowledge of metal objects from the terp-region of Westergo.[3] During the investigation of the brooch, a number of gold and silver fragments found by the amateur archaeologists drew attention. They were attributed to the Wijnaldum brooch and with this came the complete reconstruction of the original object.[4] During a meeting in the Fries Museum the different parts were compared and put together. This produced an affirmation of the findings.

During the archaeological excavations at the terp Wijnaldum-Tjitsma, additional parts of the Wijnaldum brooch were found. In the 1991 excavation these consisted of a number of thin gold sheet fragments from the reverse of the brooch. That these pieces of gold sheet really came from the ornament became apparent in the course of the season. The Bergman family also appeared to be in possession of a little piece of gold sheet from the reverse. In the excavation season of 1992 another important part of the brooch was discovered. On the first of June of that year a piece of the head-plate was discovered by the metal detector in an excavation trench. It was not instantly recognised as being a part of the brooch, because it was heavily damaged by the plough. However, comparing and matching clearly proved that it was indeed part of it. The last discovery was a knob that belonged to the brooch. It was found through the extensive survey of the dwelling mound in 1995. All parts that were found after the 1950s, including the ones found during the excavation, were found in the till-layer.

3 DESCRIPTION

Although the Wijnaldum brooch was first considered a part of a buckle, the reconstruction made it clear that we are dealing here with an example of a disc-on-bow brooch. A disc-on-bow brooch consists of a foot-plate and a head-plate. Both are linked to each other by a hinge to which a disc is attached as an added decorative element. Although many pieces of the brooch were traced, an important part is still missing, namely the disc. This is a pity, not only because the object of art appears incomplete to us, but mainly because we are still in the dark about this part of the reconstruction of the brooch.

The brooch is executed in the cloisonné technique. With this method the depiction is set out in gold lines. The intervening space, the cells, are filled up with a contrasting material. In this case the cloisonné cells are filled with red garnet and a little aragonite or mother of pearl.[5]

In the decoration of the foot-plate three distinctive elements are added. One element is composed of geometrical patterns, another element consists of a crooked line and the third element is formed by five circles. These five circles frame the entire piece. The biggest circle is situated at the foot and consists of one central knob surrounded by two circles, one composed of white aragonite and one in cloisonné. In the cloisonné strip the cells are arranged to form a St Andrew's Cross. The central knob in the bottom circle consists of two 'mushroom-figures' and two 'staircase-figures' that are cleverly pieced together. The two circles in the middle have the same design, but are somewhat smaller and the knob is a bit more plain, while the smallest circles at the top only consist of a central cell with a garnet and a surrounding field of aragonite.

The lowest part of the foot-plate shows a geometrical decoration. The different cells have been carefully incorporated into each other. The staircase is used as the principal form and is combined with the mushrooms, as is the case on the bottom knob. On both sides an ornamented border has been applied in which a staircase is also incorporated, while on the upper side a curved band links the two middle circles.

The most ingenious decoration can be found on the upper side of the foot-plate. All kinds of curved lines are interlaced into one depiction. At first sight it looks like a tangle of oblong forms, but on closer inspectation it reveals a certain coherence. The depiction is mirrored and the two edges on the outside provide the solution. In these forms a little eye is marked in gold. If one traces the lines it turns out to be a twisted snake or dragon-form. Between these lines a central figure is formed out of bands. We

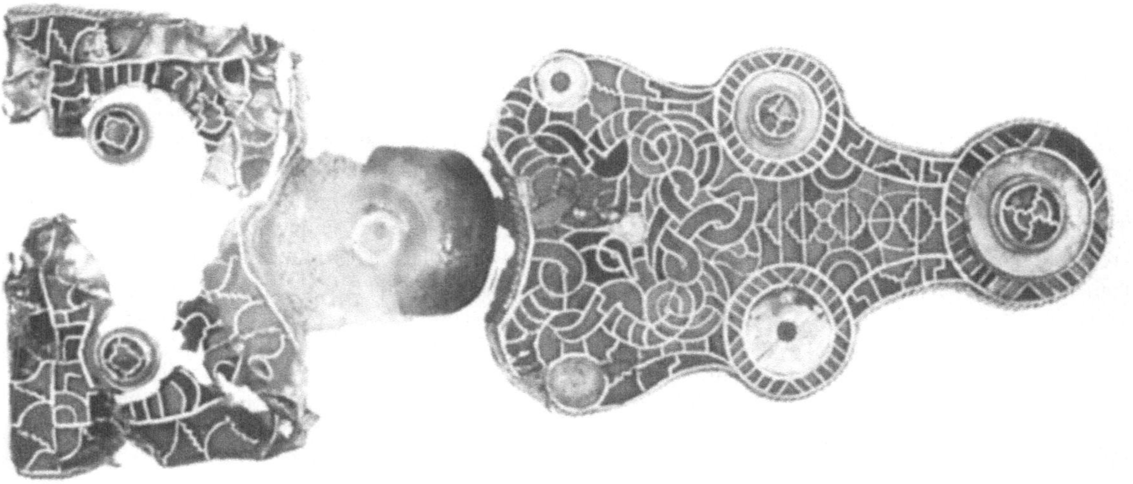

Figure 1. The disc-on-bow brooch completed with metal detector and excavation finds (photograph: Fries Museum, Leeuwarden). Scale 1:1.

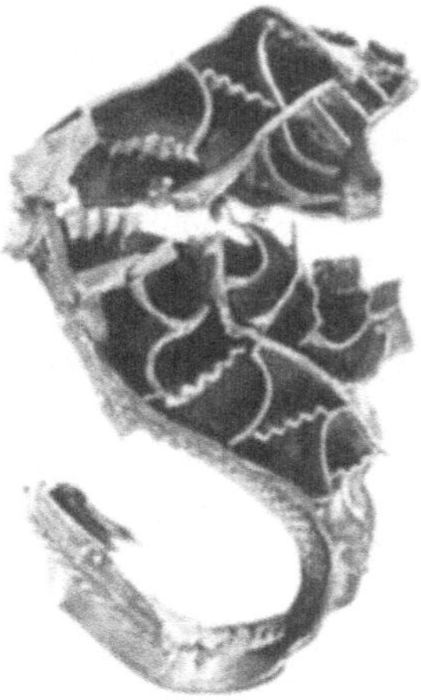

Figure 2. Fragment of the disc-on-bow brooch found during the excavation.

are dealing with a highly stylised depiction of the 'man between two animals'.[6] A more naturalistic example of this depiction is known from the purse-cover of the Sutton Hoo ship burial.

Also in the decoration of the head-plate three different elements can be distinguished. The first element consists of two circles that are juxtaposed in the centre of the rectangular headplate. These circles were probably filled up with two little knobs, that were found separately. One of these knobs originally belonged to the Zijlstra collection and was earlier ascribed to the ornamental disc. The other one was found after a long search by H. Hommes from Wijnaldum in 1995. The space surrounding the two circles is filled up in a rectangular form. In the corners quarters of circles are placed that are formed by two lines. The rest of the rectangle is filled in with straight lines. Possibly a curved line has been added here and there, but this cannot be confirmed because this part of the brooch is severely damaged. The third part, the rim, consists of alternate curved lines that are linked by 'staircase-forms'. The head-plate and foot-plate are connected by a hinge. The hinge bulges because the cloak was folded underneath. This silver hinge is covered very beautifully by a gold sheet and decorated with *S*-shapes of filigree. This hinge is encrusted with little garnets as well. The big disc is positioned on top of the hinge.

Unfortunately we are less well informed about the reverse of the Wijnaldum brooch. We know that it had a fixing pin and that it was probably decorated with a small animal head. This little head is made of filigree and provided with a little eye of garnet. This reconstruction is plausible considering other examples such as the reverse of the Kingston Brooch from Kent. The small head differs in material and craftsmanship.[7] Perhaps it was made by a second person, possibly an apprentice in the workshop. It eventually ended up at the backside of the brooch. The side of the entire brooch is finished off with braided gold wire.

4 MATERIALS

The gold brooch of Wijnaldum is completely covered with garnet, so it must have been a special ornament for the people in the early Middle Ages. This kind of stone is hard to work with and does not occur naturally in our area. Garnet is a name for a collection of gems. This group includes almandine, pyrope, spessartite and grossular. The garnet was already used as an ornament in Egypt during the pre-dynastic period (before 3300 BC).[8] In Europe it occurs as an ornament since the 4th century BC. Greek writers like Aristotle and Theophrastus make mention of the gem. Both report the principal characteristic that garnet or anthrax is not affected by fire.[9] They are right about this that garnet does not change at a temperature of 400 to 500°C. However, it does start to melt at 800°C. And if one holds the stone in the sunlight it seems to flicker like a little firecoal. In Pliny's *Historica Naturalis* the garnet is called 'carbunculus', which means glowing coal. He repeats the observations made by Theophrastus and observes that the stone is able to melt sealing-wax and that it even glows beautifully in the dark. We hear about this story again from the Christian writer Epiphanius. According to this bishop the stone is not visible during the day. Therefore it must be located at night because only then can one see even from a distance the illumination of the gem and the sparks that jump around it. If one possesses the stone it cannot be hidden because it even shines through clothes.

The Christian writers attach great value to the stone because it is the symbol of Christ. The garnet forms a part of the high priest's chest ornament. In the description of the chest ornament in *Exodus*, the red stone is mentioned as the first stone in the second row of gems and connected to the tribe of Judah. Kings like Solomon and David are descended from Judah and therefore the garnet became the symbol of secular power. Moreover it symbolised Christ, the most important descendant of the tribe of Judah. So the garnet has a rather wide symbolism, in different cultures as well as in different regions.

The garnet does not occur in our region, so the stones belonging to the Wijnaldum brooch had to be imported. Different find-spots of the gems are known from antiquity. The Old-Egyptian specimens were found in Aswan, at the first cataract in the river Nile and in the Sinai. They occurred here until the last century. From Aswan and Elephantine crystals with a diameter of 10 cm were mentioned. Besides this Theophrastus mentions Marseilles and Carthage as the place of origin. The surroundings of Marseilles seems somewhat strange, but perhaps this was a place where the stone could be obtained through commerce.

Plinius is the first to name India as a source for garnet. The garnets that came from India were so big that tableware could be manufactured from them. These garnets were of a much lighter colour than the ones from Carthage and it is quite possible that Pliny meant the big blocks of hessonite that came from Ceylon. Early-medieval writers like Epiphanius only knew Carthage as a place of origin, and not until trade with India flourished in the fifteenth and sixteenth centuries was this origin mentioned again. Marco Polo reports trade in Indian stones in the markets of Asia Minor, but he also mentioned origin elsewhere.

Modern laboratory research has shown that the garnets in early-medieval jewellery from Northern Europe belong to the group of almandine with the most likely source being Bohemia.[10] This conclusion is not based on chemical analysis alone. Stones from the same origin can sometimes have considerable differences in constitution and properties. It is mostly a technical property that turned the balance to use garnets from Bohemia, as we will see in Section 5.

5 TECHNOLOGY

With the encrusted objects made in cloisonné technique like the Wijnaldum brooch, thin slivers of garnet are being used. These slivers are obtained by cleaving the stone. Due to the crystal structure it is virtually impossible to cleave garnet. Garnet has a cubic crystal system and crystallises in virtually round crystals. These are very rich in facets. That is why garnets are mostly executed 'en cabochon'. In this way they get an elliptical shape. However, from an 18th-century description we know of a technique to cleave garnet by means of heating and cooling off quickly. This technique is only applicable to a special sort of stone from Bohemia. Only these garnets have a make-up in which the crystals lie in a sort of slate-like structure between the rough rocks. This makes it possible to obtain the thin slabs instead of obtaining grit when hammering the rocks. The small slabs are extremely even, because of the position of the layers of crystals.

After cleaving, the stones are polished.[11] This is not as labour-intensive as would appear. If necessary, it can be done in large quantities simultaneously by mass-production. For that purpose the small discs are glued on a bearer and are all polished at the same time.

Once the thin slabs of garnet are obtained, then they have to be cut into the required forms. This is done with a little carving-wheel. At that time really thin little wheels could not be made, so a *V*-shaped cut was used. This can be observed on the sides of the pieces of garnet, which have an oblique angle. Since iron is not hard enough to cut garnet, a means of grinding was used.[12] The wheel has to be turned very regularly, and then the forms can be finished with all kinds of grinding-devices. A theory exists that the different little shapes of garnet were supplied from different centres in Europe, ready to hand to the gold- an silversmiths, who subsequently made their jewellery from it. This is quite possible in the case of the geometrical part of the Wijnaldum brooch. The man between the two animals, however, is such a personal design with specially formed pieces of inlay, that this must be a personal creation of the artist, in design as well as in finish.

Cloisonné is an old technique, already used in Egypt during the Old Kingdom (circa 2850-2200 BC). In this technique, gems were put in special settings next to each other and secured with a glue made of resin and limestone. Then the whole surface was levelled by scrubbing and polishing. The upper side of the little walls of the settings would be somewhat broadened, which would set the stones even better.

This method cannot be used with garnets however. The white glue with which the stones are set has a dulling effect on garnets and will stop the beautiful glowing red light. In order to solve this problem an attempt was made to colour the glue red by adding pigment. But the best method proved to be to cover the reverse side of the garnets with gold foil. Subsequently the stones were fastened by heating a paste of quartz or calcite. A little bit of albumen was used to achieve a thickening effect, until the temperature of 100°C to 200°C was reached.

The glue that was used on the Wijnaldum brooch dissolves fairly easily in water. The last discovered fragment of the head-plate (in 1992) hardly contained any paste any more. It had been completely dissolved by the groundwater, which made the garnet sink down to the bottom of the cloisonné cells.

So in order to make the light reflect in the stones, gold foil was attached to the reverse. The thickness of the gold foil varied around 0.03 mm. To heighten the effect the foil was pressed into a wafer-patterned dye instead of using plain foil.[13] This wafer-pattern differs strongly per object, but in general around 3.5 lines per millimetre are used, forming a crosswise pattern. The number of lines used with the Wijnaldum brooch even amounts to four lines per millimetre. During the excavations of the Tjitsma terp a small dye was found, with which this wafer-pattern could be imprinted in the gold foil. Unfortunately the pattern of the dye did not match the one belonging to the brooch.

The gold plate on which the cloisonné cell walls were attached was itself supported by a silver plate. On the reverse this was again covered with a thick plate of gold foil. In this way a strong construction was created on which the pin could be attached to the back. The hinge between the foot-plate and the head-plate is decorated with filigree. The small head of an animal that possibly beautified the reverse is also embellished in this technique.

6 SYMBOLISM

The reason that the Wijnaldum brooch is completely covered with garnets is not only because of the beauty of this gem. It also has a very clear symbolic function, as mentioned above. In antiquity, garnet was always considered as a symbol of power. For this reason the Sassanides (224-651 AD) had their portraits cut especially in garnet. As we have seen this symbolism is continued in the Jewish-Christian tradition. The garnet resembles the Messiah himself, who with his clear, permeating radiance comes to the world to lighten the darkness. Therefore the stone was not only used on mass chalices and crucifixes, but was also eagerly worn because it would yield its owner much esteem. He would become brave, misfortune would be kept at bay and prosperity would smile upon the owner. Moreover it was beneficial to friendship and love. It is understandable that the garnet was also much respected in the Germanic world.[14] This was in fact a direct inheritance from the Roman Empire. Roman emperors gave their officers brooches, mostly cross-shaped, as gifts. An example of such a brooch was found in King Childeric's grave

(† 481). In the Germanic world such brooches were inlaid with garnets and each time one comes across such objects in a Germanic context one should assume that it represents the wealth of a Germanic king. The kings copied the object from the Roman Emperors as well as the habit of giving it away as a symbol of status. In contradiction to brooches, swords with inlaid garnets are much more of a rarity in the Germanic world. These are direct imitations of models from the Byzantine Empire and were also imported from that area. Only somewhere around the reign of Childeric were they manufactured in the north as well.

The distribution of objects with garnets in general is remarkable. Sword pommels and buckles for example are mainly found in Sweden. Possibly the Frankish Kings exchanged elite commodities on a royal level with these northern regions. Women's jewellery with garnets has an entirely different distribution. They are typically uncommon in Scandinavia but occur frequently in southern regions. Perhaps because marrying off as a form of diplomacy within elite-networks was not done in the north. But it is also possible that the Christian Church played a prominent role in the distribution. During Christian weddings valuable gifts may have been exchanged. Besides there were workshops that made objects with an adornment of garnets for the bishops. Or these objects may have been made at the instruction of the church as presents for Christian women. Apart from a few exceptions, these objects in a Christian context would not have reached the northern regions.

7 FREYA AND HER BRISINGAMEN

The Wijnaldum brooch, however, is a distinctive case. Although the symbolism of the garnet is clear, the shape only plays an important part in Scandinavia. Disc-on-bow brooches are found in sizes that, considering their magnitude, were no longer wearable. They possibly played a role in the cult. What the disc-on-bow brooch is supposed to represent as regards its meaning can be found out from Scandinavian literature. In this literature a beautiful jewel is described that belonged to the goddess Freya.[15] The saga of Thrymskvida tells the story of how the giant Thrym steals Thor's hammer, named Mjolnir. Thrym only wants to return the hammer if he can have Freya as his bride. Heimdall then proposes to let Thor wear Freya's wedding garment in order to approach the giant Thrym and fight him. Freya's wedding garment is described in this episode from the Thrymsong. She wears a dress that reaches to her knee, on her chest she wears a broad diamond necklace and her head is decorated with a pointed headdress. But her most beautiful jewel is the Brisingamen.

A gilt silver medallion-pendant from Aska, Hagebyhöga Östergötland, Sweden shows what is meant by this Brisingamen. The medallion portrays a female figure, seated, and with her arms folded.[16] She is wearing a characteristic dress consisting of a wide coat with underneath it a long skirt, trimmed with beads. The sleeves are also bead-trimmed. On her head she is wearing an ornament in the shape of a diadem, consisting of two parts and a double headband in between. On her chest a necklace made of four rows of beads can be observed, the lowest row being somewhat thicker than the others. But most important in this context is the object she is wearing directly below her chin: a big disc-on-bow brooch. Because of the size of the portrayed jewel it makes an almost grotesque impression. But the purpose of emphasizing this object is most of all to convince us of the importance of the ornament. This female figure with her beautiful jewels is also found in other pictures. A small female figure form Tuna in Alsike, Uppland, also depicts

Figure 3. Scandinavian pendant (original size 1.5 cm) (after Arrhenius 1969).

Freya in her long robe. She is wearing the same headdress, but because of the reproduction in profile it is now clearly visible that the cap is built up in a much more complex fashion. On her chest she again wears the necklace consisting of four rows of beads with the bigger lowest row. Directly underneath the chin a circle with a thick raised border is displayed. Birgit Arrhenius has proved that in this case too we are dealing with a disc-on-bow brooch, only indicated by the large disc. Freya is also wearing the same disc-on-bow brooch in a number of pictures of her from Helgö, Ekerö Ksp., Uppland. Here the goddess is depicted being led by the hand by a man. The large disc, the head-plate and the foot-plate are clearly indicated.

The name Brisingamen can also be connected with a jewel such as the Wijnaldum brooch. The word 'men' means necklace in Old-Scandinavian. Especially in the pictures of Freya we see that the disc-on-bow brooch is worn directly under the chin. So it is quite possible that this word defines this very object.

The first half of the compound Brisingamen, 'brisinga' is significant. It is related to the modern Norwegian word 'brisa' which means flame or blaze. In old descriptions of precious stones this term was often associated with garnets and therefore it is possible that it is a literal translation of the Latin word 'carbunculus' or the Greek word 'anthrax'. Apparently the Old-Scandinavian genitive plural 'brisinga' can be directly translated to the separate gems and is a direct indication of the very characteristic garnet cloisonné of the disc-on-bow brooch. The term gives a beautiful description of the jewel lighting up in the sun to a splendid red, underneath the goddess's chin. The garnet, which brings courage in battle and friendship and love to man, fits Freya like a glove. For Freya is not only the goddess of love, she also shares with Odin the power over the warriors that have died in battle. In her honour, wives of prominent and brave folk were given the honorary title 'Fru'.

Apart from the Thrymkvida there are some other places in Old-Scandinavian literature in which the 'Brisingamen' is mentioned.[17] In the 'Snorra Edda' by Snorri Sturlason, a passage from the Húsdrápa is quoted. This song was composed by Úlfr Uggason in Iceland in 983 and describes the woodcarving in Oláfr's new house. In this song the 'Singa' gem is described, 'singa' being related to 'brisinga'. In the Húsdrápa-song the half-god Loki steals the jewel and Heimdall tries to recover it. Loki fights Heimdall in the sea and they both turn into seals.

The Sqorla Tháttr tells the story of four dwarfs, Alfrigg, Dwalinn, Berlingr and Grerr, who manufacture a precious jewel, called the 'Brisingamen'. Freya wants to possess this jewel. She can have it, but in return she will have to spend one night with each dwarf. She agrees to the terms, but Loki hears of it and tells Odin, who then forces Loki to steal the Brisingamen. In the shape of a mosquito Loki enters Freya's locked room where she is sleeping with the Brisingamen around her neck. In his new appearance he is able to sting her, which makes her turn around so he can take away the necklace. The next morning Freya notices her necklace is stolen and wants Odin to return it to her. She can get it back on condition that she succeeds in bringing about an eternal war between two kings. Because she succeeds the Brisingamen is returned to Freya.

In the Old English epic of Beowulf, Freya's jewel is also mentioned, as the 'Brosinga mene'.[18] In this case a necklace which Beowulf receives is compared to the Brisingamen. The fact that Hama takes the Brosinga mene from Eormenric is also casually mentioned. Although this Eormenric is a well-known figure in the Old-Scandinavian heroic poems, he is never connected with the ornament.

8 DISC-ON-BOW BROOCHES

Very large-sized disc-on-bow brooches occur in the Scandinavian region. The largest one is probably that from Kåsta, Roslagsbro, Uppland in Sweden. Although it is kept in fragments, it must have been 28.5 cm long and 15 cm in height. The disc has a diameter of 10 cm. The brooch is beautiful, it is made of bronze and decorated with animal ornamentation. An almost equally big brooch was found in Näshulta, Södermannland, Sweden. This specimen is somewhat more simply decorated but still has one garnet. The undecorated parts of the ornament probably contained these gems as well. Furthermore two other fragmentary specimens are known: these were found in the burial place of Tuna, Uppland in Sweden. Such gigantic disc-on-bow brooches also occur in Norway, one was found in Rogaland and three brooches were found in northern Norway. The National Museum in Copenhagen, Denmark, has some disc-on-bow brooches in its collection. Most of them come from the island of Bornholm. As to craftsmanship and valuable finishing touch they cannot compare with the Wijnaldum brooch. Although all the above-mentioned brooches have or have had almandine decoration, such a large, all-over inlaid brooch is very extraordinary. So it is of another magnitude compared with the finds in Scandinavia.

Apart from the Wijnaldum brooch, two more examples can be given from Friesland.[19] They are not furnished with garnet cloisonné, but completely executed in gold and filigree. In Westergo, in the terp of Bessens at Wieuwerd, the foot-plate of a disc-on-bow brooch was found in 1886. We also know of a specimen from Hogebeintum in Oostergo, which is 15.5 cm long and complete. Although they are both inferior in execution as well as in height to the Wijnaldum brooch, they indicate that the shape of the disc-on-bow brooch as an important elitist or cult object was not unknown in early-medieval Frisia.

9 SUTTON HOO

The shape of the Wijnaldum brooch (disc-on-bow) links it to Scandinavia, and its material, garnet and gold, links it to the cult of Freya. The style of the brooch however leads us to another region, to East-Anglia.

Facing the town of Woodbridge on the river Deben lies a burial site which consists of fifteen tumuli. In 1938 the owner of the land had a couple of burial mounds investigated. In 1939 the biggest tumulus was excavated, and when the excavators discovered that this was no ordinary burial mound, a team of experienced archaeologists was formed to carry out the work.[20]

The discoveries were unexcelled in the history of British archaeology. In a very large boat-grave an impressive amount of objects was brought together to serve as grave goods for a king, most likely king Redwald. The ship must have been 27 m long, a tremendous length in those days. Only the rivets had been preserved in the soil and marked the outline of the ship. A large ossuary was placed amidships in which a large amount of objects was found. These consisted of gold jewellery, coins, silver plates, dishes, cauldrons, buckets and dishes made of bronze and iron, textile, leather, goblets and drinking horns and many other objects. Human remains were not found during the excavation of 1939, and not during the recent research of the 1966-1968 either. These remains had probably vanished completely into the acid soil during the course of time. Laboratory research of samples of the soil indicates this but is not conclusive. Comparing the grave of Sutton Hoo with the Wijnaldum brooch can be done by means of the gold artifacts. A description is given below of some of them that are executed in the garnet cloisonné technique.

A beautiful object is the mounting piece of a moneybag. It consists of a rim, executed in cloisonné which surrounds different images. These images are placed in two rows, three plaques on top and four plaques underneath with another four circles on the outsides and on the bottom. The upper row consists of a central image of fighting, twisted animal shapes. On both sides of this a tablet has been attached with a geometrical decoration, which, with regard to technique is first-rate. The bottom row consists of four shapes; paramount are four eagles attacking a duck, on the outside we see a man who keeps two animals under control. Next to cloisonné in gold and garnet, millefiori glass has also been used as an inlay. The central shape of the geometrically decorated hexagonal tablets in the top row is empty. Possibly an *intarsia* of millefiori glass was placed here, but it is also possible that it consisted of a paste of calcite. This was used as a glue to fasten the garnets, but has been dissolved because of the acidity of the soil.

The mounting of the moneybag shows some similarities with the Wijnaldum brooch. If one compares the geometric part of the foot-plate with the geometrically decorated tablets of the pouch, the similar shapes are striking. We see the staircase shapes and the mushrooms, the semi-circles in which the stairs are portrayed and the mushrooms that link up with the stairs. The curved line that connects the staircase shape halfway and forms an element in the rim of the head-plate of the Wijnaldum brooch is present and appears in the tablets as well. Another striking similarity is the outermost cloisonné inlay circle of the round forms. On both objects they have a cross-shaped filling, built up in straight cells. These straight cells are also seen in the rim of the pouch mounting, and in some places in the figurative images.

The human figure between the two animals that we recognised on the upper side of the foot-plate, recurs very identifiably in the lowest row of the mounting of Sutton Hoo. This image is very old and already occurs in the Sumeric art of Mesopotamia. Through the art of the steppe-peoples it is included in the range of Scandinavian art. Here we know this image from the Vendel Period in Sweden, which then influences Sutton Hoo. The fact that Wijnaldum is geographically situated exactly between both places is striking.

Bruce-Mitford sees in the meandering lines of

the Wijnaldum brooch two animal figures facing each other.[21] He regards the ribbon-shaped lines that are seen as a strongly stylised human figure as part of the abdomen of the animals. Even correct, this explanation ends up at the mounting piece of the moneybag of Sutton Hoo, because the central tablet of the upper row shows a similar image.

The other gold objects from the Sutton Hoo treasure can also be connected to the Wijnaldum brooch. The ornamental pieces of a sword from Sutton Hoo present the same style. The quatrefoil we know from the centre of the big disc belonging to the brooch, recurs in the knob, and the filigree *S*-shape in the hinge recurs in the mounting of the sword hilt. In the same style a rectangular buckle, the *T*-shaped mounting, a mounting in the shape of a buckle and rectangular mounting pieces can be mentioned.

10 THE ARTIST OF SUTTON HOO

The Wijnaldum brooch has some properties that are characteristic of 'the school of Sutton Hoo'. One of the most important and characteristic themes that recurs in almost all objects from this school, is the mushroom shape. Notable is the fact that the mushrooms on the Wijnaldum brooch touch each other with the stems and not with the hoods. Bruce-Mitford already made the link with Sutton Hoo in 1955.[22] He did not see the work of the goldsmith of Sutton Hoo himself or one of his workshop in the Wijnaldum foot-plate, but rather that of a jeweller who was trained in Suffolk or Kent and was working for a king on the European continent. Some years later, in 1974, Bruce-Mitford does contemplate ascribing the brooch to the master craftsman of Sutton Hoo, but more as an early work, a work that came into being before his creations reached maturity in East-Anglia.[23]

To make his point clear, Bruce-Mitford draws the attention to the further development of the work of Sutton Hoo compared to that of Wijnaldum. In the objects of Sutton Hoo some cloisonné cells are filled up in the millefiori technique. Moreover the artist has covered cells that belong to the background of a figurative image with gold foil in order to bring clarity of arrangement to his work. Indeed these two characteristics do not occur on the Wijnaldum brooch. But this could be caused by further developments throughout time. That no use was made of the millefiori technique is not of great significance. Several objects from the Sutton Hoo treasure are not executed in the technique. Yet on the other hand the most important pieces do have the millefiori decoration. The same goes for the parts which are covered with gold foil in order to clarify the image in the composition. The eagles consuming a duck, depicted on the mounting of the pouch, do not have this covering of cells with gold foil.

Besides, the figurative image on the Wijnaldum brooch is much more formalised than the image of Sutton Hoo. It has almost become an abstract form. It may very well be possible that an apprentice of the Sutton Hoo workshop was working for a king in Frisia. The shape of the knobs of the rounded cells forms an argument for dating the Wijnaldum brooch some years later than the objects from Sutton Hoo, dated ca 625. In Sutton Hoo they are closed off with gold just like some fragments with cloisonné decoration from Uppsala Västhög, Sweden. On the Wijnaldum brooch however, they are open and encrusted with a garnet.

Because of this the decoration comes closer to similar ribbon-shaped compositions from a somewhat later date. A good example can be found in 'The Book of Durrow'. On fol. 125v. opposite Luke's gospel, a decoration is applied which shows the same knobs as broadly executed in open work as that of Wijnaldum. The dualistic conception of the page is also striking, a curved ribbon shape facing geometric patterns. The Book of Durrow is dated some years later than the treasure of Sutton Hoo, namely circa 680 AD. Wijnaldum seems to be a good connection between both works of art. Dating the Wijnaldum brooch some years before 650 AD therefore seems more likely.

11 THE ALEMANNI

Scandinavian and British traditions are reflected in this precious ornament of Wijnaldum. But there is also a connection to be made with the Alemanni in Southern Germany. Bruce-Mitford has already pointed out close parallels with discoveries from this region in connection with the image on the helmet of Sutton Hoo.[24] Storms considers the frequent use of the mushroom-design in different places in Southern Germany striking and not fortuitous.[25] It can be dated around 600 AD and he suggests that diplomatic ties between the Alemanni and the East Anglians existed at that time. He thinks it possible that an Alemannic jeweller was employed by an Anglo-Saxon king. The fact is that the refined style and execution of many ornaments from the Alemanni bear more resemblance to the objects from Sutton Hoo and the Wijnaldum brooch, than to most Frankish and Scandinavian jewellery. This could point to a special relation between Frisia, Southeast England and South Germany. That there was a close tie between Frisia and Southeast England is beyond any doubt.[26] The relation between Frisia and Southeast Germany has not been thoroughly researched, but there are definitely indications that point to intensive contact.[27] Taking the geographical situation into account, Frisia could very well have been a sort of intermediary between the Alemanni and the Anglo-Saxons. The Rhine estuary ended in Frisian territory and the most important trade routes with South Germany ran via this river. Besides the Scandinavian (the shape) and the English, the brooch probably also represents the South-German decoration traditions, and from that point of view, it may represent a Frisian variant or hybrid. In that case the ornament itself forms a node of cultures.

12 THE DICHOTOMY

It is not sufficient to regard the Wijnaldum brooch simply as a beautiful ornament made out of gold and precious stones with a symbolic value. Considering the obvious dichotomy in the ornament's foot-plate, the use of a sacred theme in the decoration of the brooch should be taken into account as well. The main problem, however, is the unfamiliarity with preceding forms in Germanic art. That is why at most an attempt can be made to understand the dichotomy of the Wijnaldum brooch and only on the basis of contemporary Christian themes elsewhere.

The work of the master craftsman of Sutton Hoo shows that he was familiar with Christian symbolism. In support of this the cross-shape is brought forward, which appears on the pommels of the swords in the treasure.[28] 'The Book of Durrow' also shows a similar cross-shape in the 'Gospel of John'.[29]

The contradiction between the geometrical decoration and the figurative parts, as seen in the Wijnaldum brooch, is also present with objects used in Christian worship or with a Christian symbolic context. Examples of this are the 'Engerer Bursenreliquiar' with cross-shapes and animal-shapes, the Kingston brooch, also decorated with cross-shapes and animals and the brooch of Dorestad, where a cross-shape and animals can be recognised as well. Elbern sees in this scheme a 'heidnischen Geruch'. This makes the Wijnaldum brooch an intriguing pendant on the pagan side of the objects that are inspired by the newly introduced Christianity. With the Christian objects it is Christ that conquers nature in the shape of intertwined animals, on the Wijnaldum brooch it is the hero, the northern heir of the Sumeric Enkidu, who controls nature and is victorious.

13 THE TJITSMA TERP AND METALWORKING

It is of course pleasant to speculate about the location of the place where the Wijnaldum brooch was manufactured. Considering the shape, it is a Scandinavian object but judging from its decoration it is a product from Kent or Suffolk with possible influences from South Germany. Bruce-Mitford has already included a possible manufacture in Frisia.[30] This view is supported by the existence of two other precious disc-on-bow brooches in Frisia, of nearly the same specific shape as regards side-view. These pieces from Wieuwerd and Hoogebeintum are also hybrid.[31] They have a Scandinavian shape in combination with English and possibly South German decorative motives.

Later authors, e.g. Mazzo Karras, suggest one specific Frisian workshop that was probably active for some decades.[32] Other medieval jewellery found in Friesland may originate from this workshop as well. A number of artists could be employed in the workshop in the course of time. This makes it plausible that there was an important elite in Frisia. According to the discovery of prestige goods in gold-hoards, the elite of Westergo must have been wealthy and successful.[33] It is therefore no coincidence that the most precious disc-on-bow brooches (Wieuwerd and Wijnaldum) were found in this *antiquus districtus*.

The workshop mentioned above could be situated in northern Westergo, if it existed as the central workshop in Friesland at all. Wijnaldum may well have housed such a workshop. During the excavations proof was found for intensive metal industry. Not only were bronze and silver used, but a jeweller must also have been present. Melting-pots, gold-drops, small sledgehammers, touchstones with traces of gold on them and gold nails were found. One of the latest discoveries proves that we are dealing with the art of an eminent jeweller. It is a die for stamping gold foil with the wafer pattern. Because the Wijnaldum brooch was probably offered for repair to the workshop on the terp at Wijnaldum, it is a tempting thought to suppose that it was also manufactured here originally.

REFERENCES

ARNE, T.J. 1932. Ein bemerkenswerter Fund in Ostergötland, *Acta archeologica* 3: 67-112.

ARRHENIUS, B. 1969. Zum symbolischen Sinn des Almandins im früheren Mittelalter, *Frühmittelalterlichen Studien* 3: 47-59.

ARRHENIUS, B. 1971. *Granatschmuck und Gemmen aus nordischen Funden des frühen Mittelalters*, Stockholm.

ARRHENIUS, B. 1985. *Merovingian Garnet Jewellery*, Stockholm.

BERNHEIMER, R. 1931. *Romanische Tierplastik und die Ursprünge ihrer Motive*, 90-101.

BESTEMAN, J.C., J.M. BOS & H.A. HEIDINGA (eds) 1992. *Graven naar Friese koningen*, Franeker, 35-36.

BIMSON, M. 1985. Dark-Age Garnet Cutting, *Anglo-Saxon studies in archaeology and history* 4: 125-128.

BOELES, P.C.J.A. 1951. *Friesland tot de 11e eeuw*, 2nd edition, 's-Gravenhage.

BOS, J.M., & J. ZIJLSTRA 1991. Nieuwe fragmenten van de 'koninklijke' spang van Wijnaldum (Fr.) *Paleo-aktueel* 2: 97-99.

BRUCE-MITFORD, R. 1955. Gold and silver cloisonné buckle from Wynaldum, *Fries Genootschap van Geschied-, Oudheid-, en Taalkunde te Leeuwarden, 126ste Verslag (tevens verslag van het Fries Museum) over het jaar 1954*. 126: 16-17.

BRUCE-MITFORD, R. 1972. *The Sutton Hoo ship-burial*, London.

BRUCE-MITFORD, R. 1974. *Aspects of Anglo-Saxon archaeology*, London, 270-273.

CARVER, M.O.H. 1992. The Anglo-Saxon cemetery at Sutton Hoo: An Interim Report. In: M.O.H. Carver (ed.), *The Age of Sutton Hoo* Woodbridge, 343-371.

EAST, K. 1985. A study of the cross-hatched gold foils from Sutton Hoo, *Anglo-Saxon Studies in Archaeology and History* 4: 129-142.

ELBERN, V.H. 1971. Das engerer Bursenreliquiar und die Zierkunst des frühen Mittelalters, *Niederdeutsche Beiträge zur Kunstgeschichte* 10: 41-103.

EVANS, A.C. 1986. *The Sutton Hoo ship-burial*, London.

KNOL, E. 1993. De Noordnederlandse kustlanden in de Vroege Middeleeuwen, Amsterdam (thesis Vrije Universiteit).

MAZZO KARRAS, R. 1985. Seventh-Century jewellery from Frisia: A re-Examination, *Anglo-Saxon studies in archaeology and history* 4: 159-177.

LAFAURIE, J., B. JANSEN & A.N. ZADOKS-JOSEPHUS JITTA 1961. Le trésor de Wieuwerd, *Oudheidkundige medelingen uit het Rijksmuseum van Oudheden te Leiden* 42: 78-107.

LUCAS, A. 1948. *Ancient Egyptian materials and industries*, London, 451-452.

SIMEK, R. 1984. *Lexikon der germanischen Mythologie*, Stuttgart, 57-59.

STORMS, G. 1978. The Sutton Hoo ship burial; an interpretation, *Berichten van de Rijksdienst voor het Oudheidkundig Bodemonderzoek* 28: 309-344.

VRIES, J. DE 1933. *The problem of Loki*, Helsinki, 125-141.

WERCKMEISTER, O. 1967. *Irisch-northumbrische Buchmalerei des 8. Jahrhunderts und monastische Spiritualität*, Berlin.

WYATT, A.J. (ed.) & R.W. CHAMBERS (rev.) 1920. *Beowulf with the Finnsburg fragment*, Cambridge.

ZIJLSTRA, J. 1990-1994. *Friese bodemvondsten*, 5 vols., Leeuwarden.

ZIJLSTRA, J. 1991. Finn's Brooch? Belangwekkende vroegmiddeleeuwse vondsten te Wijnaldum (Fr.) *Westerheem* 40: 51-62.

NOTES

1. Elbern (1971, p. 41).
2. In a letter to Dr Wassenbergh (1955). See Besteman et al. (1992, pp 35-36).
3. Zijlstra (1990-1994). A summary of volume 1 is to be found in Zijlstra (1991, pp 51-62).
4. Bos & Zijlstra (1991, pp 97-99).
5. See Nijboer & Van Reekum (this volume).
6. Bernheimer (1931, pp 90-101).
7. See Nijboer & Van Reekum (this volume).
8. Lucas (1948, pp 451-452).
9. Arrhenius (1985, p. 23).
10. Arrhenius (1985, pp 29-36).
11. Bimson (1985, pp 125-128).
12. Arrhenius (1971, p. 45).
13. East (1985, pp 129-157).
14. Arrhenius (1985, p. 196).
15. Arrhenius (1969, pp 47-59).
16. Arne (1932, p. 73).
17. De Vries (1933, pp125-141) and Simek (1984, pp 57-59).
18. Wyatt & Chambers (1920, p. 62).
19. Lafaurie et al. (1961, pp 78-107).
20. Bruce-Mitford (1972), Evans (1986) and Carver (1992).
21. Bruce-Mitford (1974, pp 270-273).
22. Bruce-Mitford (1955, pp 16-17). Here the footplate is still regarded as a buckle.
23. Bruce-Mitford (1974, pp 270-273).
24. Bruce-Mitford (1974, p. 36). See also: Storms (1978, p. 343).
25. Storms (1978, pp 343-344).
26. Mazzo Karras (1985, pp 159-177).
27. Zijlstra (1993, pp 8-13).
28. Elbern (1971, pp 65-66).
29. Werckmeister (1967, afb. 6c).
30. Bruce-Mitford (1955, pp 16-17).
31. Boeles (1951, pl. XLII, XLIII).
32. Mazzo Karras (1985, 159-177).
33. Knol (1993, p. 241).

Scientific analysis of the gold disc-on-bow brooch

A.J. NIJBOER & J.E. VAN REEKUM

1 INTRODUCTION

The gold cloisonné disc-on-bow brooch from Wijnaldum-Tjitsma has been recently analysed and inspected for a material-scientific description. Previously the brooch was described in terms of its art history and archaeology,[1] but a technical examination was wanting. By means of various scientific methods, a description of the materials was obtained as well as a closer observation on the manufacturing- and decoration techniques of the brooch. One of the motives to analyse the composition of the metal alloys of the diverse small components, was to ascertain whether all the fragments belonged to the same brooch. Additionally, information on the history of the brooch and it's destruction is provided. After a preliminary microscopic examination of all components of the brooch, the following scientific techniques were applied:

- X-ray photography for an examination of the construction, primarily of the foot-plate,
- Scanning Electron Microscopy (SEM) for detailed images of several segments of the brooch in order to obtain more information about the manufacturing- and decoration techniques,
- X-ray Fluorescence (XRF) for an analysis of the various metal components of the brooch,
- X-ray Diffraction (XRD) for a determination of the cloisonné inlay and the white filling in the roundels of the foot-plate,
- High Pressure Liquid Chromatography (HPLC), Gas Chromatography (GCMS), Fourier Transform Infra Red (FTIR) in combination with XRD was employed for the paste underneath the gold backing foils beneath the cloisonné inlay.

A reconstruction of the main segments of the brooch is on display in the Fries Museum in Leeuwarden.[2] Figure 1 illustrates the assembled brooch, in combination with the associated fragments.[3] The numbers and letters in Figure 1 are used in this paper to indicate the different components of the brooch.

The various components of the brooch have been recovered at several stages and the archaeological history of the brooch is intricate. The footplate was found in the early 1950s, during drainage works on the Tjitsma terp near Wijnaldum. It was recovered almost intact. The majority of the gold sheet on the back, two roundels and some cloisonné inlays were missing. From the 1980s onwards, several other pieces have been found like two heavily damaged pieces of the head-plate with similar cloisonné decoration as the above mentioned foot-plate. Additionally the following fragments were recovered:

- 2 pieces of the connecting bow were recovered as one gold filigree and one silver 'filling' fragment (Fig. 1),
- 7 fragments of gold sheet (Fig. 1, no. 6 up to 12),
- 2 pieces of silver 'filling' (Fig.1, no. 2 and 3),
- 1 fragment of the gold rim with twisted beaded wire (Fig. 1, no. 13),
- 1 piece of the hinge with gold sheet and silver filling (Fig. 1, no. 5),
- 1 piece of gold filigree of a stylized animal head with a garnet eye (Fig. 1, no. 4).

The above mentioned fragments were found by amateur-archaeologists and were collected by Mr J. Zijlstra. The Fries Museum at Leeuwarden acquired most of the components of the brooch. During the excavation on the Tjitsma terp in 1991, 6 more fragments of gold sheet (Fig. 1, no. 15 up to 19) and in 1992 a piece of the head-plate were recovered.

From onwards the 1950s, all the major components of the disc-on-bow brooch from Wijnaldum have been recovered except for the disc itself. As yet no fragments of the disc have been found which could be significant for the final interpretation of the presence of the brooch on Tjitsma terp.

The paper will first describe the analyses of the singular materials which were employed during the manufacture of the brooch. The determination of the non-metals, silver alloys and the gold alloys are

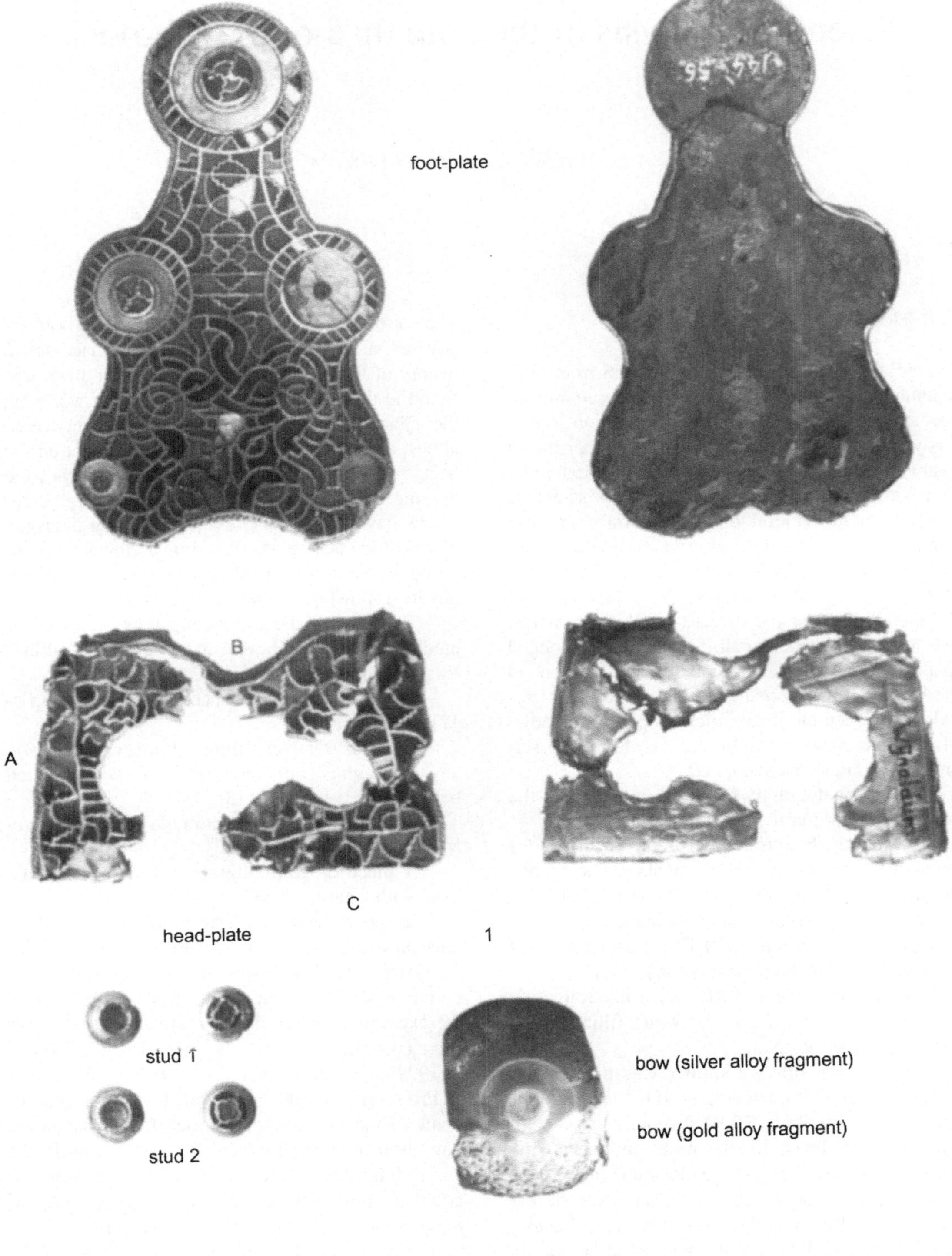

Figure 1. Photograph of the various components associated with the Wijnaldum brooch (computer graphics by H.J. Waterbolk, GIA).

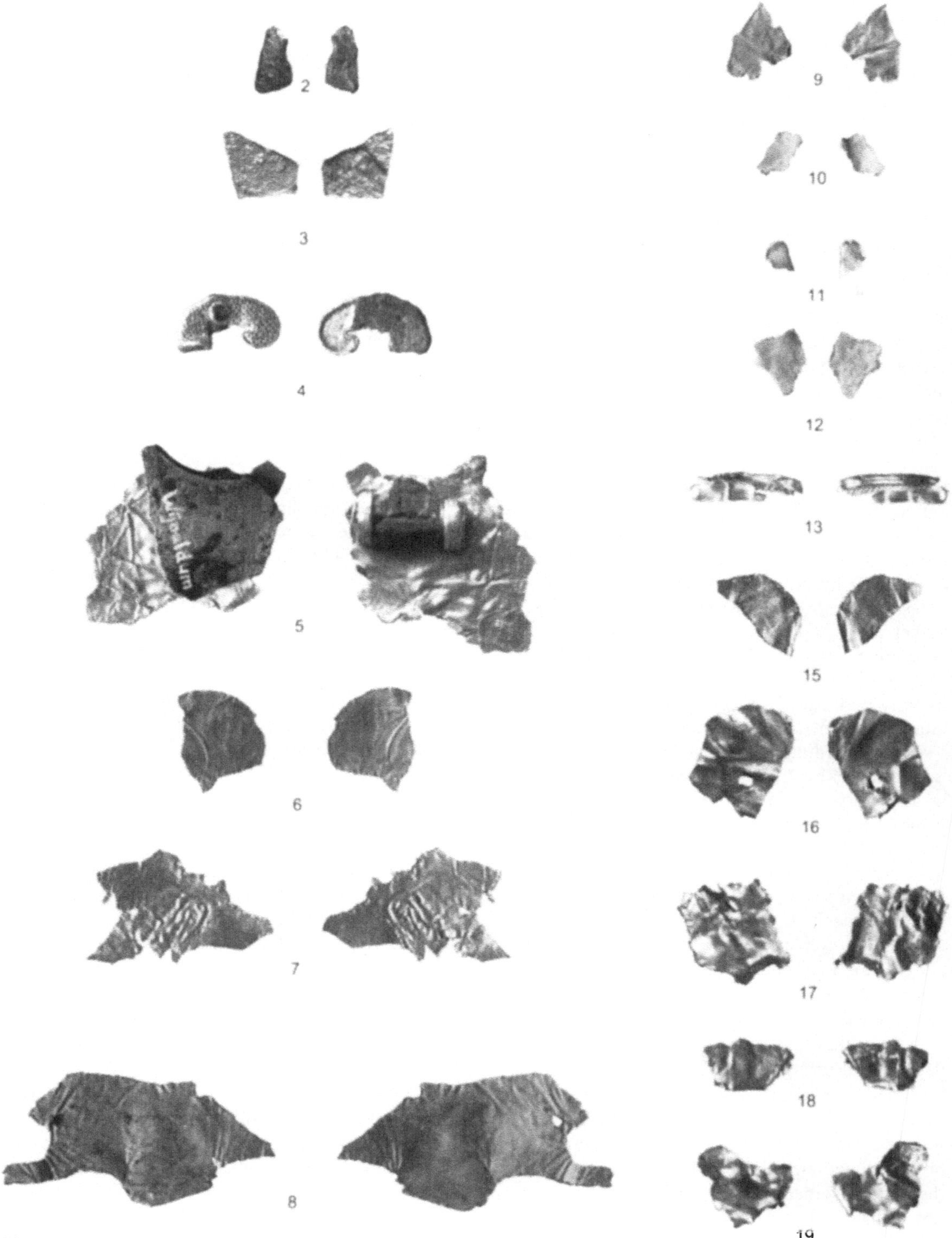

Figure 1. Continued.

presented before a discussion of the manufacturing and decoration techniques is given.

2 ANALYSES OF THE MATERIALS

2.1 *Non-metals*

The non-metals of the Wijnaldum brooch consist of garnets, white fillings in the roundels of the foot-plate and white paste underneath the cloisonné inlay.

The cloisonné surface of the foot-plate has been characterized by Bruce-Mitford as a 'a continuous carpet-like spread of garnets'.[4] The assumption that garnets were employed during the manufacture of the brooch is rational since the colour of the transparent stones is wine-red and they do not resemble glass. In order to characterize the garnet more precisely, a particle was temporarily lifted from a heavily damaged fragment of the head-plate (B) and analysed with a Debeye-Scherrer camera. The analyses demonstrated that an 'iron-aluminium garnet' or almandine, with the chemical formula $Fe_3Al_2(SiO_4)_3$ had been employed. Almandine belongs to the group of garnets. Other varieties of garnet are pyrope, spessartite and grossular. The almandine variety can be found in Sweden, Austria and Italy[5] but garnet or almandine which was used in early dark-age jewellery in Northern Europe, is thought to come from Bohemia.[6]

The white filling in the roundels of the foot-plate was analysed by XRD. It appeared to be aragonite or orthorhombic calciumcarbonate. Aragonite is valued for its colour, lustre and surface pattern and therefore frequently used as an inlay material. It is quite soft (3.5-4 on Mohs'scale) and therefore easy to cut. The appearance of the aragonite in the Wijnaldum brooch is opaque white with a greenish streak and tiny black veins. The surface displays a silky lustre and because of preservation occasionally a fibrous structure (Fig. 7). Due to post-depositional changes, the original appearance of the aragonite inlay has altered. The process of post-depositional changes of aragonite has been hardly documented which makes it difficult to identify archaeological samples of aragonite. Modern specimens can be distinguished in diverse varieties. The original aragonite of the Wijnaldum brooch could therefore have been mother of pearl or other varieties such as:[7]

– An 'alabaster variety', which have been reported in Mexico and Pakistan or,

– A snow-white coralloid or 'flos ferri' variety of aragonite which is typical for the mines of Styria in Austria.

The more fibrous variety shows a silky shine and a cat's-eye effect and is often cut into cabochon-shaped stones and beads. Besides the above mentioned provenances, aragonite is also found in Czechoslovakia, Sicily, France, England and Spain.[8] It has been rarely attested as inlay material for garnet cloisonné jewellery. An identification of the aragonite as mother of pearl is ambivalent. Of all the samples analysed by Arrhenius, only one contained aragonite as an inlay, probably as mother of pearl[9] though inlay of mother of pearl has been fairly often registered by Nerman.[10]

A small sample from the white paste underneath the gold foil, was taken from head-plate, fragment C. The sample derived from a cell without the almandine inlay. The paste was analysed by HPLC, GCMS and FTIR. Other samples of the paste came from head-plate, fragment B of which from one of its cells, the almandine inlay was temporarily lifted. These samples were analysed by XRD, HPLC, GCMS and FTIR. Analysis by XRD demonstrated the presence of a silicate combination: $H_2Si_2O_5$. The analyses by FTIR, HPLC and GCMS indicated a mixture of a silicate, a binder (all samples contained both a resin and a protein, similar to animal glue), sulphur and a sulphur compound. The samples were uniform in composition. Paste mixtures were set in the cells after which the gold foil and the almandine was positioned. This was heated slightly in order to soften the paste. After cooling, it would keep the backing foil and the almandine in it's cell at its proper height.

Diverse paste materials which came from garnet cloisonné jewellery have been analysed by Arrhenius. She made different clusters, based on the main components of the paste, like a carbonate-, silicate- gypsum- sulphur- and an organic group.[11] Within these clusters, wax was frequently a constituent of the paste mixture, though, unlike the paste of the Wijnaldum brooch, a combination of a resin and animal glue has not been reported. It appears that the paste mixture of the Wijnaldum brooch belongs to the silicate group but is dissimilar from the mixtures which were analysed by Arrhenius with regard to the binder. The presence of sulphur improves the adhesion quality of the paste and has been attested by Arrhenius as a component of some of the paste mixtures.[12] It has been reported that sulphur only occurs in volcanic regions and is hard to obtain in northern Europe.[13]

2.2 *Silver alloys*

Silver alloys were predominantly employed during the manufacture of the brooch as framework for the gold sheet on the back and the gold sheet with cloisonné on the front. Four fragments silver were analysed by XRF. The composition of the silver alloys is shown in Table 1. The high copper content of the filling of the foot-plate is probably due to corrosion of the metal since a thick copper-corrosion layer covers the back. The other three samples had been cleaned by a conservator to a silver-shiny surface.

2.3 *Gold alloys of different parts of the brooch*

The composition of the gold alloys have been analysed by XRF and are presented in Table 2. Clusters can be made such as indicated by the composition of the foot- and head-plate. Probably a slightly different gold alloy has been used for the foot-plate. Its gold and copper content is higher and its silver content is lower than the metal components of the gold alloy which was used for the manufacture of the head-plate and associated rim fragment. The gold content of the foot-plate varies around 95%. 23 Carat gold contains 95.83% of pure gold, and is supposed to be the best grade of gold alloy for decorative purposes. Because of its purity it stays bright indefinitely, but is quite soft and easily worn.[14] The gold alloy of the Wijnaldum brooch is essentially a binary alloy made by melting gold with some silver since many of the samples do contain less than 2% copper.

The rim fragment and the head-plate are correlated on account of the similar construction. The gold sheet on the back had been connected by small gold rivets and was decorated with a twisted, beaded wire. The rim of the foot-plate is not associated with a construction using small rivets. The rivets of the head plate are probably used in order to strengthen the construction. This was necessary because the hinge is likely to have been attached to the back of the head-plate. The component of the brooch with the hinge had to be sturdy because of the substantial weight of the complete brooch. The backing foil of head-plate B has the highest gold content of all samples. The manufacture of the thin foil from gold sheet requires an extremely soft and malleable alloy with a high gold content. The relatively high silver content of the front and back of the bow fragment can be correlated with the silver-content of the head plate. The difference in copper content can not be explained. It could be that the X-ray beam had been located on one of the copper corrosion spots on the front of the bow.

2.4 *Composition of the gold sheet fragments*

Table 3 presents the composition of gold sheet fragments which were analysed by XRF. The most obvious cluster consists of sheet number 10, 11 and 12 because of the copper content which is present as a trace element. The three fragments differ in appearance from the other fragments and it is likely that these sheet fragments do not belong to the Wijnaldum brooch. Two other fragments, number 7 and 19 have a relatively high copper content.

The gold content of all the gold sheet fragments is very high, which makes the alloy soft and malleable, and therefore easy to roll or stamp into such thin sheet.

Comprehensively, the gold content of the Wijnaldum brooch is high when compared with the gold jewellery from Sutton Hoo and other sites. The results from the Sutton Hoo jewellery demonstrate a range of gold content from 97 to 70% which compares with the range found in the coins, blanks and ingots in the purse (96.9 to 69.2%). Only the gold content of the sword pommel (97%) and the two shoulder clasps (93.5 and 94%) from Sutton Hoo correlates with the analyses of the gold alloys from the Wijnaldum brooch. The gold content of the Wijnaldum brooch does not reflect the general 'flight from the gold standard' which can be deduced from Kent's time-fineness graphs based on the gold content of the Merovingian tremisses during the 7th century BC. This 'flight' is also reflected in the gold content of jewellery and on account of the gold content, the manufacture of the brooch is possibly dated to the early 7th century.[15]

Table 1. Composition (in percentage) by XRF of silver alloys.

Specification	Au	Ag	Cu	Pb	Ni	Co	Cd
Backside foot-plate	1.33	69.82	26.30	1.62	0.25	–	0.04
Bow	1.41	95.07	2.54	0.80	0.13	0.01	0.04
Fragment 2	2.77	93.41	2.50	0.91	0.19	0.01	0.03
Fragment 3	2.47	93.27	2.68	1.42	0.13	0.02	0.01

Table 2. Composition (in percentage) by XRF of gold alloys used for the brooch.

Specification	Au	Ag	Cu	Pb	Ni	Co	Cd
Foot-plate rim	95.25	3.02	1.53	0.01	0.12	0.03	0.02
Foot-plate roundel	95.06	3.21	1.11	0.09	0.16	0.16	–
Foot-plate stud	95.52	3.32	0.84	0.06	0.08	0.09	–
Head-plate A (back)	94.59	4.99	0.25	–	0.09	0.03	–
Head-plate B (back)	93.45	5.99	0.34	–	0.17	0.03	0.03
Head-plate C (back)	94.03	5.56	0.26	–	0.10	0.02	0.01
Fragment rim, 13	93.80	5.20	0.84	–	0.11	0.03	–
Backing foil h-pl.B	96.85	2.69	0.28	–	0.14	0.04	–
Bow (front)	91.59	5.85	2.13	0.25	0.15	0.02	0.00
Bow (back)	92.59	6.72	0.47	–	0.19	0.02	–
Stud 1	93.69	4.22	1.94	–	0.08	0.02	–
Stud 2	93.10	4.45	2.25	–	0.09	0.04	–
Animal head, 4	96.59	2.08	1.12	0.08	0.08	0.04	0.00

Table 3. Composition (in percentage) by XRF of gold sheet fragments.

Specification	Au	Ag	Cu	Pb	Ni	Co	Cd
Sheet with hinge, 5	96.66	2.91	0.30	–	0.07	0.04	0.01
Sheet, 6	96.67	2.90	0.25	–	0.11	0.03	–
Sheet, 7	95.64	2.98	1.24	–	0.11	0.02	–
Sheet, 8	96.43	3.14	0.27	0.04	0.10	0.01	–
Sheet, 9	96.65	2.87	0.32	–	0.13	0.02	–
Sheet, 10	96.80	3.02	0.03	–	0.11	0.04	–
Sheet, 11	96.18	2.63	0.02	–	0.11	0.02	0.04
Sheet, 12	96.16	2.67	0.01	–	0.12	0.03	–
Sheet, 15	96.90	2.66	0.29	–	0.12	0.03	–
Sheet, 16	96.35	3.09	0.41	–	0.08	0.03	0.03
Sheet, 17	96.72	2.84	0.30	–	0.10	0.03	–
Sheet, 18	96.16	3.15	0.55	–	0.10	0.04	–
Sheet, 19	95.56	2.82	1.51	–	0.07	0.02	–

3 MANUFACTURING- AND DECORATION TECHNIQUES

3.1 *Head-plate*

Examination with a microscope and a scanning electron microscope revealed aspects of the production and decoration technique of the brooch. The bulk of the head- and foot-plate is obtained by a silver 'filling'. This gives the brooch its strength after which the gold sheet could be applied for the visible parts. Many Saxon silver brooches are made by casting the silver framework.[16] A soap-stone mould has been found on the farm Hinna, Hetland Parish, Rogaland. The mould shows the exact outlines of a square-headed brooch without ornamentation.[17] After casting the silver could be decorated with techniques like hammering, embossing and carving.

The silver fragments of the Wijnaldum brooch displays casting characteristics and thus the silver framework is considered to have been cast.

The cloisons are made by soldering pre-shaped gold strips on an underlayer of gold sheet. Figure 2 illustrates a soldering seam between a gold sheet surface and a cell wall. The cell wall has rounded edges where the angles of the cloison are made. This is the result from bending. Another method for shaping cell walls consists of filing and soldering but this would result in a straighter angle. Thus it is probable that the cloisons from the Wijnaldum brooch are made by bending gold strips.

Additionally Figure 2 illustrates that the cell walls have a small protruding edge to mount the almandine. This protruding edge on top of the cell walls could be made by gently hammering or rubbing the soft gold cell walls, after having placed the aimandine.

All head-plate fragments have a rim with small gold rivets. These rivets connect the two-ribbed rim of the gold sheet which is folded around the silver filling, with the gold sheet on the backside of the head-plate. Some torn pieces of gold sheet from the backside are still present on the head-plate frag-

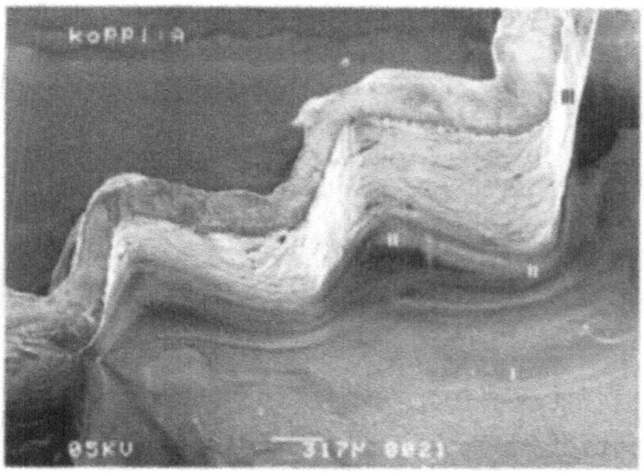

Figure 2. SEM photograph of head-plate A. Magnification 31.5 ×. I = gold sheet, II = soldering seam, and III = cell wall (photo GIA).

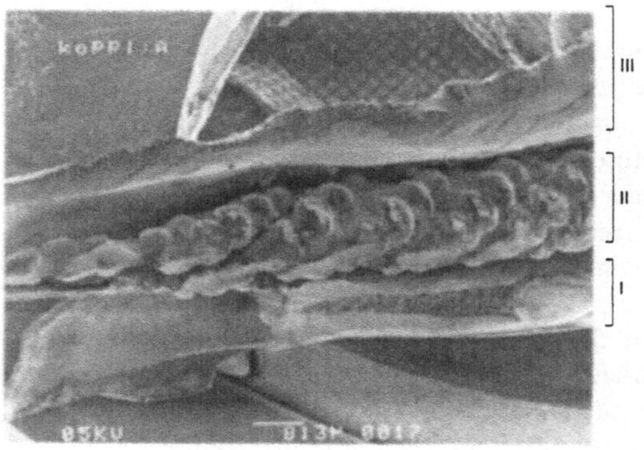

Figure 3. SEM photograph of head-plate fragment A. Magnification 12.3 ×. I = ribbed edging with one rivet at the right, II = decorative cord consisting of three beaded wires, and III = front of the headplate with cell walls and backing foil (photo GIA).

ments. Above the two-ribbed rim of the upper gold sheet, a decorative rope is soldered. This rope consists of three twisted beaded wires. Figure 3 shows the rim of head-plate fragment A, with the twisted beaded wires, one rivet and a piece of backing foil in a cloisonné cell.

Fragment A of the head-plate is heavily damaged, probably caused by ploughing. It is one of the pieces, found by amateur-archaeologists during the 1980s. The beaded wire of the head plate shows signs of wear and must originally have been in use for a considerable time. Beaded wire can be made with several devices. Duczko describes an experiment with a beading-file, made according to a description by Theophilus. The beading-file is a small iron bar with a lengthwise groove with sharp edges. The implement leaves a groove round the maximum diameter of every bead as result of the pressure exerted by the groove of the tool.[18]

Another method to make beaded wire is by using an *organarium* which has also been described by Theophilus. The *organarium* is an iron implement with an upper and a lower part with holes and grooves on both sides. When the upper iron part is lowered, beads are formed while turning the wire or rod. The holes have different diameters and one is supposed to start beading the wire in the largest hole.[19] Untracht illustrates a two-piece die which is similar to the above mentioned *organarium.*[20] The beaded wires of the Wijnaldum brooch do not have a groove around the maximum diameter of every bead, which would have been indicative for a beading file. This indicates that the beaded wire is made with an *organarium* or a die as described by Untracht.

Backing foil is present in all cloisonné cells and also underneath the garnet-eye of the stylized animal head. The backing foil (Fig. 3) is folded into shape and covers the base and partially the sides of the cell walls. The almandine is placed inside the folded foil. All visible foils in the cells without the almandine, demonstrate that the foil has folded edges. The Wijnaldum foil is similar to the 'standard foils', discussed by East and Meeks & Holmes.[21] Standard foils display a pattern with a simple uniform grid. An example of another type of foil is the 'boxed' foil where more robust, wider lines enclose small squares. The most common standard foil of the Sutton Hoo jewellery has a line spacing of 4 lines/mm. The foil of the Wijnaldum brooch exhibits 4 to 5 lines/mm. Less fine spacing of 3.5 lines/mm is more prevalent in Anglo-Saxon garnet jewellery.[22] Meeks and Holmes suggest that foils with the fine spacing of the Sutton Hoo jewellery could have been manufactured with a brass or ivory die. The pattern is stamped into the gold foil with a die which has very fine cutting grooves. Such fine and regularly spaced grooves are hard to produce without the use of an implement and thus the use of a 'mechanical die-cutting jig' has been suggested. This instrument cuts grooves of the required spacing into the flat die faces, like a pantograph. The movement of a lever on a scale is imparted to the die face in a corresponding but much smaller pattern. By experiment, Meeks and Holmes were able to produce very similar foils with the aid of a grooving jig.[23] A mechanical die-cutting jig produces negative dies. In 1996, a positive die for making boxed pattern foils was discovered in the conservation laboratory of the University of Groningen during mechanical removal of the corrosion layers on the artefact. This die was excavated at Wijnaldum and dates to the 7th century. Figure 4 presents photographs of the die and a detail of the pattern while Figure 5 illustrates the lines of the boxed pattern which could be identified after examination with a scanner and a microscope (the grey lines in Figure 5 are reconstructed). From Figure 5 it is deduced that the line pattern is regular though not as uniform as suggested by the experiments of Meeks and Holmes with a die-cutting jig. The die from Wijnaldum is slightly less mechanical than the dies produced by Meeks and Holmes. Some of the lines in Figure 5 are askew though most are aligned and straight. The subdivision of the individual boxes into squares in neither regular. The die from Wijnaldum has preserved 16-, 20- and 25-square boxes while the size of some of the boxes suggest a smaller or higher number of squares.[24]

3.2 *Bow fragment of gold*

The bow fragment of gold is decorated with several pieces of beaded wire which are bent into a S-shape with spirals. The beaded wire resembles the beaded wires discussed above. It is intermittently soldered on the gold sheet which is visible on the SEM photograph (Fig. 6) as dots where the solder is melted.

Untracht explains how filigree is soldered nowadays.[25] A mixture of (gold) solder scraping with water, flux and gum is made. This mixture is painted on the wire. At the wire ends, particles of solder can be placed in order to obtain a solid joint. The amount of solder used must be just enough to make a solid junction, too much solder will flood the beaded wires. The heating of the filigree piece, to melt the solder, requires skill of the craftsmen, because the thin wires are very easily melted themselves. Interval soldering of wire to sheet, like the s-shaped spirals of the bow, is sufficient. It is not necessary to solder the whole wire to the ground sheet.

3.3 *Hinge and pin*

The hinge fragment which comprises the remains of the pin, exists of a segment of gold sheet (no. 5, Table 3) with silver filling and a copper alloy pin-fragment. The copper alloy pin is riveted in between two upright sections of silver- or copper alloy which are covered with gold sheet. The gold sheet is decorated with a two-ribbed rim similar to the rim of the head- and foot-plate. Figure 7 illustrates a fragment of the gold sheet which covers the hinge and which is part of the gold sheet that covers the back of the brooch. This construction makes it probable that a section of the sheet is cut out and bent upwards to cover the hinge. The covering sheet of the hinge could be partially soldered.

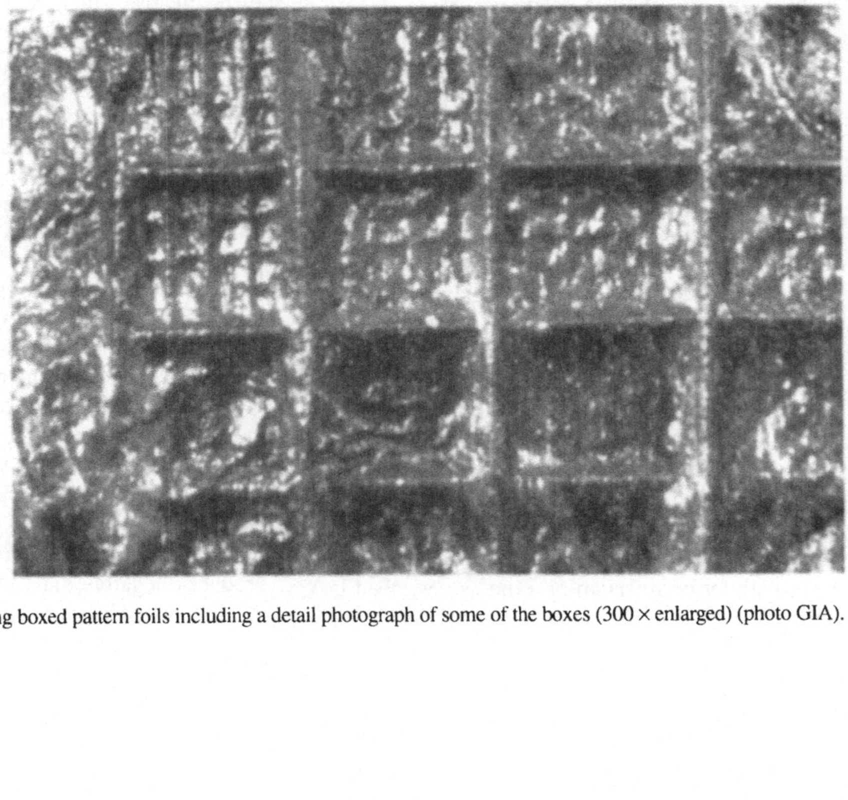

Figure 4. Positive die for making boxed pattern foils including a detail photograph of some of the boxes (300 × enlarged) (photo GIA).

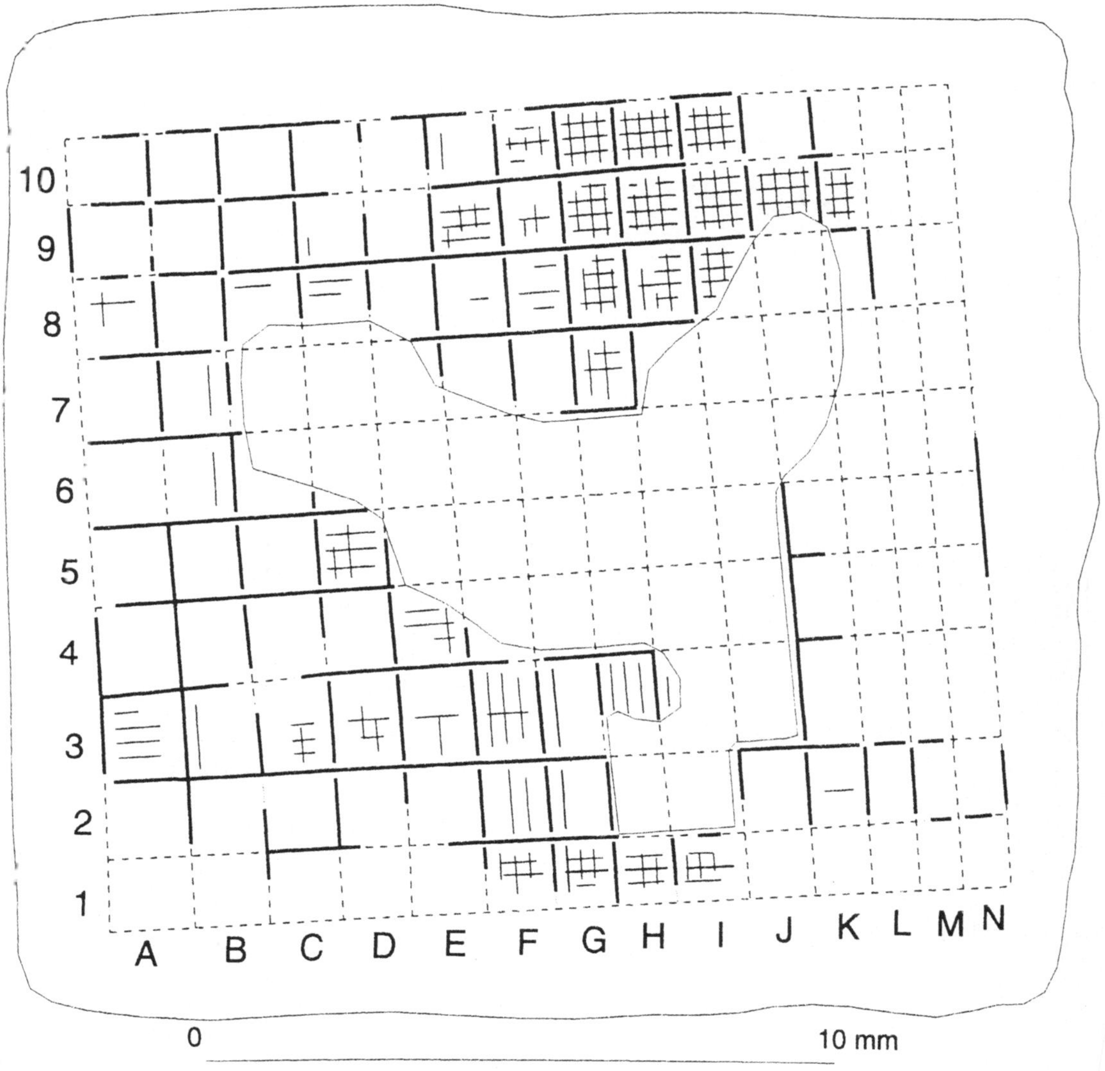

Figure 5. A line drawing of the die. The dark lines were identified by a scanner and under a microscope while the grey lines are reconstructed (drawing by H.J. Waterbolk, GIA).

3.4 *Foot-plate*

The construction of the foot-plate which consists of gold sheet with cloisons on a silver framework is similar to the fragments already discussed as is the rim which is decorated with two ribs and a rope of three beaded wires. The gold sheet at the back is attached to the brooch by folding the gold sheet from the front over the backing sheet. An X-ray photograph (Fig. 8) reveals several holes in a symmetrical pattern. These holes in the foot-plate are probably associated with the construction for fastening both the pin catch as well as the backing sheet to its framework. The hinge and catch construction for the pin can be illustrated by the gold and silver disc-on-bow brooch from Hoogebeintum of which the hinge and catch construction for the pin has been preserved. The reverse of this brooch displays a decoration of two animals which originate from the catch and continue to the top and side of the foot-plate. Mazzo Karras suggests that an animal head decorated the end of the catch.[26] The Wijnaldum brooch is comparable in size and con-

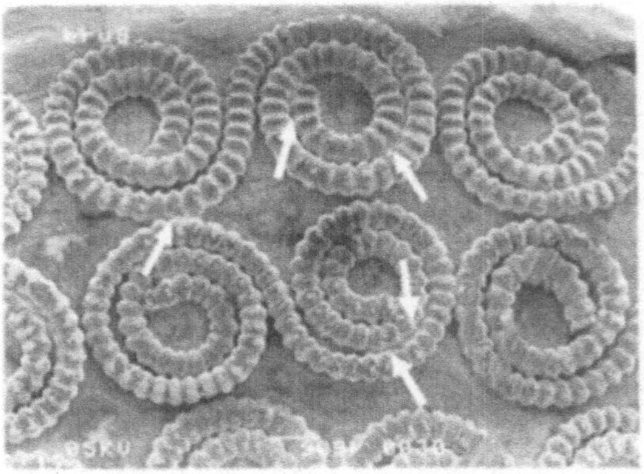

Figure 6. SEM photograph of the bow fragment of gold. The arrows indicate the soldering joints. Magnification 11 × (photo GIA).

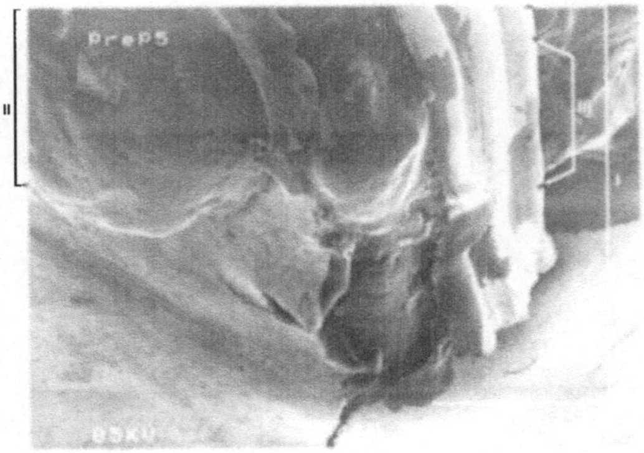

Figure 7. SEM photograph of the hinge. Magnification 13.4 ×. I = vertical section of silver- and copper alloy, II = detail of copper alloy pin, and III = ribbed gold edging covering the silver and copper alloy (photo GIA).

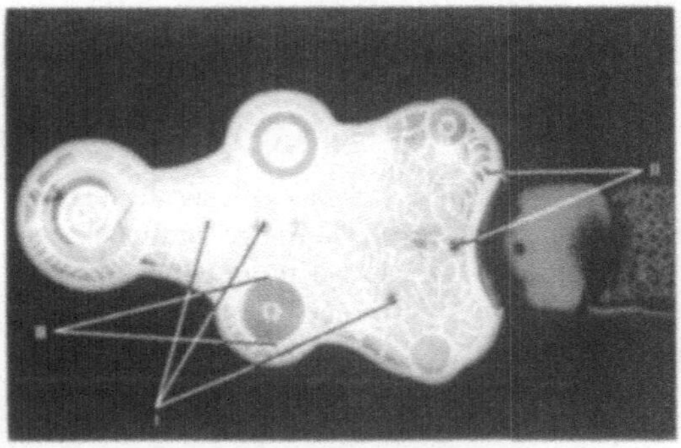

Figure 8. X-ray photograph of foot-plate and bow. I = holes in the silver framework, II = two larger holes for connecting the bow, and III = tube with several cuts in the edge (photo R. Gerritsen).

struction with the Hoogebeintum brooch which is almost complete. Both are disc-on-bow brooches with similar construction characteristics. Taking its size and weight into consideration, the Wijnaldum brooch must at least have had a similar complicated and strong catch. This could account for the number of holes which are revealed by the X-ray photograph. Especially the many small holes in between the two roundels in the centre and the roundel at the top could have been used for joining the animal decorations of the catch. Nearby the side of the foot-plate which is closest to the bow, at least two larger holes are visible on the X-ray photograph. The silver framework of the bow has three holes and therefore the bow might have been connected to the foot-plate by a row of three rivets. The attachment of the studs to the roundel and the foot-plate is revealed by the roundel without a stud. A small tube is present in the middle of the aragonite inlay. This tube has an edge with several small cuts which are revealed by the X-ray photograph. This implies that the studs are connected by folding outwards the edges of the tube. The small cuts facilitate the folding. The tube is pierced through the gold sheet but is not visible at the back of the foot-plate. The tube therefore, does not perforate the silver framework. The gold tubes which are soldered to the back of stud 1 and 2 are bigger than the small tube left in the foot-plate. This means that stud 1 or 2 are not an integral part of the foot-plate, although the same folded and cut edges are present at the end of both tubes. The circular shapes formed by the cell walls of the different head-plate fragments indicate the initial presence of two roundels in the middle of the head-plate. Both studs were probably placed in the centre of these roundels.

The top of the foot-plate is decorated with two small roundels. One of these roundels is filled with the aragonite inlay and a round, flat piece of almandine while the decoration of the other roundel has not been preserved. The gold cylinder of the roundel is soldered to the gold sheet which covers the silver framework. Figure 9 is a SEM photograph of the roundel with preserved inlay. The gold cylinder of the roundel is an element of the cloisonné structure. This cylinder encloses another gold cylinder which contains the aragonite and almandine setting. The gold cylinder with the aragonite and almandine setting is attached to the roundel by a small projecting edge which has been rubbed or hammered. This method of fastening the cylinder is similar to the setting of the almandine.

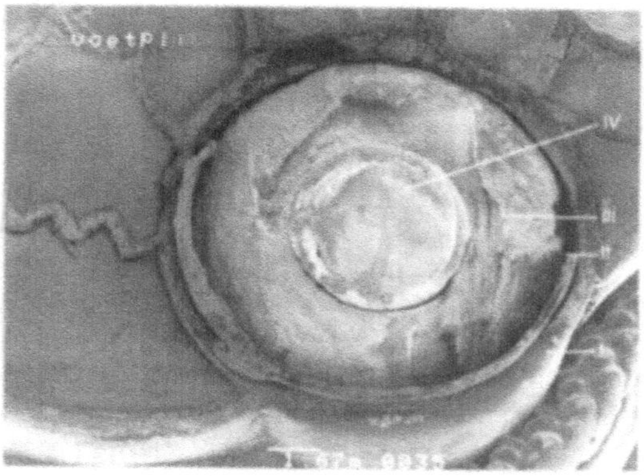

Figure 9. SEM photograph of the roundel with preserved setting at the top of the foot-plate. Magnification 6 ×. I = gold cylinder of roundel, part of the cloisonné structure, II = gold cylinder for aragonite inlay, III = aragonite, and IV = aragonite in its setting (photo GIA).

4 RECONSTRUCTION OF THE GOLD SHEET AT THE BACK OF THE FOOT-PLATE

Due to the holes in the foot-plate which were revealed by the X-ray photograph, it was possible to reassemble some fragments of the backing gold sheet. The gold sheet numbers 5, 6, 7, 8, 15 and 16 and 17 belong to the back of the brooch. The reconstruction of the gold sheets on top of a X-ray photograph is presented in Figure 10. Most of the gold sheet fragments which belonged to the back of the foot-plate, could be reassembled since they match. Several gold sheet fragments show traces of cutting. A small pair of scissors was used to shear the gold sheet at the back of the foot-plate. Especially gold sheet 8 demonstrates the shearing traces on the side to the middle of the foot-plate.

Table 3 displays the difference in copper content of gold sheet 10, 11 and 12. Since these sheet fragments also have a different appearance, it is assessed that these fragments do not belong to the brooch.

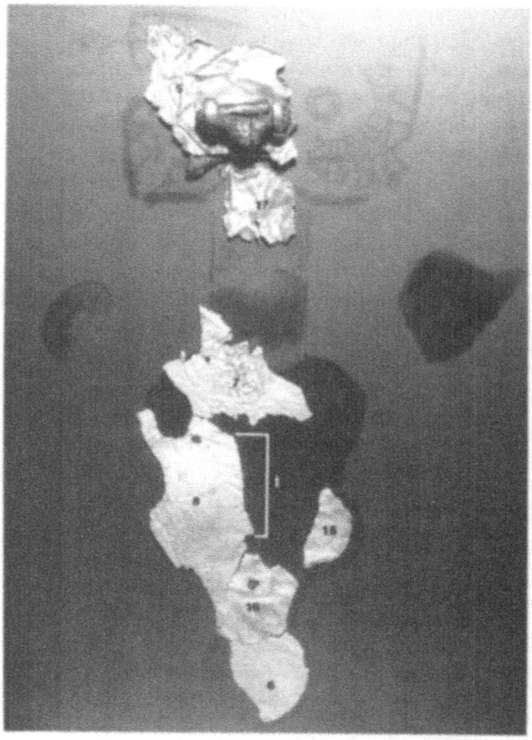

Figure 10. The reconstruction of fragments of gold sheet on a X-ray photograph. I = shearing traces, 5 to 8 and 15 to 17 are the numbers of the gold sheets as used in Figure 1 and Table 3 (photo R. Gerritsen).

5 CONCLUSIONS

The XRF analysis of the gold alloys indicates that for the various decoration- and construction techniques, slightly different gold alloys were used. All alloys which were analysed, have a very high content of precious metal. It seems that silver and to a lesser extent also copper were melted together with very high percentages of pure gold. This is probably done to make the alloy harder, because pure gold is too soft for most purposes. Especially the gold alloy of the foot-plate, the gold sheet from the back and the backing foil reveal very high gold contents. As such the gold content of the Wijnaldum brooch mirrors the 'gold standard' as revealed by the late 6th and early 7th century Merovingian tremisses.

The examination by SEM illustrates that the manufacture of the brooch involved a highly qualified craftsman as is indicated by the fine spacing of the grid on the foil. The brooch is made by combining several metalworking- and decoration techniques, which all exhibit the subtle eminence of its composer. On account of the craftsmanship, it seems unlikely that the segment of gold filigree with a stylized animal head, belongs to the brooch mainly because of the inferior quality which is displayed in the filigree itself and the plain setting of the garnet.

From an archaeological point of view it has not

been established why this brooch was present on the Wijnaldum site. Taking into consideration that many traces of metalworking have been recovered during the excavations, it is suggested that the Wijnaldum brooch is associated with the metalworking activities on the site. The context and nature of the deposition of the various fragments of the brooch remains ambivalent as well as the reason why the fragments were actually preserved at the site. Nevertheless, the positive die for making boxed pattern foils which has been recovered at Wijnaldum confirms the stature of the goldsmith who worked at the site. The evidence for shearing on the backing gold sheet and the fact that the different components which belong to the brooch, are not really torn but only disconnected like the two studs and the bow, implies that the brooch was reused or mended. Both possible interpretations of reuse or repair are associated with the many data on goldsmithing on the site such as the touchstones, a crucible for melting gold and the die. Reuse is implied by the cutting marks on the gold sheet from the back of the foot-plate as well as by the absence of two roundels of the head-plate and the disc itself. Several pieces from the brooch were found at various stages during a period covering more than 40 years. It is notable that during this period no fragments of the disc were recovered while it seems as if both roundels from the head plate were intentionally separated from the head-plate (Fig. 1). It is suggested that these roundels were taken from the head-plate and might have been used to compose a new ornament. The disk itself could have been transformed into a pendant as has been frequently attested for Anglo-Saxon disc brooches.[27] The Monkton disc brooch might even derive from a larger and grander brooch.[28] In this context it is furthermore relevant to recall the 7th century demolition segment of an almandine buckle from Dronrijp and the disconnected foot-plate of a substantial gold brooch from the Wieuwerd hoard.[29] A late example of reuse of 7th century garnet jewellery is found on the Egbert shrine at Trier. During the late 10th century, the outer roundel of a possible Frisian disc was reused in the Egbert workshop during the manufacture of the shrine.[30] On the other hand, it could be that the evidence for shearing on the gold sheet of the back of the foot-plate points to mending. The flattened beaded wire demonstrates that the brooch was extensively worn. The gold sheet of the back of the foot-plate could have been sheared in order to replace the catch of the pin. Because of the weight and dimensions of the brooch, especially the pin and catch are under a lot of strain when worn regularly. Altogether, the idea rises that the Wijnaldum brooch was in possession of a goldsmith for reuse or repair.

ACKNOWLEDGEMENTS

Gratitude is especially outstanding to drs E. Kramer from the Fries Museum in Leeuwarden who as Keeper of the archaeological department gave permission to analyse the Wijnaldum brooch.

The materials of the brooch were analysed at the Central Research Laboratory for Objects of Art and Science in Amsterdam. SEM-photographs were taken at the department for Cell-Biology and Electron Microscopy of the University of Groningen with an 'Akashi DS 130' with LABS-6 source electron microscope. We would like to thank Dr W.L. Jongebloed and Mr I. Stokroos for their kind support with regard to the electron microscope facilities. Finally, credits are due to Mr R. Gerritsen from Amsterdam who made the X-ray photographs.

REFERENCES

ARRHENIUS, B. 1971. *Granatschmuck und Gemmen aus Nordischen Funden des Fruhen Mittelalters*, Stockholm (Studies in North European archaeology).

AVENT, R. 1975. *Anglo-Saxon disc and composite brooches*, 2 vols., Oxford (British Archaeological Reports 11).

BESTEMAN, J.C., J.M. BOS & H.A. HEIDINGA 1993. *Graven naar Friese Koningen. De opgravingen te Wijnaldum*, 2nd edition, Franeker.

BIMSON, M. 1985. Dark-age garnet cutting, *Anglo-Saxon Studies in Archaeology and History* 4: 125-128.

BRUCE-MITFORD, R.L.S. 1955. Gold and silver cloisonné buckle from Wijnaldum Friesland, *Fries Genootschap van Geschied-, Oudheid-, en Taalkunde: 126ste Verslag over het jaar 1954*, 16-17.

BRUCE-MITFORD, R. 1972. *The Sutton Hoo ship-burial*, 2nd ed., London.

BRUCE-MITFORD, R. 1978. *The Sutton Hoo ship-burial*, II, London.

DRAGSTED, O. & B. KRIJGER 1978. *Edelstenen in kleur*, Baarn.

DUCZKO, W. 1985. *Birka V: The filigree and granulation work of the Viking Period: An analysis of the material from Björkö*, Stockholm.

EAST, K. 1985. A study of the cross-hatched gold foils from Sutton Hoo, *Anglo-Saxon Studies in Archaeology and History* 4: 129-142.

HAWKES, S.C. 1974. The Monkton Brooch, *Antiquaries Journal* 54: 245-256.

HAWTHORNE, J.G. & C.S. SMITH 1979. *Theophilus, On divers Arts: the foremost medieval treatise on painting, glassmaking and metalwork*, New York.

HØILUND NIELSEN, K.P. & P. VANG PETERSEN 1993. Detector finds. In: S. Hvass & B. Storgaard (eds), *Digging into the past, 25 Years of archaeology in Denmark*, Aarhus, 223-225.

HOUGEN, B. 1967. *The migration style of ornament in Norway*, 2nd edition, Oslo.

JESSUP, R. 1974. *Anglo-Saxon jewellery*, Aylesbury.

MACDONALD 1983. *The Macdonald encyclopedia of rocks and minerals*, London.

MAZZO KARRAS, R. 1985. Seventh-Century jewellery from Frisia: A re-examination, *Anglo-Saxon Studies in Archaeology and History* 4: 159-177.

MEEKS, N.D. & R. HOLMES 1985. The Sutton Hoo garnet jewellery: an examination of some backing foils and a study of their possible manufacturing techniques, *Anglo-Saxon Studies in Archaeology and History* 4: 143-157.

RONIG, F.J. (ed.) 1993. *Egbert, Erzbischof von Trier 977-993*, I, Katalog- und Tafelband, Rheinisches Landesmuseum, Trier.

SCHONEVELD, J. 1993. De grote fibula van Wijnaldum, *De Vrije Fries* 73: 7-24.

THRANE, H. 1994. Goldene Schwertteile der Völkerwanderungszeit aus Gudme auf Fünen, Dänemark. In: H. Keller & N. Staubach, (eds), *Iconologia sacra, Festschrift für Karl Hauck*, Berlin, 106-117.

UNTRACHT, O. 1975. *Metal techniques for craftsmen*, New York.

UNTRACHT, O. 1985. *Jewelry, concepts and technology*, London.

VIERCK, H. 1974. Werke des Eligius. In: G. Kossack & G. Ulbert (eds), *Festschrift für Joachim Werner zum 65. Geburtstag*, II, München, 309-381.

WESTERMANN-ANGERHAUSEN, H. 1973. *Die Goldschmiedearbeiten der Trierer Egbertwerkstatt*, Trier (Beiheft zum 36. Jahrgang der Trierer Zeitschrift).

WESTERMANN-ANGERHAUSEN, H. 1987. Spolie und Umfeld in Egberts Trier, *Zeitschrift für Kunstgeschichte* 50: 305-336.

ZIJLSTRA, J. 1991. Finns fibula? Belangwekkende vroegmiddeleeuwse vondsten te Wijnaldum, *Westerheem* 40: 51-62.

NOTES

1. Bruce-Mitford (1955, pp 16-17), Mazzo Karras (1985, pp 168-170), Schoneveld (1993, pp 7-24), and Zijlstra (1991, pp 51-55).
2. The weight of this reconstruction on perspex is 253 g.
3. The numbering of the fragments of the disc-on-bow brooch from Wijnaldum in Figure 1 agrees with the numbering of the fragments in the loan-conditions of the Fries Museum in Leeuwarden.
4. Bruce-Mitford (1955, p. 16).
5. Macdonald (1983, p. 161).
6. Schoneveld (1993, pp 13-14).
7. Macdonald (1983, p. 95).
8. Dragsted & Krijger (1978, p. 299).
9. Arrhenius (1971, p. 223).
10. The identification by Nerman of inlay as mother of pearl are presented in Arrhenius (1971, pp 231-233).
11. Arrhenius (1971, pp 216-222).
12. Arrhenius (1971, pp 90-94). Sulphur was most likely a separate component for the paste mixture of the Wijnaldum brooch since during the pre-treatment of the samples for the analyses, the smell of sulphur was noticed in the laboratory.
13. Arrhenius (1971, pp 93-94).
14. Untracht (1975, pp 8-9).
15. Bruce Mitford (1978, pp 611-613, Appendix C). The majority of the Sutton Hoo gold jewellery was analysed by atomic absorption spectrometry with an analytical precision for gold of ±2%. The analytical precision worsens as the sample weight decreases. The metal composition of the disc-on-bow brooch from Wijnaldum was analysed by XRF with an analytical precision for gold of ±0.5%.
16. Jessup (1974, p. 22).
17. Hougen (1967, p. 10).
18. Duczko (1985, pp 17-22).
19. Hawthorne & Smith (1979, pp 88-90).
20. Untracht (1985, pp 168-169).
21. East (1985, pp 129-142) and Meeks & Holmes (1985, pp 143-157).
22. East (1985, p. 137).
23. Meeks & Holmes (1985, pp 148-157).
24. We would like to thank Dr N. Meeks of the British Museum and Benner Larsen, Dianalund, Denmark for their valuable comments on the the die from Wijnaldum. Dies for making foils from excavations are extremely rare. Two positive dies for standard foils were recovered in Denmark as detector finds from Gudme and Neble (Høilund Nielsen & Vang Petersen 1993, p. 225). The die from Gudme has been associated with local goldsmithing activities by Thrane who also mentioned a die from the settlement Dalshøj on Ostbornholm (Thrane 1994, p. 110). The die from Dalshøj is correlated to the central settlement of Sorte Muld from where gold cloisons, raw and worked almandine fragments for the production of cloisonne jewellery are reported.
25. Untracht (1985, pp 178-179).
26. Mazzo Karras (1985, pp 166-168).
27. Avent (1975, p. 3 and Table 5).
28. Hawkes (1974, pp 250-252).
29. Besteman et al. (1993, pp 20-22) and Mazzo Karras (1985, pp 165-166).
30. Vierck (1974, pp 359-360). The provenance of the various spolia on the Egbert shrine is debated in e.g. Vierck (1974), Ronig, (1993, pp 36-37), and Westermann-Angerhausen (1973, 1987). Especially the reuse of the 7th century garnet jewellery has attracted much attention. The discs have been linked to workshops in England (Sutton Hoo), Frisia and Trier as well as to various Merowingian workshops.

Medieval coins from Wijnaldum

A. POL

1 REGISTRATION OF THE COINS

The medieval coins from Wijnaldum reported up to and including 1997 form an important group among the coin finds discovered in Frisian soil.[1] Wijnaldum here refers to the village centre plus surroundings including the Tjitsma terp. 87 coins from the Roman Period[2] and 95 coins from the Middle Ages originate from Wijnaldum. Although the total number of medieval coins is not unusually large in the absolute sense, the figure becomes more impressive when one realizes that there is virtually no material from the Late Middle Ages. Generally speaking, far larger numbers of coins from the periods before and after the Early Middle Ages are found than from the Early Middle Ages themselves. At Wijnaldum, the situation is reversed which points to the great importance of this place in the Early Middle Ages. This is even more the case at the Tjitsma terp, the productive site which was the focus of excavation campaigns. Apparently, at the site of the excavation this terp was no longer used for settlement after the Carolingian Period.

Prior to the excavation which started in 1991, several coins had already been found by metal detector amateurs at Tjitsma and elsewhere in the immediate vicinity of Wijnaldum. In some cases the exact origin of these coins in Wijnaldum is not certain, but they generally come from one of the other terps at Wijnaldum. In the list of finds the coins from Tjitsma are mentioned separately.

2 DATING OF THE COINS

The dating of the coin finds is based on the usual historical and numismatic information such as the reign of the issuer, year of acquisition of a mint, first emission date of a specimen from elsewhere, development of type, impression sequence, occurrence in hoards etc. In the case of the excavated finds from Wijnaldum we also occasionally have a dated archaeological context.

The commencing date of a minting period is the earliest moment at which the coin could enter the soil. As a result, a coin produces a reliable *terminus post quem* for the context in which it is found. Far more intangible is the duration of use of a coin prior to the moment when it actually entered the soil as a individual find.[3] In such a case it is impossible to trace whether the coin entered the soil shortly after production started or a long time after. The *terminus ante quem* can never be established with certainty. However, for most types of coin it is possible to establish a rough duration of circulation based on their occurrence in hoards of varying ages. The *terminus ante quem* which is sometimes postulated below is no more than a rough estimate, valid in general for the coin type to which the coin find belongs and possibly also valid for the coin find in question if one takes into account the fact that the period mentioned may have been exceeded.

2.1 *Merovingian Period – first phase: gold coins (500-675)*

In the Early Middle Ages, gold coins in Western Europe were at first only minted in imitation of the late-Roman and Byzantine examples which also circulated here on a large scale.[4]

They were more or less free imitations of solidi and especially tremisses (tremissis = 1/3 solidus = 1.4 to 1.2 g). In these copies the name of the emperor was originally used in a more or less corrupted form and for this reason they are known as pseudo-imperial coins. A little later, the imperial name was occasionally replaced by that of a Frankish king. Far more frequently the coins only bore the name of the mint and that of the *monetarius* in question. For most of the Frankish coins, the name of the mint cannot be identified by a present

Table 1. Frequency of the various types of coin at Wijnaldum and percentage compared with individual coin finds in Friesland (localized).

Period	Coin type	Tjitsma	Tjitsma excavation	Other Wijnaldum	Total Wijnaldum	Total Friesland.*	Percentage Wijn:Frsl.
Roman coins		46 = 53%		41 = 47%	87 = 100%		
Medieval coins		36 = 38%		59 = 62%	95 = 100%		
500-675	tremissis	2	5	8	15 = 16%	105	17%
675-750	denarius	1	–	2	3 = 3%		
	sceatta	7	12	17	36 = 38%	132	31%
750-900	denarius	–	15	2	17 = 18%		
	dirhem	1	1	–	2 = 2%	80	24%
900-1100	penny	–	2	4	6 = 6%	15	(40%)
Late Middle Ages		8	1	7	16 = 17%		
Total		19	36	40	95 = 100%		

* = not localized; other Wijnaldum is outside Tjitsma terp or not localized.

place-name. Moreover, there are many imitations of uncertain origin, some of which are regarded as Frisian, and Ostrogothic, Burgundian and Visigothic coins were also used here, though on a lesser scale, and, more rarely, Anglo-Saxon, Alemannic and Lombard coins. In the Low Countries, nearby Frankish workshops such as Huy and Maastricht flourished shortly after 600, as did the production of the so-called Dronrijp type, situated in Frisia. Coins from the productive city of Mainz are also often found in the Netherlands. A short time later, Dorestad also became one of the important mints. Summing up, we may state that in the 6th-7th century a decreasing number of Byzantine and an increasing number of Frankish gold coins were in circulation, together with a growing number of coins of Frisian origin.

The series of medieval coins from Wijnaldum-Tjitsma starts with 7 tremisses from the Merovingian Period. Solidi are frequently found in Friesland, not only in the hoards of Midlum (ca 525) and Wieuwerd (ca 625), but also as individual finds; in this case mainly coins minted before 600. No solidi have so far been found at Wijnaldum. Among the tremisses from Wijnaldum are two 6th-century imitations of Byzantine coins, one of which is a carefully executed specimen in the name of emperor Anastasius (491-518). The place or district of production cannot be identified but the fine workmanship may indicate Italy (cat. 2). The second specimen is also pseudo-imperial but has no legible imperial name; it belongs to a type thought to be produced in Frisia (cat. 4). The oldest coin from the excavation also dates from the 6th century, but little can be said about its origin (due to the lack of legible circumscriptions or findspots of similar specimens). The obverse shows a rather distorted bust, while the reverse has a figure of a Victoria in profile with a long cross as found on solidi of various emperors from the 5th and 6th centuries but never on their tremisses. However, this depiction is found on Western imitations of tremisses of the emperors Libius Severus (461-465) and Zeno (476-491); these tremisses were probably struck around 500 by the Visigoths in Southern Gaul.[5] It would be going too far, though, to seek the origin of this Wijnaldum specimen there, for this coin is made of gilded copper and may be regarded as a fake. It may have been made anywhere the coin was known. It is perforated, so it may have been worn as an ornament (cat. 1, Fig. 1). Of the 7th-century tremisses, one specimen was struck at Bannassac, an extremely productive mint in Central France, coins from which are often found in these parts; the coin in question dates from ca 650 (cat. 12, Fig. 2). A second tremissis was struck around 640 and originates from the Eastern French mint of Gizia (Jura) – coins from this place are rare and have not previously been found in the Netherlands (cat. 9, Fig. 3). The third coin is from Dorestad and is in the name of the *monetarius* Madelinus, and also dates from ca 640 (cat. 11, Fig. 4). On stylistic grounds, this specimen may be designated an original product of Madelinus, thus belonging to the relatively small group which is separate from the mass of imitations of this type, the pseudo-Dorestad/Madelinus tremisses. This series ends with an anonymous tremissis from the same period (cat. 10, Fig. 5). Its exterior alone (obverse unrecognizable, reverse cross; no circumscriptions) does

FS BS

Figure 1. Germanic false tremissis ca 500?, design copied from Visigothic prototype; gilt copper, pierced (catalogue no. 1). Scale 2:1 (photo KPK).

FS BS

Figure 5. Frisians?, tremissis ca 630 (gold) (catalogue no. 10). Scale 2:1 (photo KPK).

FS BS

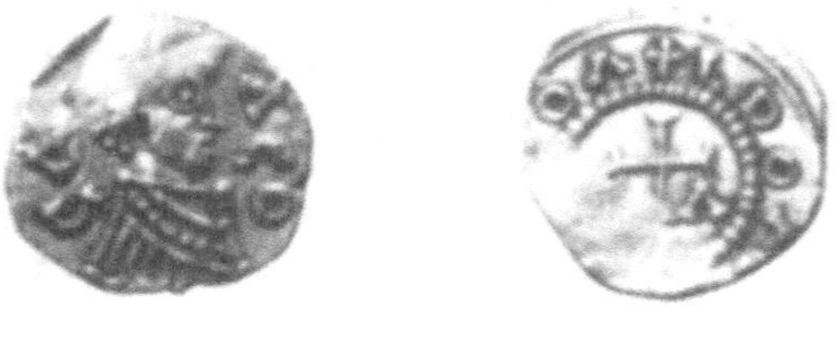

Figure 2. Franks, tremissis ca 650, Bannassac (gold) (catalogue no. 12). Scale 2:1 (photo KPK).

FS BS

Figure 3. Franks, tremissis ca 625, Gizia/Andoaldus (gold) (catalogue no. 9). Scale 2:1 (photo KPK).

FS BS

Figure 4. Franks, tremissis ca 640, Dorestat/Madelinus (gold) (catalogue no. 11). Scale 2:1 (photo KPK).

not provide a basis for identification, but typologically related items were found at Crondall (GB), Coddenham (GB), Nietap, Wijk bij Duurstede, Remmerden and Hemer (Germany). From this distribution it may be assumed that this type was probably minted in the Lower Rhine area or even in Frisia itself; only the Dutch hoards of Nietap and Remmerden contain more than one specimen. On the basis of its occurrence in these and the other hoards mentioned, the production of this tremissis must be dated to ca 630. Of the remaining 8 Merovingian gold coins from Wijnaldum, 7 come from the other terps at Wijnaldum; the exact findspot of the last coin is unknown. This is just as varied a group as that of Tjitsma. The series starts with four 6th-century pseudo-imperial tremisses, one of which is certainly Visigothic and two others possibly Frisian, while the last coin remains obscure due to the deficiency of the find report. The Visigothic tremissis bears the name of emperor Justinian (527-565) and the designation for Constantinople, but may, on stylistic grounds, be ascribed to the Visigoths who, in the 6th century, had settled mainly in Spain and in a small part of Southern France. After ca 580, King Leovigild (575-586) set his own name on the coins, as did his successors. This coin does not bear a king's name, so it dates from before this period; it bears the name of the emperor Justinian (527-565) and was therefore struck after the beginning of his reign, though it is not possible to date this Wijnaldum coin more precisely than the middle of the 6th century (cat. 3). The two tremisses of possible Frisian origin are less easy to identify (cat. 5). Their attribution to Frisia is based on the relatively large number of findspots of this type of coin in the region, and on their virtual absence elsewhere. The smaller of the two belongs to a small group clearly related to a group of pseudo-imperial tremisses already ascribed to Frisia by Boeles (cat. 6).[6] The last of the four pseudo-imperial tremisses was found in or before 1912 and published without illustration; its present whereabouts are unknown, and its exact appearance is uncertain so that it is impossible to find out whether it is an original Justinian tremissis or an imitation.

The remaining four tremisses belong to the group of *monetarius* coins. The oldest is probably the specimen struck by the *monetarius* Alebodus in Sully-sur-Loire (dep. Loiret) in the first decennia of the 7th century (cat. 8). A number of coins of this type with *appendice perlée* and *croix ancrée* were already found in the Escharen hoard buried in about 600, but the type was in use for a long period. The other three tremisses, all Dorestad/Madelinus, are of later date. These are, however, not originals, but imitations of various origin and age, probably struck in the decades around the middle of the 7th century in workshops in the Netherlands which have not been localized. One specimen is still of reasonably good quality (cat. 13), the two others have deteriorated stylistically and are of a lesser quality, i.e. their gold content is lower (cat. 14, 15).

The tremisses from Wijnaldum give a good impression of what is found in Frisia in the way of gold coins from the Merovingian Period. In the find material from this one village there are 6th- and 7th-century coins, some exotic material in addition to the Frankish and Frisian finds, of pseudo-imperial and urban type, struck in what is now France as well as in the Netherlands. They are mostly actual specimens and imitations of reasonably good quality plus one fake coin, all without special traces of use apart from the one perforated coin. However, no solidi or Byzantine coins were found at Wijnaldum, though these do occur in other parts of Friesland.

From the composition of this group of coin finds or from their varying dates and origins no special conclusion can be drawn regarding the position of Tjitsma in particular or Wijnaldum in general. The at this site is, however, remarkable. All in all, Wijnaldum produced 15 tremisses. For the province of number of coin finds from the Merovingian Period Friesland which is so rich in Merovingian coins – I know of some 250 gold coins from the period 500-675 – this would seem a modest number, but nothing could be further from the truth. The total of Merovingian coins (solidi and tremisses) from Friesland which do not belong to the three hoards (Midlum 1925, Wieuwerd 1866 and Dronrijp 1876) amounts to over 160. If we disregard the 55 specimens which cannot be localized, Wijnaldum's proportion is 17% (see Table 1). There are only a few places in Friesland where more than one coin has been found, and in these cases the number of finds remains below four finds per site. Half of the Wijnaldum tremisses come from the Tjitsma terp, making this terp certainly one of the richest sites regarding individual coin finds from the Merovingian Period.

2.2 *Merovingian Period – second phase: silver coins (675-750)*

The next phase of early-medieval coinage is characterized by the introduction of silver coins: after 675 gold is no longer minted in France and the transition to silver as bullion in the Low Countries probably took place at about the same time. While the Franks struck silver denarii of ca 1.2 g, the Frisians and Anglo-Saxons minted similar pieces of silver referred to by the modern name of sceatta. The representations on the Frankish coins partly correspond to those on the previous gold coins, but the bust was increasingly replaced by symbols, monograms and other decorative elements. Nor was a circumscription always added. The latter phenomenon occurred even more frequently in the Frisian and Anglo-Saxon types: these are virtually always anepigraphical. This phase of small, thick pieces of silver continued in the Frankish empire until the middle of the 8th century, when Pippin I announced a coinage reform (751). By that time, Frisia had been conquered by the Franks, so we may assume that the second phase in the Frisian territory ended at the same time,[7] although there is a possibility that the minting of sceattas continued until into the second half of the 8th century.

At Wijnaldum-Tjitsma, 20 coins from this period were excavated. These include 1 Frankish denarius of a type attributed to Paris, struck in the decades around 700 (cat. 16). The other pieces are sceattas of 5 different types. The most common, with 10 specimens, is the so-called porcupine/standard type (BMC type 4-5) which, with all kinds of variations, has a badly distorted head on the obverse and a square containing several geometrical figures on the reverse. This type is generally regarded as Frisian and was probably in production in the first half of the 8th century. A more precise dating of individual variants is not yet possible. In addition, Tjitsma produced 5 more specimens of the so-called continental rune type (BMC type 2c) and 1 specimen of a hybrid of both types (BMC type 8. The origin of the continental rune type is uncertain.

Figure 6. Franks, denarius ca 700, Marseille (silver) (catalogue no. 17). Scale 2:1 (photo KPK).

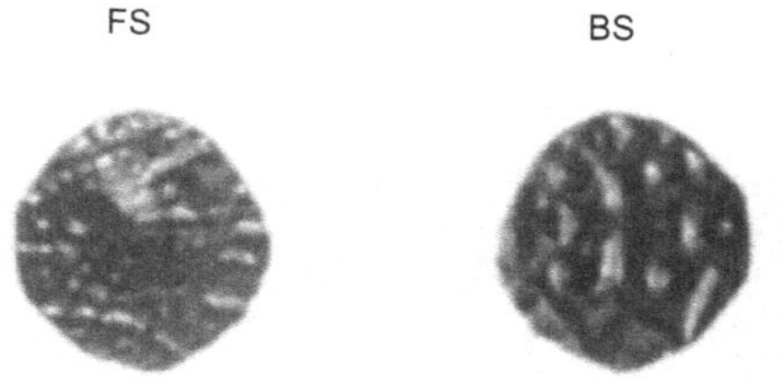

Figure 7. Franks?, Frisians?, sceatta ca 700-720, 'continental runic' (silver) (catalogue no. 23). Scale 2:1 (photo KPK).

Since the nucleus of the distribution area is clearly outside Frisia, it is assumed that the centre of production was in the area around the great rivers (Rhine and Meuse) or the Scheldt estuary (Domburg?), or perhaps even further south. The Remmerden hoard (near Rhenen) contained some 150 specimens of this type.[8] This sceatta is an imitation of an English coin dating from about 700. Since the continental rune type does not appear in the around the 730's hidden Frisian hoards it is assumed that the minting of these coins only lasted a short time: despite the many variants, no longer than two decades, from 700 to 720. Finally, 3 sceattas are reported from Tjitsma of the so-called wodan/monster type (BMC type 31), with a full face head on the obverse and an animal figure on the reverse (cat. 50-52). This type used to be generally attributed to Frisia, but in recent years there is a tendency to ascribe it to Denmark. The large number found since 1990 at the excavation of the 8th-century marketplace of Ribe is one of the main arguments for this, but the discussion continues.[9] The minting of this type took place between 720-760.

In addition to the above-mentioned coins from Tjitsma, 2 Frankish denarii and 17 sceattas were found in and around Wijnaldum. This brings the number of coins from the whole of Wijnaldum from the period 675-750 to 39. Apart from the coins in the three sceatta hoards (Terwispel 1863, Hallum 1866 and Franeker 1868) a total of 139 individual finds of denarii and sceattas have been registered in the province of Friesland, 131 of which can be localized. The percentage from Wijnaldum was 30%. No other site has reached such a high percentage. The localizable Frisian coin finds from the period 675-750 are distributed over 29 different toponyms and again, Wijnaldum shows up favourably.[10] Apart from Dongjum with 16 coin finds, all the other findspots stop at 10 or fewer.

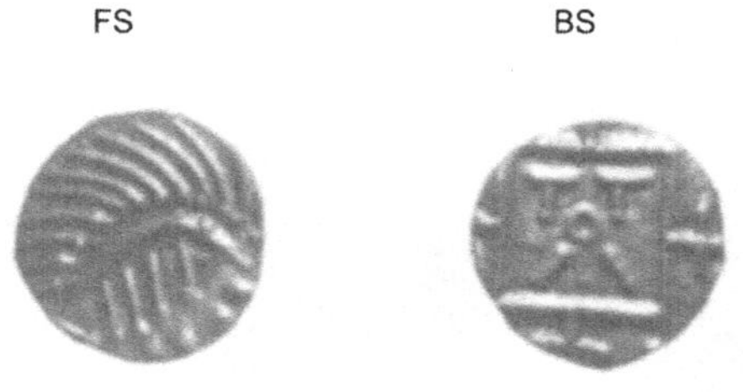

Figure 8. Frisians, sceatta ca 700-750, 'porcupine' (silver) (catalogue no. 39). Scale 2:1 (photo KPK).

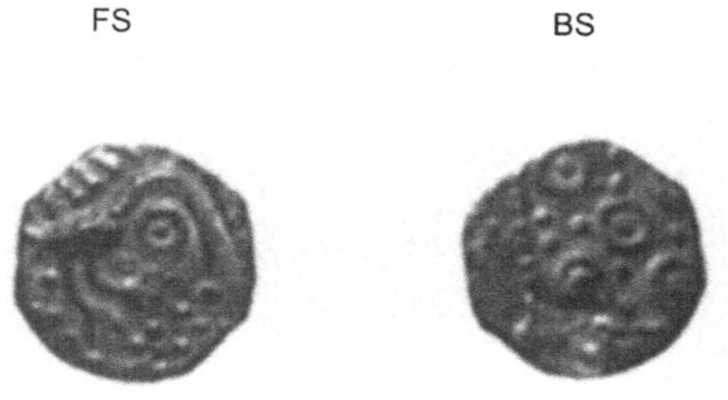

Figure 9. Frisians, sceatta ca 700-750, 'porcupine' (silver) (catalogue no. 45). Scale 2:1 (photo KPK).

2.3 *Carolingian Period (750-900)*

Silver coins are struck almost exclusively in the Carolingian Period. The silver denarii reformed by Pippin in 751 are of more or less the same mass as the coins from the preceding period, but of a larger size and therefore much thinner. The repertoire of heads is very limited and apart from the Christian cross and a temple or church other things are rarely found: a portrait, a ship, a gateway or a set of dies. Carolingian coins are first and foremost coins with a text in the form of circumscriptions, inscriptions and monograms. The greater degree of royal control over coin production resulted in a restriction on

Figure 10. Frisians?, Danes?, sceatta ca 710-750, 'woden/ monster' (silver) (catalogue no. 52). Scale 2:1 (photo KPK).

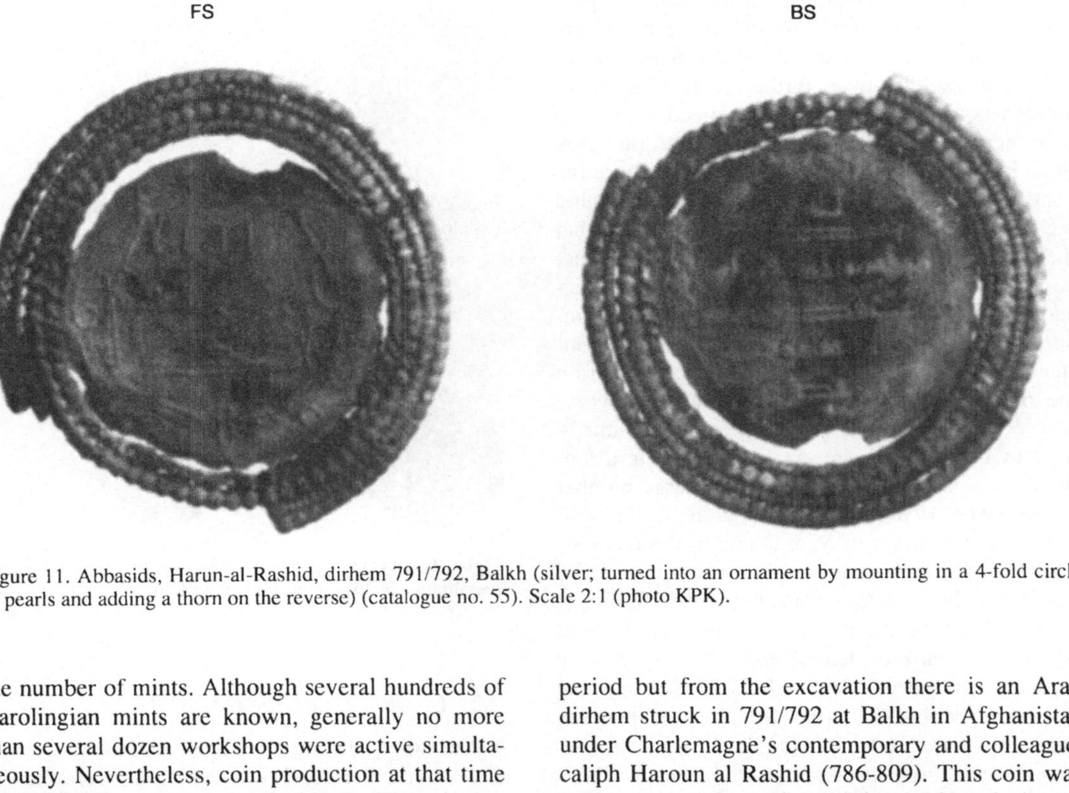

Figure 11. Abbasids, Harun-al-Rashid, dirhem 791/792, Balkh (silver; turned into an ornament by mounting in a 4-fold circle of pearls and adding a thorn on the reverse) (catalogue no. 55). Scale 2:1 (photo KPK).

the number of mints. Although several hundreds of Carolingian mints are known, generally no more than several dozen workshops were active simultaneously. Nevertheless, coin production at that time was probably much greater than in the Merovingian Period.

15 Carolingian denarii were discovered at Wijnaldum-Tjitsma, all but two during excavation. The series opens with 5 coins of Louis the Pious (814-840) followed by 8 coins of Lotharius I (840-855) and ends with 2 coins of Louis III (899-911) from Strasbourg (cat. 72-73). From Wijnaldum and surroundings 2 further Carolingian denarii are reported, including a coin of Charlemagne (768-814) struck at Laon (cat. 57). From the same period but from the excavation there is an Arab dirhem struck in 791/792 at Balkh in Afghanistan under Charlemagne's contemporary and colleague, caliph Haroun al Rashid (786-809). This coin was once set in a decorative edging and made into a brooch (cat. 55). Arab coins from the 8th-9th century are extremely rare here and are occasionally found as ornaments. The recent hoard of Westerklief 1995/1996 contained 3 of such coin ornaments as well as 78 Carolingian denarii, 7 silver armrings, 1 neck-ring and 16 silver ingots.[11] The assumed Viking context immediately indicates how these oriental coins usually end up here: probably via Russia and Scandinavia. A second dirhem was also found at Tjitsma, outside the excavation, a

FS BS

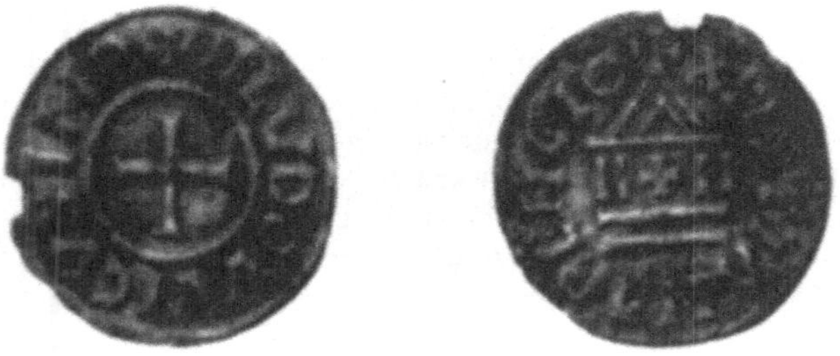

Figure 12. Franks, Louis the Pious, denarius 822-840 (silver) (catalogue no. 60). Scale 2:1 (photo KPK).

halved specimen which could not be fully identified.[12] The total number of 18 (19) coin finds from the Carolingian Period at Wijnaldum does not appear unusually high: in fact it sinks into insignificance when compared to the gigantic hoard from neighbouring Tzummarum (1991) which consists of almost 2800 items.[13] Nevertheless, the relative importance of Wijnaldum is not inconsiderable. For Friesland a preliminary total of some 80 individual coin finds from the Carolingian Period has been estimated, distributed over 31 toponyms.[14] Among them, Wijnaldum is yet again the best represented site, with more than 20% {24%} of the total. From most of the 31 places, only one Carolingian coin is reported and in none were more than 5 found, not counting the hoards.

2.4 *Ottonian-Salian Period (900-1125) and later Middle Ages*

After the Carolingian Period another major change takes place in the coinage system. This time it does not affect the overall exterior of the object (metal, size and weight remain more or less the same for the time being), but the decentralization of coinage under the influence of the shifting power relations in Europe at that time. After the disintegration of the enormous Carolingian empire and the rise of smaller territorial kingdoms, the new rulers took control of many royal rights including mintage rights which were exercised on a local/regional level. The increasing political and territorial disintegration also affected the appearance (heads) and quality (weight, value) of the coins. In the Low Countries, coins were not struck again until the end of the 10th century, the so-called *penning* (penny) over 1 g.[15] This mint at Tiel was quickly followed in the 11th century by Deventer, Utrecht and Groningen.[16] In Friesland, a not insignificant minting by count Wigman III apparently took place in about 1000 at an unspecified mint, followed by extensive minting activities especially at Bolsward, Dokkum, Leeuwarden and Stavoren.[17] Of all the coins struck in large numbers in this period, most were exported to the Baltic countries. Very little of this mintage has been found at Wijnaldum or elsewhere in Friesland or even in the rest of the Netherlands.

The Wijnaldum excavation yielded 2 pennies from this period: 1 of the above-mentioned Wigman (994-1016) and 1 probable late 11th-century anonymous *pfennig* from Münster (cat. 74, 75). From the other Wijnaldum terps came 3 Frisian pennies of Bruno III (1038-1057) and a specimen from Utrecht of Conrad of Swabia (1076-1099) (cat. 76-79). The number of hoards from this period in Friesland has risen to 6, the number of individual finds to 15 (from 12 toponyms). Again Wijnaldum proves to be the site with the greatest number of individual finds in the province. The small number of coins for this period, however, does not really permit a quantitative comparison.

Not at all representative is the picture of late-medieval coin finds at Wijnaldum. Apart from several very light small Frisian pennies from the 12th-13th century (cat. 81-84)[18], these include a pair of Utrecht pennies and two Flemish ones from the same period. Hardly any other later coins were discovered, neither native nor foreign. Although coin production boomed in the 14th century and after, this cannot be seen at Wijnaldum. Worth mentioning as curiosities are a Serbian dinar and a fragment of a local imitation of an Almohad dinar, both from the 13th century (cat. 89-90).[19] Coins like these are virtually never found in the Netherlands.

3 CONCLUSION

Before excavation started at Wijnaldum, there were great expectations regarding the early-medieval coins which one hoped to find there. Especially the great gold fibula contributed to the image of Tjitsma as an extremely rich site, and that is perhaps why the resulting numbers of coin finds were somewhat disappointing. However, closer analysis of the coins found before and during excavation shows that Wijnaldum certainly is an important site. One should not forget, though, that the picture may be biased, for Wijnaldum is the only large-scale excavation at which systematic use was made of the metal detector. On the other hand, even if the objects from the excavation itself are excluded and only the individual finds are compared, Wijnaldum still remains the most productive site in Friesland with regard to early-medieval coins.

Almost all the medieval coins from the Tjitsma excavation can be dated on numismatical grounds. The mintage period is certain and therefore also a reliable *terminus post quem*. A large proportion of the find material from Tjitsma has an archaeologically datable context, e.g. by means of the Harris matrix. The *terminus post quem* provided by the coins in the find material corresponds in nearly all cases with the archaeological datings and is before or around the archaeological date. When this date is later than the *terminus post quem* it still falls within the duration of use acceptable for that type of coin. This is an important conclusion when determining the value of the archaeological dates.

In this way it can be stated that the medieval coins were always 'in the correct place' in the profile. In contrast to what Van der Vin has established for a number of Roman coins from the excavation,[20] it is impossible to conclude that the early-medieval coins were still occasionally used a very long time after their minting period. The fact that the archaeological date and numismatical *terminus post quem* of the early-medieval coins correspond shows that Van der Vins conclusion that Roman coins were used in the Early Middle Ages is correct: it is certainly not a matter of dating the archaeological context incorrectly or too broadly.

Moreover, in the period between production (1st-2nd century) and final use (6th-7th century) no new copper and silver coins were imported, nor were these produced locally. Perhaps it was out of sheer necessity that the old coins were used – and luckily they were still available.

CATALOGUE OF MEDIEVAL COINS FROM WIJNALDUM

The right-hand column shows:

- The find number if the object was found during the excavation of the Tjitsma terp,
- The letter T if the object was found outside the excavation but on the same terp,
- No mark if the object was found within Wijnaldum or surroundings other than Tjitsma or if the location within Wijnaldum and surroundings is unknown.

The dates are dates of emission unless marked with * for reigns of rulers; if a date has the designation *ca* a margin of about 10 years before and after this date must be taken into account.

No.	Origin	Coin	Date	Ruler/mint/mintmaster/type/(reference)	Remarks	Find o.
MEROVINGIAN PERIOD – FIRST PHASE (gold) 500-675						
1	Germanic	tremissis	500?	Pseudo-imperial after Visigothic prototype	false, gilt copper, pierced	4534
2	Franks	tremissis	500-520	Pseudo-imperial, name of Anastasius (Werner 177+179)		T
3	Visigoths	tremissis	ca 550	Name of Justinian (Tomasini 6; Grierson/ Blackburn 192)		
4	Frisians	tremissis	ca 550-600	Pseudo-imperial type (Belfort 5385)		T
5	Frisians	tremissis	ca 550-600	Pseudo-imperial type (Belfort 5385var)		
6	Frisians	tremissis	ca 550-600	Pseudo-imperial type (Belfort 5387var)		
7	Franks?	tremissis	ca 500-600	Pseudo-imperial type [old find, exact type unknown]		
8	Franks	tremissis	ca 600-620	Sully-sur-Loire/Alebodus (Belfort 4164 var ?, Prou 667var)		
9	Franks	tremissis	ca 625	Gizia/Andoaldus (Belfort 1939var)		748
10	Frisians?	tremissis	ca 630	Nietap-type (Belfort 5517var)		7046
11	Franks	tremissis	ca 640	Dorestad / Madelinus (Belfort 1760)		4630
12	Franks	tremissis	ca 650	Bannassac (Belfort 749)		467
13	Fra/Fri?	tremissis	640-650	Dorestad/Madelinus-type (Belfort)		
14	Fra/Fri?	tremissis	650-660	Dorestad/Madelinus-type (Belfort 1784)		
15	Fra/Fri?	tremissis	650-660	Dorestad/Madelinus-type (Belfort)		

MEROVINGIAN PERIOD – SECOND PHASE (silver) 675-750

16	Franks	denarius	ca 700	Parijs? (Belfort 6346)		T
17	Franks	denarius	ca 700	Marseille (Belfort 3057-3058)		
18	Franks	denarius	ca 700	Bourges? Poitiers? (cf. Belfort 853, 5636, 6366)		
19	Fra/Fri?	sceatta	ca 700-720	continental runic-type (BMC type 2c)		T
20	Fra/Fri?	sceatta	ca 700-720	continental runic-type (BMC type 2c)		T
21	Fra/Fri?	sceatta	ca 700-720	continental runic-type(BMC type 2c)		T
22	Fra/Fri?	sceatta	ca 700-720	continental runic-type (BMC type 2c)		5184
23	Fra/Fri?	sceatta	ca 700-720	continental runic-type (BMC type 2c)		10166
24	Fra/Fri?	sceatta	ca 700-720	continental runic-type (BMC type 2c)		
25	Fra/Fri?	sceatta	ca 700-720	continental runic-type (BMC type 2c)		
26	Fra/Fri?	sceatta	ca 700-720	continental runic-type (BMC type 2c)		
27	Fra/Fri?	sceatta	ca 700-720	continental runic-type (BMC type 2c)		
28	Fra/Fri?	sceatta	ca 700-720	continental runic-type (BMC type 2c)		
29	Fra/Fri?	sceatta	ca 700-720	continental runic-type (BMC type 2c)		
30	Fra/Fri?	sceatta	ca 700-720	continental runic-type (BMC type 2c)		
31	Fra/Fri?	sceatta	ca 700-720	continental runic-type (BMC type 2c)		
32	Frisians	sceatta	ca 700-750	porcupine-type (BMC type 4-5)		T
33	Frisians	sceatta	ca 700-750	porcupine-type (BMC type 4-5)		T
34	Frisians	sceatta	ca 700-750	porcupine-type (BMC type 4-5)		T
35	Frisians	sceatta	ca 700-750	porcupine-type (BMC type 4-5)		2
36	Frisians	sceatta	ca 700-750	porcupine-type (BMC type 4-5)		860
37	Frisians	sceatta	ca 700-750	porcupine-type (BMC type 4-5)		1095
38	Frisians	sceatta	ca 700-750	porcupine-type (BMC type 4-5)		2640
39	Frisians	sceatta	ca 700-750	porcupine-type (BMC type 4-5)		2727
40	Frisians	sceatta	ca 700-750	porcupine-type (BMC type 4-5)		2773
41	Frisians	sceatta	ca 700-750	porcupine-type (BMC type 4-5)		7553
42	Frisians	sceatta	ca 700-750	porcupine-type (BMC type 4-5		
43	Frisians	sceatta	ca 700-750	porcupine-type (BMC type 4-5)		
44	Frisians	sceatta	ca 700-750	porcupine-type (BMC type 4-5)		
45	Frisians	sceatta	ca 700-750	porcupine-type (BMC type 6)		
46	Frisians	sceatta	ca 700-750	porcupine-type (BMC type 6)		
47	Frisians?	sceatta	ca 700-750	(BMC type 8)		10901
48	Frisians?	sceatta	ca 700-750	(BMC type 8)		
49	Frisians?	sceatta	ca 700-750	(BMC type 10)		
50	Fri/Danes?	sceatta	ca 710-750	woden/monster (BMC type 31)		T
51	Fri/Danes?	sceatta	ca 710-750	woden/monster (BMC type 31)		T
52	Fri/Danes?	sceatta	ca 710-750	woden/monster (BMC type 31)		4504
53	Fri/Danes?	sceatta	ca 700-750	woden/monster (BMC type 31)		
54	Frisians?	sceatta	ca 700-750	[unidentifiable]		

CAROLINGIAN PERIOD – 750-900

55	Abassids	dirhem	791/792	Harun-al-Rashid/Balkh (Mitchiner 188)	mounted: 4-fold circle of pearls, thorn	5338
56	Abbass.?	dirhem	8/9th cent.	[unidentifiable]	cut	T
57	Franks	denarius	793-812	Charlemagne/Laon (Morrison/Grunthal 130var)		
58	Franks	denarius	819-822	Louis the Pious (Morrison/Grunthal 332-336)		73
59	Franks	denarius	822-840	Louis the Pious (Morrison/Grunthal 472.ff)		7325
60	Franks	denarius	822-840	Louis the Pious (Morrison/Grunthal 472.ff)		
61	Franks	denarius	822-840	Louis the Pious (Morrison/Grunthal 483 ?)		3176
62	Franks	denarius	822-840	Louis the Pious (Morrison/Grunthal 488 ?)		87.a
63	Franks	denarius	822-840	Louis the Pious (Morrison/Grunthal 496-498)		4001.a
64	Franks	denarius	840-855*	Lothar I (Morrison/Grunthal 525-530)		87.b
65	Franks	denarius	840-855*	Lothar I (Morrison/Grunthal 525-530)		1965
66	Franks	denarius	840-855*	Lothar I (Morrison/Grunthal 525-530)		2673
67	Franks	denarius	840-855*	Lothar I (Morrison/Grunthal 525-530)		2780
68	Franks	denarius	840-855*	Lothar I (Morrison/Grunthal 525-530)		3176
69	Franks	denarius	840-855*	Lothar I (Morrison/Grunthal 525-530)		4001.b
70	Franks	denarius	840-855*	Lothar I (Morrison/Grunthal 525-530)		5601
71	Franks	denarius	840-855*	Lothar I (Morrison/Grunthal ?)		4001.c
72	Franks	denarius	900-911	Louis III/Strassbourg (Grierson/Blackburn 830-831)		1502
73	Franks	denarius	900-911	Louis III/Strassbourg (Grierson/Blackburn 830-831)		5616

OTTONIAN-SALIAN PERIOD – 900-1150						
74	Frisia	penny	994-1016*	Wichman III (Dannenberg 1229)		2271
75	Münster?	penny	ca 1100?	Mimigardeford? (Ilisch III.1-13 ?)	minute fragment	5706
76	Frisia	penny	1038-1057*	Bruno III/Stavoren (Dannenberg 503)		
77	Frisia	penny	1038-1057*	Bruno III/Dokkum (Dannenberg 499-501)		
78	Frisia	penny	1038-1057 *	Bruno III/Dokkum (Dannenberg 499-501)		
79	Utrecht	penny	1076-1099*	Conrad of Swabia (Van der Chijs IV.7)		
LATE MIDDLE AGES – after 1150						
80	Utrecht	penny	1150-1156*	Herman (Van der Chijs V.1-7)		T
81	Frisia	penny	ca 1175-1200	(Scholten 20)		T
82	Frisia	penny	ca 1175-1200	(Scholten 21)		T
83	Frisia	half penny	ca 1250-1275	(Scholten cf 19 = penny)		T
84	Frisia	penny	ca 1250-1275	(Scholten 28)		T
85	Holland	penny	1190-1203*	Dirk VII (Van Hengel C8.2)		T
86	Utrecht	penny	1198-1212 *	Dirk van der Ahr/Utrecht (Van der Chijs VI..4)		
87	Utrecht	penny	1198-1212 *	Dirk van der Ahr/Utrecht (Van der Chijs VI..4)		
88	Utrecht	penny	1198-1212*	Dirk van der Ahr/Deventer (Van der Chijs XXXI. 5)		1707
89	Frisian?	dinar	1200-1300	Almohad imitation (Lanting/Molema 1993/1994, no. 4)	gilt copper?, fragment	T
90	Serviё	dinar	1243-1276*	Urosh I (Metcalf 7.2)		
91	Flanders	penny	1259-c.1290	Gent (Ghyssens 465)		
92	Flanders	penny	1259-c.1290	Gent (Ghyssens 481)		
93	Flanders	eagle-groat	1275-1280	Margarete of Constantinople/Aalst (Gaillard 144-147)	cut	
94	Hainaut	rider-groat	1275-ca 1290	Margarete of Constantinople/Valenciennes (Chalon 13-16)		
95	Liège	postulate florin	15th century	[unidentifiable]	fragment	T

REFERENCES

ALBRECHT, G. 1959. *Das Münzwesen im niederlothringischen und friesischen Raum vom 10. bis zum beginnenden 12. Jahrhundert*, Hamburg (Numismatische Studien 6).

BESTEMAN, J.C. 1997. De vondst van Westerklief, gemeente Wieringen: een zilverschat uit de Vikingperiode, *Oudheidkundige Mededelingen uit het Rijksmuseum van Oudheden* 77: 199-226.

BOELES, P.C.J.A. 1951. *Friesland tot de elfde eeuw*, s-Gravenhage.

GRIERSON, P. & M. BLACKBURN 1986. *Medieval European coinage, I: The early middle ages, 5th-10th centuries*, Cambridge.

HAERTLE, C.M. 1997. *Karolingische Münzfunde aus dem 9. Jahrhundert*, Köln.

ILISCH, P. 1990. Friesische Prägungen des Grafen Wichmann, *Jaarboek voor Munt- en Penningkunde* 77: 5-25.

JACOBI, H.W. 1984. Twee muntvondsten uit de 13de eeuw: Hemrik 1984, Leeuwarden 1983, *Jaarboek voor Munt- en Penningkunde* 71: 133-144.

KENT, J.P.C. 1994. *The Roman imperial coinage, X: The divided empire and the fall of the Western parts, AD 395-491*, London.

KLUGE, B. 1991. *Deutsche Münzgeschichte von der späten Karolingerzeit bis zum Ende der Salier, ca 900 bis 1125*, Sigmaringen (Römisch-Germanisches Zentralmuseum, Monographien 29).

KOERS, J.P., J.N. LANTING & J. MOLEMA 1990. De muntfibula van een almohadische dobla uit Scheemda: vondstomstandigheden, parallellen en historische context, *Palaeohistoria* 32: 331-338.

LANTING, J.N. & J. MOLEMA 1993/1994. Nogmaals gouden muntfibula's uit de 12e-13e eeuw, *Palaeohistoria* 35/36: 323-328.

METCALF, D.M. 1996. Viking Age numismatics, 2: Coinage in the Northern lands in Merovingian and Carolingian times, *Numismatic chronicle* 156: 399-428.

VELDE, W. OP DEN 1982. De in Nederland voorkomende sceattas, *De Beeldenaar* 6: 40-52, 83-96.

POL, A. 1989. Remmerden 1988, Een vondst van vroegmiddeleeuwse munten bij Rhenen, *De Beeldenaar* 13: 39-47.

POL, A. 1992. Spectaculaire schatvondst uit de 9e eeuw, *De Beeldenaar* 16: 66-71.

POL, A. 1992. Münzfunde und Fundmünzen der Jahrtausendwende in den Niederlanden. In: B. Kluge (ed.), *Fernhandel und Geldwirtschaft. Beiträge zum deutschen Münzwesen in sächsischer und salischer Zeit; Ergebnisse des Dannenberg-Kolloquiums 1990*, Mainz (Römisch-Germanisches Zentralmuseum, Monographien 31 = Berliner Numismatische Forschungen, Neue Folge 1), 189-204.

VIN, J.P.A. VAN DER 1992. *Fundmünzen der römischen Zeit in den Niederlanden, I: Provinz Friesland*, Berlin.

VÖLCKERS, H.H. 1963. *Karolingische Münzfunde der Frühzeit, 751-800*, Göttingen (Abhandlungen der Akademie der Wissenschaften in Göttingen, Philologisch-Historische Klasse, 3F, 61).

ZIJLSTRA, J. 1990. Friese bodemvondsten, I: Finns fibula, Leeuwarden.

NOTES

1. Reports of individual finds have increased enormously since the use of the metal detector became popular in the 1980s. Numerous new reports of thousands of coins have provided a valuable contribution to the source material on the history of coins in the Netherlands. If these data are systematically examined they may produce new insights.

   I am grateful to my colleague Drs B.J. van der Veen who has been responsible for the registration of new coin finds at the Rijksmuseum Het Koninklijk Penningkabinet (KPK) since 1996 and has also undertaken the retrospective registration of older reports present in the KPK archive, for his assistance in producing the quantitative data in this article.
2. Published in Van der Vin (this volume).
3. The term 'individual find' is not used here in the archaeological sense as a find whose (archaeological) context is unknown, but in the numismatic sense of a find of one single coin and no more (in contrast to the term 'hoard' which refers to a find of several coins which entered the soil on one and the same occasion). From a numismatic point of view, the medieval coins from Wijnaldum are all individual finds, including those found during the excavation.
4. For a detailed survey of Western European coinage in the Early Middle Ages, see Grierson & Blackburn (1986).
5. Kent (1994 nos 3756-3763 and 3770-3775) and Grierson & Balckburn (1986, no. 173). For a better, gold specimen of Libius Severus from Frisia see Boeles (1951, p. 504, no. 25 = Van der Vin 1992, p. 143, no. 249.39).
6. Boeles (1951, especially pp 266-268).
7. Op den Velde (1985).
8. Pol (1989).
9. Metcalf (1996, pp 402-409).
10. In this survey the 8 coins from an unknown findspot in the province count as 1 toponym.
11. Besteman (1997).
12. Zijlstra (1990, pl. VIII, no. 33).
13. Pol (1992).
14. Völckers (1963), Haertle (1997) and KPK register of finds; Haertle's catalogue was completed in 1991, but the finds reported since have all been included in the electronic coin finds register of the KPK.
15. This is a translation of the Latin *denarius* (cf. English penny and German *pfennig*, and also the French *denier* and Italian *dinero*) a term generally meaning coin. A more precise indication was unnecessary since there was only one kind of coin, one denomination.
16. Albrecht (1958) Kluge (1991) and Ilisch (in press). For an inventory of Dutch finds, Pol (1993).
17. Ilisch (1990), since then the hoard of Franeker/Witmarsum 1980 (46 items) has been discovered.
18. Jacobi (1984).
19. Koers et al. (1990) and Lanting & Molema (1993/1994, especially p. 327, no. 4, ill. 5).
20. See Van der Vin (this volume).

The glass vessel finds

Y. SABLEROLLES

1 INTRODUCTION[1]

The excavations of the settlement on the Tjitsma terp at Wijnaldum yielded almost 200 fragments of glass vessels which show a wide variety of forms. The retrieval of this surprisingly large and varied assemblage of vessel glass from Wijnaldum results from the fact that for the first time a Dutch terp was thoroughly excavated on a large scale accompanied by systematic waterscreen sieving using a 3 mm mesh. In comparison, other excavated terps along the coast of the provinces of Friesland and Groningen only yielded a small range of more or less complete vessels, which mostly originate from graves. This picture provided by burials appears not to be representative of the glass in use during the full length of occupation of a terp-site.[2]

The excavation of part of the settlement at Wijnaldum enables us for the first time to catch a glimpse of the range of glass vessel types in use and the composition of glass assemblages by period of habitation of a terp. This, in turn, could give us an indication of the importance which glass vessels played in the material culture, and possibly the drinking culture, of a terp population over a period of almost 8th centuries.

The first part of this section focuses on the glass finds from Wijnaldum and is followed by a discussion in which all the evidence is presented in a broader context so as to try and interpret the significance of the finds. The second part, a catalogue with a detailed description of the diagnostic vessel fragments, is provided as an appendix.

2 METHODS OF QUANTIFICATION

One of the results of the sieving activities is that even the smallest glass fragments and splinters were retrieved. Consequently, the percentage of undiagnostic fragments is rather high (72%) and only 55 out of 194 fragments were found to be diagnostic (Table 1).

When we want to compare certain groups within the Wijnaldum glass assemblage with each other or compare the Wijnaldum assemblage to other assemblages, quantification by simple fragment totals as used for pottery quantifications is unsatisfactory. For one thing, thinly blown vessels such as Carolingian funnel beakers, are more likely to break into a larger number of small pieces than thick-walled vessels like Roman square bottles; any further fragmentation after the initial breakage and disposal would depend on post-depositional processes.

Therefore an attempt has been made to estimate the minimum number of glass vessels (EMN) in the Wijnaldum assemblage. This method takes into account all the fragments of one colour, including body fragments, identified as coming from a particular vessel type, and then makes a conservative estimate of the minimum number of different vessels they might have come from.[3] Contrary to quantification methods used for pottery, which involve rim and base fragments, body fragments are specifically included, since they tend to be equally or often more diagnostic than rim or base fragments.

By using the EMN method for the vessel glass from Wijnaldum, it was possible to establish that 38 fragments represent a minimum number of 25 vessels of a specific form or type. These include three vessels dating to the Roman Period, seven vessels dating to the Migration Period, four to the Merovingian Period and eleven to the late Merovingian/Carolingian Period (Table 1). The vessel form of further 17 body fragments representing at least 12 vessels cannot be identified with certainty. They can, however, be grouped into periods with a high degree of confidence, because of specific characteristics such as the quality and colour of the metal and/or the technique of decoration used. This adds another 2 vessels to the Roman Period, three to the Migration Period, three to the Merovingian Period and four to the late Merovingian/Carolingian Period, bringing the total EMN to 37 (Table 1).

As is usual in settlements, many Wijnaldum vessels are represented by only one fragment: after breaking a vessel, the fragments were apparently carefully collected and disposed of, making sure that no sharp fragments were lying around. Some vessels which are represented by more than one fragment are either very thinly blown or made of – generally brittle – colourless glass. Their fragmentation is likely to be due to post-depositional processes (differential conservation, redeposition, excavating and sieving) rather than to the initial breakage of the glass.

The extent to which the glass vessels which were recovered represent the total number which were in use at Wijnaldum at any one time remains open to speculation, though it can obviously be only a very small part of the actual assemblage.

Table 1. Wijnaldum – Number of glass Fragments (NFrs) and Estimated Minimum Number of glass vessels (EMN) per period. Glass fragments from cremation burials are not included.

Period	Nfrs	%	EMN	%
Diagnostic				
Form certain:				
– Roman	3	8	3	12
– (Late Roman)/Migration	13	34	7	28
– Merovingian	8	21	4	16
– Late Merovingian/Carolingian	14	37	11	44
– Subtotal	38	100	25	100
Form uncertain:				
– Roman	4	24	2	17
– Migration Period	4	24	3	25
– Merovingian	3	18	3	25
– Late Merov./Carolingian	6	35	4	33
– Subtotal	17	101	12	100
Total diagnostic	55	28	37	100
Undiagnostic				
Decorated	10			
Undecorated:				
– Colourless	1			
– Translucent blue	2			
– Translucent purple	1			
– Translucent greenish*	125			
Subtotal	139	72		
Total	194	100		

*Plain fragments of translucent greenish glass are not included in the catalogue.

3 COLOURS USED IN WIJNALDUM GLASSES

The majority of the fragments found at Wijnaldum are made of translucent, 'naturally' coloured greenish glass in different shades. The greenish colour arises from the presence of impurities in the raw materials used for making glass, for instance iron oxide which occurs naturally in sand; the precise shade, bluish, yellowish or pale green, depends on the atmospheric conditions in the furnace during the melting process of the glass batch.[4]

During the first half of the 1st century AD, the Roman glass service is dominated by strongly coloured moulded and free-blown tablewares. Before the mid 1st century AD a major change occurred with the introduction of cheaply produced, mould-blown containers of ‘naturally coloured’ blue-green glass (2-3) used for transport and storage. The most successful of these were square bottles, which were mass-produced throughout the 2nd century AD (3).

This blue-green glass is also increasingly used for the production of tablewares (1). From the late 1st century onwards, tablewares are made of blue-green glass or of (almost) perfectly colourless glass and less frequently of lightly tinted greenish glass. Initially used for the more luxurious end of the tableware production such as facet-cut beakers, colourless glass became more widely used in the course of the 2nd century. Some perfectly colourless fragments from Wijnaldum are almost certainly Roman in date and may have belonged to such common colourless tablewares, for instance cylindrical drinking cups on a double base-ring which were the dominant drinking vessel of the late 2nd and early 3rd centuries throughout the north-western provinces (32-34).

Blue-green and colourless glass still dominated Roman glass production during the 3rd century. Lightly tinted glasses were still less frequent, but by the end of the 3rd century a lightly-tinted and bubbly glass had evolved which was to dominate the entire 4th century. This typical 4th-century bubbly glass ranged in colour from greenish colourless to light green and yellowish green.

During the late Roman Period strongly coloured glass was mainly used for purposes of decoration, for instance the application of blue blobs on late Roman drinking vessels.

In the late 4th and first half of the 5th century lightly tinted greenish glass continued to be used, but more often strong shades of yellow-green (4-5) and olive (6) occurred. In general, the metal tends to contain many small and large bubbles as well as inclusions such as small fragments of quartz.

Besides yellowish green and olive, bluish-green was again frequently used for glass vessels from the 2nd half of the 5th century onwards; shades of brown are frequently seen in the 6th and early 7th centuries (16, 37). These colours result from the presence of oxides of iron and manganese in the raw materials and are achieved when melting the raw materials under reducing circumstances in the furnace; they therefore do not arise from deliberately added metal oxides.[5] Deliberately coloured glass was still mainly used for decorational purposes, though some cobalt blue glasses were made during the 5th-7th centuries (44-45).

In the late 7th/8th century, the technique of colouring glasses by deliberately adding metal oxides – other than cobalt oxide for deep blue glass – was re-introduced. A small fragment of translucent red glass from Wijnaldum (38) most likely derives from a late Merovingian or Carolingian glass vessel, but could also have been part of an inlay imitating almandine. Translucent red glasses are rare and the few existing vessel fragments of this colour do not pre-date the 8th century, whereas there is evidence that translucent 'almandine' red glass inlays were used as early as the first half of the 7th century in Anglo-Saxon and continental gold jewellery.[6]

Two vessel fragments of translucent, deep turquoise glass were both found in contexts dating to the first half of the 8th century (46, 47); they were decorated with a thin opaque white spiral trail (see also below). Translucent turquoise glass was found at the 9th century glass working site at Glastonbury in the form of a small piece of translucent deep turquoise glass with one drawn end, a thin rod of glass of the same colour and a translucent turquoise and opaque white bichrome cable. This turquoise glass is believed to have been imported and worked on the site for decorational purposes.[7]

In the 5th century and later attempts may have been made to produce colourless glass vessels, but this always resulted in a yellowish or greenish-tinged metal. Only during the Carolingian 'Renaissance' do perfectly colourless glasses appear again. In contrast to the Roman Period, the technique of producing colourless glass seems to have been reserved for highly decorated vessels. The only Carolingian fragment from Wijnaldum which is made of nearly colourless glass with a slight greenish tinge had a separately added blue rim and probably belonged to a funnel (26). Most of the ordinary funnel beakers from Wijnaldum are made of blue-green glass (18, 20, 24, 25, 27-29), some are made of a lightly tinged (yellowish) greenish glass (19, 21, 23), whereas a stronger yellowish-green is rather uncommon for funnels (22).

4 TECHNIQUES OF DECORATION USED IN WIJNALDUM GLASSES

Many Roman decoration techniques were discontinued in the 5th century, specifically 'cold' techniques. These are techniques which are executed after the vessel is blown and cooled: engraving and cutting, painting, enamelling, gilding and the application of gold-foil. It was only in the 8th century that one such 'cold' technique, viz. the application of gold-foil, was re-introduced.

Two decorative techniques which were very popular in the Roman Period and continued to be used throughout the largest part of the early medieval period, trailing and optic blowing, can also be seen on early-medieval vessel glass from Wijnaldum.

The technique of trailing threads onto a vessel was very popular during the entire early medieval period. During the Migration Period the application of self-coloured threads predominates, mostly in the form of spiral trails applied to rims, sometimes accompanied by vertical loops on the body. This kind of decoration can be seen on a range of different vessel forms from this period, in the case of Wijnaldum on rim and body fragments belonging to 5th-century cone beakers (6-8).

Opaque white trailing was also used during the entire early medieval period, but seems to have been most sophisticated during the 5th-6th centuries. After application, the white trailed threads were often combed into intricate patterns and marvered into the vessel by rolling it over a flat surface. Cones and bowls of the 5th and early 6th centuries were decorated with opaque white spiral trails, vertical loops and combed feather, rosette and arcade patterns. At Wijnaldum four fragments with opaque white decoration were identified as coming from bowls (9-12); three undiagnostic fragments with similar decoration may also have belonged to such bowls (43-45).

In the late 7th/8th century opaque yellow thread decoration, which occurs occasionally during the 5th to 7th centuries, became more fashionable and soon gained popularity over opaque white trailing. In contrast to the previous periods, white and yellow trails are no longer (completely) marvered into the surface and tend to have a shiny 'enamel-like' appearance (27, 48). A rare example from Wijnaldum combines part of an opaque white trail, which was probably combed into an arcaded pattern, below an opaque yellow spiral trail on a rim fragment of a late Merovingian/Carolingian vessel, probably a pouch-shaped or cylindrical beaker (30).

Two translucent turquoise fragments with a thin, opaque white spiral trail from Wijnaldum (46-47) may belong to a group of deep blue vessels decorated with applied, thin opaque white trails. These are mostly found on German sites in (later) 10th to 11th/mid 12th-century contexts, although some earlier 8th/9th-century finds are known from Dorestad.[8] An undiagnostic body fragment of deep blue glass decorated with a half-marvered, shiny white spiral trail (62) may also belong to this group, although its find context suggests a Merovingian date.

Spiral trails were sometimes combined with bichrome cables, made by twisting threads of two different colours into slender, twisted rods. They occur mostly on squat jars/beakers and bowls and occasionally on funnel beakers.[9] At Wijnaldum no examples of these more luxuriously decorated vessels were recovered.

A characteristic feature on squat jars/beakers and funnels are rims of translucent purple, deep green and deep blue glass (26). These coloured rims are sometimes referred to as trails or coils, but are in fact 'strips' separately fused to the edge of the vessel and then fire-rounded.

The second main decorative technique is achieved by optic blowing, the process by which a gather of glass is first blown into a mould with a sharply ribbed pattern in it and then free-blown or blown in a second mould so that the pattern in relief becomes faint; by twisting the vessel at the end of the process the ribbing may become diagonal. Faint, diagonal ribbing occurs mostly on 5th-6th-century vessels such as cones and bell beakers (14, 39-40); the latter also show faint, vertical ribbing. Stronger vertical ribbing is found on 7th-century palm cups (16) and on 8th-century palm funnels which may also be represented at Wijnaldum (41-42). In the 9th-10th centuries the technique of optic blowing seems to have gone out of fashion.

5 VESSEL GLASS OF THE ROMAN PERIOD

Glass vessel finds from the Roman Period are relatively few at Wijnaldum and represent tablewares and containers.

Contexts

It is remarkable that fragments of two 1st/2nd century diagnostic Roman vessels (2-3) were found in contexts dating between 650-750 (occupation Period V). A third fragment (1), which belongs almost certainly to an early 1st-century cast form, was found in a 3rd-century context. Of the perfectly colourless fragments which are almost certainly Roman in date, two (32, 34) derive from post-Roman contexts and one (33) from a context dating to the first half of the 4th century.

This seems to indicate that the Roman vessel fragments, with the possible exception of one (33), did not reach the settlement at Wijnaldum in the form of complete vessels during the Roman occupation of the terp, but only much later. The same phenomenon was noticed for other Roman finds at Wijnaldum.[10]

Tablewares

A fragment of a base with what seems to be a low base-ring is wheel-polished on both sides (1). It most likely comes from an early-Roman cast vessel, probably a dish, of the 1st century AD.

Some undiagnostic body fragments of perfectly colourless glass containing a few pinprick bubbles (32-35), almost certainly belonged to Roman tablewares. Colourless tablewares were made from the late 1st century into the late Roman Period, but became much rarer in the early 4th century. Therefore, the only colourless fragments to have been found in a Roman context (33) dating to the first half of the 4th century, are more likely to have belonged to an earlier vessel which may have reached the site sometime in the 3rd century. One possible form from which these fragments could derive is a cylindrical cup on a double base ring (Isings form 85b), which is found on numerous sites throughout the north-western provinces during the later 2nd and (first half of the) 3rd centuries.[11]

Containers

These include a fragment of collared rim (2) which is most likely to have come from an ovoid or globular jar type of Isings form 67b/c, though the rim form also occurs on jars with square bodies of Isings form 62 (Fig. 1). Of these, optic blown ribbed jars of Isings form 67c were the most popular type of jar in the north-western provinces.[12] The main period of use for globular ribbed jars was

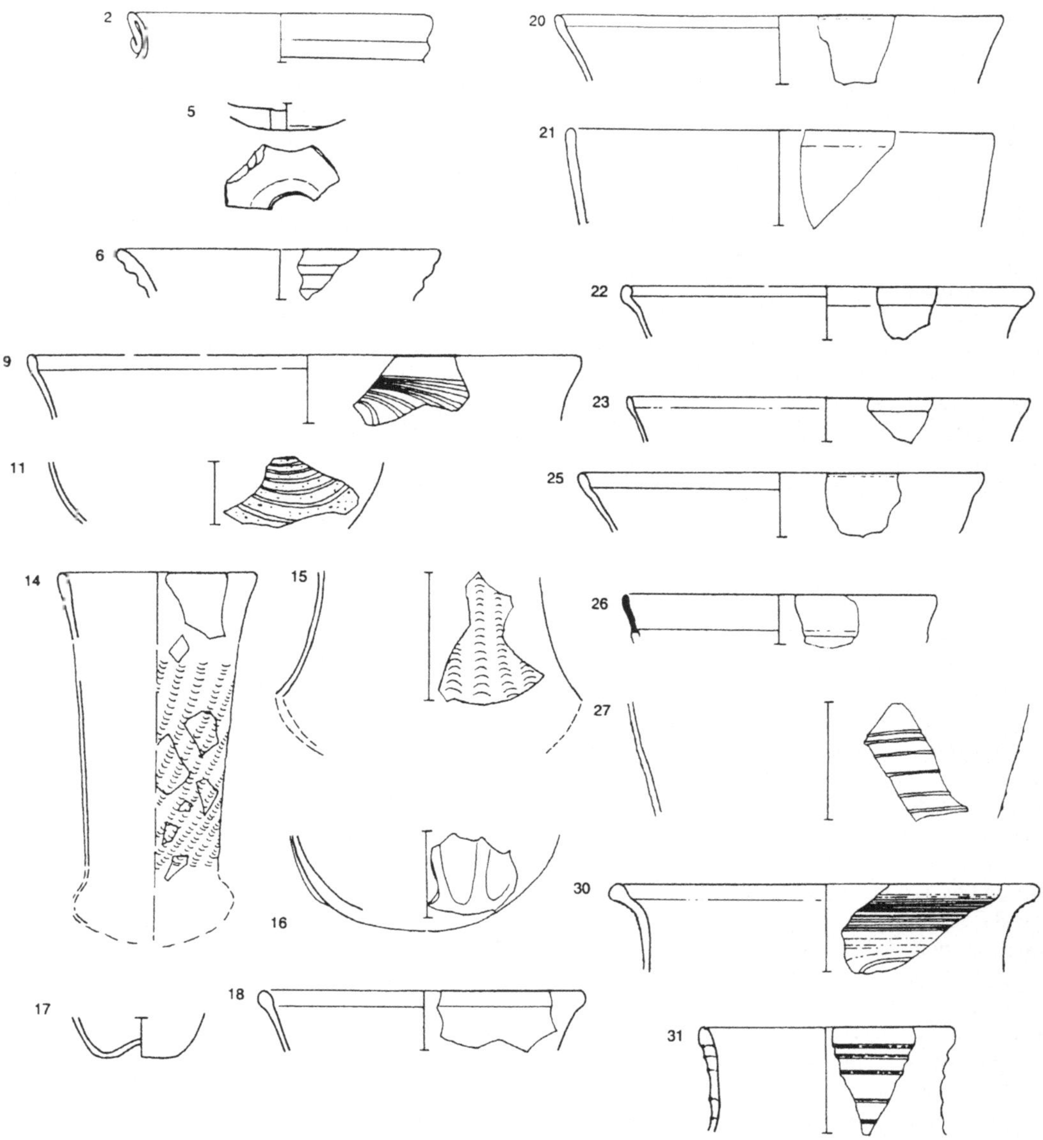

Figure 1. Glass vessels from Wijnaldum with catalogue numbers. Scale 1:2,5.

from the mid 1st century to the early 2nd century. Plain ovoid jars were certainly in use during the second half of the 1st century, but it is not clear how long they continued in use due to their apparent scarcity.[13] Collared jars with square bodies are even rarer and therefore more difficult to date, but the rim form would argue for a similar date range as for the globular and ovoid jars.

The second fragment almost certainly belongs to a square bottle of Isings form 50 (3). Square bottles were mass-produced in the later part of the 1st and the entire 2nd century; they were probably still in use in the early 3rd century.[14]

6 VESSEL GLASS OF THE LATE ROMAN/MIGRATION PERIOD

Glass from the Migration Period is relatively well represented at Wijnaldum with a minimum number of ten vessels. When looking at the Wijnaldum glass assemblage from this period, one thing that immediately attracts the attention is the absence of unworked, 'knocked-off' rims. While rims on Roman glass vessels were usually finished by heat-treating or tooling, there is a tendency in the course of the 4th century to leave the rims unworked after the vessel is knocked-off from the blowpipe, and this becomes the distinctive feature on drinking vessels during the late 4th and first half of the 5th century. Almost all of the Migration Period glass from Wijnaldum can be attributed to cones and bowls with fire-rounded rims of the second half of the 5th and early 6th century (5-13).

Glass vessels from the Migration Period are still frequently found in graves and are also well represented in settlements. In the Netherlands one of the largest glass assemblages of this period (ca 1000 fragments) comes from a settlement at Gennep on the Niers (Prov. of Limburg) which is dated from the late 4th to the late 5th/early 6th century.[15] Numerous glass vessel fragments from this period come from various settlement contexts in Maastricht.[16] Outside the borders of the Roman empire some 30 fragments of vessels of the late 4th-early 5th centuries were recovered from the late-Roman settlement at Wijster (Prov. of Drenthe).[17] Glass of this period has also found its way up north into Scandinavia by way of Denmark; on Jutland, numerous fragments of glass vessels, predominantly of the second half of the 5th/early 6th century, were found in the settlements at Dankirke and Dejberg where they are believed to have been traded.[18]

Apart from Wijnaldum, there are no glass vessels from the late Roman and Migration Period known from settlement or burial contexts in Friesland or Groningen. This is probably partially caused by the burial customs in this region at the time: inhumation without grave goods and cremation of the deceased together with the grave goods. A recent survey of Roman cemeteries in neighbouring Lower Saxony and Schleswig-Holstein has revealed that molten glass vessels frequently occur in cremation burials, often in association with late-Roman militaria.[19]

At Wijnaldum a cremation burial with two individuals from the second quarter of the 5th century contained part of at least one molten, bluish green glass vessel, which probably stood on a base-ring (63) and possibly a second vessel of colourless glass (64). A molten fragment of a blue-green vessel containing calcinated bone (65) was found in a settlement context (as part of sod structure 32).

Contexts

Most fragments attributed to vessels from the Migration Period come from contexts dating to the second half of the 5th century.

Two fragments of bowls were found in later contexts. Bowls rapidly went out of production in the early 6th century and are still frequently found in graves dating to the 2nd half of the 6th century. A fragment of an undecorated bowl (13) comes from the same ditch (550-600) as fragments of a Merovingian bell beaker (38) and probably represents a treasured old piece which was discarded together with the bell beaker. A tiny rim fragment of a bowl (10) from a waste deposit dated between 600 and 700 is probably residual. One fragment of a conical beaker type of the second half of the 4th and the early/first half of the 5th century (5) comes from a context dating to the second half of the 5th century and probably represents the earliest Migration Period glass import on the terp.

Cup or beaker on a base-ring

A fragment of a small, tubular base ring of bubbly yellow-green glass (4) probably belonged to a late-Roman drinking vessel, probably a hemispherical cup (Isings form 108) or a conical beaker (Isings form 109) on a base ring. These are both forms typical of the later 4th and first half of the 5th century which are often found in late-Roman graves in northern Gaul, e.g. at Spontin, Furfooz, Samson, Eprave etc.[20] By the mid 5th century only two vessel forms, bell shaped beakers and claw beakers, still had a base in the form of a small base ring. The colour and quality of the fragment argue for a date in the late 4th or the first half of the 5th century, although the size of the base ring could also indicate a date after the mid 5th century.

Conical beaker

A concave base fragment of yellowish green glass (5) probably belonged to a slim conical beaker of Isings form 106b (Fig. 1). The relatively narrow base make it a transitional form between relatively broad 4th-century Roman conical beaker types and pointed 5th-century 'Frankish' cones which could not stand upright. The rims on this type of beaker were usually left unworked and were sometimes lightly wheel-polished. The observation that the base lacks a pontil scar is a strong indication that the Wijnaldum beaker had a knocked-off rim. This type of beaker can be dated from around the mid 4th to the early 5th century.

Cones

During the 1st half of the 5th-century conical beakers gradually became slimmer until their bases were so narrow that they could not stand upright any more. These forms are known as cones and represent the typical drinking vessel of the 5th and early 6th centuries. Cones have fire-rounded and fire-thickened, straight or everted rims. They are usually decorated with spiral trails below the rim, while the body can be plain, trailed or ribbed by means of optic blowing.

At Wijnaldum fragments of at least two cones can be identified. A fragment of a fire-rounded, everted rim with a self-coloured spiral trail (6) of yellowish olive glass belonged to a splayed cone (Fig. 1). The colour and the thickness of the rim (0.40 cm) may indicate this is an earlier type of cone dating to (just) before the mid 5th century, but this is not certain since the type cannot be established.[21]

Two thinly blown body fragments of a bluish green cone decorated respectively with a self-coloured spiral trail below the rim and apparent vertical loops on the body (7) most likely belonged to a cone of the Kempston type, so called after the findspot of one of the finest examples in Bedfordshire, England.[22] Three further similarly decorated fragments (8) may have belonged to the same cone, but come from a different context.

With their average height of 25 cm Kempston cones are by far the largest type of cone. They first appear in the mid 5th century and quickly went out of production in the first quarter of the 6th century. Find concentrations of greenish Kempston cones along the Middle Rhine and of nearly colourless, yellowish specimens in the Belgian-French area suggest different production zones, probably in the Middle Rhine and the Middle Meuse areas.[23]

In her survey of the occurrence of Kempston cones on the continent, Koch points out that they are predominantly found in Frankish, Saxon, Alamannic and Thuringian graves of high-ranked, fully-armed warriors, horsemen and noblemen and only occasionally form part of the set of grave goods of aristocratic women.[24] In these graves Kempston cones are regularly accompanied by scales for precious metals, spits for roasting and bronze basins which were probably used as hand-wash basins.[25] In the Netherlands, one complete Kempston cone from a grave in Aalden (Prov. of Drenthe) was published without further mention of the grave contents.[26]

Over the last years more information has become available on Kempston cones from settlement contexts. They occur in settlements as far south as Baden-Württemberg, where as many as 26 Kempston cones were recovered from the hill-settlement the Runde Berg near Urach, where an Alamannic military elite resided from the second half of the 3rd to the early 6th century,[27] and as far north as Scandinavia, for instance at Dankirke and Dejberg on Jutland.[28] In the Netherlands, some 15 specimens come from a Frankish settlement at Gennep in Dutch Limburg.[29] Located close to the confluence of the Niers and the Meuse, this settlement was probably built here in the late 4th century to protect a nearby, yet undetected, stronghold. Surprisingly, no specimens were identified amongst the glass assemblages from two settlement sites in Maastricht.[30] Fragments of two Kempston cones were found in a gully of the Old Rhine running through the early-medieval settlement at Valkenburg-de Woerd (Prov. of South Holland; unpubl. data), which appears to have functioned as a riparian market during the Merovingian Period. Far more inland, a sunken hut in a late-Roman settlement at Elsen, Markelo (Prov. of Overijssel) yielded a base fragment of a Kempston cone.[31] For the area, this settlement is surprisingly rich in luxury wheel-thrown pottery, mostly footed beakers of terra nigra, and yielded traces of iron production and bronze working.[32]

Bowls

Hemispherical cups and truncated conical bowls with knocked-off rims, which were very popular drinking vessels during the late Roman Period, were discontinued sometime in the first half of the 5th century. Around the mid 5th-century bowls with fire-rounded, everted rims and mostly convex-curved sides appeared. These were produced in large series during the second half of the 5th century and often were high-quality products with intricately trailed-on and marvered opaque white decoration, such as spiral, feather, arcade and rosette patterns. The function of these modestly sized bowls (rim diameter between 11-15 cm) may have been that of drinking vessels, but they were (also) used to serve food in. This is suggested by the find of a bowl decorated with two opaque white spiral zones in a rich 5th-century burial of a three and a half-year-old boy under the St. Severin Church in Cologne. The boy was buried in a tuff sarcophagus with many grave goods, amongst others an iron francisca interpreted as a symbol of his special status in an aristocratic-warrior world. The bowl contained the remains of a bird and certain pollen and herbs, which suggests it contained a dish prepared in a Mediterranean tradition with fresh herbs and honey.[33]

At Wijnaldum a minimum number of four bowls decorated with opaque white marvered trailing were identified. A rim fragment (9) and a body fragment (11) of bluish-green glass, representing at least one bowl, are decorated with an opaque white marvered feather pattern (Koch bowl form IV L) (Fig. 1). Two blue fragments with opaque white decoration (44, 45) may also have belonged to (a) bowl(s) of this type. Although there are no parallels known to the author for bowls with a white feather pattern of blue glass, there are examples of contemporary vessel types of blue glass with opaque white decoration, e.g. a blue Kempston cone with an opaque white spiral trail and self-coloured vertical loops from Dankirke in Jutland (Den.)[34] and (fragments of) a blue bowl with a similar decoration from Maastricht.[35]

A rim fragment with one turn of a spiral trail (10) probably comes from a bowl decorated with either one or more zones of white spiral trails (Koch form H/I) or from a bowl with a white spiral trail and arcades (Koch form K). A body fragment (12) is probably to belong to a bowl decorated with arcades (Koch form IV K). An undiagnostic body fragment (43) is too small to attribute with certainty to a bowl with feather or arcaded pattern.

Although there is not much dating evidence available for bowls decorated with opaque white trails, grave finds seem to indicate that most bowls decorated with opaque white decoration can be dated in the second half of the 5th and early 6th century with one possible exception of bowls with a feather pattern which seem to be exclusively of an early 6th-century date, making these the youngest type in this bowl series.[36]

The distribution of bowls with opaque white decoration shows different find concentrations for different types of bowls.[37] Bowls with one spiral trail below the rim occur predominantly on the Middle Meuse and Sambre and the Middle Rhine and Lower Moselle, whereas bowls with more spiral trails were found along the Oise, Marne and the Upper Moselle. Bowls with an arcaded pattern show the widest distribution: from the Middle Meuse and the area to the west of it, along the Oise, Marne and Upper Moselle, to the Middle Rhine region. Bowls with a feather pattern are more frequent along the Middle Rhine and less frequent along the Middle Meuse than bowls with arcades and do not occur to the west of the Meuse. Bowls with rosettes on the base, which are not represented at Wijnaldum, are almost exclusively found in the Middle Meuse area to the north of the Ardennes.

A pale green rim fragment (13) probably belonged to an undecorated bowl (Koch form IV E). Undecorated bowls are less numerous than bowls with a self-coloured or opaque white decoration and are concentrated in the Rhineland, from Cologne upwards. They are found in both rich graves of women and children and modest graves of men dated to the late 5th/first half of the 6th century.[38]

7 VESSEL GLASS OF THE MEROVINGIAN PERIOD

At Wijnaldum, glass from the Merovingian Period is relatively poorly represented. The two most characteristic types of Merovingian drinking vessels, the bell beaker and the palm cup, are each represented with a minimum of two vessels.

In comparison to the 4th and 5th century, Merovingian glass vessels are found less frequently in graves of the 6th and 7th centuries. For example, the largest *Reihengräberfeld* in the Netherlands excavated at Rhenen counted ca 1100 graves, ranging in date between 375 and 750 with most graves belonging in the 6th and 7th centuries; some 820 graves contained grave goods, of which only 36 graves contained glass vessels, some of which were 'antiques'.[39]

This phenomenon is often ascribed to the decline of production in the Merovingian Period. As an alternative explanation it has been suggested that this burial custom weakens in the last part of the Merovingian Period under the influence of Christianity.[40] Others, however, have pointed out the reduction of glass vessels in graves, while ceramics still regularly occur as grave goods.[41]

Contexts

Both bell beakers fit well into the contexts in which they were found, respectively a ditch (14) and a truncated sod layer (15), which are both dated to the second half of the 6th century. Fragments of one bell beaker (14) were found in the same ditch as the rim fragment of the undecorated bowl (13) described in section 6.

The same can be said for the palm cups. A fragment belonging to a typologically early type of palm cup (16) was found in a truncated sod layer dated to the second half of the 6th century, whereas a base fragment (17) which probably belongs to a typologically late type was found in a sod layer which is dated to the second half of the 7th/first half of the 8th century.

Bell beakers

In the course of the first half of the 6th-century, cones, footed beakers and bowls fell into disuse and were replaced by bell beakers which were produced in large series during the 6th into the first half of the 7th century. Up to this day, Böhner's typo-chronology for bell beakers is still being used.[42] The earliest, rather squat bell beakers with splayed rims and splayed sides originate in the first half of the 6th century (form A). A taller, somewhat later development with almost straight, conical sides (form B) is typical of the largest part of the 6th into the early 7th century (Stufe III, early Stufe IV). A rare variety of form B has short, conical sides (form C). A tall, slim type of bell beaker with almost cylindrical sides (form D) is dated similarly to bell beakers of form B. The latter lacks the marked carination to the lower body characteristic of the earlier forms. Recently, Siegmund has distinguished a 4th form, a slim variety of Böhner form B, which is typical of the late 6th and early 7th century on the Lower Rhine (Niederrhein Phasen 7 and 8a).[43]

Over the years another typological distinction has been noticed between the Belgian/northern-French and the Rhenish types. The former were generally found to have a small knob at the centre of the base, whereas the Rhenish types tend to have a simple rounded base.[44] A recent survey of bell beakers in this area by Koch makes it clear that an even finer distinction can be made.[45] A rounded base is typical for the, often ribbed, bell beakers found along the Rhine, while ribbed beakers with a pointed, drawn base predominate between the Meuse and the Rhine. Plain beakers with a pointed, drawn base predominate in the Sambre-Oise-Marne area and beakers of which the pointed base is emphasized by adding a drop of glass are found to the west of the Meuse; the latter often have a decoration of opaque white spiral trails. Evidence for production of the third type of bell beaker with opaque white decoration comes from a possible glass production site at Macquenoise (Belgium).[46] All types of bell beakers are characterized by simple, fire-rounded and thickened rims which are usually straight and sometimes slightly in-turned. The quality of the glass varies considerably, from a clear, transparent metal with few small bubbles to a low quality metal clouded with many bubbles, impurities and streaks.

Bell beakers are frequently encountered in the numerous *Reihengräberfelder* along the Rhine as well as in cemeteries in northern France and Belgium. In the Netherlands the popularity of the bell beaker is obvious in the *Reihengräberfeld* at Rhenen (Prov. of Utrecht), where as many as 17 out of 36 graves with glass grave goods contained a bell beaker.

Bell beakers do not seem to be related to gender or the quality of grave goods to the extent that was seen during the Migration Period when distinct glass vessel types seem to have specifically selected as part of specific grave contents.

Bell beakers are given as grave goods to males, sometimes fitted out with full armour, females and children. They occur not only in rich graves, accompanied by many other grave goods, but also in modestly furbished graves in which they sometimes represent the only luxury item. The reduction in glass vessel forms in the 6th century apparently led to less diversification in the burial ritual with regards to the deposition of glass vessels in graves.

The shape of the first bell beaker from Wijnaldum, which is represented by a rim fragment and several delicately, diagonally ribbed body fragments (14), cannot be identified (Fig. 1). However, the colour and quality of the metal and the fine mould-blown pattern on the body fragments argue for a date in the 6th century rather than in first half of the 7th century. The diameter of the rim (5 cm) indicates a rather small beaker of ca 9 cm in height.[47]

The second bell beaker is represented by a vertically ribbed lower body fragment (15) which is concave-curved, indicating it belonged to a bell beaker with splayed or conical sides and a marked carination to the lower body, rather than to a type of bell beaker with straight, cylindrical sides (Fig. 1).

Because no base fragments of the Wijnaldum beakers are preserved, it is impossible to specify the exact type of bell beaker, but a Rhenish production or production in the Rhine-Meuse region seems probably on the basis of the optic blown, ribbed decoration of both Wijnaldum specimens.

Palm cups

Palm cups represent a somewhat later development than bell beakers. The earliest specimens are from the late 6th century, but the majority date to the 7th century. Rather than following Böhner's typology which is based on the form of palm cup, viz. palm cups with almost conical profile (form A) and palm cups with a splayed profile (form B),[48] a typology on the basis of the rim seems more useful. A survey of palm cups in Merovingian graves in Belgium shows that almost cylindrical cups with fire-rounded and fire-thickened or inward folded rims with vertical ribs are dated to the late 6th and early 7th century.[49] More conical, wider cups with in-folded rim are encountered during most part of the 7th century, while wide cups with deeply outward folded rims belong to the last phase of the production of the type. Siegmund's typo-chronology for the lower Rhine region shows that palm cups with a simple, fire-rounded and fire-thickened rim are typical of the late 6th and early 7th century (ca 585-610) and that palm cups with deeply outward folded rims are typical of the middle part of the 7th century (640-670).[50]

Palm cups can be made in shades of olive, brown and (blue-) green glass, although late palm cups with an outward folded rim are frequently bluish (green) or greenish blue. The metal tends to be bubbly with many inclusions and impurities such as dark and red/purplish streaks, although the latter are regarded by some as intentional.[51]

The rather low quality is often interpreted as a feature of the 'mass-production' of these cups, but there is no reason to assume that they were produced in far greater numbers than bell beakers or their successors the palm funnels, which are generally of a much better quality. Perhaps the low quality is a reflection of the level of recycling and the availability of raw materials for glass making in the 7th century.

Palm cups have the same distribution as bell beakers and occur in *Reihengräberfelder*; in the Netherlands, for instance, from Meerveldhoven (grave 24) and Bergeijk (grave 53). However, because of their greater uniformity in shape, it is more difficult to establish possible production centres on the basis of specific production features; palm cups with outward-folded rims of blue-green/greenish blue glass, for instance, are very frequent both in the Rhineland and in northern France.[52]

At Wijnaldum a vertically ribbed body fragment of brownish glass (16) belongs to a palm cup of the late 6th or early 7th century (Fig. 1). A thick-walled blue-green base fragment with dark streaks (17) almost certainly belongs to a later 7th-century type of palm cup with an outward folded rim (Fig. 10). A convex-curved blue-green body fragment with red streaks (36) probably belongs to a similar type of palm cup.

8 VESSEL GLASS OF THE LATE MEROVINGIAN AND CAROLINGIAN PERIOD

Glass of the late Merovingian/Carolingian Period is the best represented category at Wijnaldum with a minimum number of 15 vessels. Most fragments are identified as belonging to (palm-) funnels (18-29), wile two other fragments probably belonged to a bulbous beaker or squat jar (30) and a pouch-shaped or cylindrical beaker (31). These are the most frequently occurring vessel forms during the 8th-9th/10th centuries.

An important source of information on the typo-chronology of Carolingian glass vessels are Viking Age cemeteries in Scandinavia, where the impact of Christianity was not felt for a further 300 years. The rich burials from Uppland and Södermanland in Sweden, e.g. the well-known Viking cemetery at Birka, provide the main corpus of vessels known from the 8th-10th centuries.

The rise of the *emporia* or large trade centres and the expansion of the markets stimulated the production of luxury glass wares for the privileged. In their important survey of medieval glass, Baumgartner and Krüger divided the find contexts of fragmentary Carolingian vessel glass in north-western Europe into *emporia*, aristocratic seats, ecclesiastical buildings and glass workshops associated with any of these contexts.[53]

These are all high-status contexts in which highly-decorated glass vessels are frequently found. A combination of coloured trails and bichrome cables is predominantly used to decorate bulbous or pouch-shaped beakers (squat jars) and bowls, whereas gold-foil decoration is mostly found on funnels of (almost) colourless glass. Most fragments from Wijnaldum, however, belong to undecorated, plain wares and only a few are decorated with simple spiral trails of opaque yellow and white glass.

Contexts

Most fragments identified as belonging to glass vessels of the late Merovingian/Carolingian Period come from contexts dated to (the first half of) the 9th century. A fragment of a relatively wide base (28) probably comes from a palm funnel and is dated accordingly in the first half of the 8th century. A fragment of a rolled-in rim from a palm funnel or an early funnel (18) was recovered from a ditch dated between 800-900 and was probably in use sometime in the second half of the 8th century.

Two blue-green body fragments with strong vertical ribbing (41, 42) may have belonged to palm funnels and are dated by their contexts to respectively the second half of the 8th and the last quarter of the 8th/first half of the 9th century.

Fragments of vessels which have a trailed decoration and/or are deliberately coloured fragments were predominantly found in contexts dating to the first half of the 8th century: a rim fragment decorated with opaque yellow and white trails (30), two translucent turquoise body fragments with opaque white spiral trailing (46, 47) and a fragment of translucent red glass (38). A rim fragment from a bulbous beaker (squat jar) with a self-coloured spiral trail (31) found in a well dated between 800-900, and a body fragment with opaque white 'enamel-like' trailing (48) from a well dated to the first half of the 9th century may be residual and well have been in use during the first half of the 8th century.

Palm funnels/funnels

In the late 7th/early 8th century the Merovingian palm cup developed into a deeper shape with splayed sides. This form is christened 'palm funnel' because it represents a transitional form between the palm cup and the later funnel beaker. As Christianity gained influence in the northern part of the Frankish empire, palm funnels are amongst the latest grave goods which are represented in only a few graves. Some remarkable palm funnels originate in the Marne area and are richly decorated with mould-blown patterns, including illegible inscriptions and what seem to be Christian crosses surrounded by festoons.[54] On the lower Rhine palm funnels (*Glockentummler*) are found in graves belonging to the last phase in this region which is still characterized by the practice of giving grave goods (Lower Rhine phase 11, ca 705-740).[55] Palm funnels are occasionally found in weapon graves which make their appearance in the 8th century throughout Frisia. Heidinga suggests that these should be seen as a symbol of mental resistance of the newly established, Frankish order.[56] Fragmentary palm funnels, mostly rims and/or bases, are known from trade centres such as Dorestad (Wijk bij Duurstede),[57] Medemblik,[58] Hamwih (Southampton)[59] and Ribe (Denmark).[60]

Palm funnels tend to be of a rather good quality, predominantly bluish (green) or greenish blue glass. They mostly have rolled-in rims, often with a small cavity. Some have a deeply outward or inward folded rim or a simple, fire-thickened and slightly inturned rim. The latter also occurs on funnels which can make it difficult to distinguish both forms from each other; the body of palm funnels is more splayed though (swings in more) below the rim than is the case with funnels.[61] They display a wide variety of shapes and decoration.[62]

Palm funnels gradually evolved into elongated funnels with only slightly thickened, inturned rims and a narrow, tubular base.[63] The funnel *(Trichterbecher)* was to be the dominant glass drinking vessel of the 9th and 10th centuries. Funnels went out of fashion sometime in the late 10th century, probably because of changing drinking habits, since beakers without a true base are completely absent from the range of vessel forms in use during the following ages.[64]

Most funnels come from trade centres and ecclesiastical and aristocratic sites. The distribution of funnels ranges from the trade centres of Quentovic[65] in northern France and Dorestad[66] in the Netherlands to Haithabu in northern Germany,[67] Hamwih (Southampton) in Anglo-Saxon England[68] and Kaupang, Helgö and Birka in Scandinavia.[69] Funnels are also known from 9th-10th/11th-century contexts at Saint-Denis in northern France[70] and Liège (Place Saint-Lambert) in Belgium.[71]

Evidence for the production of funnels is, until now, only known from ecclesiastical and aristocratic sites in Germany: from a monastery at Augsburg (Benediktinerstift St. Ulrich und Afra),[72] the site of Cordel near Trier, which is believed to have been associated with a monastery,[73] and a royal palace (*Pfalz*) at Paderborn.[74]

Funnels are probably amongst the products of glass houses in the Rhine area and were exported as far afield as Borg on the Lofoten (northern Norway), where fragments of funnels were found in association with the latest phase (890-1030) of a chieftain's farm.[75] No examples of 9th-10th-century funnels from Frisia are known to the author.

Most of the Wijnaldum fragments from this period can confidently be identified as belonging to plain (palm-) funnels (19-25, 28-29). A fragment of a rolled-in rim (18) belongs to a palm funnel or an early funnel (Fig. 1). Only two fragments belonged to simply decorated funnels. One has a separately added dark blue rim (26; Fig. 1) for which there are parallels from Dorestad (8th/9th century)[76] and Birka (9th century).[77] A concave-curved body fragment decorated with an opaque yellow spiral trail (27) probably belonged to a funnel, although a bulbous beaker (squat jar) cannot be excluded (Fig. 19); opaque yellow spiral decoration on funnels has not until now been recorded outside Hamwih.[78]

Very thin-walled, delicate rim fragments of pale, greenish tinged glass (19, 21, 23) may have belonged to palm funnels with a ring just below the middle of the body (Fig. 1). An almost complete example with a fire-rounded, hardly thickened rim and tooled-out ring below the middle of the body comes from Dorestad and is very thinly blown of almost colourless glass with a greenish tinge and very small bubbles.[79] Palm funnels with tooled-out rings are probably also well represented at 8th-century Ribe (Denmark), where rim fragments of a similar quality are thought to be associated with body fragments of palm funnels with tooled rings.[80] The 9th-century context dates of the Wijnaldum rim fragments (21, 23), however, do not agree with an identification as palm funnels.

Bulbous and cylindrical/pouch-shaped beakers
Two rim fragments belong to vessels other than (palm-) funnels. One fragment is decorated with a half-marvered, self-coloured spiral trail with a thin dark thread in the centre causing a brown-yellowish discolouration (31) and probably belongs to a bulbous beaker or squat jar (Fig. 1). The second fragment is decorated with an opaque yellow spiral trail above the upper part of an opaque white arcaded trail (30; Fig. 1). The white arcaded trailing was probably quite similar to an intricate pattern of 'marvered opaque white trails combed into festoons' on a blue fragment from Hamwih which belonged to a small globular vessel or jar.[81] These beakers may have been decorated with bichrome cables, examples of which were found, for instance, in Dorestad and Birka,[82] but not a single vessel fragment decorated with bichrome cables has been recorded at Wijnaldum.

9 DISCUSSION AND CONCLUSIONS

The glass vessel assemblage from the Tjitsma terp at Wijnaldum represents a wide range of vessel forms ranging in date from the 1st to the 9th/10th centuries. In total, 194 fragments of vessel glass were recovered, a high proportion of which are undiagnostic (72%) due to the systematic sieving carried out using a sieve with a 3 mm mesh. The diagnostic fragments ($n = 55$) represent a minimum number of 37 glass vessels, 25 of which could be attributed to a specific vessel form or type. Although it was not possible to establish the vessel form of a further 12 vessels, the quality and colour of the metal and the decoration of the body fragments involved make their attribution to a specific period very likely.

None of the fragments of early-to-mid Roman vessel glass were recovered from Roman contexts; most were found in early-medieval contexts. It is possible that glass fragments were inadvertently mixed-in with scrap metal and other Roman *spolia* which reached the terp at Wijnaldum during the early medieval period. Fragments of glass vessels found on early-medieval 'barbarian' sites, e.g. in the late Roman Iron Age in Germania and Scandinavia, are often interpreted as cullet to be melted down and worked into glass beads, although the evidence is scanty.[83] There are certainly no indications that this is the case at Wijnaldum where only very few Roman glass vessel fragments were found. It is possible that a collared rim from a 1st/2nd-century Roman jar (2), recovered from a late-Merovingian context, was specially selected to be worn as a bead. There are parallels for the re-use of rim fragments of Roman and early-medieval glass vessels in necklaces from the Scandinavian world, Anglo-Saxon England and Germany.[84] In the Netherlands, an unpublished necklace from a grave in a Merovingian cemetery at Rijnsburg includes a fragment of a collared rim from a 1st century-AD Roman storage jar.[85]

Per period of habitation, all the most characteristic early-medieval glass vessel types, predominantly drinking vessels, are represented at Wijnaldum: cones (6-8) and bowls (9-13) of the Migration Period, bell beakers (14-15) and palm cups (16-17) of the Merovingian Period and (palm) funnels (18-29), bulbous and cylindrical/pouch-shaped beakers (30, 31) of the late Merovingian and Carolingian Periods.

The discontinuation or sharp decline in the population on the terp during the second half of the 4th and first quarter of the 5th century is well reflected by the Migration Period glass from Wijnaldum. Most fragments come from cones and bowls typical of the mid 5th to early 6th century. Only one fragment, a base of a conical beaker (5), can with certainty be ascribed to a late-Roman vessel type typical of the second half of the 4th and first half of the 5th century. Since it was recovered from a context dating to the second half of the 5th century, it probably represents the earliest glass vessel which reached the site before the mid 5th century.

If we are able to suggest that the estimated minimum number of glass vessels (EMN) reflects the ratios between the vessels in use at Wijnaldum during the different periods of habitation, and considering the length of each period of habitation, it seems possible to say that no glass vessels reached the site during the Roman Period, that they first reached the site during the Migration Period, probably starting around the mid 5th century and increasing in the course of the second half of the 5th/early 6th century, only to reach a low point during the (second half of the) 6th and 7th century, after which they picked up again in the 8th century reaching a 'peak' – relatively speaking – during the (first half of the) 9th century. This shows that during the entire early-medieval occupation of the terp, the site was continuously embedded in (an) exchange system(s) which enabled the inhabitants to obtain such relatively luxurious commodities as glass vessels.

The Migration Period glass from Wijnaldum includes fragments of at least one Kempston cone and one bowl decorated with an opaque white feather pattern, which are probably imports from the Rhineland. It has been suggested that the popularity of the large Kempston cones with warriors may simply mean that they were owned by those who could afford more beer or wine.[86] This may well be true, and rather than associating Kempston cones with military aspects of 5th-century society on the basis of finds in warrior graves, it seems more likely that they formed part of a package of commodities symbolic of an elite life style. From their (presumed) production centres in the Middle Rhine region, these products found their way up north into areas peripheral to the Frankish empire, as far as Scandinavia; it seems the greater the distance from the source, the higher the associated value.

It is known from Anglo-Saxon sources that cut horns were the ceremonial drinking vessel for those of high status all through the Anglo-Saxon Period[87] and they also feature prominently in Viking Age epic poems and iconography.[88] It seems likely that the glass horn, a rare elite drinking vessel, was a costly imitation in glass of the Germanic drinking horn. Large Kempston cones which have the same content as glass horns and are associated with high status, warrior burials may represent the 'next step down' in the hierarchy of glass drinking vessels. The technical skills involved to produce the regular pattern of spirals and elongated loops on the thinly blown, tall Kempston cones undoubtedly added to the perceived value of these vessels. Smaller cones characterized by a simply executed, mould-blown ribbed decoration were produced in large series and perhaps represent a popular, 'watered-down' version of these elite drinking vessels.

Viking Age sagas reveal the 'down-in one ritual' of drinking without pausing for breath; it is likely that this ritual is reflected in the morphology of early-medieval drinking vessels which lack a base and presumably had to be drunk empty before being put down.[89] The obvious close relationship between alcohol consumption and martial behaviour has been pointed out by many. Drinking ceremonies also helped create and maintain clientship through drink; large quantities of wine enabled a king to maintain vast numbers of retainers. Such ceremonies may well have been held on a terp such as Wijnaldum, especially since the *Beowulf* mentions the existence of a drinking-hall of the (presumably) Frisian King Finn.[90]

The most prestigious drink which features in the Anglo-Saxon and Viking epic poems is wine, followed by mead, cider and beer, and these are the probably drinks to be associated with glass drinking vessels. Ale, on the other hand, was the everyday drink of common people and was probably mostly drunk from wooden and ceramic beakers. The prominent place of alcoholic drinks in Germanic society is almost a cliche and, as such, was probably often intentionally exaggerated by religious leaders/missionaries. Anglo-Saxon sources often refer to excessive drinking and intoxication and regularly mention warriors who carry on drinking till they fall on the floor, drunkenness during marriages and even during funerals.[91] It is likely that such behaviour was also experienced by the Anglo-Saxon missionary Boniface in Frisia who saw drunkenness as being peculiar to the heathen and the English: 'For neither the Franks, nor the Gauls, nor the Lombards, nor the Romans, nor the Greeks have it.'[92]

The following Merovingian Period shows a low point in glass imports. This seems contradictory because habitation Period IV (550-650) is characterized by large numbers of imported, wheel-thrown pottery (64%), almost exclusively *Wölbwandtöpfe* from Mayen.[93] The area around Mayen is one of the probable candidates for the continuation of glass production from the (late) Roman into the early medieval period[94] and it is also a possible origin for the Merovingian glass vessels from Wijnaldum. The argument of trade routes to Wijnaldum being temporarily hampered does not, therefore, seem to be a valid assumption: the glass vessels could easily have arrived through the same channels as the imported pottery from Mayen.

The observation that the numbers of glass vessels fell in Merovingian graves has already been discussed (see above) and it may be that excavations of Merovingian settlements will provide a different picture. However, a very similar pattern to that in Wijnaldum is observed in the early-medieval settlement at Valkenburg-de Woerd (Prov. of South-Holland) located on a gully of the Old Rhine (unpublished data). During the Merovingian Period, the settlement probably served as a riparian market and was well connected to the hinterland and presumed centres of glass production at the time. The same can be said for several settlements on Jutland (Denmark) which have yielded fairly large quantities of vessel glass of the 5th and first half of the 6th centuries; for the following 150 years or so nothing much is reported and only in the early 8th century do large numbers of fragmentary vessel glass appear again in trade centres such as Ribe.

If early-medieval production did rely heavily on recycling, resulting in low quality vessels such as seen in many palm cups, then a decline in glass production may indeed have taken place during the Merovingian Period; this being a time when the availability of imported raw materials – mineral soda from Egypt – was restricted and experimentation with the exploitation of new, local raw materials (potash) had not yet or only just begun.[95] For the time being, the small proportion of 6th-7th-century vessel glass at Wijnaldum is seen as a reflection of the level of Merovingian glass production, rather than of the status of the site itself. The special status of Wijnaldum may also be confirmed by the fact that glass beads were being produced at the site sometime in the late 6th/early 7th century.[96]

It is remarkable that the 5th-7th-century vessels from Wijnaldum are all 'Frankish' imports. There are no fragments of glass vessels which are considered typical products of Kentish glass workshops and which are also found in Scandinavia.[97] As far as the glass vessels are concerned, the excavations at Wijnaldum have not been able to substantiate the idea that Frisia played an intermediary role between Scandinavia and Anglo-Saxon England.

The way in which glass vessels reached the terp during the 5th-7th centuries can only be hypothesized, but commercial trade by way of the most important waterways certainly seems a possibility. In Frisia there is evidence for the minting of coins as early as ca 600 and in Dorestad in the first half of the 7th century. This further underlines the economic importance of coins in early-medieval society; a role which has long been underestimated or even entirely negated.[98]

The glass vessels could have reached Wijnaldum together with imported pottery from Mayen which was imported to the site during the 5th-7th centuries (in small numbers during habitation Period III (425-475) and in large numbers during habitation Period IV (550-650). Ships with a cargo of pottery and glass from the Middle Rhine region may have stopped over on the Frisian coast on their way to Jutland, where fairly large quantities of 5th/early 6th-century glass were found in 'trade-houses' (*Handelshuset/Glashus*) in Dankirke and Dejberg.

The distribution of relatively luxurious goods such as glass vessels does not, of course, necessarily indicate organized trade; their distribution can also be the result of other exchange mechanisms such as ceremonial gift exchange, dowry tradition, plunder, or the occasional visit of a wandering trader.

Known settlement find contexts for 8th-10th-century vessel glass include aristocratic and ecclesiastical sites and large trade centres. The terp at Wijnaldum therefore offers a rare and interesting insight in the glass assemblage used at the time in an entirely different type of settlement. In this respect it is interesting to see that no luxury glasses are present amongst the Carolingian glass. Most fragments of this period belong to plain (palm-) funnels; only one funnel had a separately added blue rim and another one was decorated with a simple opaque yellow spiral.

There is no evidence for the use of vessels decorated with bichrome cables, although it cannot be excluded that a bulbous and a cylindrical/pouch-shaped beaker from Wijnaldum were decorated with bichrome cables. Nor is there any evidence for the presence of vessels decorated with gold foil at Wijnaldum. Fragments of gold foil decorated vessels are rare and Dorestad, where fragments of five specimens were found, has one of the highest concentrations.[99] Based on their find distribution, gold foil decorated vessels may have been produced in Germany, possibly at Paderborn, and were exported as far north as Scandinavia. Chemical analysis of a colourless funnel beaker with gold-foil decoration found at a chieftain's farm at Borg on the Lofoten (northern Norway) has pointed out that – in contrast to plain greenish funnels – especially pure raw materials had been selected for its production, which undoubtedly added to the value of the beaker.[100] The late-Merovingian/Carolingian glass assemblage from Wijnaldum is again very similar to that of the earlier mentioned settlement at Valkenburg-De Woerd, which at this time probably functioned as a minor trading place in the catchment area of Dorestad.[101] This suggests that the expansion of the markets in the Carolingian Period not only included the production of glass vessels – some of which are highly decorated – for the elite: the same development probably also made plain vessel glass more widely available to 'commoners' than was previously the case.

Whereas the elongated forms of cones and bell beakers suggest that their principal function may have been for beer drinking, palm funnels and funnels are generally considered to be wine glasses, something which is suggested by their wide, open (splayed) shape. With the rise of large trade centres and the increase in long-distance trade, wine was increasingly becoming a more widely available commodity. The presence of one barrel of silver pine and another one of oak (from the Danube and Middle-Rhine regions resp.), both used on the terp in the 9th century to line wells, is evidence that wine was imported to the site at that time.[102] Wine was probably taken directly from the barrel into earthenware pitchers or jugs and drunk from wooden, ceramic and glass drinking vessels in the provincial-Roman tradition.[103] A probable drinking set of the 9th century consists of Tating ware jugs and glass funnel beakers, both of which are represented at Wijnaldum. It has also been pointed out that the geometric designs of the tin-foil decoration on Tating ware jugs are very similar to those of the gold-foil decoration on funnel beakers.[104]

The 'peak' in the number of glass vessels in use on the terp in the 1st half of the 9th century matches well with the fact that the settlement was most densely populated during habitation period VII. It is very likely that glass vessels were traded from the Rhineland to the terp by way of Dorestad. Although luxury glasses were available from Dorestad, these do not seem to have been in use on the terp. Thus the quality of the glass vessel assemblage of the late Merovingian and Carolingian Periods suggests that by this time the heyday of the terp as an 'elite centre' was over.

CATALOGUE OF VESSEL GLASS FROM WIJNALDUM

FindNo = find number; ContNo = context number; Cont = nature of context (di = ditch; cl = cultivation layer; os = occupation surface; ph = post hole; sf = stray find; sh = sunken hut; sl = sod layer; trl = truncated layer; ts = top soil; wa = waste deposit); DatA/DatB = earliest/latest possible date of context; Rel = reliability of date (1 = very reliable, 2 = unreliable, 3 = under certain conditions reliable, 4 = reliable); H/L = maximum height of rim and body fragments/maximum length of base fragments (cm); Diam = diameter of the fragment (cm); Thri = thickness of the rim (cm); Thbo = thickness of the body (cm); Thba = thickness of the base (cm); (ContNo)rel = related context number, when context uncertain.

I GLASS OF THE ROMAN PERIOD

FindNo	ContNo	Cont	DatA	DatB	Rel	H/L	Diam	Thri	Thbo	Thba

Tablewares

1. Fragment of a flat base with what seems to be part of a low base-ring; wheel-polished on the interior and exterior. Most likely of an early-Roman cast vessel, a dish or plate. Translucent pale blue-green glass with some bubbles, iridescent all over.

FindNo	ContNo	Cont	DatA	DatB	Rel	H/L	Diam	Thri	Thbo	Thba
2379	44	ph	200	300	1	3.58	–	–	–	0.43

Containers

2. Fragment of an outward folded cavity rim. Most likely of an oval/bulbous jar of Isings type 67 b/c or possibly of a square jar of Isings type 62. Translucent blue-green glass with many horizontally elongated bubbles (Fig. 1).

FindNo	ContNo	Cont	DatA	DatB	Rel	H/L	Diam	Thri	Thbo	Thba
9025	159	os	650	750	4	1.28	7.50	0.48	–	–

3. Small body fragment with two sides at right angles to each other. Most likely of a prismatic bottle, probably a square bottle of Isings type 50. Translucent, bright blue-green glass, saturated with small bubbles.

FindNo	ContNo	Cont	DatA	DatB	Rel	H/L	Diam	Thri	Thbo	Thba
11184	1220	di	650	750	3	1.53	–	–	0.35	–

II GLASS OF THE LATE ROMAN/MIGRATION PERIOD

FindNo	ContNo	Cont	DatA	DatB	Rel	H/L	Diam	Thri	Thbo	Thba

Cup or beaker

4. Small fragment of a tubular base-ring. Translucent yellow-green glass with many bubbles. Probably of a hemispherical cup on a base-ring of Isings form 108 or of a conical beaker on a base-ring of Isings form 109; a later form, viz. footed beaker or claw beaker, cannot be excluded.

11060	1471	trl/wa	450	500	1	0.74	–	–	–	–

Beakers

5. Complete, pushed-in base with part of lower body; almost straight side sloping in steeply towards the base. The absence of a pontil mark on the base implies an unworked rim. Probably of a conical beaker with knocked-off rim of Isings type 106b. Translucent yellowish green glass with many small bubbles, some whitish weathering.

5208	801	di	450	500	1	1.30	2.00	–	0.11	0.15

6. Small rim fragment; edge fire-rounded and fire-thickened, curved-out rim decorated with a self-coloured spiral trail of which two half-marvered turns are preserved. Most likely belonging to a cone beaker with a spayed rim, type undiagnostic. Translucent yellowish olive glass with a few small horizontally elongated bubbles (Fig. 1).

1	–	ts	–	–	–	1.34	8.00	0.40	–	–

7. A concave-curved upper body fragment decorated with a self-coloured spiral trail of which three half-marvered turns are preserved; broken off below a splayed rim; a straight lower body fragment with three applied vertical trails, two of which are slightly converging. Probably belonging to a large Kempston cone, decorated with vertical loops of Koch type III N. Translucent pale, bluish green glass with small elongated bubbles; slightly iridescent.

9782	2351	os	450	500	1	0.84	–	–	0.07	–
9782	2351	os	450	500	1	0.80	–	–	0.05	–

8. Three small body fragments: a concave-curved, upper body fragment from below a splayed rim shows two turns of an applied, self-coloured spiral trail; two straight body fragments shows two half-marvered turns of the same self-coloured spiral trail. Possibly belonging to the same Kempston cone as cat no. 7. Translucent pale, slightly bluish green glass with small elongated bubbles.

9897	2355	sh	450	500	1	0.64	–	–	0.05	–
9897	2355	sh	450	500	1	1.04	–	–	0.06	–
9897	2355	sh	450	500	1	0.5	–	–	0.05	–

Bowls

9. Rim fragment; edge fire-rounded and fire-thickened, slightly inturned rim, decorated with a marvered opaque white feather-pattern, of a bowl of Koch type IV L. Translucent bluish green glass with many horizontally elongated bubbles, slight whitish weathering (Fig. 1).

6794	1346	wa?	450	500	1	1.55	14.0	0.22	0.09	–

10. Very small rim fragment; edge fire-rounded and fire-thickened, slightly inturned rim, with one turn of a marvered, opaque white, fine spiral trail. Belonging to a bowl with an opaque white decoration, type undiagnostic. Translucent, pale bluish green glass, dully weathered.

10802	2565	os/wa	600	700	3	0.65	–	0.21	–	–

11. Fragment of lower body with slightly convex-curved side; decorated with seven broadly marvered, opaque white threads, combed into a feather pattern. Very likely of a bowl of Koch type IV L. Translucent bluish-green, iridescent, glass with many small bubbles.(Fig. 1).

9902	2353	os	450	500	1	1.77	8.50	–	0.12	–

12. Fragment of lower body with slightly convex-curved side decorated with three turns of a marvered, opaque white, fine spiral trail and a broadly marvered opaque white thread, combed into an arcade or feather pattern. Most likely belonging to a bowl decorated with an opaque white arcade pattern of Koch type IV K. Translucent bluish green, slightly iridescent glass with many small bubbles.

6993	962	cl	450	500	1	1.90		–	0.19	–

13. Small rim fragment; edge fire-thickened and fire-rounded, slightly inturned rim. Translucent pale-green glass with small bubbles; slightly molten. Probably of an undecorated bowl.

7448	1233	di	550	600	3	1.53	–	0.30	0.19	–

III GLASS FROM THE MEROVINGIAN PERIOD

Bell beakers

14. Small rim fragment; edge fire-rounded and fire-thickened, straight rim. Four small body fragments with a delicate obliquely ribbed pattern. Twelve very small undecorated body fragments (splinters: no measurements). Most likely of a small, optic blown bell beaker. Translucent pale greenish olive glass with pinprick bubbles, dully weathered (Fig. 1).

FindNo	ContNo	Cont	DatA	DatB	DatC	H/L	Diam	Thri	Thbo	Thba
7448	1233	di	550	600	3	1.72	5.00	0.26	0.11	–
7448	1233	di	550	600	3	1.40	–	–	0.10	–
7448	1233	di	550	600	3	1.12	–	–	0.08	–
6704	1233	di	550	600	3	1.50	–	–	0.09	–
6704	1233	di	550	600	3	1.44	–	–	0.08	–

15. Body fragment with concave side (curving in two planes), decorated with a faint vertically ribbed pattern. Belonging to an optic blown bell beaker. Translucent yellowish green, slightly iridescent glass with many vertically elongated bubbles (Fig. 1).

FindNo	ContNo	Cont	DatA	DatB	DatC	H/L	Diam	Thri	Thbo	Thba
6256	1287	trl/sl	550	600/650	1	4.10	7.50	–	–	0.12

Palm cups

16. Lower body fragment; convex-curved, with three remaining, pronounced, vertical ribs, interconnected by 'shoulders'. Belonging to a palm cup. Translucent, slightly iridescent, yellowish brown glass with darker brown streaks and many small bubbles (Fig. 1).

FindNo	ContNo	Cont	DatA	DatB	DatC	H/L	Diam	Thri	Thbo	Thba
6210	1287	trl/sl	550	600/650	1	2.16	6.50	–	0.11	0.32

17. Base fragment; irregular, thick-walled, rounded (convex-curved) base. Most likely belonging to a palm cup. Translucent, iridescent, blue-green glass with dark streaks and numerous small bubbles (Fig. 1)0.

FindNo	ContNo	Cont	DatA	DatB	DatC	H/L	Diam	Thri	Thbo	Thba
10142	2219	sl	650	750	4	2.95	3.50	–	0.21	0.51

IV GLASS OF THE LATE MEROVINGIAN/CAROLINGIAN PERIOD

(Palm) funnels

18. Rim fragment; edge rolled-in without a cavity and slightly splayed. Most likely of a palm funnel. Translucent blue-green glass with dark inclusions and many horizontally elongated bubbles; some whitish weathering (Fig. 1).

FindNo	ContNo	Cont	DatA	DatB	Rel	H/L	Diam	Thri	Thbo	Thba
2020	3539rel	di	800	900	4	1.55	8.50	0.38	0.10	–

19. Rim fragment; edge fire-rounded and fire-thickened, slightly inturned, splayed rim of a funnel. Translucent, pale green glass with pinprick bubbles; slight whitish weathering.

FindNo	ContNo	Cont	DatA	DatB	Rel	H/L	Diam	Thri	Thbo	Thba
1	–	ts	–	–	–	1.83	13.00	0.23	0.08	–

20. Rim fragment; edge fire-rounded and slightly fire-thickened, slightly inturned, splayed rim rim of a funnel. Translucent blue-green glass with many small and large, horizontally elongated bubbles; some whitish weathering (Fig. 1).

FindNo	ContNo	Cont	DatA	DatB	Rel	H/L	Diam	Thri	Thbo	Thba
3695	2605	trl/wa	800	850	1	2.10	11.50	0.24	0.10	– –

21. Rim fragment; edge fire-rounded and hardly thickened, rather straight rim (faint ridge below edge of exterior rim). Most likely of a funnel. Translucent pale yellowish green, slightly iridescent glass with numerous horizontally elongated, small and large bubbles (Fig. 1).

FindNo	ContNo	Cont	DatA	DatB	Rel	H/L	Diam	Thri	Thbo	Thba
7359	1636	trl/wa	800	900	1	2.40	11.00	0.23	0.18	–

22. Rim fragment; edge fire-rounded and slightly thickened, inturned, splayed rim of a funnel. Translucent yellowish green glass with many horizontally elongated bubbles, slight whitish weathering on the inside (Fig. 1).

FindNo	ContNo	Cont	DatA	DatB	Rel	H/L	Diam	Thri	Thbo	Thba
7507	3296	trl/os	775	850	1	1.35	10.50	0.27	0.11	–

23. Rim fragment, edge fire-rounded and hardly fire-thickened, slightly inturned, splayed rim of a funnel. Translucent, almost colourless glass with a greenish tinge and and many pinprick bubbles, slight whitish weathering on the inside (Fig. 1).

FindNo	ContNo	Cont	DatA	DatB	Rel	H/L	Diam	Thri	Thbo	Thba
7877	3358	trl/os	800	850	3	1.30	10.00	0.17	0.09	–

24. Tiny rim fragment; edge fire-rounded and slightly fire-thickened, slightly inturned rim. Probably of a funnel. Translucent blue-green glass with horizontally elongated bubbles.

FindNo	ContNo	Cont	DatA	DatB	Rel	H/L	Diam	Thri	Thbo	Thba
7960	3324	trl/wa	800	850	3	0.50	–	0.20	–	–

25. Rim fragment, edge fire-rounded and slightly fire-thickened, slightly inturned and slightly splayed rim of a funnel. Translucent blue-green glass with many small and larger, horizontally elongated bubbles; some whitish weathering (Fig. 1).
11871 2957 well12 850 900 3 2.00 10.50 0.22 0.10 –

26. Small fragment of a separately added rim with small part of upper body; edge fire-rounded and slightly thickened, slightly inturned rim. The separately added rim of translucent dark blue glass with small, horizontally elongated bubbles is broken off at the point of fusion to an almost colourless vessel, most likely a funnel. The glass is dully weathered (Fig. 1).
10906 340rel os 770 900 1 1.25 – 0.15 – –

27. Body fragment, slightly concave-curved, decorated with six half-marvered, irregular turns of an opaque yellow spiral trail. Translucent deep blue-green, slightly iridescent glass with many small bubbles. Most likely belonging to a funnel (Fig. 1).
1423 – stray – – – 3.15 10.00 – 0.10 –

28. Body fragment with almost straight side, most likely of the relatively wide, lower end of a palm-funnel. Translucent blue-green glass with many vertically elongated bubbles.
1543 1166 trl/sl 700 750 2 2.21 3.00 – 0.10 –

29. Small body fragment, most likely of the narrow, tube-like lower end of a funnel. Translucent, slightly iridescent blue-green glass with two dark vertical streaks and many small, vertically elongated bubbles.
7510 3277 trl/os 800 850 1 0.90 – – 0.25 – –

Beakers (bulbous/pouch-shaped/cylindrical)

30. Rim fragment, edge fire-rounded and fire-thickened, slightly inturned, everted rim; straight-sided upper body. The rim is decorated with ten turns of a marvered, opaque yellow spiral trail; below the spiral trail the upper part of a marvered opaque white trail remains, which was probably part of a combed, arcaded pattern. Possibly of a pouch-shaped or a cylindrical beaker. Translucent blue-green glass with numerous small bubbles, some dark inclusions; dark and red streaks in the metal are visible between the opaque yellow and white decoration (Fig. 1).
1526 1114 sl 700 750 1 2.31 11.00 0.42 0.19 –

31. Rim fragment; edge fire-rounded and thickened on the outside (folded out?) and smoothed on the inside; straight-sided neck beginning to curve out. The fragment preserves five irregular turns of a half-marvered, self-coloured spiral trail, which has a very thin dark thread in the centre causing a yellowish-brown (dis-) colouration. Probably belonging to a bulbous beaker (squat jar). Translucent blue-green glass with horizontally elongated bubbles and pale inclusions, slight whitish weathering (Fig. 1).
5359 997 well7 800 900 3 2.81 6.50 0.39 0.18 –

V VESSEL FORM UNCERTAIN

FindNo ContNo Cont DatA DatB Rel H/L Diam Thri Thbo Thba

Undecorated

32. Four small, rather thickwalled fragments of perfectly colourless glass with some pinprick bubbles and internal cracks. Most likely of a Roman date.
1478 621 os/wa 575 625 1 0.73 – – 0.27 –

33. Four small, rather thickwalled fragments of perfectly colourless glass with some pinprick bubbles and internal cracks. Most likely of a Roman date.
5155 10 trl 300 350 1 0.73 – – 0.28 –

34. Small, perfectly colourless, thickwalled fragment, with pinprick bubbles and internal cracks. Most likely of a Roman date.
5995 997 well7 800 900 3 0.81 – – 0.43 –

35. Base fragment; thickwalled, pushed-in, concave base. Translucent blue-green glass with few pinprick bubbles, partially molten. Probably of a Roman date.
5296 ? ? – – – 4.16 – – – 0.57

36. Body fragment, lower body with convex-curved side. Translucent blue-green glass containing numerous bubbles, pale inclusions and 'flamed' red streaks. Most likely of a palm cup.
10786 514 di 750 770 3 2.18 – – 0.22 –

37. Small body fragment, concave-curved. Translucent olive brown glass with darker brown streaks and small, vertically elongated bubbles. Most likely of a Merovingian date, probably of a bell beaker.
7600 1328 di 500 550 1 1.62 – – 0.08 –

38. Tiny thickwalled fragment with internal cracks. Translucent 'wine' red glass.
1526 1144 sl 700 750 1 0.90 – – 0.29 –

Decorated
39. Small body fragment, straight with faint, narrow, diagonal ribs; broken off below everted rim. Translucent, pale blue-green glass. Probably belonging to a bell beaker or possibly a cone.
3391 1206 ph? 550 650 1 1.50 – – 0.15 –

40. Small body fragment, straight side with faint, narrow, diagonal ribs. Translucent, yellow-green glass. Possibly belonging to a cone.
9723 139 cl 450 525 3 1.12 – – 0.10 – –

41. Small body fragment, almost straight side with two, pronounced, broad, vertical ribs. Translucent blue-green glass. Probably of an optic blown palm funnel.
10793 2550 di 750 800 1 1.05 – – 0.13 –

42. Small body fragment, almost straight side with one broad, vertical rib. Most likely of an optic blown palm funnel. Translucent blue-green glass with wavy dark streaks and one straight dark 'thread' and many small bubbles. From the same context as cat no. 22.
7507 3296 trl/os 775 850 1 0000 – – 0000 –

43. Body fragment, slightly convex-curved side decorated with a broadly marvered, curved, opaque white thread. Translucent bluish green glass. Possibly of a bowl decorated with an opaque white arcade or feather pattern of Koch bowl type IV K or L.
5928 – sf? 450 750 2? 1.62 – – 0.16 –

44. Small body fragment, slightly convex-curved side with four regularly interspaced, broadly marvered, opaque white threads, probably forming a feather pattern. Most likely belonging to a bowl of Koch type IV L. Translucent blue, slightly iridescent glass with many small elongated bubbles.
5182 821 sl? 450 500 1 1.53 – – 0.15 –

45. Small body fragment, slightly convex-curved side with three regularly interspaced, marvered, broad, opaque white threads, probably forming a feather pattern. Most likely belonging to a bowl of Koch type IV L. Translucent blue glass with many small elongated bubbles; slightly iridescent.
6430 861 trl/sl 450 500 1 1.03 – – 0.15 –

46. Tiny body fragment with two turns of an applied thin opaque white spiral trail. Translucent turqoise glass with small bubbles.
1526 1144 sl 700 750 1 0.42 – – 0.12 – –

47. Small body fragment, slightly convex-curved with two turns of an applied thin opaque white spiral trail. Translucent turquoise glass with many small and big, round bubbles.
10074 2219 sl 650 750 4 1.43 – – 0.12 –

48. Small body fragment, concave-curved with six turns of a half-marvered, opaque white spiral trail which is shiny ('enamel-like') in appearance; very thin dark 'threads', which are incorporated in the metal, are visible at the back of the opaque white trails. Translucent bluish green glass with small bubbbles. Probably belonging to jar or possibly to a funnel.
11518 2587 well14 800 850 1 1.25 – – 0.17 –

VI UNDIAGNOSTIC, (DE-)COLOURED/DECORATED

FindNo ContNo Cont DatA DatB Rel H/L Diam Thri Thbo Thba

Undecorated, (de-)coloured
49. Small fragment of perfectly colourless glass; heat-affected.
2809 625 met/wa 575 625 3 1.13 – – 0.10 –

50. Small convex-curved body fragment. Translucent medium blue glass; strong white weathering.
1254 373 os 700 750 3 1.40 – – 0.17 –

51. Small curved convex-curved body fragment. Translucent blue glass with elongated bubbles.
5405 512 pit 650 700 3 1.34 – – 0.15 –

52. Small, almost straight, body fragment. Translucent purple glass with many small elongated bubbles.
9902 2353 os 450 500 1 1.26 – – 0.09 –

Decorated
53. Tiny body fragment with one self-coloured, thick vertical thread (vertical loop ?). Translucent, pale yellowish green glass.
5300 793 cl 500 550 1 0.87 – – 0.16 –

54. Small body fragment with two turns of a self-coloured spiral trail. Translucent, pale bluish green glass.
5812 2098 os/wa 450 000 1 0.72 – – 0.06 –

55. Tiny body fragment with one self-coloured, thin thread. Translucent, pale yellowish green glass.
5865 793 cl 500 550 1 0.76 – – 0.06 –

56. Small body fragment with one half-marvered, self-coloured, thick vertical thread (vertical loop?). Translucent, pale yellowish green glass.
7545 1328 di 500 550 1 0.98 – – 0.12 –

57. Small, straight body fragment with two self-coloured vertical trails (vertical loop ?). Translucent, pale bluish green glass.
9608 213 pit 425 550 3 0.79 – – 0.08 –

58. Small body fragment with three, of which two diverging, half-marvered, self-coloured vertical trails (loops ?). Translucent, pale bluish green glass.
9781 0.58 – – 0.08 – – – –

59. Small body fragment with one self-coloured vertical trail. Translucent, yellow-green glass.
9737 2341 os/sl 450 500 1 0.88 – – 0.13 –

60. Small body fragment, convex-curved, with a half-marvered, self-coloured, vertical thread. Translucent, pale yellowish green glass with some small bubbles. From the same context as catno. 8. Possibly belonging to a bowl decorated with vertical loops of Koch type IV M.
9897 2355 sh 450 500 1 1.08 – – 0.08 – –

61. Body fragment, concave-curved, broken off below everted rim, with the 'negative' of three turns of a weathered opaque white spiral trail. Possibly of a cone or bell beaker.
5095 2222r sl 650 700 4 1.50 – – 0.08 –

62. Small body fragment with three regular turns of an opaque white spiral trail, which is not completely marvered into the surface and is shiny ('enamel-like') in appearance. Translucent, bright blue glass with small bubbles.
7448 1233 di 550 600 3 0.96 – – 0.17 –

VII UNDIAGNOSTIC, CREMATIONS

FindNo ContNo Cont DatA DatB Rel H/L Diam Thri Thbo Thba

63. Large, molten fragment of bluish green glass, probably of a vessel with a (tubular) base-ring.
4570 448 crem 425 450 1

64. One molten and one almost intact fragment of perfectly colourless glass, slightly iridescent. Possibly of a thickwalled (late Roman) vessel.
4570 448 crem 425 450 1

65. Fragment of molten, bluish-green vessel glass; fragments of calcinated bone incorporated.
6502 3067 os 775 850 3

REFERENCES

ALÉNUS-LECERF, J. 1993. Le verre mérovingien dans les tombes de Wallonie. In: *Trésors de Wallonie. Les verres mérovingiens.* (Exhibition catalogue, Musée Ourthe-Amblève, Comblain-au-Pont 1 juillet-31 0ctobre 1993), 27-50.

ALÉNUS-LECERF, J. 1995. Contribution à l'étude des verres provenant des tombes mérovingiennes de Belqique. In:

D. Foy (ed.), *Le Verre de l'Antiquite Tardive et du Haut Moyen Age. Typologie – Chronologie – Diffusion*, 57-83.

AMENT, H. 1976. *Die fränkischen Grabfunde aus Mayen und der Pellenz*, Berlin (Germanische Denkmäler der Völkerwanderungszeit B9).

ARBMAN, H. 1937. Schweden und das Karolingische Reich, Stockholm.

CHAMBON, R. & H. ARBMAN 1952. Deux fours à verre d'époque mérovingienne à Macquenoise (Belgique), *Meddelanden fran Lunds Universitets Historiska Museum*, 199-232.

ARBMAN, H. 1967. *Birka, 1: Die Gräber*, Uppsala 1940-1943 (2 vol.). Baume, P. la, *Das fränkische Gräberfeld von Junkersdorf bei Köln*, Berlin.

BAUMGARTNER, E. & I. KRÜGER 1988. Phoenix aus Sand und Asche. Glas des Mittelalters, München.

BAYLEY, J. 1991. Le verrerie en Angleterre pendant l'époque anglo-saxonne. In: D. Foy & G. Sennequier (eds), *Ateliers de verriers de l'antiquité à la période pré-industrielle*, Rouen, 31-34.

BESTEMAN, J.C. 1974. Carolingian Medemblik, Berichten van de Rijksdienst voor Oudheidkundig Bodemonderzoek 24: 43-106.

Bloemers, J.H.F., L.P. Louwe Kooijmans & H. Sarfatij 1986. *Verleden Land*, Amsterdam.

BÖNER, K. 1958. Die fränkischen Altertümer des Trierer Landes, Berlin.

CABART, H. 1993. Les gobelets mérovingiens à inscriptions du Musée de Châlons-sur-Marne (France), *Annales du 12è Congrès de l'Association Internationale pour l'Histoire de Verre*, 225-234.

CABART, H. & J.-Y. FEYEUX 1995. *Verres de Champagne. Le verre à l'époque mérovingienne en Champagne-Ardenne*, Reims (Bulletin de la Société Archéologie Champenoise).

COOL, H.E.M. & J. PRICE 1995. *Roman vessel glass from excavations in Colchester, 1971-1985*, Colchester (Colchester Archaeological Report 8).

DÉKOWNA, M. 1976. Glasbecher. In: M. Müller-Wille (ed.), *Das Bootkammergrab von Haithabu*, Neumünster, 63-66 (Berichte über die Ausgrabungen in Haithabu Bericht 8).

ELLMERS, D. 1964/1965. Zum trinkgeschirr der Wikingerzeit, *Offa* 21/22: 21-43.

ERDRICH, M. & H.-U.VOSS 1997. Die Perlen der Germanen des 1.-5. Jahrhunderst in Mecklenburg-Vorpommern, Schleswig-Holstein und Niedersachsen. In: U. von Freeden & A. Wieczorek (eds), *Perlen: Archäologie, Techniken, Analysen*, 77-94 (Römisch Germanische Kommission Kolloquien zur Vor- und Frühgeschichten, Band 1).

ES, W.A. VAN 1967. Wijster: A native village beyond the imperial frontier 150-425 AD (Palaeohistoria 11).

ES, W.A. VAN 1990. Dorestad centred. In: J.C. Besteman, J.M. Bos & H.A. Heidinga (eds), *Medieval Archaeology in the Netherlands, Studies presented to H.H. van Regteren Altena*, Assen/Maastricht, 151-182.

EVISON, V. 1972. Glass cone beakers of the 'Kempston' type, *Journal of Glass Studies* 14: 48-64.

EVISON, V. 1982. Bichrome glass vessels of the seventh and eighth centuries, *Studien zur Sachsenforschung* 3: 7-21.

EVISON, V. 1983. Some distinctive glass vessels of the post-Roman period, *Journal of Glass Studies* 25: 87-93.

EVISON, V. 1988. Vieux-Marché, Place Saint-Lambert, Liège – The glass. In: M. Otte (ed.), *Les fouilles de la Place Saint-Lambert à Liège 2. Le vieux marché*, 215-219 (Etudes et recherches Archéologiques de l'Université de Liège).

EVISON, V. 1989. Le verre carolingien. In: D. Foy & G. Sennequier (eds), *A travers le verre du moyen âge á la renaissance*, 137-144 (Exhibition catalogue, le Musée des Antiquitées de Seine-Maritime à Rouen).

FEYEUX, J.-Y. 1995. La typologie de la verrerie mérovingienne du nord de la France. In: D. Foy (ed.), *Le Verre de l'Antiquite Tardive et du Haut Moyen Age. Typologie – Chronologie – Diffusion*, 109-138.

FREMERSDORF, F. 1941. Zwei germanische Grabfunde des frühen 5. Jahrhunderts aus Köln, *Germania* 25: 180-188.

GIFFEN, A.E. VAN 1953. De Romeinse castella in de dorpsheuvel te Valkenburg aan de Rijn (Z.H.). Praetorium Agrippinae II: de opgravingen in 1942-43 en 1946-50, *Jaarverslag voor de Vereniging voor Terpenonderzoek* 33-37: 1-209.

GLAZEMA, P. & J. YPEY 1956. *Merovingische ambachtskunst.*

HABEREY, W. 1942. Spätantike Gläser aus Gräbern von Mayen, *Bonner Jahrbücher* 147: 249-284.

HAEVERNICK, T.E. 1972. Zu Trierer Glasfragmenten, *Trierer Zeitschrift* 35: 215-217.

HAGEN, A. 1995. A Second Handbook on Anglo-Saxon Food & Drink. Production and Distribution.

HANSEN, T.E. 1993. Glashus. *Skalk* 5: 3-7.

HEIDINGA, H.A. 1997. *Frisia in the first millennium. An Outline*, Utrecht.

HENDERSON, J. 1985. The raw materials of early glass production, *Oxford Journal of Archaeology* 4(3): 267-291.

HENDERSON, J. 1993. Aspects of Early Medieval glass production in Britain. In: *Annales du 12e Congrès de l'Association Internationale pour l'Histoire de Verre*, 247-260.

HENDERSON, J. & I. HOLAND 1992. The Glass from Borg: an Early Medieval Chieftain's Farm in Northern Norway, *Medieval Archaeology* 36: 29-58.

Henricson, L.G. 1995. Broken glass beakers re-used as glass beads. In: M.S. Rasmussen, U. Lund Hansen, U. Näsman (eds), *Glass beads. Cultural History, Technology, Experiment and Analogy*, Lejre, 13-17 (Studies in Technology and Culture 20).

HUNTER, J. 1980. The glass. In: Ph. Holdsworth, *Excavations at Melbourne Street, Southampton, 1971-1976*, 59-71 (Southampton Archaeological Research Committee Report I. CBA Research Report 33).

ISINGS, C. 1957. *Roman glass from dated finds*, Groningen (Archaeologica Traiectina).

ISINGS, C. 1959. Merovingisch glas uit Nederland. In: J.E. Bogaers, W. Glasbergen, P. Glazema et al. (eds), *Honderd eeuwen Nederland*, 212-223.

ISINGS, C. 1980. Glass finds from Dorestad, Hoogstraat I. In: W.A. van Es & W.J.H. Verwers (eds), *Excavations at Dorestad 1. The Harbour: Hoogstraat I*, 225-237.

JENSEN, S. 1991. *The Vikings of Ribe*, Ribe.

KNOL, E. 1993. *De Noordnederlandse kustlanden in de Vroege Middeleeuwen*, Groningen.

KOCH, U. 1977. *Das Reihengräberfeld bei Schretzheim*, Berlin (Germanische Denkmäler der Völkerwanderungszeit A 13).

KOCH, U. 1987. *Die Glas- und Edelsteinfunde aus den Plangrabungen 1967-1983. Teil 1: Text. Der Runde Berg Berg bei Urach*, Heidelberg.

KOCH, U. 1996. Glas – Luxus der Wohlhabenden. In: A. Wieczorek et al. (eds), *Die Franken – Wegbereiter Europas. Vor 1500 Jahren: König Chlodwig und seine Erben*, 605-617.

KOCH, U. 1998. Glas – Merowingerzeit. In: H. Beck, H. Steuer & D. Timpe (eds), *Reallexikon der Germanischen Altertumskunde*, Band 12: 153-164.

LITH, S.M.E. VAN 1987. Late Roman and Early Merovingian glass from a Settlement Site at Maastricht (Dutch South Limburg), I, *Journal of Glass Studies* 29: 47-59

LITH, S.M.E. VAN 1988, Late Roman and Early Merovingian glass from a Settlement Site at Maastricht (Dutch South Limburg), II, *Journal of Glass Studies* 30: 62-76.

LITH, S.M.E. VAN & Y. SABLEROLLES 1995. Verres du IVe et du Ve siècles des sites d'habitat aux Pays-Bas. In: D. Foy (ed.), *Le Verre de l'Antiquite Tardive et du Haut Moyen Age. Typologie – Chronologie – Diffusion*, 41-50.

LUND HANSEN, U. 1984. Dankirke, Gläser. In: J. Hoops (ed.), *Reallexikon der Germanischen Altertumskunde* 5: 249-252.

LUND HANSEN, U. 1993. Neues aus der Glasforschung in Skandinavien, 4. bis 8. Jh.. In: *Annales du 12e Congrès de l'Association Internationale pour l'Histoire de Verre*, 235-246.

LUNDSTRÖM, A. 1971. Cuppa vitrea auro ornata, *Early Medieval Studies* 3: 52-68 (Antikvariskt Arkiv 40).

MÜLLER-WILLE, M. 1996. Königtum und Adel im Spiegel der Grabfunde. In: A. Wieczorek et al. (eds), *Die Franken – Wegbereiter Europas. Vor 1500 Jahren: König Chlodwig und seine Erben*, 206-221.

NÄSMAN, U. 1984. Vendel period glass from Eketorp-II, Öland, Sweden, *Acta Archaeologica* 55 (Copenhagen 1986), 55-116.

NIEUWENHUIS, M. 1990. Het glaswerk uit het grafveld van Rhenen, unpubl. IPP report, University of Amsterdam, Amsterdam.

PÄFFGEN, B. & S. RISTOW 1996. Die Römerstadt Köln zur Merowingerzet. In: A. Wieczorek et al. (eds), *Die Franken – Wegbereiter Europas. Vor 1500 Jahren: König Chlodwig und seine Erben*, Mannheim, 145-159.

PÉRIN, P. 1989. Le verre mérovingien en France d'après les trouvailles funéraires. In: *A travers le verre du moyen âge á la renaissance*, (Exhibition catalogue, le Musée des antiquitées de Seine-Maritime à Rouen), 125-129.

POHL, G., et al. 1972. Frühmittelalterliche Glaswerkstatt bei St. Ulrich und Afra in Augsburg. In: *Bayerische Vorgeschichtsblätter* 37: 60-68.

POHL, G. 1977. Frühmittelalterliche Grubenhütte mit Resten frühmittelalterlicher Glasherstellung. In: J. Werner (ed.), *Die Ausgrabungen in St. Ulrich und Afra in Augsburg 1961-1968*, 465-470 (Münchener Beiträge zur Vor- und Frügeschichte 23).

RADEMACHER, F. 1942. Fränkische Gläser aus dem Rheinland, *Bonner Jahrbücher* 147: 285-344.

ROES, A. 1965. *Vondsten van Dorestad*, Groningen (Archaeologica Traiectina 7).

RÜTTI, B. 1991. *Die römischen Gläser aus Augst und Kaiseraugst*, 2 vol., Augst (Forschungen in Augst 13 1/2).

SABLEROLLES, Y. 1992. Het glas van Gennep, De glasvondsten van een nederzetting uit de volksverhuizingstijd te Gennep (Limburg), Amsterdam (unpubl. IPP-Report, University of Amsterdam).

SABLEROLLES, Y. 1993. A Dark-Age glass complex from a Frankish settlement at Gennep (Dutch Limburg), *Annales du 12e Congrès de l'Association Internationale pour l'Histoire du Verre*, 1993, 197-206.

SCHMID, P. 1970. Das frühmittelalterliche Gräberfeld von Dunum, Kreis Wittmund (Ostfriesland) (Grabung 1967-1968), *Neue Ausgrabungen und Forschungen in Niedersachsen* 5: 39-62.

SCHOTTEN, J. & B.J. GROENEWOUDT 1997. *Halverwege Wijster en Gennep, vierde- en vijfde-eeuwse nederzettingsvondsten uit Elsen, gemeente Markelo*, Amersfoort (Interne Rapporten ROB 34).

STEIN, F. 1967. *Adelsgräber des achten Jahrhunderts in Deutschland,* Berlin (Germanische Denkmäler der Völkerwanderungszeit A9).

STEPPUHN, P. 1998. *Die Glasfunde von Haithabu*, Neumünster (Berichte über die Ausgrabungen in Haithabu 32).

TALBOT DONALDSON, E. 1966. *Beowulf* (a new prose translation), New York etc.

THOMSEN, P.O. 1995. The question of beadmaking in the late roman Iron Age at Lundeborg, Denamark. In: M.S. Rasmussen, U. Lund Hansen & U. Näsman (eds), *Glass beads. Cultural History, Technology, Experiment and Analogy*. Lejre, 19-24 (Studies in Technology and Culture 20).

TRÉSORS DE WALLONIE 1993. *Les Verres Mérovingiens*. (Exhibition catalogue, Musée Ourthe-Amblève, Comblain-au-Pont, 1 juillet-31 octobre).

WINKELMANN, W. 1977. Archäologische Zeugnisse zum frühmittelalterlichen Handwerk in Westfalen, 3. Die Glaswerkstatt in der karolingischen Pfalz Paderborn, *Frühmittelalterlichen Studien* 11: 92-126.

WOOLF, A. & R. ELDRIDGE 1994. Sharing a Drink with Marcel Mauss: the Uses and Abuses of Alcohol in Early Medieval Europe, *Journal of European Archaeology* 2.2: 327-340.

YPEY, J. 1962-1963. Die Funde aus dem frühmittelalterlichen Gräberfeld Huinerveld bei Putten im Museum Nairac in Barneveld, *Berichten van de Rijksdienst voor Oudheidkundig Bodemonderzoek* 12-13: 99-152.

YPEY, J. 1968. Vroeg-middeleeuws glas. De verzameling Van Bogaerts-Wauters te Hamme aan de Durme, België, *Antiek* 8: 376-386.

YPEY, J. 1973. Das fränkische Gräberfeld zu Rhenen, Prov. Utrecht, *Berichten van de Rijksdienst voor Oudheidkundig Bodemonderzoek* 23: 289-335.

NOTES

1. The author wishes to thank Dr M. Stiff (Oxford) and Mrs L. Lund-Faveile (Ribe) for giving advice on the identification and dating of some of the fragments. Further, Dr S.M.E. van Lith (University of Amsterdam) and Dr J. Henderson (University of Nottingham) for reading through several drafts and giving useful comments.

2. Knol (1993, pp 78-79 and pp 181-188).
3. Cool & Price (1995, pp 9-10).
4. Henderson (1985, p. 278 and pp 283-284).
5. Henderson (1985, p. 283).
6. Evison (1982, pp 8-9).
7. Bailey (1991, p. 32).
8. Baumgartner & Krüger (1988, pp 77-80).
9. See for instance Baumgartner & Krüger (1988, pp 70-76, cat. nos. 12-23 – only cat. nos. 19-21 belong to funnels).
10. See contributions by Erdrich (this volume), Volkers (this volume), Galestin (this volume) and Van der Vin (this volume).
11. Rütti (1991, 13/1, AR Form 98.1, 34, Figs 21, 49, 35) and Cool & Price (1995, pp 82-85).
12. Isings (1957, p. 88).
13. Rütti (1991, 13/1, AR Form 118.1 (plain globular) and AR Form 118.2 (ribbed globular), 36, Figs 27, 51, 35) and Cool & Price (1995, pp 106-109).
14. Rütti (1991, 13/1, AR Form 156, 37, Figs 29, 54-55, 35) and Price & Cool (1995, pp 190-191, Fig. 11.6).
15. Sablerolles (1993, pp 197-206).
16. Van Lith (1987, pp 47-59), Van Lith (1988, pp 62-76) and Van Lith & Sablerolles (1995, pp 41-50).
17. Van Es (1967, pp 153-156 and Fig. 76).
18. For the distribution of 4th-8th-century glass vessels in Scandinavia, see Lund Hansen (1993, pp 235-246), for Dankirke see also Lund Hansen (1984, pp 248-253) and for Dejberg see Hansen (1993, pp 3-7).
19. Erdrich & Voss (1997, pp 90-91 and Table 1).
20. Isings (1957, pp 134-135 and pp 136-138).
21. Cf. Koch (1987, pp 117-119, 124 cat. no. 113 and Fig. 49).
22. Evison (1972, p. 48, cat. no. 1).
23. Evison (1972, pp 59 and 62) and Koch (1987, pp 117-118 and Fig. 47).
24. Koch (1987, pp 118-120).
25. Ellmers (1964/1965, p. 38, note 101) and Müller-Wille (1996, pp 215 ff., Figs 153-157).
26. *Van Friezen, Franken en Saksen* (1959/1960 – catalogue, cat. no. 40, Fig. VIII) and Evison (1972, cat. no. 46).
27. Koch (1987, p. 116 and pp 122-149).
28. See note 18.
29. Sablerolles (1992 – catalogue, pp 49-54, plate 4.1 – cat. no. 110.1, plate 2.8 – cat. no. 107.1) and Sablerolles (1993, pp 200-203, Fig. 3.1).
30. Van Lith & Sablerolles (1995, p. 47, Fig. 3).
31. Schotten & Groenewoudt (1997, p. 21, Fig. 4.6).
32. Schotten & Groenewoudt (1997, pp 16 ff).
33. Fremersdorf (1941, p. 184) and Päffgen & Ristow (1996, pp 147-148).
34. Lund Hansen (1984, p. 251).
35. Van Lith (1988, p. 76, Fig. 48).
36. Koch (1987, pp 208-236).
37. Koch (1987, Figs 88, 93, 96, 99). For an update on finds of bowls with opaque white decoration in Belgium, see Alénus-Lecerf (1995, Fig. 6).
38. Koch (1987, pp 200-201).
39. Nieuwenhuis (1990, 2 cat. nos. 1-36).
40. For northern France, see Périn (1989, pp 126-127).
41. For instance by Koch (1987, p. 153) in the context of the Merovingian cemetery at Schretzheim.
42. Böhner (1958, pp 228-230).
43. Siegmund (1998, pp 170-172, Fig. 77).
44. For an inventory of bell beakers from graves in the Walloon provinces in Belgium, see Alénus-Lecerf (1993, p. 46, cat. nos. 37-41, 44-45). For northern France, see Feyeux (1995, p. 116, pl. 10 and 11) and Cabart (1995, pp 61-81, Figs 32-45). For the Rhine region (i.e. the Rhineland, the Rhine-Moselle and the Rhine-Neckar regions) see, for instance, Böhner (1958, pp 228-231 – *Trierer Land*), la Baume (1967, pp 79-81 – *Junkersdorf bei Köln*), Ament (1976, pp 55-56 – *Mayen und der Pellenz*) and Koch (1977, pp 153-154 – *Schretzheim*).
45. Koch (1996, pp 615-617, Figs 469-470) and Koch (1998, p. 160).
46. Arbmann (1952, p. 72 and Fig. 3.10).
47. For a similarly small bell beaker from Limet, Vierset-Barse, Modave, province of Liège (Belgium), see Alénus-Lecerf (1993, p. 128, no. 42).
48. Böhner (1985, pp 226-227).
49. Alénus-Lecerf (1995, p. 68, Fig. 17).
50. Siegmund (1998, pp 165-166).
51. For instance, Böhner (1958, p. 227) mentions a palm cup with outward folded rim from Hohenfels grave 20 with purplish streaks *(violett-bräunlicher Färbung)*.
52. See for instance Ament (1957, p. 57 – Rhineland) and Cabart & Feyeux (1995, pp 88-94, Figs 52, 54, 55 – Champagne).
53. Baumgartner & Krüger (1988, pp 58-59).
54. For inscriptions, see Cabart (1993, pp 225-233, Figs 1, 2, 6), Cabart (1995, pp 101-103 cat. nos, pp 277-280, Figs 60-61). For crosses, see Ypey (1968, pp 383-385, Figs 14, 15 – collection Van Bogaerts-Wauters, Hamme, Belgium) and Cabart (1995, pp 99-100 cat. no. 270, Fig. 59).
55. Siegmund (1998, pp 165-166).
56. Heidinga (1997, p. 48).
57. Isings (1980, p. 230 cat. nos. 25, 26, Fig. 153: 24, 25; p. 231 cat. no. 38, Fig. 153: 33).
58. Besteman (1974, pp 17 and 21: Fig. 1.20).
59. Hunter (1980, pp 59-69, e.g. Fig. 11.2: 11, 12, 14; Fig. 11, 3.3, 8; Fig. 11.4: 5; Fig. 11, 5: 5, 7, 12).
60. Jensen (1991, pp 14-15).
61. Cf. Hunter (1980, p. 69) and Koch (1987, p. 257).
62. For example, an undecorated palm funnel with a deeply outward folded rim from grave 30 at Katwijk-Binnen (Prov. of South-Holland) clearly shows its origins in later palm cups with deeply outward folded rims (Stein 1967, pp 383-384, plate 67.18). A squat palm funnel of pale blue-green glass, a stray find associated with a grave, from Valkenburg-village (Prov. of South-Holland) has an inward folded cavity rim and self-coloured, thick arcaded trails on the lower part of the body (Van Giffen 1953, pp 47-48; drawing kindly provided by M. Dijkstra, AAC/IPP). More elongated palm funnels with a fire-thickened rim and self-coloured, arcaded trails extending beyond the lower half of the body from grave 1, Pingjum-village terp (Westergo, Prov. of Friesland; Knol 1993, pp 181 and 184: Fig. 3) and from Medemblik (Besteman 1989, p. 21, Fig. 1.20) probably represent a somewhat later development of the type. From grave 1, Ferwerd-Burmania I (Oostergo, Prov. of Friesland) comes a beaker with a tooled ring

below the middle of the body (Knol 1993, pp 181 and 186: Fig. 13) which is similar to a specimen from Dorestad dated to the 8th/9th century (Baumgartner & Krüger 1987, pp 62-63 cat. no. 3). An elongated palm funnel with a fire-rounded, but hardly thickened rim and a tooled ring above the middle of the body from grave 1, Westergeest-Weerdeburen (Oostergo, Prov. of Friesland) may represent the latest development in the production of the type, although the (scanty) available evidence from this cemetery does not suggest a date beyond the early 8th century (Knol 1993, p. 166, Table 13; pp 182 and 187: Fig. 19). A similar beaker from Dunum (Ostfriesland) is dated to the second half of the 8th century (Schmid 1970, p. 52, plate 24).

63. The typo-chronological development of palm cups into (palm-)funnels published by Ypey (1964, p. 145, Fig. 40) is still useful and regularly referred to.
64. Baumgartner & Krüger (1988, p. 60).
65. Evison (1989, p. 140).
66. Roes (1965, p. 38, Figs 124 and 125) and Isings (1980, pp 227-232, Figs 153, 154.2).
67. Dékowna (1976, pp 63-66, Fig. 29 – complete funnel from a grave), Baumgartner & Krüger (1988, pp 60-64, Figs 14/5) and Steppuhn (1998, pp 59-60, plates 10.1-11; 11.1-3).
68. Hunter (1980, pp 59-72).
69. For a listing of Scandinavian find-places and references, Dekówna (1976, pp 63-66).
70. Evison in: Foy & Sennequier (1989, pp 139-140, pl. VI, 60).
71. Evison in: Otte (1988, pp 215-219, Fig. 140.4 – gold-foil funnel).
72. Pohl et al. (1972, pp 60-70) and Pohl (1977, pp 465-470).
73. Arbman (1937, pp 26-36, Figs 1-2) and Haevernick (1972, pp 215-217). On chemical analysis of glass from Kordel, see Henderson (1993, p. 255).
74. Winkelmann (1977, pp 92-126, esp. pp 123-126) and Winkelmann (1985, 30 ff).
75. Henderson & Holand (1992, pp 44-53, Figs 5 and 6).
76. Isings (1980, p. 227, Fig. 153.4-6, 10 – one has a pale blue rim (10), the other three have a dark bluish-green rim and one a pale blue rim (4-6).
77. Arbman (1937, 39 ff.), Arbman (1940/1943, vol. 1, 161 ff., vol. 2, plate 190, 1) and Baumgartner & Krüger (1988, pp 61-62 cat. no. 2).
78. Cf. Hunter (1980, p. 70).
79. Baumgartner & Krüger (1988, pp 62-63, no. 3).
80. Pers. comm. L. Lund-Faveile, Archaeological Museum Ribe, Denmark.
81. Hunter (1980, Fig. 11, 2.10).
82. Isings (1980, p. 231, Fig. 154.1 and 154.3-5) and Baumgartner & Krüger (1988, pp 71-74, cat. nos. 13, 14, 16, 17 – Dorestad, 15 – Birka).
83. Thomsen (1995, pp 19-24, Figs 3-5).
84. Henricson (1995, pp 13-17).
85. The Merovingian necklace is made up of glass beads and a fragment of a collared rim of a Roman olla of Isings type 67 b/c or type 62 and is exhibited in Museum Rijnsburg, Rijnsburg (Prov. of South Holland).
86. Cf. Evison (1972, p. 62) and Siegmund (1998, pp 172-173) calculated the contents for cones and bell beakers in the lower Rhine region and found that most specimens contain between 175 and 335 ml (average 245 ml), while Kempston cones, drinking horns and claw beakers contain between 495 and 695 ml.
87. Hagen (1995, p. 243).
88. Ellmers (1967, pp 24-25, Fig. 2.1; pp 36-37, Fig. 5; 40ff).
89. Woolf & Eldridge (1994, p. 337).
90. Talbot Donaldson (1966, pp 20-21).
91. Talbot Donaldson (1966, p. 239).
92. Hagen (1996, p. 247).
93. See also Gerrets & De Koning (this volume).
94. Haberey (1942, pp 260-261).
95. For early-medieval mixed-alkali and high-potash glasses, see Henderson (1993, pp 247-260).
96. See also the contribution of the author on beads in this volume.
97. Evison (1983, pp 87-93) and Näsman (1986, pp 69-76 and pp 82-85). However, against the idea of English glass vessel imports in Scandinavia, see Lund Hansen (1993, pp 242-244).
98. Cf. Heidinga (1997, p. 22 and pp 27-30).
99. Isings (1980, p. 230, no. 27, Fig. 153.26) and Baumgartner & Krüger (1988, p. 66, nos. 8-10). An almost complete, colourless funnel beaker with gold-foil decoration from Dorestad has not been published yet. For a recent survey of Carolingian gold-foil glasses, see Henderson & Holand (1992, pp 50-51).
100. Henderson & Holand (1993, pp 39-41, Fig. 4a-b and pp 48-51, Fig. 6) also for an update on the distribution of gold-foil vessels.
101. Van Es (1990, pp 171-172) argues that Dorestad was at the top of a hierarchy of trading towns located on the lower reaches of the Rhine and Meuse. As part of the catchment area of Dorestad, the type of settlement probably ranged from specialized trade settlements to agrarian settlements with mooring facilities for ships, of which the Carolingian settlement at Valkenburg-De Woerd may be an example.
102. See Gerrets & De Koning (this volume).
103. Ellmers (1964, p. 30, note 51).
104. Lundström (1971, 52 ff).

Beads of glass, faience, amber, baked clay and metal, including production waste from glass and amber bead making

Y. SABLEROLLES

1 INTRODUCTION

The systematic waterscreen sieving (3 mm mesh) carried out during the excavations at the terp Tjitsma at Wijnaldum also positively influenced the retrieval of beads.[1] In total, 214 (fragments of) beads were found which are mostly made of glass (82%); the rest is made up of amber and a few beads of faience, baked clay and metal (Table 1). Beads made of shell (*Muschelscheibchen*), sapropelite (*Gagat*), i.e. a biogenous black stone, or rock cristae, which are occasionally found in graves in the Frisian terp-area, are absent.[2]

Most of the beads were found in settlement contexts and are single finds, probably originating from accidentally broken necklaces. Some may be waste from glass bead production though, which will be discussed below.

It should be noted that the beads found in sunken huts (3, 8, 9, 64, 181, 184, 185, 188, 189) are from the secondary fill of the huts; they were not lost in the sunken huts and cannot – at least not directly – be related to any (female) activities carried out in them. Many beads were recovered from one of the main features of the settlement during habitation Period IV (550-650), a large accumulation of burnt and collapsed 5th/6th century houses which probably marks the southernmost limit of the settlement. The feature is therefore largely made up of layers which contain domestic refuse.

From an inhumation burial comes a complete necklace consisting of 19 glass beads (140-158), 23 amber beads (191-213) and one metal bead (214). The beads were in part found in the area of the neck and upper breast of a young woman of about 18 or 19 years old.[3] Remains of 11 (partially) molten glass beads (159-169) were found in a cremation burial which contained two individuals, an infant between 3 years (+/– 12 months) and 5 years (+/– 16 months) old and an adult between 20 and 34 years old;[4] besides the glass beads, other grave goods such as a decorated antler spinning whorl and a fragment of a broken comb case, may indicate the adult was female.

As is the case with glass vessels, glass and amber beads from the Frisian terp region are mainly known from grave contexts rather than from settlements.[5] Grave finds of beads are certainly helpful in that, for instance, necklace ensembles can sometimes provide a more precise date than accompanying grave goods. It is, however, only through thorough excavations of settlement sites that other aspects such as production techniques of glass beads can be examined: production waste of glass and amber bead production from the terp site at Wijnaldum has brought to light the first evidence for the production of glass and amber beads in the Dutch terp region during the early medieval period.

2 THE GLASS BEADS

When looking at the type range of the glass bead assemblage from Wijnaldum, it is obvious that the majority of the beads fit well into established bead typologies developed in Germany and northern France on the basis of 5th-7th century grave finds.[6]

Indeed, the settlement find contexts in which the beads were found confirm this and even allow for a tighter date range. No beads were recovered from habitation Period I (175-250), only a few beads (4%) derive from contexts dating to habitation Period II (250-300/350), almost a quarter (22%) dates to habitation Period III (425-550) with a marked concentration in the second half of the 5th century, more than half (54%) come from Period IV (550-650) of which almost all date to the second half of the 6th century, while numbers have dropped considerably during Periods V-VII (650-850), reaching a low point in Period VIII (850-900/950) with only one bead (118); the remaining beads fall within a wider date range of

Table 1 Wijnaldum: number of beads listed by material and according to find context (inhum.= inhumation burial, crem.= cremation burial, tr.= translucent, op.= opaque, cylindr.= cylindrical, undiagn.= undiagnostic).

BEAD-TYPE	other	inhum.	crem contexts	total	n	%
GLASS						
Wound, monochrome						
irregular, 'black'	1	2	–	3		
annular	8	3	–	11		
globular	55	2				
disc-shaped	2	2				
cylindrical	21	–				
barrel	4	–				
biconical	6	–				
'segmented'	5	–				
prismatic	4	–				
long cylindr., tr. blue	–	1				
long cylindr.?, op. red	1	–				
semi-conical/semi-square	3	–				
melon	2	–	–	2		
Subtotal	112	10	6	128	73%	60%
Wound, bichrome						
tr. matrix & trails	2	–	–	2		
translucent matrix & dots	–	1	–	1		
opaque matrix & trails	8	–	–	8		
opaque matrix & dots	2	–	–	2		
opaque matrix, undiagn.	–	–	2	2		
Subtotal	12	1	2	15	9%	7%
Wound, undiagnostic	6	–	2	8	5%	4%
Drawn						
thin section, gold foil	1	–	–	1		
globular, metal foil	6	–	1	7		
barrel-shaped, hberfang	4	–	–	4		
tiny globular, tr. blue	–	8	–	8		
oblate, tr. blue	1	–	–	1		
globular, bichrome	1	–	–	1		
Subtotal	13	8	1	22	13%	10%
Millefiori						
barrel-shaped	1	–	–	1		
hexagonal	1	–	–	1		
Subtotal	2	–	–	2	1%	1%
Total glass beads	145	19	111	175	101%	82%
FAIENCE	2	–	–	2	1%	
BAKED CLAY	2	–	–	2	1%	
AMBER	11	23	–	34	16%	
METAL	–	1	–	1	0.5%	
TOTAL	157	43	11	214	100.5%	

habitation or cannot be dated/are stray finds (Table 2).

In addition, some Merovingian beads are recovered from younger contexts (after 650) and were either 'antiques' when lost or are residual, for example a typical Merovingian *millefiori* bead (139) which was found in a well in use between 750-850.

Nothing much out of the ordinary marks out the range of glass beads from Wijnaldum, whether in terms of forms, colours or decorations. The major-

Table 2. Wijnaldum – number of single glass beads per dated context.

Contexts dated between:	n	%
AD 175-250	–	–
250-350	6	4
425/450-500	21	16
500-550	8	6
550-600	54	40
575-625	10	7
550-650	7	5
600-650	1	1
650	1	1
550-800	1	1
600-700	3	2
640/50-750/75	9	7
750-800	1	1
750/70/75-850	7	5
800-850	3	2
775-900	1	1
>800	1	1
875-900	1	1
Subtotal	135	101
Date unknown	2	
Not dated (stray)	8	
Total	145	

ity of the glass beads are simply wound, monochrome beads (73%). The decoration of bichrome, wound beads (9%) is limited to the usual spiral and crossing trails (double swags) and dots. There are only two beads which show a more complex, *millefiori* design. Drawn beads, some of which enclose metal-foil, are represented in small numbers (13%) (Table 1).[7] The absence of any elaborate Merovingian drum-shaped beads with combed decoration or complex *reticella* designs should be noted.[8]

2.1 *Wound, monochrome and bichrome beads of translucent glass*

During the 4th and 5th centuries necklaces are dominated by beads made of translucent glass. This preference is clearly illustrated by translucent beads from Wijnaldum which mainly derive from contexts dated between AD 250-350 and 450-550. In addition, all (wound) beads which form part of the necklace in the inhumation burial are made of translucent glass (140-150) and all but one (160), possibly two (166) of the beads from the cremation burial are made of translucent glass (159-169).

Many of the small, undecorated bead types of translucent glass were very long-lived and are therefore difficult to date.

Small, translucent blue annular (1-4) and globular beads (162), for example, were first produced in the Iron Age, but the overwhelming proportion comes from provincial-Roman sites. They are still regularly found in post-Roman contexts, for example in the mid 5th century necklace of the 'princess' of Zweeloo.[9]

Small annular and globular beads of translucent blue-green glass (10-12, 143, 144, 159, 160) are very rare during the Iron Age and are predominantly Roman in date; small beads of (faintly) translucent or opaque green glass (9, 141, 142) tend to be 3rd century or later.[10] Guido suggests that beads of blue-green glass could have been made from recycled Roman 'bottle' glass both in the Roman and post-Roman Period.[11]

Small, sharply biconical beads of translucent, mostly (purplish) blue, glass (91-93, 163, 164) are typically Roman.[12] On the Lower Rhine they are frequently found with small annular and (flattened) globular beads of translucent blue and green glass in late Roman graves, for instance at Krefeld-Gellep.[13]

A recent study of Merovingian grave contents in the Lower Rhine region shows that small biconical beads of translucent blue glass and miniature (globular or short-cylindrical) beads (8) occur in necklace ensembles which are typical of roughly the first half of the 5th century (Lower Rhine Phase 1).[14] Further, that tube-like, long cylindrical beads of translucent blue glass such as the one included in the inhumation necklace (147) first occur during Lower Rhine Phase 1 and are typical of necklace ensembles during Phases 2-4 (440-555).[15] In northern France these beads are most typical of necklace ensembles dated between (450/480-560/570).[16]

With regard to the decoration of irregular yellow dots on a translucent blue bead from the Wijnaldum necklace (148), Koch refers to blue beads decorated with opaque yellow dots from a 4th-century women's grave at the *villa rustica* in Bondorf.[17] An annular bead of translucent 'prussian' blue glass with yellow dots comes from Rittersdorf grave 11 and is dated to Böhners Stufe 3 (525-600).[18]

Medium-sized beads of translucent lightly tinted glass, e.g. yellowish olive (6) are mainly 5th to early 6th century in date.[19] The annular yellowish olive specimen from Wijnaldum is closely parallelled by a bead from a late 4th/5th-century settlement at Gennep (Dutch Prov. of Limburg).[20] Annular beads of olive glass are represented in graves in the area around Trier between 450-600 (*Stufen* II-III) and in the cemetery at Krefeld-Gellep during

Figure 1. Location of the glass and bead finds.

Figure 1. Continued.

Figure 2. Molten and heat-affected glass beads from cremation burial (feature no. 448) including catalogue numbers 159-169. Scale 1:1 (scan Y. Sablerolles)

the 5th and 6th centuries *(Stufen* I-III).[21] It is likely that some of the annular beads of blue glass (3, 4) are of a similar date. Beads of apparently 'black' glass, which is in fact a deeply coloured olive, dark green or dark blue became popular in the late Roman Period and were popular from mid 4th into the 6th century. On the Lower Rhine small 'black' beads (102) are typical for necklace groups C and D which fall within Lower Rhine Phases 3-5 (485-570).[22]

A slightly larger variety are irregularly wound, more or less cylindrical 'black' beads which still clearly show the windings of the glass thread. They are included in the inhumation necklace (149, 150) and are typical of necklace ensembles of the late 4th and early 5th century at the Saxon cemetery at Liebenau in Lower Saxony.[23] They still occur regularly in later necklace ensembles, for instance in the mid 5th-century necklace of the 'princess of Zweeloo',[24] and in necklaces of the first half of the 6th century, for instance at Klepsau in Baden-Württemberg.[25] In northern French cemeteries small 'spiral black' beads are characteristic of necklaces of the second half of the 5th century and still occur during the first half of the 6th century.[26]

Bichrome beads of olive or yellow-green glass decorated with coloured trails (125, 126) were probably first produced in the 5th century.[27] Translucid olive green beads with opaque white and/or red decoration are found in cremation graves of the later 5th century at Liebenau.[28] Medium-sized, flattened globular beads of translucent yellow-green glass decorated with opaque yellow or white trails, sometimes accompanied by opaque red dots, are again found in the necklace of the 'princess of Zweeloo'. This necklace also includes beads of translucent glass with an opaque red spiral trail running circumferential to the perforation (127), as well as a flattened globular, almost disc-shaped, bead of 'ice' blue glass (64).[29]

Only three translucent beads come from contexts dated after 550. A small, blue globular bead (61) comes from an unreliable context (650-750) and is probably of a 4th/5th-century date. A dark blue, bead which is semi-conical/semi-square in section (108) is dated to 700; a similar bead is a stray find (109). These beads probably belong to a group of broad biconical beads of opaque or faintly translucent green and (greenish) blue glass which occasionally occur around the mid 7th century, but are most popular in the later 7th and early 8th century.[30] A long biconical bead of translucent deep greenish blue ('petrol') glass (96) is dated between 800-850. Long biconical beads of blue glass first occur in the later 6th and first half of the 7th century and are popular again in the 9th century, for instance at Haithabu where they were made.[31]

2.2 *Wound, monochrome and bichrome beads of opaque glass*

In the early 6th-century translucent beads are still regularly included in necklaces, but in course of the 6th century the number of opaque beads in necklaces increases considerably. This trend is also reflected by the beads from Wijnaldum which derive from contexts dated between 550-600/650.

The preference for opaque yellow beads at Wijnaldum, almost half of all opaque beads (Table 3), fits in with a more general pattern. Opaque yellow

Table 3. Wijnaldum – matrix colours used for single, monochrome and bichrome glass beads.

Colours	N	%	%
Colourless/metal foil	7		5
Translucent 'black'	2		2
Translucent blue(-green), olive	30		22
Opaque yellow	48	49	35
Opaque red	29	30	21
Opaque white	15	15	11
Opaque orange	5	5	4
Opaque bright green	1	1	1
Total opaque	98	100	72
Total	137		101

was the dominant colour in all categories which could be distinguished amongst the production waste of glass beads recovered from a late 6th/early 7th-century pit in Maastricht.[32] In Germany monochrome beads of opaque yellow glass occur in substantial numbers in Merovingian necklaces in the western part of southern Germany, in the Rheinpfalz and generally in the Rhineland.[33]

In the Rhineland opaque yellow beads frequently make up 45% or more in necklace ensembles during a large part of the 6th and early 7th century with a peak in the later 6th century.[34] The proportion of opaque red beads at Wijnaldum (30%) is also consistent with the pattern established for the Rhineland, where the proportion of opaque red beads remains rather constant at 20-30% from the mid 6th century onwards and is only significantly lower in the 5th and first half of the 6th century. Orange beads, which only make up 5% at Wijnaldum, first appear in the late 6th century and become more numerous (under 10%) in the course of the 7th century, possibly replacing opaque red beads which decrease slightly in number during this period. The proportion of opaque white beads, 15% at Wijnaldum, is generally low in early necklaces and increases from ca 10% in the late 6th century to ca 20% in the later 7th century. In contrast to the material from Maastricht and the Rhineland, where faintly translucent/opaque green accounts for 10-20% in the later 6th and first half of the 7th century, this colour is almost absent at Wij-naldum.[35]

Many of the opaque bead types represented at Wijnaldum are long-lived types and were already produced in the (late) Roman Period such as (flattened) globular beads, long and short cylindrical beads, long and short square-sectioned beads and small ribbed beads.[36]

Almost half of the monochrome, opaque beads from Wijnaldum are small, (flattened) globular beads, which are made of opaque (greenish) yellow (13-44), red (45-55), white (56-58) and bright green (59) glass. The next best represented group are short cylindrical beads of opaque yellow (66-72), white (73-78) and red (79-86) glass (Table 1).

Small globular and short cylindrical beads derive mostly from contexts dated between 550-600 and 575-625. They are mostly very similar in shape and quality to small globular and cylindrical beads produced in Maastricht in the late 6th/early 7th century.[37] In the Rhineland opaque yellow globular beads are represented in Merovingian necklaces between 530-670 with a peak in the later 6th and roughly the first half of the 7th century.[38]

Globular beads were also in use in the 4th-5th centuries; the mid 5th-century necklace of the 'princess' of Zweeloo, for instance, contained some 12 small opaque yellow specimens.[39] At Wijnaldum the necklace from the cremation burial (161), which is dated to the second quarter of the 5th century, includes a medium-sized, flattened globular bead of opaque red glass. Two small (flattened) globular beads of opaque yellow (13) and opaque red glass (48), which are both dated between 450-550, are clearly distinctive from the later specimens in that they are made of a better quality glass with a denser and more glossy surface, and are more regularly shaped with smaller perforations. Chemical analysis of a medium-sized globular bead of opaque bright red glass (45), which was found in a well dated between 800-850, shows that its chemical composition is very different from any of the other opaque red beads from Wijnaldum. It is, in fact, consistent with an early Islamic glass composition and it may therefore represent an example of the influx of oriental beads in Europe during the first half of the 9th century.[40]

The remainder of the Merovingian beads at Wijnaldum is made up of a disc-shaped, opaque yellow bead (65), barrel-shaped beads of opaque white (78) and orange (88-90) glass, medium-sized, (faintly) biconical beads of opaque yellow (94) and white (95) glass, 'segmented' beads of opaque yellow (97-100) and bluish white (101) glass, short rectangular beads of opaque red (103) and white (104) glass, a long cylindrical (?) bead of opaque red glass (107) and ribbed beads of opaque red (111) and yellow (112) glass.

Disc-shaped beads (65) are a long lived type which seems to be slightly earlier than the bulk of the small globular beads, starting before the mid 6th century.[41] Segmented beads (97-101), on the other hand, are somewhat later than small globular beads as is confirmed by their find contexts at Wijnaldum. On the lower Rhine they are frequent in necklaces of the later 6th and first three quarters of the 7th century.[42] Segmented beads with two and three globular segments are also amongst the products of the glass bead maker in the Merovingian settlement at Rijnsburg.[43]

In south-western Germany barrel-shaped and broad biconical beads of opaque orange, red, white and yellow glass (88-90) are typical of necklaces in the period between 610-650/670.[44] On the lower Rhine broad biconical beads of opaque white, red and orange glass first occur in the late 6th century, followed by opaque yellow beads in the early 7th which increase in numbers in the later 7th/early 8th century.[45] In the Saxon cemetery at Liebenau barrel-shaped and broad cylindrical beads, especially of opaque orange glass, dominate necklaces of the late 6th and early/first half of the 7th century.[46]

Long cylindrical beads of opaque red glass (107) can be found in necklaces from the mid 5th to the early/first half of the 7th century,[47] e.g. in a necklace from Beetgum-Besseburen.[48] However, in view of its context date (750-800) this bead may represent a Carolingian type which is frequently found in necklaces from 8th-9th-century Saxon and Frisian cemeteries, for instance on the terp at Oldorf.[49] They are also well represented in the trade centre at Haithabu.[50] Carolingian long cylindrical beads are predominantly made of translucent blue glass, although they are very occasionally also made of other colours. Knol gives examples of opaque red, brown, blue and green long cylindrical beads from Putten, Ferwerd-Burmania I, Dunum and Beers ranging in date from the 7th/8th to the 9th century.[51]

In general, decorated, bichrome beads of opaque glass dominate necklace ensembles of the later 6th and a large part of the 7th century.[52] Bichrome, opaque beads from Wijnaldum are limited in number and type (Table 1).

Six beads belong to a very popular type of globular or barrel-shaped bead with crossing trails or swags (128-133). At Schretzheim Koch distinguishes an early variety with wide crossing trails (128) which is typical of *Stufen* 1-3 (525-590/600) and still occurs in *Stufe* 4 (590/600-620/630); blue-green and green decorations (128) seem to be confined to the late 6th and early 7th century.[53] Beads with narrow crossing trails (129-133) represent a somewhat later variety of which mass production sets in around 600.[54]

In the Merovingian cemeteries at Eichstetten and Weingarten beads with wide crossing trails (128) are characteristic of the period between 530-570, while beads with opaque white beads with blue narrow crossing trails (129, 130) are typical of the later 6th century and all other beads with narrow crossing trails (131-133) are typical of the late 6th/early 7th century.[55] Opaque red (segmented) beads with narrow crossing white trails were amongst the products of the beadmaker in Rijnsburg.[56]

Cylindrical, opaque red beads with an opaque white (134) or yellow spiral decoration are a long-lived type, but seem to be most characteristic of the later 6th and first half of the 7th century.[57]

No parallels were found for two small, globular beads of opaque pale orange glass decorated with marvered, small and irregular yellow dots (135-136). In view of their more or less consistent context dates (respectively 775-850 and 700-775) an 8th-century date seems likely.

A small, globular bead of translucent orange glass with an opaque white, cylindrical core (137) is a stray find. There are no vertical *striae* to indicate that it is a drawn bead (see below). It is possible that this is a post-medieval seed bead; beads of this type were mass produced in the 19th century in Europe, especially Italy, and were very popular as Americo-Indian trade beads.[58] A close parallel of 'opaque cream and orange polychrome glass' comes from a 6th-century burial in an Anglo-Saxon cemetery at Norton, Cleveland (GB), although its context is unfortunately 'much disturbed'.[59]

2.3 *Drawn, segmented beads*

True segmented beads with one or more globular or barrel-shaped segments are known throughout the entire Roman Period and continued well into post-Roman times. The beads are made from hollow, drawn glass tubes which are crimped at regular intervals, after which the segments can then be broken off at the point of easiest fracture. Consequently, many of the beads have a thin, broken collar around the perforation(s). As a result of the drawing technique the glass metal often shows streaks running parallel to the perforation.[60] The beads can be made of translucent, opaque and colourless glass, which sometimes contains gold or silver foil.

At Wijnaldum the following types of segmented beads are represented.

2.3.1 *Beads of translucent coloured glass*

The necklace from the inhumation burial counts six tiny segmented beads with barrel-shaped segments of translucent blue glass (151-156). Segmented beads of translucent blue glass occasionally occur in the early Roman Period, for instance at the Tiberian/early-Claudian harbour-fort at Velsen (Prov. of North Holland).[61] The tiny variety seems to have been firstly introduced in the late Roman Period. Necklace ensembles from graves in northern France show that tiny segmented beads of mainly translucent, colourless, translucent pale yellowish and blue glass are typical of roughly the second half of the 5th and the first decades of the 6th century and still occur, far less frequently, in the remaining part of the 6th century.[62]

An oblate (*spitz-oval*), hollow bead of translucent blue glass (124) was found in a context dated to the second half of the 5th century and probably belongs to the same production phase as the tiny barrel-shaped segmented beads in the necklace.

2.3.2 *Beads of opaque überfang glass*

The Wijnaldum segmented beads of opaque *überfang* glass have a core of colourless glass covered with a thin layer of opaque red glass (120-122) which contains black streaks running parallel to the perforations. One bead appears 'black' (123), possibly as a result of numerous black streaks in the opaque red metal. The beads are made up of one or two barrel-shaped segments which have small glass collars around their perforations, indicating they are crimped and broken rather than cut off the glass tube from which they were made.

Small, segmented globular or barrel-shaped beads of predominantly opaque, 'black', red, yellow, green or blue glass, are known from Gallo-Roman contexts from as early as the 1st century AD, but they are most common in the late 3rd and 4th century.[63]

In the Merovingian Period opaque red *überfang* beads are most frequent in the second half of the 6th century and less so in the first decades of the 7th century, for instance at the Frankish cemetery of Klepsau[64] and the Saxon cemetery of Liebenau.[65]

The context dates for the opaque red segmented beads from Wijnaldum are variable, but two finds from very reliable contexts seem to confirm a 6th century date. A 'black' bead (122) found in a well (750-850) may have been an old piece at the time it was lost.

2.3.3 *Beads of colourless glass, mostly containing metal foil*

Colourless metal foil beads are represented in Wijnaldum in the late Roman Period by a thin section from a relatively large tube of glass containing gold foil (113). Segmented metal foil beads are represented during the Migration and early Merovingian Periods (114, 116, 117, 119); a globular segmented bead (114) does not contain metal foil, but seems to have a core made of faintly translucent, apparently yellowish glass.[66] After a gap in the 7th century, they reappear on the terp in the last quarter of the 8th century in the form of a small, bi-globular bead of colourless glass with what seems to be gold-foil (115) which is rather patchy; its surface incorporates some opaque white 'streaks', possibly bubbles, as well as small drops of spilt opaque red glass. A relatively large globule of colourless glass still contains some silver-foil where the upper layer of colourless glass is still intact (118); it comes from a ditch dated to the last quarter of the 9th century.

The necklace from the inhumation burial includes two single globular segments with silver-foil (157-158) and the beads from the cremation burial possibly include one bi-globular colourless bead without metal foil (169).

The segmented beads have thin glass collars around one or both perforations, which are typical of the production technique whereby they are broken off a crimped glass tube.

In the north-western provinces of the Roman empire segmented colourless beads with or without metal foil are known from military sites as early as the first decades of the first century AD, e.g. at the harbour-fort at Velsen (Prov. of North-Holland),[67] where two bi-globular gold-foil beads probably represent imports from Egypt.[68] They became far more popular in the late Roman Period and are frequently found in graves as part of late and post-Roman necklaces.[69] One of the necklaces found in the mid 5th century grave of the 'princess of Zweeloo' includes 16 relatively large, globular beads of colourless glass, which are apparently made by the drawing technique as is indicated by the thin collars around both their perforations.[70] Analysis of several Merovingian cemeteries in southwestern Germany, viz. Klepsau, Weingarten and Eichstetten, show that metal-foil beads occur especially frequently in necklaces from the middle part of the 6th century; at Weingarten and Eichstetten metal-foil beads occur in necklace ensembles between 530-570 in combination with small, globular beads of predominantly opaque yellow and opaque brown glass and opaque beads with wide crossing trails.[71]

Metal-foil beads which are found in Merovingian *Reihengräber* are considered imports from the Byzantine provinces of the Near East.[72] They completely disappear from the *Reihengräber* territory around 600 and after the beginning of the 7th century they seem to be totally absent from western, central and northern Europe, only to be imported again from the now Islamic Near East into southern, central and western Europe in the later 8th and first half of the 9th century together with *millefiori* beads (*Mosaikaugenperlen*, see below).[73]

Knol has listed the occurrence of metal-foil beads in graves of the second half of the 8th to the first half of the 9th century in the Dutch terp region, viz. Kimswerd (Friesland), and in northern Germany, viz. Dunum and Buchholz-Vaensen,[74] to which Oldorf (Ostfriesland)[75] can now be added.

However, this picture does not seem to match completely with evidence from Saxon cemeteries. At the Saxon cemetery at Liebenau (Lower Saxony), undamaged silver-foil beads of up to 4 segments which can be clearly distinguished from the earlier beads, are introduced in the second half of the 7th century. These apparently 'new' silver-foil beads then become more popular in Liebenau necklace ensembles of the turn of the century and continue to be included in necklaces throughout the 8th century.[76] In Saxon cemeteries in northern Germany metal-foil beads are found in graves in combination with drawn, hexagonal lead-glass beads as early as the first half of the 8th century.[77] They may have reached Lower Saxony as a result of trade contacts with Eastern Europe, where oriental silver-foil beads and so-called gold-foil beads (i.e. silver-foil covered with brownish or yellowish glass) continued to be imported throughout the 8th century.[78] Perhaps metal-foil beads were produced in northern Europe as early as the 7th/8th century; they were certainly being made at Haithabu in the 9th century of glass with a typical Islamic glass composition.[79] It is possible that a bi-globular gold-foil bead from Wijnaldum from a context dated between 775 and 850 is such a product.

Chemical analysis shows that the colourless glass of the youngest silver foil bead (118) is of a typical Islamic (post 800) glass composition.[80] It is impossible to say whether this is a European product of presumably imported Islamic glass or an example of another influx of oriental beads in northern Europe at the end of the 9th and the beginning of the 10th century.[81]

2.4 *(Blättchen-) millefiori beads*

At Wijnaldum only two *millefiori* beads were found. They both show typically Merovingian designs of white encircled red dots (138, 139) and flowers composed of four double opaque yellow petals (138) – hence the name *Blättchenmillefiori* – in the centre of translucent fields.[82] Koch's bead-typology based on the beads from the cemetery at Schretzheim offer an exact parallel for the barrel-shaped bead (138),[83] while the hexagonal bead (139) represents a short variety of a bead with a more complex design represented at Schretzheim.[84]

These are long-lived bead types which first occur before the mid 6th century, became very popular during the second half of the 6th century and rapidly decline in the course of the 7th century.[85] The hexagonal bead (138) which was found in a well dated between 750-850 may have been a much treasured old piece when lost.

Millefiori beads are often considered to be rather special, expensive beads, not in the least because of their labour-intensive production techniques. Koch, however, remarks that their presence in female graves at Schretzheim does not seem to be dependent on wealth or status; in the early phases of the cemetery *millefiori* beads are even completely absent in the richest graves, while they regularly occur in poorer graves.[86]

Millefiori beads flooded the mid-European markets in the second half of the 6th century and are believed to be imports from the Byzantine eastern Mediterranean, probably Egypt, and so are complex herringbone-*reticella* beads, metal-foil beads and melon beads of translucent glass.[87] Their distribution suggests an eastern Frankish land connection over the Alps rather than a western Frankish connection by way of the Rhône.[88]

Judging from the way in which identical elements of *millefiori* and *reticella* are used in varying combinations on different forms of beads, Evison is of the opinion that the millefiori and reticella rods were manufactured at specialized centres, after which the finished rods were then exported to other districts where operators with less expertise worked them up into beads.[89] This could maybe explain the occurrence of low quality beads, such as the barrel-shaped bead from Wijnaldum (138) with a failed flower design – which would probably have been discarded if made in a specialised workshop – and the fact that new excavations constantly bring to light new varieties in design,[90] such as seen on the hexagonal bead from Wijnaldum (139).

There is certainly evidence for the local production of millefiori beads in the 8th-9th/10th centuries in Ireland, England, Scandinavia and northern Germany.[91] On the other hand, evidence for Merovingian glass bead production, however scanty it may be, until now only supports the production of simply wound beads.[92]

An inventory of Merovingian *millefiori* beads by Knol shows that they are very rare in the Dutch Frisian terp region with only one *Blättchenmillefiori* bead known from a necklace found in a grave at Beetgum-Besseburen.[93]

Carolingian *millefiori* beads of the 8th and 9th centuries with designs of checkered patterns

(*Schachbrettmuster*), 'stars' and mosaic eyes (*Mosaikaugen*) are also considered imports from the eastern Mediterranean, but there is evidence to suggest that at least some of these types were made in Scandinavia.[94] They are equally rare in the Frisian terp region as their Merovingian counterparts: only one such bead is documented from Godlinze.[95] Surprisingly, they are completely absent at Wijnaldum. They are well documented, though, in the neighbouring areas of Drenthe, Ostfriesland and Lower Saxony as well as in the Carolingian heartland.[96]

3 FAIENCE MELON BEADS

Two halves of heavily weathered faience melon beads, both broken along their perforation, were recovered. One is a stray find from the topsoil (176) and the other is dated with reasonable reliability to the 7th century (177).

No faience melon beads are known from cemeteries in the terp-region or from cemeteries in Wijster and Zweeloo in neighbouring Drenthe.

In the past it has been suggested that all later faience melon beads are re-used early specimens, since they were mass-produced during the early Roman Period.[97] Voss & Erdrich recently pointed out the differences in the dating of faience beads in different regions of barbarian central Europe and within the Roman empire.[98] It therefore seems more likely that production centres in different areas and different periods were involved in the production of faience beads.

However, in view of the heavy wear on both broken Wijnaldum beads, they are probably early Roman in date and were brought to the terp in the early medieval period, as were fragments of early Roman vessel glass.[99] The same is suggested for two halves of faience melon beads from a late 4th/5th century settlement at Gennep (Dutch Prov. of Limburg).[100] This settlement also yielded early Roman scrap metal, vessel and window glass.[101]

Within the Roman Empire single or just a few faience beads are found in female and children's graves, suggesting a function as amulet. Settlement finds of faience beads which show heavy wear, indicating their suspension from a belt, seem to support this theory.[102] It is possible that their amulet function is continued into the early medieval period, since Merovingian necklace ensembles sometimes include a single faience melon bead.[103]

4 BEADS OF BAKED CLAY

Two small, short cylindrical beads of baked clay (178,179) were recovered from a context dating to the second half of the 5th century and from a ditch dating to the second half of the 6th/first half of the 7th century. Their regular shape suggests they may have been moulded rather than hand-formed.

Obviously clay beads are very fragile and it may be that many more were originally included in Merovingian necklaces.[104] To the author's knowledge no clay beads from the terp region have been published.

5 METAL BEAD

The necklace from the inhumation grave contains one disc-shaped bead (214). Chemical analysis demonstrated that the bead is made of a lead-rich metal.[105]

Since the bead is heavily worn with deep grooves, it may be an old, perhaps re-worked, piece.[106]

6 VARIA

The excavations yielded one fragment of a late 1st century BC La Tène bangle with D-shaped profile of translucent purple glass decorated with an applied opaque yellow zigzag trail of Haevernick type 3b (215).[107] It was found in the same large boundary-ditch (Feature 1233) as the majority of the Merovingian glass beads.

Fragments of La Tène bracelets are found more often in both Roman and early medieval settlement contexts, for instance in the Alamannic hill-settlement *'der runde Berg'*. Here seven fragments were found which show a similar find distribution to that of the late Merovingian glass beads.[108]

It is known from Merovingian grave finds that women collected fragments of highly coloured, preferably dark blue, Celtic glass bangles. They kept these fragments in their pouches, probably as amulets.[109]

7 THE EVIDENCE FOR GLASS BEAD PRODUCTION

Although the evidence is scanty, it is sufficient to

prove glass was being worked into beads on the site. The evidence for the production of glass beads at Wijnaldum includes a complete opaque yellow tessera (216), a breaking splinter from an opaque (greenish) white glass rod (217), a translucent deep blue-green fragment of 'punty' glass (218) and a fragment of baked clay with opaque yellow glass adhering to it (219).

A crucible fragment with yellowish grey glassy deposits (220) from a dump of metal waste dated between 250-350 was initially believed to be a glass-working crucible; chemical analyses have proven this to be a slaggy material resulting from metal-working.[110]

The fragment of baked clay with opaque yellow glass (219) was found amongst waste from metal-working by a blacksmith/bronze-caster. The dump is dated very closely to the last quarter of the 6th and the first quarter of the 7th century. Two small, globular beads of opaque yellow (14) and opaque white (56) were also recovered from the dump.

The fragment contains a thick layer of opaque yellow glass (1.0-1.3 cm) which has permeated through the porous, pinkish orange fabric. It may have been part of a working tray on which the glass was softened before being worked into beads. Another possibility, although less likely, is that it was part of a glass furnace or kiln floor on which a crucible with molten glass fell over.

Until now, hardly any evidence for the use of furnaces on early-medieval glass bead production sites has come to light. At Ribe, where beads were made in the first half of the 8th century, small concentrations of red-burnt clay with charcoal and glass were found (25-30 × 50-60 cm in size). Bead-making experiments were carried out using a small, shallow hearth-pit of a similar size surrounded by stones plastered with sandy clay. However, the results of the experiments did not corroborate the excavated evidence and a taller, more enclosed construction in which the heat gathers at the top was suggested as an alternative.[111]

The source for glass working at Wijnaldum was probably glass rods which are found on many early medieval glass bead-making sites.[112] It is assumed that the rods were imported to bead-making sites, where they were broken up to make their handling easier. When breaking the rods into sections, a special kind of breaking splinter occurs of which an example of opaque bluish white glass was found at Wijnaldum (213). Although context indicates a 5th century date (reliability 1), a later date is more likely since the use of opaque white glass is very limited before the late 6th century. Besides, almost all opaque white beads from Wijnaldum, amongst them a segmented bead of bluish white opaque glass (101), are context-dated in the second half of the 6th or in the last quarter of the 6th to first quarter of the 7th century (56-58, 73-75, 77, 78, 87).

The opaque yellow glass on the baked clay fragment is of a similar lead-tin oxide composition as four small, (flattened) globular beads (18, 19, 41, 59), a short cylindrical bead (71) and a broad biconical bead (94) of opaque (greenish) yellow glass.[113] It is therefore very likely that these simply wound, monochrome bead types were amongst the products of the Wijnaldum glass bead-maker.

The winding technique involves transferring molten or pre-heated glass to a solid iron 'pontil' rod, which can then be drawn out with tweezers and wound around a second, tapering iron rod or 'mandril'.[114] The Wijnaldum beads clearly illustrate the use of such a tool: apart from a few exceptions, they invariably have a conical perforation (see catalogue, P1 and P2) with reduced, greyish-black iron oxide scale on the inside.

The final shape of the bead is simply achieved either by pressing the wound glass on the mandril into the desired shape with a form-iron[115] or possibly by using moulding-tongs.[116] 'Segmented' beads (97-101, 131) are made by fusing two separate globular or barrel-shaped segments together. When finished, the beads are pushed off the rod at which critical point they can break in two halves along their perforation.[117] This is a very common production error which may account for the occurrence of broken beads at Wijnaldum, e.g. several halves of short cylindrical beads recovered from the large boundary 'ditch' (F 1233) of the second half of the 6th century (70, 77, 81, 84-86); most of the fractures are iridescent and therefore probably old. A small ribbed bead (*vierpassförmig*) of opaque red glass from a 6th/7th century context has a square perforation (111). From a 6th century bead-making site in Maastricht (Rijksarchief) comes an iron rod which is round in section at one end and square in section at the other, its diameter matching with the perforations of the associated (failed) beads.[118] It is possible that a similar instrument was used to make this bead, but whether it was made on the terp at Wijnaldum is debatable since its chemical composition is different from all the other opaque red beads analysed.[119]

The inside perforations of a short cylindrical (70) and a square-sectioned bead (106) from 6th/7th century contexts, are pinkish/purplish in

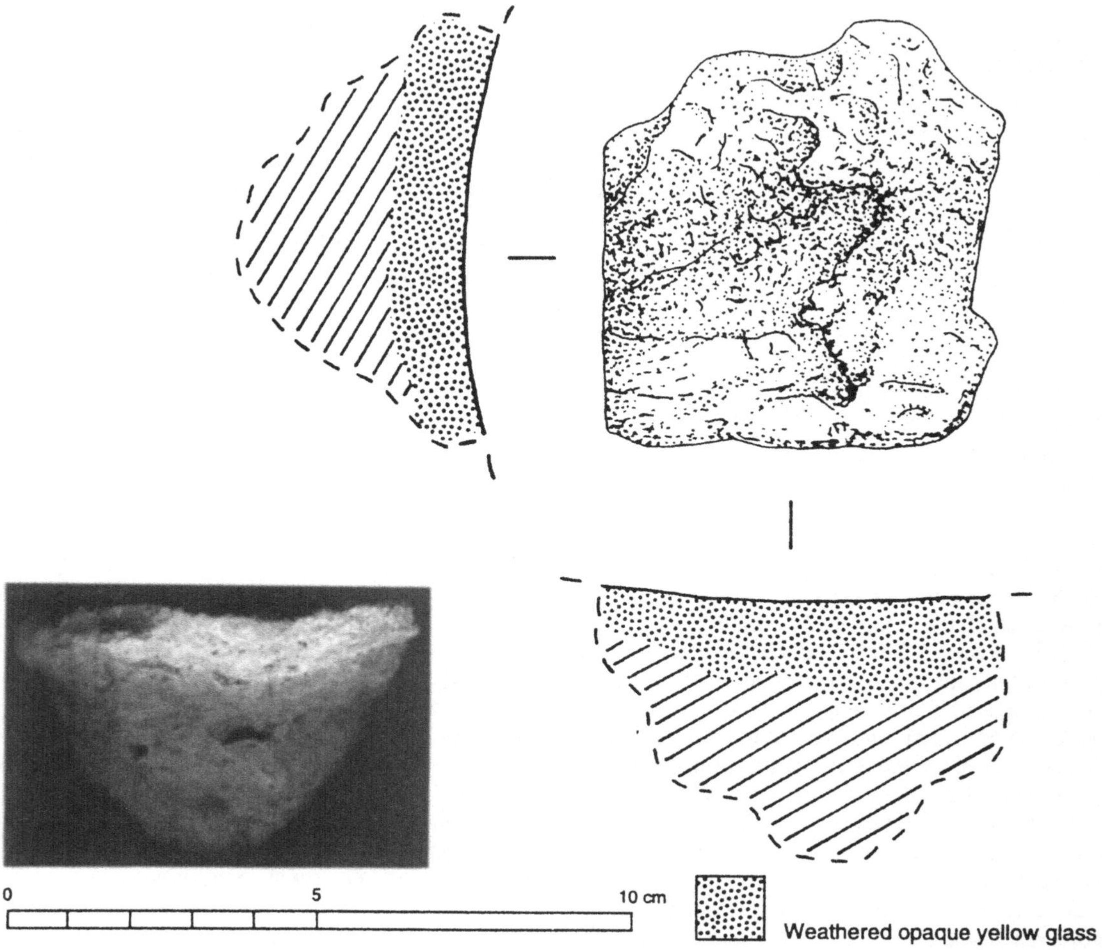

Figure 3. Probable fragment of furnace wall with opaque yellow glass; the highly mobile laed-rich glass has penetrated into the fabric (drawing D. Taylor, Dept. of Archaeology, University of Nottingham).

colour. This is most likely reduced copper implying the use of a copper or bronze tool. The inside of a fragment of 'punty' glass from Wijnaldum (218) also shows this colour. Punty glass is defined as the glass collar that remains behind on the pontil after the last bead was made: it has interacted with the pontil and cannot not be reused. When this glass around the pontil is removed, iron oxide scale is visible on the concave side of the punty glass.[120]

In the absence of blue-green beads, the punty glass fragment may be waste from decorating beads, since the winding technique is also employed for trailing. There are four beads decorated with blue-green crossing trails (128, 131-133), all of which were found in the large 'ditch' (F 1233) dated to the second half of the 6th century, except for one stray find (131). On the other hand, the context date (reliability 4) of the punty glass fragment (218) could indicate that beads were still/again being made on the terp in the (second half of the) 8th century, when blue and (bluish) green were fashionable colours for beads.

This brings us to the last piece of evidence in the form of an opaque yellow *tessera* (209) from a ditch dated between 750-770 (reliability 3). Chemical analysis demonstrated the use of lead-antimonate as opacifier instead of the lead-stannate used in the earlier opaque yellow beads and the glass from the baked clay fragment. This composition is consistent with that of opaque yellow tesserae found on Scandinavian bead-making sites and with that of the opaque yellow trailing on an 8th

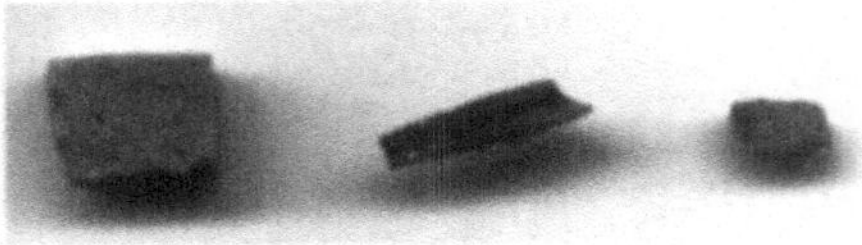

Figure 4. Tessera of opaque yellow glass (cat. no. 216), fragment of translucent deep blue-green 'punty' glass (cat.no. 218), breaking splinter of an opaque bluish white glass rod (cat.no. 217). Scale 1:1.

century cylindrical beaker with opaque white and yellow trailing from Wijnaldum (30).[121]

Glass *tesserae* are found in a great number at Viking Age sites as well as on some Anglo-Saxon and continental, Carolingian sites which yield evidence for glass bead production.[122] They were crushed and melted down on these sites as 'raw materials' and/or as colourants for the production of glass beads. Obviously, the find of one tessera at Wijnaldum is not sufficient evidence for glass bead production on the terp in the 8th century.

Glass *tesserae* are not found on bead-making sites before the 8th century and their presence should be seen in the context of the rise of the large trade centres (emporia) and long-distance trade. *Tesserae* were probably imported from the southwest, perhaps from Byzantine territory in northern Italy (Ravenna, Torcello).[123] In the monastery of San Vincenzo (Italy), where a wide range of glass objects was being made in the 9th century, *tesserae* were certainly used either as 'raw material' and/or colourants since a *tessera* was found in a half-molten state inside a crucible.[124] At Åhus (Sweden), it was observed that in some cases mortar was still adhering to the *tesserae*, which indicated that at least some were robbed from mosaics.[125]

8 AMBER BEADS AND PRODUCTION WASTE OF AMBER

Two small lumps of unworked amber (221, 222), some small waste fragments and splinters (223-228) as well as a re-worked bead (185) and unfinished beads with failed perforations (187, 188) indicate that amber was probably worked on the terp in the second half of the 5th and the 6th century.

One lump was found together with a small, roughly globular bead (181) and a re-worked, disc-shaped bead which was originally rectangular in section (185) in a sunken hut dated to the second half of the 5th century; small beads like these are also included in the necklace from the inhumation burial and it seems likely that the amber beads from the necklace were made on the terp.

In general, amber beads of the 4th and 5th centuries are regularly hand-cut or turned on a lathe, often disc-shaped, whereas crudely hand-cut, irregular beads appear in the course of the 6th century.[126]

All amber beads from Wijnaldum, including the beads from the necklace in the inhumation burial, are relatively small beads which are carefully cut by hand, with one possible exception of a medium-sized, perfectly flattened globular bead (182). This bead may have been lathe-turned, but the weathering has long since erased any traces of lathe-cutting. The perforations of the beads are cylindrical (see catalogue) so they are perhaps drilled by hand from both ends, eventually creating a cylinder.

In the terp region amber beads are known from many early-medieval graves.[127] In comparison to Wijnaldum, where 34 amber beads make up 16% of a total of 214 beads (Table 1), the cemetery and settlement at Oosterbeintum yielded 12 amber beads (6%), 195 glass beads and one bead of rock crystal. The 12 amber beads from Oosterbeintum were divided over more than 5 necklaces, whereas the Wijnaldum necklace alone included 23 amber beads as well as 19 glass beads and one metal bead.[128] Perhaps the relatively high number of amber beads in the Wijnaldum necklace can be attributed to the occurrence of amber working on the terp.

Further afield, in the lower Rhine region amber beads occur in small numbers (under 5%) in necklaces from the 5th to the 7th century, but they briefly peak in the late 5th and first half of the 6th century, when the proportion of amber beads in necklaces can form up to ca 20% of the total.[129]

The amber worked on the terp at Wijnaldum was probably collected on the North Sea sandy shores and (former) tidal flats.[130] Unfortunately, until now it has not been possible to chemically distinguish Baltic from North Sea amber. However, judging from the small quantity of amber at Wijnaldum which shows great variety in colour and surface appearance, local collection seems more likely than importation from the Baltic, which would undoubtedly have occurred in some quantity.

9 SUMMARY AND CONCLUSIONS

On the Tjitsma terp at Wijnaldum systematic water-screen sieving (3 mm mesh) greatly increased the recovery of beads. In total, 214 (fragments of) beads were found which are mostly made of glass (82%) and amber (16%). Furthermore, there are two faience beads and two beads of baked clay, constituting the remaining 2% of beads recovered. Apart from glass and amber beads, the necklace from the inhumation burial also includes a bead of (lead rich) metal. No (fragments of) beads or armlets of sapropelite (*Gagat*) which are occasionally found in late Roman graves in the terp region, were recovered; neither were beads of shell (*Muschelscheibchen)* and rock crystal which are sometimes included in necklaces of the (second half of the) 7th century.

The majority of glass beads from Wijnaldum are typical Merovingian types, mainly deriving from contexts dating to the second half of the 6th century (Table 2). A typical Wijnaldum necklace from this period would probably have been dominated by small beads, especially small flattened globular beads of opaque yellow glass, and would have included several short cylindrical beads of opaque yellow, red and white glass and a few large barrel-shaped or broad biconical beads of opaque yellow, white, red or orange glass. A perhaps more luxurious version would also have included some bichrome opaque beads with narrow crossing trails, some drawn segmented beads of opaque red *überfang* glass, and the occasional segmented metal foil and/or *Blättchenmillefiori* bead.

Glass beads from late Roman contexts and those from Migration Period contexts are characterized by the use of translucent blue (-green), olive green and 'black' glass. The glass beads from the cremation burial are late Roman types, whereas the beads included in the necklace from the inhumation burial can be dated somewhat later, in the (second half of the) 5th to the early/first half of the 6th century.

There are only a few beads in Wijnaldum which date to the late Merovingian and Carolingian Periods. Beads of this period probably include a long biconical bead of greenish blue glass (96), two beads (semi-square/semi-conical in section) of translucent blue glass (108, 109) and what is probably a long cylindrical bead of opaque red glass (107). A monochrome opaque red bead (45) and a segmented bead containing silver foil have a typical, post 800, Islamic glass composition.[131] Metal foil beads with a similar Islamic composition were made at Haithabu in the 9th century.[132]

It is surprising that other 8th-9th century bead types which are frequently combined in Saxon and Frisian necklaces with metal-foil beads are absent from Wijnaldum, e.g. drawn, hexagonal beads of green lead-rich glass, long cylindrical beads and segmented beads of blue glass and millefiori beads with checkered patterns and mosaic eyes (*Mosaikaugenperlen*).

Most of the amber beads come from two sunken huts and an associated hearth dating to the second half of the 5th/first half of the 6th century. They have more or less regular shapes with cylindrical perforations and are carefully hand-cut. Two small lumps of unworked amber (221, 222), some small waste fragments and splinters (223-228) as well as a re-worked bead (185) and unfinished beads with failed perforations (187, 188) indicate that amber was probably worked on the terp in the second half of the 5th and (first half of the) 6th century. Some of the beads which occur separately are very similar to the amber beads included in the necklace from the inhumation burial. It seems that amber working at Wijnaldum took place on a small scale to meet a local demand. The amber was probably collected on the North Sea sandy shores and (former) tidal flats.

The production of glass beads seems to have taken place somewhat later. The most important piece of evidence is a thick-walled fragment of baked clay with a thick layer of opaque yellow glass (1.0-1.3 cm) adhering to it (215). It was found together with two, small, globular beads of opaque yellow (14) and opaque white (56) glass amidst a metal-working dump of a blacksmith/bronze caster which is dated to the last quarter of the 6th to early 7th century. The fragment may have been part of a working tray on which the opaque yellow glass was softened before being worked into beads. An alternative, but less likely interpretation, is that it was part of a (rudimentary) furnace or kiln floor on which the glass was spilt.

It is probably no coincidence that this fragment was found amongst a dump of waste from metal working. It is quite common to find various crafts involving high temperatures together on the same location during the early medieval period.[133] In Merovingian Maastricht, for example, the combined occurrence of various high-temperature crafts, e.g. glass-working in combination with copper-alloy and/or iron working or pottery production, was established at five different locations.[134] It seems likely that glass bead-making was not an exclusive specialization, but that it was carried out as a secondary activity to other high-temperature crafts.

Since the early Merovingian bead assemblage from Wijnaldum forms a typologically homogenous group, it may well represent the products from one single production phase. Perhaps the beads were made by the bronze caster or perhaps by the gold- and silversmith, whose occasional presence on the terp seems certain.[135]

The products are most likely to have been simple wound, monochrome beads of opaque, predominantly opaque yellow, glass. Most popular amongst these were small (flattened) globular beads followed by short cylindrical beads. Other locally produced types may be broad biconical, barrel-shaped, 'segmented' beads and bichrome opaque beads decorated with spirals and narrow crossing trails. Most of the beads which were probably produced on the terp come from contexts dating to the second half of the 6th to first quarter of the 7th century, in particular from the large boundary 'ditch' (F1233). Some of the beads from this context are split in half along their perforation which is a common production flaw in early-medieval beads; complete beads, especially the small (flattened) globular ones, may have been discarded as seconds.

The fact that the opaque yellow glass adhering to the baked clay fragment has a similar lead-tin composition as the analysed small globular, cylindrical and biconical beads of opaque yellow glass strengthens the case for arguing that opaque yellow glass was worked into beads on the terp.[136]

It is remarkable that early-medieval bead-making sites almost exclusively yield crucible fragments with opaque yellow glass, while glass waste and (failed) beads of other colours are also present. Merovingian glass bead-making sites at Maastricht and Rijnsburg, for example, yielded many crucible fragments with opaque yellow glass and some with translucent greenish glass; opaque yellow is also the dominating colour amongst the waste fragments and the (failed) beads.[137]

The abundance of opaque yellow glass beads during most of the 6th and first half of the 7th centuries, peaking in the later 6th century, may indicate that opaque yellow glass was made in Europe either from raw materials or by adding a colourant to a base glass which could have been made by remelting greenish vessel glass. Whether the preference for opaque yellow beads triggered a greater production of opaque yellow glass or whether it was somehow easier to produce opaque yellow glass, causing an increase in the use of opaque yellow beads, is unclear.[138]

Merovingian beads with a more complex, (*Blättchen-*) *millefiori* design (138, 139) as well as the drawn, segmented metal foil beads from Merovingian contexts (144, 116) are probably Byzantine imports, which flooded the mid-European markets in the mid 6th century; the latter may also be re-used Roman beads. Merovingian *Blättchenmillefiori* beads are rare in the terp region and the only known parallel comes from Beetgum-Besseburen.[139] Chemical analysis of the opaque red glass in the Wijnaldum *Blättchenmillefiori* bead (138) shows that it is of the same essentially Roman glass composition as the opaque red beads which were most likely made on the terp. This implies that the latter were either made of recycled Roman glass or, more likely, that red glass of a Roman composition was being made from raw materials, which were probably imported, possibly from the Byzantine area.[140]

Other evidence for bead production on the terp consists of a breaking splinter of an opaque bluish white rod (218). Glass rods are the source for glass working on many early-medieval bead-production sites. The rods were probably imported and consequently broken into smaller pieces to make their handling easier, thus creating typical breaking splinters. A 'segmented' bead of bluish white glass (101) may have been made from such a rod.

A fragment of deep blue-green 'punty' glass is probably waste from trailing opaque beads with blue-green crossing trails, although its context date may indicate that beads were still/again made on the terp in the second half of the 8th century. Evidence for glass bead production in (early) 7th century Rijnsburg includes fragments of blue-green rods, a fragment of blue-green punty glass, blue-green beads and an opaque white bead with narrow crossing blue-green trails.[141]

Glass *tesserae*, which were probably mostly imported from northern Italy, are found on many 8th-9th century bead-making sites, where they were melted down and worked into glass beads.[142] The find of an opaque yellow *tessera* (of a lead-antimonate composition) from a context dated between 750-770 is in itself not sufficient evidence that beads were still/again made on the terp in the 8th century. Interestingly, Arbman mentions several small blue and green tesserae, as well as glass beads and a fragment of a (palm?) funnel from a terp at 'Wierum Schoun' (Prov. of Groningen), which is perhaps Wierumerschouw, as evidence for glass bead-production.[143]

It is remarkable that glass beads were being produced on the terp at Wijnaldum at a time when the site apparently enjoyed a high status. The evidence for glass bead making in Wijnaldum, although scarce, is very similar to waste from Merovingian glass bead production in Maastricht, an important religious, political and economic centre during the Merovingian Period,[144] and Rijnsburg which probably functioned as a small, regional trade-centre during the 7th century.[145] However, just because beads were made during a period when Wijnaldum was a high status site, does not make the processes involved in glass bead production a high status activity.

When the site had lost this status during the late Merovingian/Carolingian Period, the occurrence of glass vessels reached its peak, albeit mainly plain ones; at this same time few glass beads were found on the terp. This may either illustrate a change in 'consumption' or it may show that glass beads and vessels were obtained from different sources and/or were embedded in exchange systems which operated in different ways.

CATALOGUE OF BEADS (GLASS, AMBER, FAYENCE, BAKED CLAY, METAL), GLASS BANGLE AND PRODUCTION WASTE OF GLASS AND AMBER BEAD MAKING

Key to catalogue: CatNo = catalogue number; FindNo = find number; ContNo = context number; Contno+r = uncertain context to be related to given context number; Cont = nature of context (cl = cultivation layer; crem = cremation; di = ditch; met = area with metal waste; os = occupation surface; ph = posthole; sh = sunken hut, sl = sod layer; sstru = sod structure (houses); trl = truncated layer; ts = top soil; wd = waste deposit); DatA/DatB = earliest/latest possible date of context; Rel = reliability of date (1 = very reliable, 2 = unreliable, 3 = under certain conditions reliable, 4 = reliable); Diam = diameter of bead; P1/P2 = smaller/larger end of perforation; A = angled perforation; S = square perforation; F = failed bead; * = incomplete bead; – = insufficient to make a measurement; R = reconstructed measurement.

Round beads: globular = the height is greater than 1/2 the diameter; annular = the height is smaller than 1/2 diameter & the central hole is wider than the height of the bead; disc-shaped: the central hole is narrower than the height of the bead.

1 GLASS BEADS

1.1 *Monochrome beads (separate)*

CatNo	FindNo	ContNo	Cont	DatA	DatB	Rel	Diam	Height	P1	P2
a) Annular										
1.	9303	175r	–	250	350	?	1.05	0.30	0.40	0.40
2.	9667	–	ts	250?	350?	?	0.99	0.47	0.46	0.48
3.	9934	2355	sh	450	500	?	0.98	0.40	0.42	0.45
small, translucent dark blue										
4.	10191*	1191	sl	550	650	?	1.05	0.41	0.37	0.40
small, opaque red										
5.	5812*	2098	os/wd	450	0	1	1.10R	–	0.30R	–
medium-sized, translucent yellowish olive green										
6.	9364*	2315	os	450	500	1	1.50R	0.61	0.66R	0.68R
b) Globular										
medium-sized (flattened), translucent blue-green										
7.	1	–	ts	–	–	–	1.17	0.72	0.26	0.31
tiny, translucent (?) bluish green (turquoise)										
8.	9897	2355	sh	450	500	1	0.26	0.20	0.08	0.10
small (flattened), faintly translucent bluish green (turquoise) with red streaks (resembling spiral decoration)										
9.	9934	2355	sh	450	500	–	0.65	0.38	0.21	0.32
small (flattened), translucent deep blue-green										
10.	1284	–	ts	–	–	?	0.89	0.54	0.22	0.36
11.	9361	2340	trl/os	450	500	1	0.44	0.45	0.22	0.30
12.	10782	2416	sstru	450	500	?	0.68	0.42	0.15	0.24
small (flattened), opaque yellow										
13.	6242	801	di	450	500	1	0.58	0.41	0.10	0.20
14.	1356	625	met/wd	575	625	3	0.52	0.41	0.14	0.23
15.	1428	608	well9	575	625	1	0.65	0.40	0.22	0.25
16.	1462	1167	sl	600	700	1	0.46	0.28	0.18	0.21

17.	2194	–	ts	800?	>	?	0.46	0.36	0.17	0.21
18.	3901	574	di	575	625	1	0.47	0.33	0.17	0.18
19.	3901	574	di	575	625	1	0.47	0.40	0.17	0.20
20.	3901	574	di	575	625	1	0.45	0.31	0.17	0.20
21.	5534	1233	di	500	550	3	0.85	0.64	0.23	0.28
22.	5972	2165	os	550	600	1	0.50	0.30	0.25	0.27
23.	6208	1233	os	550	600	3	0.57	0.44	0.22	0.25
24.	6208*	1233	di	550	600	3	0.46R	0.43	0.18	0.21R
25.	6704	1233	di	550	600	3	0.66	0.45	0.24	0.29
26.	6704	1233	di	550	600	3	0.72	0.42	0.19	0.25
27.	6704	1233	di	550	600	3	0.51	0.41	0.18	0.22
28.	6704	1233	di	550	600	3	0.46	0.35	0.16	0.22
29.	6704	1233	di	550	600	3	0.53	0.39	0.16	0.22
30.	6704	1233	di	550	600	3	0.43	0.43	0.13	0.16
31.	6704	1233	di	550	600	3	0.44	0.36	0.15	0.17
32.	6704	1233	di	550	600	3	0.44	0.38	0.17	0.18
33.	6704	1233	di	550	600	3	0.49	0.33	0.20	0.23
34.	6704	1233	di	550	600	3	0.46	0.34	0.10	0.15
35.	6704	1233	di	550	600	3	0.44	0.42	0.12	0.16
36.	7029	527	os/wd	575	625	1	0.50	0.38	0.17	0.22
37.	7060	1567	wd?	575	625	4	0.40	0.36	0.13	0.15
38.	7066*	1547	os	575	625	1	0.47	0.43	0.16	0.22R
39.	7448	1233	di	550	600	3	0.47	0.30	0.16	0.20
40.	7448	1233	di	550	600	3	0.50	0.40	0.18	0.23
41.	7448	1233	di	550	600	3	0.61	0.35	0.26	0.27
42.	7448	1233	di	550	600	3	0.50	0.30	0.12	0.17
43.	11154	–	ts	175?	>	?	0.50	0.31	0.13	0.16
44.	11580	2614	well18	800	>	?	0.50	0.32	0.25	0.30
medium-sized, opaque red										
45.	3501*	2587	well14	800	850	1	0.75	0.59	0.27R	0.34
small (flattened), opaque red										
46.	5812	2098	os/wd	450	000	1	0.65	0.41	0.21	0.23
47.	6712	1330	os	450	550	4	0.69	0.40	0.30	0.35
48.	9737	2341	os/sl	450	500	1	0.50	0.32	0.10	0.15
49.	5534	1233	di	550	600	3	0.85	0.64	0.23	0.28
50.	6208	1233	di	550	600	3	0.57	0.44	0.22A	0.25
51.	6208	1233	gr	550	600	3	0.62	0.56	0.24	0.26
52.	6253	2062	os/sl	500	550	1	0.76	0.58	0.27	0.34
53.	7448	1233	di	550	600	3	0.73	0.44	0.22	0.31
54.	7448	1233	di	550	600	3	0.72	0.49	0.18	0.25
55.	11173	?	?	?	?	?	0.76	0.43	0.29	0.31
small (flattened), opaque white										
56.	2655	625	met/wd	575	625	3	0.79	0.42	0.30	0.35
57.	5632	975	cl	550	560	1	0.61	0.33	0.15	0.19
58.	6704	1233	di	550	600	3	0.59	0.44	0.17	0.22
small (flattened), dirty opaque yellow/'apple' green										
59.	1024	1079	sl	640	750	1	0.64	0.64	0.10R	0.15R
small (flattened), translucent dark blue										
60.	5534	1233	di	550	600	3	0.55	0.72	0.23	0.30R
61.	5974	2160	os	650	750	3	0.67	0.43	0.25	0.27
62.	6704	1233	di	550	600	3	0.69	0.40	0.20	0.25
63.	7448*	1233	di	550	600	3	–	0.47	0.19R	0.26R
c) Disc-shaped										
small, faintly translucent 'sky' blue										
64.	9897	2355	sh	450	500	1	0.84	0.41	0.13	0.20
small, dirty opaque yellow										
65.	6253*	2062	os/sl	500	550	1	0.78	0.31	0.15R	0.22
d) Cylindrical										
short cylinder, opaque yellow										
66.	1461	548	os/wd	550	600	1	0.62	0.70	0.21	0.33
67.	1464	548	os/wd	550	600	4	0.90	0.54	0.17	0.23

68.	3911	729	os/wd	550	600	3	0.73	0.50	0.23	0.24
69.	5358	1230	trl/wd	550	600	1	0.85	0.58	0.17	0.22
70.	5666*	1232	di	550	650	3	0.72R	0.63	0.20R	0.33
71.	7045*	1578	di	770	850	1	0.67	0.54	0.26	0.34
72.	7448	1233	di	550	600	3	0.66	0.64	0.23	0.30

short cylinder, opaque white

73.	5666	1232	di	550	650	3	1.04	0.59	0.28	0.32
74.	6208	1233	di	550	600	3	0.94	0.65	0.28	0.36
75.	6208	1233	di	550	600	3	0.82	0.61	0.24	0.32
76.	7988	3352	os/wd	775	850	1	0.85	0.71	0.30	0.35
77.	7448*	1233	di	550	600	3	0.85R	0.63	–	–
78.	7448	1233	di	550	600	3	1.00	0.63	0.25	0.31

short cylinder, opaque red

79.	1	–	ts	–	–	?	0.86	0.56	0.31	0.36
80.	2454*	558	cl	500	550	?	0.80	0.67	0.26	0.30R
81.	10824*	1384	trl/wd	650	750	1	0.68R	0.55	–	–
82.	11090	2546	di	650	000	3	0.78	0.66	0.35	0.40
83.	6884*	575	well8	770	850	3	0.56	0.57	0.24R	0.28
84.	7448*	1233	di	550	600	3	0.70	0.77	0.24	0.29R
85.	7448*	1233	di	550	600	3	0.72	0.65	–	0.30
86.	7448	1233	di	550	600	3	0.68	0.78	0.24	0.30

e) Barrel-shaped

opaque white

87.	5409*	574	di	575	625	1	–	0.76	0.28R	–

opaque orange

88.	1024*	1079	sl	640	750	1	–	0.80	0.25R	0.33R
89.	6813*	1261	di	550	650	1	0.78	0.67	0.25R	0.33
90.	110C4*		pit	550	800	?	0.93	0.75	0.45	0.50R

f) Biconical

very small, sharply biconical, ultramarine blue

91.	6208	1233	di	550	600	3	0.60	0.36	0.13	0.22

small, sharply biconical, translucent dark blue

92.	5155	10	trl	300	350	1	0.72	0.44	0.16	0.18

small, sharply biconical, translucent purple

93.	6811	1325	os	450	500	1	0.67	0.44	0.16	0.21

medium-sized, broad biconical, opaque yellow

94.	4401*	329	pit	775	900	3	1.01	0.64	0.50	0.54

medium-sized, broad biconical, opaque white

95.	6472	1848	di/trl	600	700	1	0.90	0.77	0.25	0.29

large, long biconical, faintly translucent greenish blue ('petrol') with red streaks, almost like a spiral decoration, and large, elongated bubbles circumferential to the perforation

96.	7994	3356	os	800	850	1	1.01	1.22	0.32	0.39

g) 'Segmented' (bi-globular/2 barrel-shaped segments)

globular segments, opaque yellow

97.	1024(2)	1079	sl	640	750	1	0.51	0.72	0.16	0.22
98.	5405(1)*	512	pit	650	700	3	0.41	0.40	0.13	0.17
99.	6704(2	1233	di	550	600	3	0.43	0.66	0.16	0.20
100.	7069(2	1542	trl/os	600	700	1	0.49	0.86	0.16	0.20

barrel-shaped segments, opaque bluish white

101.	6208(2)	1233	di	550	600	3	0.64	1.02	0.30	0.33

h) Irregularly wound

short cylinder/barrel-shaped, 'black'

102.	6562(1)	2064	sl	500	550	1	0.37	0.47	0.16	0.21

i) Prismatic

short rectangle, opaque red

103.	11694	1233	di	550	600	3	0.46	0.70	0.20	0.23

short rectangle, opaque white

104.	9847*	2342	os	450	500	1	–	0.85	–	

small cube(?), faintly translucent blue-green										
105.	6704*F	1233	di	550	600	3	0.80	0.73	0.30R	0.38R
106.	6704*F	1233	di	550	600	3	0.83	0.64	0.31R	0.38R

j) Llong cylindrical (?)

opaque red, heat-affected										
107.	9800F	2466	trl/wd	750	800	1	0.48	1.24	0.16	0.21

k) Semi-square/semi-conical in section

medium-sized, translucent dark blue with red streaks running circumferential to the perforation										
108.	4316	1993	trl/wd	700	0	4	0.77	1.12	0.15	0.32
small, translucent dark blue										
109.	1	–	ts	–	–	?	0.63	0.70	0.22	0.30
small, opaque bright ('apple') green										
110.	6704	1233	di	550	600	3	0.72	0.65	0.21	0.30

l) Melon

small with 4 lobes (*vierpassförmig*), opaque red										
111.	4421	–	–	550	650	?	0.78	0.46	0.20S	0.25S
medium-sized with 7 hardly pronounced lobes, opaque yellow with dark streaks										
112.	9737	2341	os/sl	450	500	1	1.07	0.58	0.26	0.31

1.2 *Drawn beads (separate)*

CatNo	FindNo	ContNo	Cont	DatA	DatB	Rel	Diam	Height	P1	P2
sliced segment (thinly cut 'slice'), enclosing gold foil										
113.	6258	59	trl	300	350	1	0.60	0.24	0.21	0.22
globular segment, colourless										
114.	6704(2)	1233	di	550	600	3	0.32	0.67	0.13	0.15
globular segments, colourless enclosing gold foil										
115.	3316(2)	3532	pit?	775	850	1	0.49	0.80	0.10	0.13
116.	6562(1)	2064	sl	500	550	1	0.48	0.40	0.11	0.11
117.	9778(1)	2315	os	450	500	1	0.45	0.42	0.13	0.15
globular segments, colourless enclosing silver foil										
118.	3326(1)	3542	di/os?	875	900	1	0.60	0.61	0.10	0.10
119.	9737(1)	2341	os/sl	450	500	1	0.46	0.43	0.13	0.16
barrel-shaped segments, *überfang* opaque red with dark streaks										
120.	2557(1)	67	sl/wd	550	600	1	0.45	0.35	0.10	0.13
121.	6253(1)	1062	os/sl	500	550	1	0.43	0.32	0.10	0.11
122.	10608(1)	100	well6	750	850	3	0.45	0.35	0.12	0.12
'black'										
123.	5553(1)	125	sl	650	750	3	0.42	0.28	0.10	0.10
oblate (*spitz-oval*), hollow, translucent dark blue										
124.	10004*	2355	os	450	500	1	0.48	1.28	<0.15	<0.15

1.3 *Wound, bichrome beads (separate)*

CatNo	FindNo	ContNo	Cont	DatA	DatB	Rel	Diam	Height	P1	P2

a) Translucent matrix with irregular trails

barrel-shaped, translucent deep olive green matrix and opaque white trailing (zigzag between two bands ?)										
125.	6404	791	di	450	500	1	1.20	0.87	0.22	0.29
medium-sized, flattened globular, translucent deep olive green matrix and opaque white trailing (2 crossing trails?)										
126.	6405	864	di	425	550	–	1.00	0.56	0.26	0.35

b) Translucent matrix with single spiral trail

disc-shaped, faintly translucent blue-green matrix and opaque red trailing										
127.	6561	2063	os	500	550	1	0.70	0.25	0.22	0.25

c) Opaque matrix with two wide crossing trails (double swag)

globular, opaque yellow matrix and translucent (blue-)green trails (Koch 1997, Württemberg/Nordbaden group 33,20)										
128.	6704	1233	di	550	600	3	0.72	0.60	0.20	0.23

d) Opaque matrix with two narrow crossing trails (double swag)

flattened globular, opaque white matrix and translucent blue decoration (Koch 1977, Schretzheim group 34,1/2/5; Koch 1997, Württemberg/Nordbaden group 34,41)

129.	1043	605	ph	600	650	4	0.99	0.65	0.35	0.40
130.	5972*	2165	os	550	600	1	–	–	–	–

segmented (2 barrel-shaped segments), opaque white matrix and translucent pale blue-green decoration (Koch 1977, variety Schretzheim group 34,6)

131.	–	–	stray	–	–	–	0.70	1.06	0.24	0.33

undiagnostic shape, opaque red matrix and opaque white decoration (Koch 1977, Schretzheim group 34)

132.	6704*	1233	di	550	600	3	–	–	–	–
133.	7035	1233	di	550	600	3	0.82	–	–	0.30

e) Opaque matrix with a single spiral trail

long cylinder, opaque red matrix and opaque white trailing (Koch 1977, Schretzheim group 42,12/13)

134.	5170*	–	–	–	–	–	0.72R	1.40	0.22	–

f) Opaque matrix with irregular small dots

small globular bead with pale orange matrix and opaque yellow dots encircled by thin 'black' lines

135.	1542*	1043	di	775	850	1	0.64	0.60	0.26	0.28

small globular bead with pale orange matrix and opaque yellow dots

136.	5971*	2163	di	700	775	1	0.62R	0.56	0.21A	0.23

g) Small, globular, bichrome

translucent orange glass with an opaque white, cylindrical core

137.	1	–	ts	–	–	–	0.44	0.28	0.08	0.11

1.4 *Millefiori beads (separate)*

CatNo	FindNo	ContNo	Cont	DatA	DatB	Rel	Diam	Height	P1	P2

a) Barrel-shaped (*Blättchenmillefiori*)

the bead is composed of two fields of almost translucent pinkish glass, each with a white encircled, opaque red dot ('eye') in the centre, and two translucent green fields, one with a flower in the centre made up of four opaque yellow, double petals in the centre and the other with a failed flower of only five opaque yellow 'streaks'; two colourless end-strips covered with a thin layer of opaque red glass strips are wrapped around the perforations. (Schretzheim Group M32)

138.	6255	975	cl	550	650	1	1.22	1.03	0.36	0.42

b) Long hexagonal

matrix of predominantly faintly translucent dark blue glass containing two white encircled, opaque red dots and a strip of opaque pale green glass at one end (P2), which extends over three quarters of the bead on one side. Strong, whitish weathering. (Small and simplified variety of *Schretzheim* Group M 62)

139.	10436	2471	well27	750	850	3	0.48	1.50	0.15	0.24

1.5 *Inhumation burial: monochrome and bichrome beads*

CatNo	FindNo	ContNo	Cont	DatA	DatB	Rel	Diam	Height	P1	P2

a) (Nearly) annular

small, translucent brown

140.	11702	1804	grave	550	600	1	0.76	0.37	0.24	0.35

small, faintly translucent/opaque green, red streaks

141.	11702	1804	grave	550	600	1	0.59	0.29	0.20	0.26

small, faintly translucent dark bluish green

142.	11702	1804	grave	550	600	1	0.67	0.29	0.24	0.28

b) Globular

small (flattened), translucent blue-green

143.	11702	1804	grave	550	600	1	0.89	0.44	0.22	0.29
144.	11702	1804	grave	550	600	1	0.80	0.49	0.20	0.27

c) Disc-shaped

translucent blue-green

145.	11702	1804	grave	550	600	1	0.78	0.32	0.19	0.26

faintly translucent bluish green with self-crossing opaque white trail, red streak parallel to white trail; weathered surface with bubbles

146.	11702	1804	grave	550	600	1	0.76	0.36	0.22	0.26

d) Cylindrical

long cylinder, translucent dark blue

147.	11702	1804	grave	550	600	1	0.44	1.36	0.18	0.20

short cylinder (*'segment'*), translucent dark blue matrix with approximately 7 marvered, irregular opaque yellow dots (blobs); weathered surface with many bubbles

148.	11702	1804	grave	550	600	1	1.42	0.73	0.45	0.46

e) Irregularly wound

roughly cylindrical, 'black'

149.	11702	1804	grave	550	600	1	0.52	0.60	0.13	0.17
150.	11702	1804	grave	550	600	1	0.42	0.49	0.13	0.18

f) Drawn, segmented

tiny barrel-shaped segments, translucent dark blue

151.	11702(2)	1804	grave	550	600	1	0.31	0.51	0.06	0.07
152.	11702(2)	1804	grave	550	600	1	0.26	0.50	0.07	0.09
153.	11702(2)	1804	grave	550	600	1	0.39	0.52	0.07	0.07
154.	11702(1)	1804	grave	550	600	1	0.32	0.21	0.05	0.06
155.	11702(1)	1804	grave	550	600	1	0.31	0.25	0.09	0.10
156.	11702(1)	1804	grave	550	600	1	0.28	0.22	0.07	0.08

barrel-shaped segments, enclosing silver foil

157.	11702(1)	1804	grave	550	600	1	0.39	0.45	0.12	0.13
158.	11702(1)	1804	grave	550	600	1	0.43	0.46	0.12	0.13

1.6 *Cremation burial: monochrome & bichrome beads*

CatNo	FindNo	ContNo	Cont	DatA	DatB	Rel	Diam	Height	P1	P2

a) Globular

small, translucent blue-green (slightly molten)

159.	4570	448	crem	425	450	1	0.40	0.30R	–	–

small, translucent dark bluish green (slightly molten)

160.	4570	448	crem	425	450	1	0.45	0.32	0.12	–

medium-sized, flattened, opaque red (intact)

161.	4570	448	crem	425	450	1	0.85	0.59	0.16	0.32

(flattened), translucent dark blue (deformed by fire)

162.	4570	448	crem	425	450	1	0.84R	0.55R	–	–

b) Biconical

small, sharply biconical, translucent dark blue

163.	4570	448	crem	425	450	1	0.68	0.41	0.12	0.14
164.	4570	448	crem	425	450	1	0.71	0.44	0.14	0.17

c) Undiagnostic (molten)

remains of a translucent 'sky' blue bead?

165.	4570	448	crem	425	450	1				

large blob of opaque white glass, streaked with tr. blue-green glass, probably a bead

166.	4570	448	crem	425	450	1	–	–	–	–

blob of translucent brown glass with opaque white glass (circles?), probably a bead

167.	4570	448	crem	425	450	1	–	–	–	–

small fragments of a translucent blue bead?

168.	4840	448	crem	425	450	1	–	–	–	–

fragments of colourless glass with what seems to be iron oxide scale inside a perforation, probably of a bead

169.	4671(2)	448	crem	425	450	–	0.50R	0.75	–	–

1.7 *Undiagnostic beads*

CatNo	FindNo	ContNo	Cont	DatA	DatB	Rel	Diam	Height	P1	P2

fragment of an opaque yellow bead decorated with translucent blue glass										
170.	5632	975	cl	550	650	1	–	–	–	–
splinters of olive green glass, probably of a bead										
171.	6208	1233	di	550	600	3	–	–	–	–
fragment of a translucent yellowish green bead										
172.	7441	907	pit	250	350	1	–	–	–	–
splinters of translucent blue glass, possibly of a bead										
173.	7441	907	pit	250	350	1	–	–	–	–
fragment of opaque red bead, possibly globular										
174.	9847	2342	os	450	500	1	–	–	–	–
fragment of an opaque red bead, possibly segmented										
175.	11659	2792	os/cl?	800	850	3	–	–	–	–

2 FAIENCE BEADS

CatNo	FindNo	ContNo	Cont	DatA	DatB	Rel	Diam	Height	P1	P2
a) Melon-bead with faint ribs, turquoise glazed exterior heavily weathered										
176.	1*	–	ts	–	–	–	1.33	0.95	0.75R	–
177.	1522*	1163	trl/os	600	700	4	–	–	–	–

3 BEADS OF BAKED CLAY

CatNo	FindNo	ContNo	Cont	DatA	DatB	Rel	Diam	Height	P1	P2
a) Short cylinder										
orange-brown, pitted surface										
178.	5425	811	trl	450	500	–	0.62	0.46	0.11	0.13
greyish exterior, pinkish interior										
179.	6257*	1232	di	550	650	–	0.65	0.54	0.15	–

4 AMBER BEADS

4.1 *Amber beads (separate)*

CatNo	FindNo	ContNo	Cont	DatA	DatB	Rel	Diam	Height	P1	P2	
a) Globular											
small, flattened globular											
180.	7600	1328	di	500	550	–	0.52	0.37	0.12	0.14	
small, roughly globular, translucent red-orange											
181.	9850	2354	sh8	450	500	3	0.51	0.45	0.12	0.13	
medium-sized, flattened globular, translucent red-orange; weathered											
182.	9652	2241	hearth	500	550	1	1.10	0.65	0.24	0.28	
				(ass. with sh8!)							
b) Roughly rectanguar/square/triangular in section											
large bead, roughly rectangular in section at one end, triangular in section at other end, translucent red-orange											
183.	6663	3152	trl/os	770	800	–	1.08	0.93	1.06	0.35	0.39
half bead, probably rectangular in section, translucent red-orange											
184.	10004*	2355	sh8	450	500	1	0.57	–	0.55	0.13	0.15
originally probably half a (failed) bead with rectangular section, re-worked into disc-shaped bead with new perforation at right angles with and through the centre of the old perforation, translucent red-brown											
185.	9850F	2354	sh8	450	500	3	0.56	0.67	0.27	0.16	0.17
half bead, probably roughly square in section with rounded corners, translucent orange											
186.	7545*F	1328	di	500	550	1	0.72	–	>0.83	0.22	–
c) Half-fabricates											
half of a roughly angular-shaped, large bead with a failed perforation, translucent red-orange											
187.	6273*F	–	ts	–	–	–	1.28	–	1.43	–	–

small asymmetrical fragment with failed perforation, opaque orange-yellow amber

188.	9772*F	2341	sh7	450	500	1	0.94	–	–	–	–

d) Undiagnostic

small fragment of a bead, possibly with a triangular section, translucent orange

189.	5548	2226	sh	650	750	–	–	–	–	–	–

small fragment with apparent perforation of an undiagnostic bead, translucent red-orange

190.	9902	2355	sh8	450	500	4	–	–	–	–	–

4.2 *Amber beads from inhumation burial*

CatNo	FindNo	ContNo	Cont	DatA	DatB	Rel	Width	Thickn	Height	P1	P2
a) roughly rectangular in section with rounded edges translucent red-orange											
191.	11702	1804	grave	550	600	1	1.09	0.95	1.10	0.17	0.17
192.	11702	1804	grave	550	600	1	1.01	0.82	1.00	0.14	0.15
193.	11702	1804	grave	550	600	1	0.80	0.50	0.83	0.15	0.16
194.	11702	1804	grave	550	600	1	0.62	0.36	0.70	0.15	0.16
195.	11702	1804	grave	550	600	1	0.79	0.53	0.80	0.17	0.18
196.	11702F	1804	grave	550	600	1	0.50	0.30	0.44	0.15	0.17
197.	11702	1804	grave	550	600	1	0.76	0.54	0.77	0.17	0.18
198.	11702	1804	grave	550	600	1	0.74	0.66	0.75	0.16	0.17
199.	11702	1804	grave	550	600	1	0.75	0.58	0.68	0.17	0.18
200.	11702	1804	grave	550	600	1	0.83	0.71	0.81	0.17	0.18
b) Roughly triangular in section with rounded edges translucent red-orange											
201.	11702	1804	grave	550	600	1	0.65	0.57	0.54	0.15	0.16
202.	11702	1804	grave	550	600	1	0.58	0.50	0.53	0.16	0.16
203.	11702	1804	grave	550	600	1	0.74	0.65	0.71	0.18	0.19
204.	11702F	1804	grave	550	600	1	0.67	0.47	0.65	0.20	–
205.	11702	1804	grave	550	600	1	0.48	0.37	0.46	0.15	0.16
c) Roughly square in section with rounded edges/almost globular translucent red-orange											
206.	11702	1804	grave	550	600	1	0.70	0.63	0.67	0.15	0.16
207.	11702	1804	grave	550	600	1	0.66	0.64	0.65	0.17	0.18
208.	11702	1804	grave	550	600	1	0.58	0.57	0.57	0.16	0.17
209.	11702	1804	grave	550	600	1	0.50	0.50	0.46	0.16	0.18
210.	11702	1804	grave	550	600	1	0.40	0.40	0.38	0.15	0.16
d) Oblate (*spitz-oval*) translucent red-orange											
211.	11702	1804	grave	550	600	1	0.54	0.41	1.32	0.15	–
e) Asymmetrical translucent red-orange											
212.	11702	1804	grave	550	600	1	1.15	0.65	0.33	0.22	0.24
f) Undiagnostic splinters, translucent orange											
213.	11702	1804	grave	550	600	1	–	–	–	–	–

5 METAL BEAD FROM INHUMATION BURIAL

CatNo	FindNo	ContNo	Cont	DatA	DatB	Rel	Diam	Height	P1	P2
a) Disc-shaped, heavily weathered bead of lead-rich metal										
214.	11702	1804	grave	550	600	1	0.82	0.41	0.13	0.16

6 VARIA

CatNo	FindNo	ContNo	Cont	DatA	DatB	Rel	Length	Width	Thickn.
a) Bangle, fragment of a La Tène bangle with D-shaped profile of translucent purple glass with part of an applied opaque yellow zigzag trail (Haevernic type3b)									
215.	6704	1233	di	550	600	3	8.00	0.51	1.67

7 WASTE FROM GLASS BEAD MAKING

CatNo	FindNo	ContNo	Cont	DatA	DatB	Rel	Height	Width	Length
a) Glass tessera, complete tessera, opaque yellow with colourless streaks									
216.	10736	514	di	750	770	3	0.81	0.97	1.25
b) Glass rod, breaking splinter, opaque greenish white									
217.	3829	696	os	425	500	1	0.55	0.72	0.40
c) 'Punty' glass, hollow splinter of translucent deep blue-green (turquoise) glass with reduced copper of a pinkish-purplish colour on the concave interior									
218.	4601	2817	trl/os	750	800	4	0.21	0.40	1.59
d) Fragment of baked clay (tray/furnace) with glass adhering, thick-walled fragment of baked clay of a relatively soft and porous, pinkish orange matrix with a thick layer of weathered, opaque yellow glass (thickness 1.0-1.3 cm), which has permeated throughout the matrix									
219.	2943	625	met	575	625	3	4.30	3.30	6.80
e) Metal-working crucible, base fragment of a thin-walled crucible of hard-baked, greyish white fabric with yellowish grey 'glassy' deposits on the inside, the fractures and the outside									
220.	7824	36	met	250	350	1	0.82	0.55	4.33

8 UNWORKED AMBER/WASTE FROM AMBER BEAD MAKING

CatNo	FindNo	ContNo	Cont	DatA	DatB	Rel	Height	Width	Length
a) Unworked amber									
irregular lump of weathered, unworked, faintly translucent red-brown amber with one fractured side									
221.	826	143	sl	500	550	–	1.24	1.56	2.04
small lump of weathered, unworked, translucent red-orange amber									
222.	9850	2354	sh8	450	500	3	0.68	1.01	1.09
b) Waste fragments									
splinters of translucent red-orange amber									
223.	997	1072	?	670	750	–	–	–	–
splinters of translucent orange amber (possibly from a bead)									
224.	5537	1261	di	550	650	–	–	–	
small fragment of translucent pale orange amber									
225.	6414	822	–	450	525	–	0.46	0.72	1.14
fragments of opaque yellow/pale orange amber									
226.	7448	1233	di	550	600	3	0.73	0.92	1.36
227.	7448	1233	di	550	600	3	0.36	1.20	1.61
228.	7448	1233	di	550	600	3	0.30	0.63	1.20

APPENDIX: NECKLACE ENSEMBLE FROM INHUMATION GRAVE (FEATURE 1804)

1 GLASS BEADS

CatNo	FindNo	ContNo	Cont	DatA	DatB	Rel	Diam	Height	P1	P2
a) Aannular										
small, translucent brown										
140.	11702	1804	grave	550	600	1	0.76	0.37	0.24	0.35
small, faintly translucent (?) green, red streaks										
141.	11702	1804	grave	550	600	1	0.59	0.29	0.20	0.26
small, faintly translucent dark blue-green (turquoise)										
142.	11702	1804	grave	550	600	1	0.67	0.29	0.24	0.28

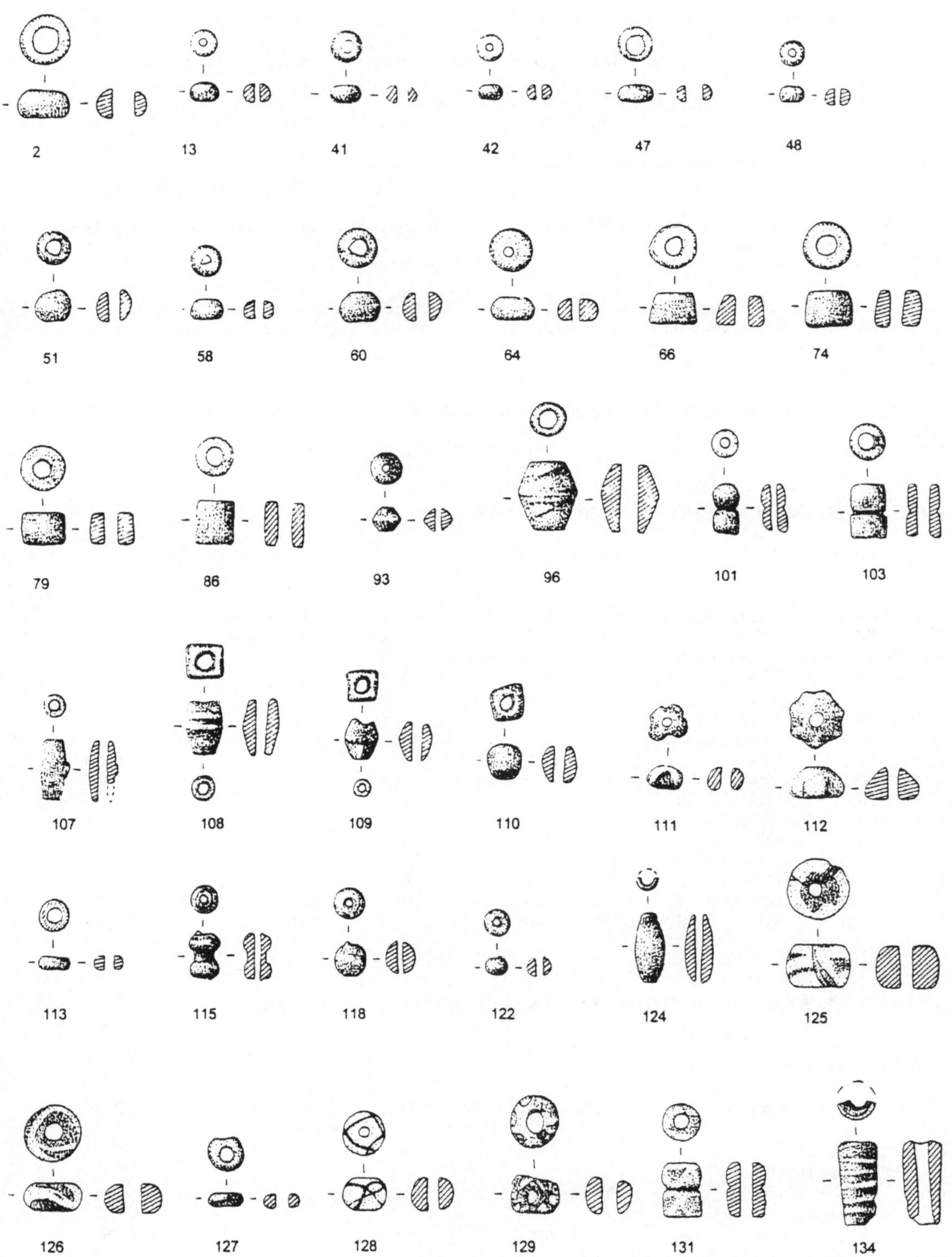

Figure 5. Beads from Wijnaldum with catalogue numbers. Scale 1:1 (drawing D. Taylor, Dept. of Archaeology, University of Nottingham).

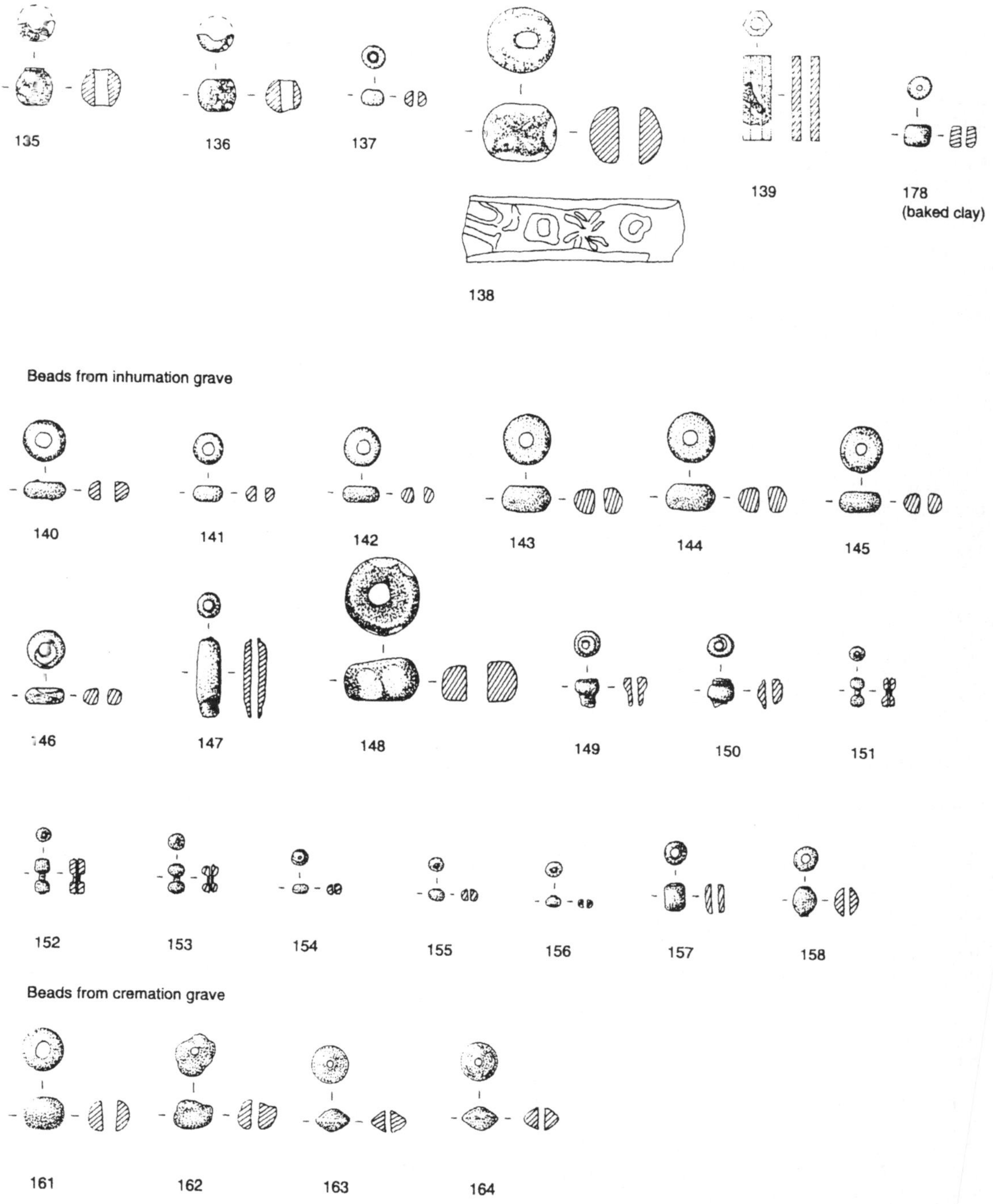

Figure 6. Beads from Wijnaldum with catalogue numbers. Scale 1:1 (drawing D. Taylor, Dept. of Archaeology, University of Nottingham).

b) Globular, small (flattened), translucent blue-green

143.	11702	1804	grave	550	600	1	0.89	0.44	0.22	0.29
144.	11702	1804	grave	550	600	1	0.80	0.49	0.20	0.27

c) Disc-shaped

translucent blue-green

145.	11702	1804	grave	550	600	1	0.78	0.32	0.19	0.26

faintly translucent (?) bluish green with self-crossing opaque white trail, red streak parallel to white trail; weathered surface with bubbles

146.	11702	1804	grave	550	600	1	0.76	0.36	0.22	0.26

d) Cylindrical

long cylinder, translucent dark blue

147.	11702	1804	grave	550	600	1	0.44	1.36	0.18	0.20

short cylinder ('*segment*'), translucent dark blue matrix with approximately 7 marvered, irregular opaque yellow dots (blobs); weathered surface with many bubbles.

148.	11702	1804	grave	550	600	1	1.42	0.73	0.45	0.46

e) Irregularly wound 'black'

149.	11702	1804	grave	550	600	1	0.52	0.60	0.13	0.17
150.	11702	1804	grave	550	600	1	0.42	0.49	0.13	0.18

f) Drawn, segmented

tiny barrel-shaped segments, translucent dark blue

151.	11702(2)	1804	grave	550	600	1	0.31	0.51	0.06	0.07
152.	11702(2)	1804	grave	550	600	1	0.26	0.50	0.07	0.09
153.	11702(2)	1804	grave	550	600	1	0.39	0.52	0.07	0.07
154.	11702(1)	1804	grave	550	600	1	0.32	0.21	0.05	0.06
155.	11702(1)	1804	grave	550	600	1	0.31	0.25	0.09	0.10
156.	11702(1)	1804	grave	550	600	1	0.28	0.22	0.07	0.08

barrel-shaped segments, enclosing silver foil

157.	11702(1)	1804	grave	550	600	1	0.39	0.45	0.12	0.13
158.	11702(1)	1804	grave	550	600	1	0.43	0.46	0.12	0.13

2 AMBER BEADS

CatNo	FindNo	ContNo	Cont	DatA	DatB	Rel	Width	Thickn	Height	P1	P2
a) Roughly rectangular in section with rounded edges translucent red-orange											
191.	11702	1804	grave	550	600	1	1.09	0.95	1.10	0.17	0.17
192.	11702	1804	grave	550	600	1	1.01	0.82	1.00	0.14	0.15
193.	11702	1804	grave	550	600	1	0.80	0.50	0.83	0.15	0.16
194.	11702	1804	grave	550	600	1	0.62	0.36	0.70	0.15	0.16
195.	11702	1804	grave	550	600	1	0.79	0.53	0.80	0.17	0.18
196.	11702F	1804	grave	550	600	1	0.50	0.30	0.44	0.15	0.17
197.	11702	1804	grave	550	600	1	0.76	0.54	0.77	0.17	0.18
198.	11702	1804	grave	550	600	1	0.74	0.66	0.75	0.16	0.17
199.	11702	1804	grave	550	600	1	0.75	0.58	0.68	0.17	0.18
200.	11702	1804	grave	550	600	1	0.83	0.71	0.81	0.17	0.18
b) Roughly triangular in section with rounded edges translucent red-orange											
201.	11702	1804	grave	550	600	1	0.65	0.57	0.54	0.15	0.16
202.	11702	1804	grave	550	600	1	0.58	0.50	0.53	0.16	0.16
203.	11702	1804	grave	550	600	1	0.74	0.65	0.71	0.18	0.19
204.	11702F	1804	grave	550	600	1	0.67	0.47	0.65	0.20	–
205.	11702	1804	grave	550	600	1	0.48	0.37	0.46	0.15	0.16
c) Roughly square in section with rounded edges translucent red-orange											
206.	11702	1804	grave	550	600	1	0.70	0.63	0.67	0.15	0.16
207.	11702	1804	grave	550	600	1	0.66	0.64	0.65	0.17	0.18
208.	11702	1804	grave	550	600	1	0.58	0.57	0.57	0.16	0.17
209.	11702	1804	grave	550	600	1	0.50	0.50	0.46	0.16	0.18

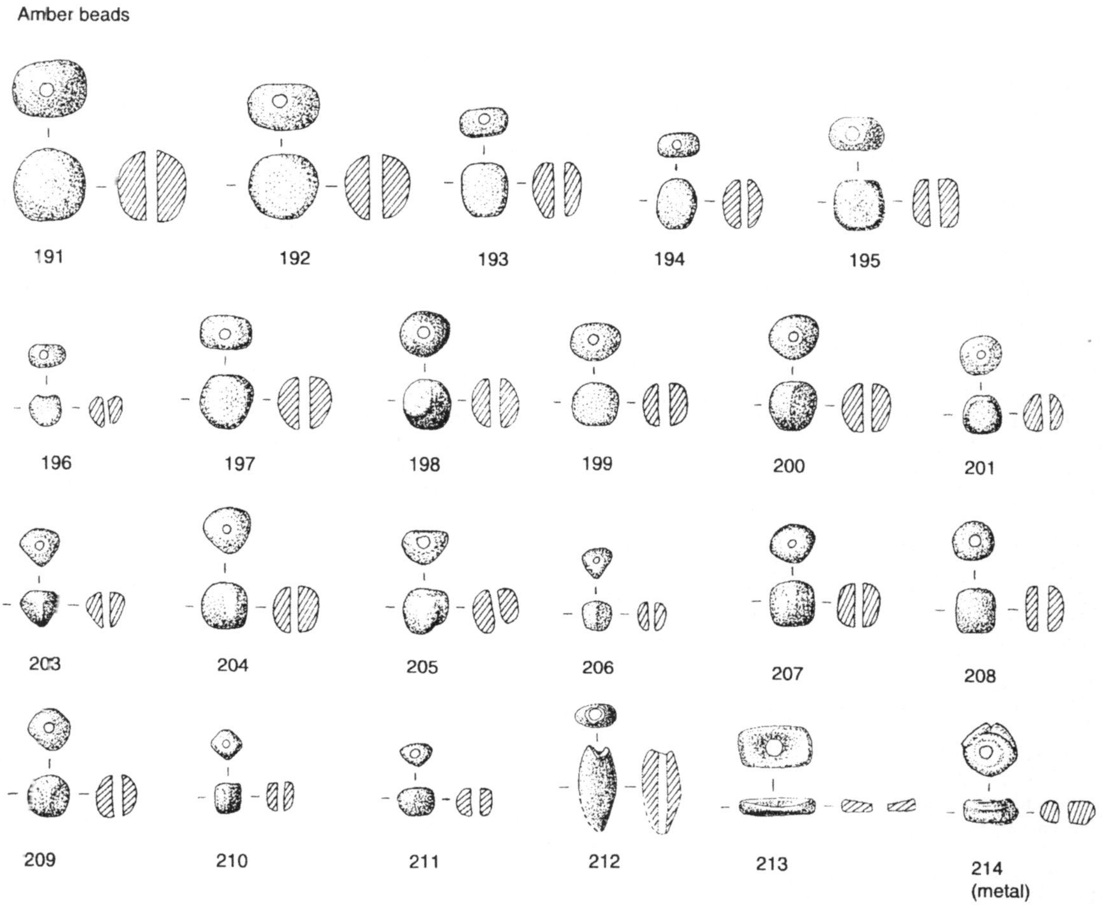

Figure 7. Beads from Wijnaldum with catalogue numbers. Scale 1:1 (drawing D. Taylor, Dept. of Archaeology, University of Nottingham).

210.	11702	1804	grave	550	600	1	0.40	0.40	0.38	0.15	0.16

d) Oblate (spitz-oval) translucent red-orange

211.	11702	1804	grave	550	600	1	0.54	0.41	1.32	0.15	–

e) Asymmetrical translucent red-orange

212.	11702	1804	grave	550	600	1	1.15	0.65	0.33	0.22	0.24

f) Undiagnostic splinters, translucent orange

213.	11702	1804	grave	550	600	1	–	–	–	–	–

3 METAL BEAD

CatNo	FindNo	ContNo	Cont	DatA	DatB	Rel	Diam	Height	P1	P2
a) Disc-shaped heavily weathered bead of lead-rich metal										
214.	11702	1804	grave	550	600	1	0.82	0.41	0.13	0.16

REFERENCES

ANDRAE, R. 1973. Mosaikaugenperlen. Untersuchungen zur Verbreitung und Datierung Karolingerzeitlicher Millefioriperlen in Europe, *Acta Praehistorica et Archaeologica* 4: 101-198.

BÖHME, H.W. 1974. *Germanische Grabfunde des 4. bis 5. Jahrhunderts zwischen unterer Elbe und Loire*, (Münchener Beiträge zur Vor- und Frühgeschichte 19).

BÖHNER, K. 1958. *Die fränkischen Altertümer des Trierer Landes*, (Germanische Denkmäler der Völkerwanderungszeit B10).

BLOEMERS, J.H.F., L.P. LOUWE KOOIJMANS & H. SARFATIJ 1986. *Verleden Land*, Amsterdam.

CALLMER, J. 1995. The influx of oriental beads into Europe during the 8th century AD. In: M. Rasmussen, U. Lund Hansen & U. Näsman (eds), *Glass beads. Cultural History, Technology, Experiment and Analogy*, 49-54 (Studies in Technology and Culture vol. 2).

CALLMER, J. 1997. Beads and bead production in Scandinavia and the Baltic region ca 600 AD-100: a general outline. In: U. von Freeden & A. Wieczorek (eds), *Perlen: Archäologie, Techniken, Analysen*, 197-202 (RGK Kolloquien zur Vor- und Frühgeschichten 1).

CALLMER, J. & J. HENDERSON 1991. Glassworking at Åhus, S. Sweden (eight century AD), *Laborativ Arkeologi* 5: 143-154.

ERDRICH, M. & H.-U. VOSS 1997. Die Perlen der Germanen des 1.-5. Jahrhundert in Mecklenburg-Vorpommern, Schleswig-Holstein und Niedersachsen. In: U. von Freeden & A. Wieczorek (eds), *Perlen: Archäologie, Techniken, Analysen*, 77-94 (RGK Kolloquien zur Vor- und Frühgeschichten 1).

ES, VAN W.A. & J. YPEY 1977. Das Grab der 'Prinzessin von Zweeloo' und seine Bedeutung im Rahmen des Gräberfeldes, *Studien zur Sachsenforschung* 1: 97-126.

EVISON, V. 1987. *Dover: Buckland Anglo-Saxon cemetery*, London (Historical Buildings and Monuments, Archaeological Report no. 3).

GAM, T. 1990. Prehistoric Glass Technology. Experiments and Analyses, *Journal of Danish Archaeology* 9: 203-213.

HAEVERNICK, TH.E. 1960. *Die Glasarmringe und Ringperlen der Mittel- und Spätlatènezeit auf dem Europäischen Festland*, Bonn.

HENDERSON, J. 1992. Dunmisk and Glass-making in Early Christian Ireland, *Antiquity* 66: 52-64.

HENNING, J. 1996. Handel, Verkehrswege und Beförderungsmittel im Merowingerreich. In: A. Wieczorek et al. (eds), *Die Franken-Wegbereiter Europas. Vor 1500 Jahren: König Chlodwig und seine Erben*, Mainz.

HODGES, R. 1991. A fetishism for commodities: ninth-century glass-making at San Vincenzo al Volturno. In: M. Mendera (ed.), *Archeologia e storia della produzione del vetro preindustriale*, Florence, 67-90.

HULST, R.A. 1992. Opgravingen Rijksarchief te Maastricht, *Archeologie in Limburg*, 90.

JENSEN, S. 1991. *The Vikings of Ribe*, Ribe.

JÓNSSON, M. & HUNNER, P. 1995. Gold-foil beads. In: M. Rasmussen, U. Lund Hansen & U. Näsman (eds), *Glass beads. Cultural History, Technology, Experiment and Analogy*, 113-116 (Studies in Technology and Culture vol. 2).

KATZAMEYER, TH. 1997. Verbreitungsbilder ausgewählter Perlentypen des Frühmittelalters in Süd- und Westdeutschland. In: U. von Freeden & A. Wieczorek (eds), *Perlen: Archäologie, Techniken, Analysen*, 149-160 (RGK Kolloquien zur Vor- und Frühgeschichten 1).

KNOL, E. 1993. *De Noordnederlandse kustlanden in de Vroege Middeleeuwen*, Groningen.

KNOL, E., W. PRUMMEL, W., H.T. UYTTERSCHAUT et al. 1996. Het vroeg-middeleeuwse grafveld van Oosterbeintum, *Jaarverslagen van de Vereniging voor Terpenonderzoek* 73-74, Groningen.

KOCH, U. 1977. *Das Reihengräberfeld bei Schretzheim*, Berlijn (Germ. Denkmäler Völkerwanderungszeit A 13).

KOCH, U. 1982. *Die fränkischen Gräberfelder von Bargen und Berghausen in Nordbaden*, Stuttgart (Forschungen und Berichte zur Vor- und Frühgeschichte in Baden-Württemberg 12).

KOCH, U. 1987. *Die Glas- und Edelsteinfunde aus den Plangrabungen 1967-1983. Teil 1: Text. Der Runde Berg Berg bei Urach*, Heidelberg 1987.

KOCH, U. 1990. *Das fränkische Gräberfeld von Klepsau im Hohenlohekreis*, Stuttgart (Forschungen und Berichten zur Vor- und Frühgeschichte in Baden-Württemberg 38).

KOCH, U. 1997. Polychrome Perlen in Württemberg/Nordbaden. In: U. von Freeden & A. Wieczorek (eds), *Perlen: Archäologie, Techniken, Analysen*, 143-148 (RGK Kolloquien zur Vor- und Frühgeschichten 1).

LEGOUX, R. 1994. Objets de parure. De la typologie à la chronologie. In: *Verre et merveilles. Mille ans de verre dans le nord-ouest de la Gaule. Exhibition catalogue of the 'Musée Archéologique Départemental du Val-d'Oise' 1993-1994*, Guiry-en-Vexin, 103-115.

LITH, VAN S.M.E. 1977. Römisches Glas aus Velsen, *Oudheidkundige Mededelingen van het Rijksmuseum voor Oudheden te Leiden* LVIII: 1-62.

LUNDSTRÓM, A. 1976. Bead Making in Scandinavia in the Early Middle Ages, *Early Medieval Studies* 9: 3-19.

LUNDSTRÓM, A. 1981. Survey of the Glass from Helgö. In: A. Lundström & H. Clarke (eds), *Excavations at Helgö VII. Glass-Iron-Clay*, Stockholm, 1-38.

MIEDEMA, M. 1983. Vijfentwintig eeuwen bewoning in het terpenland ten noordwesten van Groningen, dissertation VU, Amsterdam.

PIRLING, R. 1966. *Das Römisch-Fränkische Gräberfeld von Krefeld-Gellep*, Berlin (Germanische Denkmäler Völkerwanderungszeit B 2).

SABLEROLLES, Y. 1991. Het glas van Gennep. De glasvondsten van een nederzetting uit de Volksverhuizing te Gennep (Limburg) (Text and Catalogue), unpublished IPP-report, Amsterdam.

SABLEROLLES, Y., J. HENDERSON & W. DIJKMAN 1997. Early medieval glass bead making in Maastricht (Jodenstraat 30), The Netherlands. An archaeological and scientific investigation. In: U. von Freeden & A. Wieczorek (eds), *Perlen. Archäologie, Techniken, Analysen*, 293-314 (RGK Kolloquien zur Vor- und Frühgeschichten 1).

SCHMID, P. 1994. Oldorf (Niedersachsen), ein frühmittelalterliche Wurtsiedlung, *Germania* 72 (1): 231-267.

SIEGMANN, M. 1997. Die Perlen des frühmittelalterlichen Gräberfeldes von Liebenau, Kr. Nienburg/Weser (Niedersachsen). In: U. von Freeden & A. Wieczorek (eds), *Perlen. Archäologie, Techniken, Analysen*, 133-142

(RGK Kolloquien zur Vor- und Frühgeschichten 1).

STEPPUHN, P. 1997. Bleiglasperlen des frühen und hohen Mittelalters in Nordeuropa. In: U. von Freeden & A. Wieczorek Analysen (eds), *Perlen. Archäologie, Techniken, Analysen*, 203-210 (RGK Kolloquien zur Vor- und Frühgeschichten 1).

STEPPUHN, P. 1998. *Die Glasfunde von Haithabu*, Neumünster (Berichte über die Ausgrabungen in Haithabu 32).

THOMSEN, P.O. 1995. The question of bead making in the late Roman Iron Age at Lundeborg, Denmark. In: M. Rasmussen, U. Lund Hansen & U. Näsman (eds), *Glass beads. Cultural History, Technology, Experiment and Analogy*, 19-24 (Studies in Technology and Culture vol. 2).

WATERBOLK, H.J. & H.T. WATERBOLK 1991. Amber on the Coast of the Netherlands. In: H. Thoen et al. (eds), *Studia Archaeologica Liber Amicorum Jacques A.E. Nenquin*, 201-209

NOTES

1. See the introduction of Sablerolles on the glass vessel finds (this volume).
2. See Knol (1993: necklaces with glass and shell beads come from graves at Hogebeintum pp 85-87, Fig. 86, Cornjum-Wijdeveld pp 207-208, Fig. 66 and Aalzum-village terp p 213, Fig. 710), Knol has been able to add 6 finds to the distribution map of shell beads in the northern Dutch coastal area (p. 207, Fig. 67). Two beads of rock crystal were found in graves at Beetgum-Besseburen and Oosterbeintum-village terp (p. 177, note 1062). A necklace with beads made of sapropelite comes from a late Roman grave at Welsrijp-village terp (p. 190, Fig. 54.3).
3. See contribution by Cuijpers et al. (this volume).
4. See previous note.
5. Necklaces with glass and amber beads were for instance found in graves at Corjum-Wijdeveld, Aalzum-village terp, Beers, Hogebeintum-village terp and Oosterbeintum, (Knol 1993, p. 208, Fig. 66, p. 213, Fig. 69 4, p. 215, Fig. 71, p. 216, Fig. 72, p. 221, Fig. 77,). For Oosterbeintum: see also Knol et al. (1996, pp 77-78).
6. For Germany see: Koch (1977, pp 71-218, colour plates 1-6 – Schretzheim), Koch (1997, pp 143-148, colour plates 5-13 – Württemberg/Nordbaden), Sasse & Theune (1996, pp 187-231 – Weingarten & Eichstetten, Beilage 2) and Siegmund (1998, pp 57-78, plates 1 and 2 – Niederrhein). For northern France, see: Legoux in: Verre et Merveilles (1993, pp 103-115, 2 colour plates).
7. For the production technique of drawn beads, see: Guido (1978, p. 93).
8. Cf. Koch (1977, colour plates 4-6).
9. Van Es & Ypey (1977, pp 113-114, Fig. 8c).
10. Guido (1978, pp 65-71, small, undecorated annular beads (diameter under 1.5 cm) of Group 6.IIb (translucent blue-green), 6.IVa (translucent/opaque blue) and Group 7.II (translucent blue-green) and small, undecorated globular beads of group 7.III (translucent/opaque green) and 7.IV (translucent/opaque blue), colour plate II).
11. Guido (1978, pp 65-71, small undecorated annular beads (diameter under 1.5 cm) of Group 6.IIb (translucent blue-green), 6.Iva (translucent/opaque blue) and Group 7.II (translucent blue-green), 7.IV (translucent/opaque blue), colour plate II).
12. Guido (1978, pp 97-98, small biconical beads, blue (diameter 0.5-0.7 cm), Fig. 37.12).
13. Cf. Koch 1987, 320.
14. Siegmund (1998, p. 75, Perlen 47.1, 2 Kombinationsgruppe A, colour plate 1).
15. Siegmund (1998, 64, Perlen 1.2, Kombinationsgruppen B & C, colour plate 1).
16. Legoux in: Verre et merveilles (1993, pp 105-106 code 5, colour plate).
17. Koch (1987, p. 324 – note 60).
18. Böhner (1958, p. 74 – note 24).
19. Cf. Koch (1987, pp 320-321, Gruppe C – Mittelgrosse transluzide Perlen; diameter 0.95-1.55 cm).
20. Sablerolles 1991 (catalogue), 82 & colour plate 4,5 (cat no. 206).
21. Respectively Böhner (1958, p. 72) and Pirling (1966, p. 156).
22. Siegmund (1998, p. 67, Perle 31.1, colour plate 1).
23. Siegmann (1997, p. 136).
24. Van Es & Ypey (1977, p. 113, Fig. 8g).
25. Koch (1990, p. 120).
26. Legoux in: Verres et Merveilles (1993, pp 105-106 – perle code 2: petite perle spiralée, opaque, de couleur noire ou marron).
27. Böhme (1974, p. 41).
28. Siegmann (1997, pp 136-137).
29. Van Es & Ypey (1977, pp 113-115, Fig. 8l-p).
30. Sasse & Theune (1996, pp 215-216, Kombinationsgruppe G, Typengruppe 5, Liste 1 – Perlentypen 26,27), Siegmann (1997, p. 137, plate 4.3) and Siegmund (1998, p. 59, Kombinationsgruppe I – Perlentypen 37.1, 47.3, colour plate 1).
31. Siegmund (1998, p. 75, Perlentyp 47.5, colour plates 1 and 2) and Steppuhn (1998, p. 38, Form 15, colour plate 76).
32. Sablerolles et al. (1997, pp 297-298, Table 1).
33. Katzameyer (1997, pp 149-160, Figs 4 and 5).
34. Siegmund (1998, pp 58-64, Figs 15 and 16, Lower Rhine Phases 4-7/early 8 (530-610/620)).
35. See notes 31 and 33.
36. Guido (1978, pp 92-99, Figs 37.4, 37.5-37.7 and 37.19).
37. For an overview of dating evidence of monochrome small, globular and cylindrical beads (see Sablerolles et al. 1997, p. 302).
38. Siegmund (1998, p. 69, Perle 33.3, colour plate 1).
39. Van Es & Ypey (1977, pp 113-114, Fig. 8b).
40. See contribution Henderson (this volume).
41. Sasse & Theune (1996, p. 211, Liste 1 – Perlentyp 16, Liste 3 – Typengruppe 2b) and Siegmund (1998, pp 69-70, Perlen 33.2, 35.5, colour plate 1).
42. Siegmund (1998, p. 70, Perle 33.6, colour plate 1).
43. Bloemers et al. (1986, p. 132).
44. Sasse & Theune (1996, pp 213-214, Liste 1 – Perlentypen 18 (Tonne orange), Perlentyp 26 (Doppelkonus), Liste 3 Typengruppe 4b and Typengruppe 5).
45. Siegmund (1998, pp 69-70, Perlen 33.5 and 43.1, colour plates 1 and 2).

46. Siegmann (1997, p. 137, colour plate 4.2).
47. Siegmund (1998, p. 70, Perlentyp 35.1).
48. Knol (1993, p. 80, Fig. 15.1).
49. Schmid 1994, 253-254, Fig. 11.
50. Steppuhn (1998, p. 36, Form 11, plate 5.14-23 (5.22: opaque red)).
51. Knol (1993, p. 79, notes 681-683).
52. Siegmund (1998, pp 58-64, Figs 15-16).
53. Koch (1977, p. 206, Gruppe 33 – Perlen mit weiter Schleifenauflage) and Koch (1997, colour plate 9, Gruppe 33.20).
54. Koch (1977, p. 207, Gruppe 34 – Perlen mit enger Schleifenauflage, colour plate 3).
55. Sasse & Theune (1996, p. 211, Kombinationsgruppen C – D1, 2, Liste 1 –Perlentypen 38-40, Liste 4 – Typengruppe 3.17, 4a.27, 4b.38).
56. Bloemers et al. (1986, p. 132).
57. Koch (1977, p. 209, Gruppe 42, colour plate 4), Sasse & Theune (1996, p. 213, Kombinationsgruppe E, Fig. 15.54, 55, Liste 1 – Perlentyp 44, Liste 4 – Typengruppe 4a.26) and Siegmund (1998, p. 72, Perlentyp 35.13, colour plates 1 and 2).
58. On a photograph issued by the Arabia Steamboat Museum, Kansas City (Missouri, US) different kind of seed beads are shown, amongst others tiny orange and white specimens very similar to the Wijnaldum bead.
59. Sherlock & Welch (1992, pp 176-177, Fig. 58: grave 82).
The author wishes to thank Birte Brugmann for kindly pointing out this reference.
60. For production techniques of segmented beads see Steppuhn (1998, p. 27, Fig. 9a), for segmented beads containing metal foil (see Guido 1978, p. 93) and Jönsson & Hunner (1995, p. 6).
61. Van Lith (1977, p. 49, plate 5.190).
62. Legoux in: Verres et Merveilles (1993, pp 103-106, colour plate – perle code 3).
63. Guido (1978, pp 91-93).
64. Koch (1990, p. 119).
65. Siegmann (1997, 136-137, colour plate 3 and 2).
66. Cf. Guido (1978, p. 94) and Jönnsson & Hunner (1995, p. 115).
67. Unpublished data.
68. Cf. Guido (1978, p. 93).
69. Guido (1978, pp 93-94, Fig. 37.3).
70. Van Es & Ypey (1977, p. 113, Fig. 115.(a?),d,e).
71. For Klepsau (see Koch 1990, p. 119), for Weingarten and Eichstetten (see Sasse & Theune 1996, p. 211 – Kombinationsgruppe C, Fig. 13, Liste 1 – Perlentyp 4, Liste 4 – Typengruppe 2a).
72. Callmer (1995, p. 49). Contra: Henning (1996, note 91).
73. Ibid.
74. Knol (1993, p. 79, notes 686-688, Fig. 15.2).
75. Schmid (1994, p. 254, Fig. 11 – graves 3, 7, 9, 11).
76. Siegmann (1997, pp 135-138, Figs colour plates 3.1-3.3 and 4.1).
77. Steppuhn (1997, p. 205).
78. Callmer (1995, pp 50-51) points out that oriental silver-foil beads continued to be imported into eastern Europe throughout the 8th century and that the first wave of oriental beads did not reach northern Europe before the second half of the 8th century and that the second wave did not reach western Europe before the late 8th century.
79. Steppuhn (1998, pp 40-41).
80. See Henderson (this volume).
81. Callmer (1997, p. 199, colour plate 17.B40).
82. Koch (1977, pp 215-218, colour plate 6).
83. For instance at Schretzheim, Millefiori group M32 (Koch 1977, p. 215, colour plate 6);
84. Respectively Schretzheim Millefiori Gruppe M32 and M62 (Koch 1977, p. 216, colour plate 6).
85. Andrae (1975, pp 162-163), Koch (1977, p. 218) and Sasse & Theune (1996, p. 212 – Kombinationsgruppe D (570-610), Liste 1 – Perlentypen 51,52, Liste 4 – Typengruppe 3.16,19).
86. Koch (1977, p. 218).
87. Andrae (1973, pp 158-160) and Koch (1997, p. 147).
88. Henning in: Wieczorek et al. (1996, pp 794-796).
89. Evison (1987, p. 65).
90. See for instance Koch (1997, colour plate 12).
91. For a summary of the evidence for 8-11th-century centres where glass was being worked, including *millefiori*-rods for bead-making (see Steppuhn 1998, pp 97-98, 106-110).
92. Sablerolles et al. (1997, pp 293-312, esp. 304 – note 54).
93. Knol (1993, p. 205 – note 1207).
94. Callmer (1997, pp 197-201, colour plates 15 A10, 16 A13-15, B24-25, 17 B25).
95. Knol (1993, pp 205-206, note 1208).
96. For a summary of the evidence, see Knol (1993, p. 207, note 1208) and Callmer (1995, pp 52-53). Newly published: Siegmann (1997, pp 137-138, Fig. 16.7,11, colour plate 3.4).
97. See for instance Böhme (1987, p. 289).
98. Voss & Erdrich (1997, pp 90-91 and Finds list 4).
99. See Sablerolles on glass vessels (this volume).
100. Sablerolles (1992 – catalogue, pp 80-81, colour plate 4, 1-3).
101. Ibid. (note ..), 6.
102. Böhme (1987, p. 289).
103. See for instance Böhner (1958, p. 72) and Pirling (1966, p. 158).
104. Cf. Knol et al. (1996, p. 77).
105. Qualitative analysis using XRF-analysis in the Dept. of Archaeology, University of Nottingham.
106. Perhaps made from Roman scrap metal, pers. comm. M. Erdrich.
107. Haevernick (1960, pp 75-94).
108. Koch (1987, p. 341).
109. Haevernick (1981, pp 180-187).
110. See contribution Henderson (this volume).
111. Gam (1990, p. 209) and Jensen (1991, p. 36).
112. Lundström (1976, pp 7-10), Callmer & Henderson (1991, p. 144 and Table 1 DC.6) and Sableroles et al. (1997, p. 298 and Fig. 2.1).
113. See contribution Henderson (this volume).
114. Gam (1990, pp 204-206).
115. Sode (1997, p. 323, Figs 2-5).
116. Cf. Sableroles et al. (1995, p. 303).
117. The same was observed amongst the waste of glass bead making from the Jodenstraat in Maastricht. In: Sableroles et al. (1997, p. 303 – note 45, Fig. 2.6, colour plate 26). Experiments by J. Henderson have demonstrated that when a bead is reheated a number of

times, the glass becomes stickier; the more viscous the glass becomes, the more difficult it is to remove the bead from a tapering iron rod without causing it to fracture.

118. Hulst (1992, p. 90).
119. See contribution Henderson (this volume).
120. Gam (1990, pp 204-206).
121. See Henderson (this volume).
122. For a summary of the occurrence of *tesserae* on bead-making sites (see Steppuhn 1998, pp 86-87).
123. Lundström (1981, p. 17) and Callmer (1995, p. 50).
124. Hodges (1991, p. 81).
125. Callmer & Henderson (1991, pp 144-145).
126. Koch (1987, p. 348, Fig. 138).
127. Knol (1993, p. 87).
128. Knol et al. (1996, p. 77).
129. Siegmund (1998, pp 61-62, Fig. 15).
130. Waterbolk & Waterbolk (1991, pp 201-209).
131. See Henderson (this volume).
132. See contribution Henderson (this volume).
133. See for instance Callmer (1997, p. 197).
134. Sableroles et al. (1997, pp 293-294, Fig. 1.2,3,4,7,8).
135. Gerrets & De Koning (this volume) and Tulp (next volume).
136. See Henderson (this volume).
137. For Maastricht, see Sableroles et al. (1997, pp 297-298, Table 1; pp 304-305, plate 25.4,5) and Rijnsburg (unpubl. data).
138. See also Henderson (this volume).
139. Knol (1993, p. 80, Fig. 15.1).
140. See contribution Henderson (this volume).
141. Bloemers et al. (1981, p. 132, unpubl. data).
142. For a summary of the occurrence of *tesserae* on bead-making sites (see Steppuhn 1998, pp 86-87).
143. Arbman (1937, p. 67): 'Province of Groningen. Wierum Schoun. Terpfund. Mehrere kleine blaue Würfelchen, mit einer Scherbe eines Trichterbechers, s.o. S. 47, Fragmenten eines Kugeltopfes, Perlen u.a. gefunden.' The alternative 'Wierumerschouw', a 'wierde' (terp) located just above Groningen, is suggested to the author by D. Gerrets. However, no such finds are known from the Groninger Museum today (pers. comm. E. Knol, curator Groninger Museum). Neither are they included in Miedema's 1983 inventory of finds in the terp-region to the north-west of the city of Groningen.
144. Sableroles et al. (1997, pp 293/294).
145. Personal communication M. Dijkstra, University of Amsterdam.

Scientific analysis of the glass and the glass-bearing artefacts: Technique, raw materials used and archaeological interpretation

J. HENDERSON

1 INTRODUCTION

The reasons for carrying out a scientific investigation of the glass from Wijnaldum are both technological and archaeological. The chemical compositions of ancient glasses can provide an indication of the kinds of primary raw materials used to make the glass, the impurities which were introduced with the raw materials and the means by which colours were achieved. In almost every case it is possible to use the chemical composition of the glass as a means of characterising the glass to the time of production and very occasionally to a specific region or workshop site in which it was made. As more Early Medieval glass from Europe is analysed, it has become apparent that there are clear technological traditions which can be distinguished from each other amongst the glass which are sometimes correlated to the context date of the glass and to the vessel forms[1]; this suggests that different vessel types (and the glasses themselves) were made in a range of production centres.[2] This evidence also suggests strongly that glass recycling did not occur widely, or these fine compositional distinctions would be lost. Early medieval glass can also sometimes be distinguished compositionally from that made in the preceding 'Roman' technological tradition. Glass production sites are a rarity in Early Medieval Europe (even if the analytical data suggests otherwise); the evidence for the fusion of glass from its primary raw materials (fritting) is especially rare. There is some evidence for the manufacture of glass cables that were used for decorating beads and vessels, such as at Barking, Essex, England and at Åhus, Sweden and also for making glass beads themselves by working glass. A far more common situation, however, is that production zones inferred from distributions of early medieval glass vessels based on vessel forms or their colours can not be related to industrial evidence for their production. At least at Wijnaldum there is primary evidence for *working* opaque yellow glass: such glass has been found attached to a thick layer of baked ceramic material and was probably used for making simple wound glass beads. On the basis of this evidence we can not, however, suggest that glass vessels were manufactured from hot glass or that glass was fused from raw materials at Wijnaldum.

The range of vessel and bead forms that date from the Migration (including 450-550), Merovingian (ca 550-750) and Carolingian (ca 750-950) periods at Wijnaldum were chemically analysed. It is of special interest to note the compositional trends in the glass used over this period because of the possibility that glass may have been fused from primary raw materials in early medieval Europe (though direct industrial evidence is scanty to say the least). The other possible contribution to the glass available for glass-working is recycled early medieval and Roman glass. One might expect the changes in late Merovingian and Carolingian society and economy to be reflected in changes in the use of glass raw materials and perhaps changes in the organisation of the glass technology over this period, so this is one area which will be examined in relation to the glass chemical analyses. One other interesting question to be asked is whether any of the glasses could have originated in the Islamic world; this also ought to be reflected in their chemical compositions.

In a broader context the technology behind the production of a glass with a specific chemical composition is an objective means of providing a way of assessing its relative value. Naturally glass value can be assessed in several different ways and the type of archaeological context in which it is found is another essential parameter to be taken into account. Nevertheless, if a glass bead, for example, is highly decorated it will have taken longer to manufacture it and will have involved a much wider range of (possibly imported) raw materials than a simple undecorated bead. It may also have been a 'one-off' product which will add to its value. It is obviously important to attempt to establish whether particular bead types were mass-produced because this will have affected their value which may be reflected in the raw materials used.

It was with these technological and archaeological considerations in mind that the glass from Wijnaldum was sampled.

2 THE SAMPLING STRATEGY BEHIND THE GLASSES ANALYSED

Samples were removed from glass vessels and beads. They are listed in Table 1, where their object type, sample number, catalogue number,small find number (Vondst Nr), colour and date are given.[3] Thirty-seven samples of glass were taken, and one sample of slightly contaminated vitreous material

Table 1. List of glasses sampled from Wijnaldum.

Sample/ Cat. no.	Find no	Sample type	Sample colour	Context date
1/31	5359	Squat jar	Blue green	800-900 (3) (form 8th C.)
2/30	1526	Cylindrical beaker	Blue green	700-750 (1)
3/30	1526	The same	Opaque yellow decoration	700-750 (1)
4/27	1423	Funnel, yellow decoration	Blue green	Stray find
5/22	7507	Funnel, inturned rim	Yellowish-green	775-850 (1)
6/18	2020	Early funnel inturned rim	Blue green	800-900 (4) (form 8th C. ?)
7/20	3695	Funnel, inturned rim	Blue green	800-850 (1)
8/23	7877	Funnel rim	Almost colourless	800-850 (3) (form 8th C. ?)
9/16	6210	Palm cup-optic blown base	Yellow-brown	550-650 (1)
10/15	6256	Bell beaker, body	Pale green	550-600 (1)
11/9	6794	Rim bowl-white feather decoration	Bluish-green	450-/500 (1)
12/5	5208	Cone beaker, base	Yellowish-green	450-500 (1)
13/111	4421	Ribbed bead	Opaque red	550-650
14/133	7035	Globular/ barrel-shaped bead (opaque white decoration)	Opaque red	550-600
15/83	6884	Cylindrical bead	Opaque red	770-850 (3)
16/45	3501	Globular bead	Opaque red	800-850 (1)
17/89	6813	Small barrel-shaped bead	Opaque orange	550-650 (1)
18/59	1024	Small globular bead	Opaque yellow-green	640-750
19/134	5170	Cylindrical bead with spiral opaque white decoration	Opaque red	Stray: late 6th/ early 7th C. type
20/134	5170	Decoration of Sample 19	Opaque white	
21/72	7448	Cylindrical	Opaque yellow	550-600 (3)
22/41	7448	Small globular	Opaque yellow	550-600 (3)
23/18	3901	Small globular	Opaque greenish-yellow	575-625 (1)
24/19	3901	Small globular	Opaque yellow	575-625 (1)
25/94	4401	Broad biconical	Opaque yellow	775-900 (3)
26/71	7045	Short cylindrical	Opaque yellow	770-850
27/76	7988	Short cylindrical	Opaque white	775-850 (1)
28/95	6472	Broad biconical	Opaque white	600-700 (1)
29/74	6208	Short cylindrical	Opaque white	550-600 (3)
30/101	6208	Segmented	Opaque bluish-white	550-600 (3)
31/96	7994	Large long biconical	Translucent greenish-blue	800-850 (1)
32/118	3326	Drawn, globular, segmented with silver foil	Colourless	875-900 (1)
33/126	6405	Globular with cross-swag opaque white decoration	Deep translucent olive green	425-550
34/129	1043	Globular with cross-swag decoration in translucent blue	Opaque white	600-650 (4)
35/138	6255	Barrel-shaped Blättchen-millefiori design, colourless strips around perforation, covered with red glass	Colourless	550-650 (1)
36/138	6255	Part of 35	Opaque red	
37/220	7824	Opaque grey blob on thin-walled crucible	Opaque grey	250-350 (1)
38/219	2943	Glass on ceramic surface	Opaque yellow	575-62 (3)

from a crucible. The widest range of object type and colour were selected for chemical analysis from a range of dated contexts including Migration Table, Merovingian and late Merovingian/Carolingian glass. Thus the glass vessels included a cone beaker, a bowl, a bell beaker, a palm cup, a squat jar and funnel beakers. The beads analysed ranged in shape from annular, cylindrical and globular to ribbed, biconical and segmented. The types of decorated beads analysed were cross-swag and millefiori-decorated. An opaque yellow glass sample was removed and analysed from an object considered to be part of a thick ceramic tray or possibly part of the floor of a rudimentary kiln/furnace. Finally, a small fragment of fuel-ash slag from a thin-walled crucible was analysed qualitatively so the result is not included in Table 2. This turned out to be composed of a kind of vitreous material which is formed in almost all high temperature environments where plant ash and silica are present. Silica is a very common constituent of ceramics, including crucibles.

3 THE ANALYTICAL TECHNIQUE USED

Micro-samples of glass were removed from each of the objects and mounted in epoxy resin. The samples in the epoxy block were polished flat using a variety of grades of diamond paste down to 0.25 micron grade. This process exposes a flat unweathered glass surface which is then coated with a thin layer of carbon. The carbon conducts the electron beam and prevents it from distortion and deflection. The epoxy block with glass samples mounted in it was then placed in an electron-microprobe (a Cambridge mark 9 probe with two crystal spectrometers). Each sample was analysed several times and the results averaged; the beam was defocused to 80 microns so as to minimise the lost of sodium due to volatilisation under the electron beam. The probe was calibrated using European Science Foundation and Corning Museum glass standards. By cross-analysing with a reliable standard of proven durability once the machine had been calibrated it was possible to establish the level of accuracy that could be obtained: Typical levels of detection in p.p.m. for a low lead glass (95.5% probability level) are: Na^2O 760; K^2O 250; CaO 170; Fe^2O3 640; CuO 1200; PbO 200.

The relative analytical accuracy for major oxides under the conditions of analysis used were 4% for soda (Na_2O), 1% for silica (SiO_2) 3% for potassium oxide (K_2O), 5% for calcium oxide (CaO) and 2% for lead oxide (PbO). For minor elements the levels of accuracy were 2% for magnesia (MgO), 1% for aluminia (Al_2O_3), 3% for phosphorus pentoxide (P_2O_5), 10% for ferric oxide (Fe_2O_3) and 20% for chlorine (Cl).

Further details of the analytical conditions used are published elsewhere.[4]

4 THE RESULTS

The chemical analyses of the glasses are given in Table 2. Many of the glasses are of a soda-lime-silica composition, or SLS, and, in some, lead oxide was added deliberately forming part of the colourant. These glasses would probably have been made from sand as a source of silica, calcium oxide, perhaps introduced as fragments of sea shell in the sand, and soda which may have derived from a mineral source known as natron found at Wadi el-Natrun in Egypt.[5] However, in addition, the glasses contained the expected impurity levels of magnesia, aluminia, phosphorus pentoxide, sulphur trioxide, chlorine and potassium oxide. These are thought to be the impurities associated with the use of sand and the alkali natron. There are also glasses which have more unusual impurities. These are discussed below.

5 TECHNOLOGY OF THE TRANSLUCENT GLASSES ANALYSED

The translucent glass samples analysed (analyses 1, 2, 4-12; 31, 33 and 35; Table 2) are basically all of a soda-lime composition, with few impurities other than the oxides of iron and manganese. This, it might be suggested, is because the glass represents recycled late Roman glass that also tends to have these characteristics. However, although it is normally anticipated that such glass would have magnesia and potassium oxide levels of around 0.5%, there are examples of glass chemical compositions here which have unusual compositional features. Both numbers 10 and 11, which are translucent pale green and bluish green in colour contain higher magnesia levels than the others, and it is no coincidence that they both also contain 2.0% and 2.2% iron oxide giving the green colour; both samples are from vessels which are distinctively early – a bell beaker of 6th century date and a bowl of the second half of the 5th century/early 6xth century in

Table 2. Chemical composition of Wijnaldum glass, expresses as weight % oxide.

Sample number Colour	1 blue green	2 blue green	3 opaque yellow	4 blue green	5 yellow green	6 blue green
Na_2O	13.2	15.8	14.5	17.0	17.0	16.9
MgO	0.6	0.8	0.8	0.8	1.3	0.7
Al_2O_3	3.1	2.6	2.6	2.4	2.2	2.6
SiO_2	72.5	70.6	68.9	70.1	69.8	67.3
P_2O_5	0.1	0.2	0.2	0.2	0.2	0.1
SO_3	ND	0.2	0.2	0.3	0.3	0.3
Cl	0.9	0.9	0.8	1.0	0.9	0.8
K_2O	0.5	0.9	1.0	0.7	0.9	0.9
CaO	7.7	7.4	7.3	6.8	7.4	8.0
TiO_2	0.1	0.1	0.1	0.1	0.2	0.1
Cr_2O_3	ND	ND	ND	ND	ND	0.1
MnO	ND	0.5	0.5	0.5	1.2	1.1
Fe_2O_3	0.5	0.9	1.0	0.8	1.0	0.5
CoO	ND	ND	ND	ND	ND	ND
NiO	ND	ND	ND	ND	ND	ND
CuO	0.1	0.2	0.3	0.7	ND	0.2
ZnO	ND	ND	ND	ND	ND	ND
As_2O_3	ND	ND	ND	ND	ND	ND
SnO_2	ND	0.1	0.1	0.1	ND	ND
Sb_2O_3	ND	0.3	0.2	1.1	0.1	0.2
BaO	0.1	0.1	0.1	0.1	0.1	0.1
PbO	ND	0.5	4.6	0.6	0.2	0.2

Sample number Colour	7 blue green	8 colour-less	9 yellow-brown	10 pale green	11 blue green	12 yellow green	13 opaque red	14 opaque red	15 opaque red	16 opaque red	17 opaque orange
Na_2O	14.8	17.0	14.0	18.4	18.5	18.5	12.4	16.3	16.4	13.3	5.1
MgO	0.8	0.7	0.6	1.4	1.4	0.9	0.9	1.3	1.3	2.3	1.8
Al_2O_3	2.3	2.1	3.0	3.1	3.0	2.7	2.6	2.4	2.0	1.9	1.4
SiO_2	68.4	68.9	70.1	64.9	63.5	66.9	56.0	59.3	60.5	60.1	3.9
P_2O_5	0.1	0.1	0.1	0.2	0.2	0.1	0.3	0.2	0.2	0.9	0.4
SO_3	0.1	2.0	0.1	0.3	0.4	0.2	0.2	0.4	0.2	0.2	0.2
Cl	1.1	1.0	1.0	1.0	1.0	0.9	0.7	0.9	0.8	1.0	0.8
K_2O	0.6	0.7	0.7	0.9	1.0	1.0	0.7	0.9	0.9	2.4	1.6
CaO	9.4	7.3	8.9	7.1	6.7	6.4	4.9	6.4	5.8	11.7	6.5
TiO_2	0.2	0.1	0.1	0.1	ND	0.3	0.2	0.1	0.2	0.1	0.1
Cr_2O_3	ND	ND	ND	ND	ND	ND	ND	ND	ND	ND	ND
MnO	0.2	0.8	ND	0.6	0.8	1.7	2.0	1.2	0.9	0.3	0.1
Fe_2O_3	1.1	0.7	0.5	2.0	2.2	1.5	4.6	4.1	4.5	1.1	0.9
CoO	ND	0.1	ND	ND	ND	ND	ND	ND	ND	ND	ND
NiO	ND	ND	ND	ND	ND	ND	0.1	ND	ND	ND	ND
CuO	ND	ND	ND	ND	ND	ND	1.8	1.4	2.4	2.1	1.1
ZnO	ND	ND	ND	ND	ND	ND	ND	0.4	0.8	ND	ND
As_2O_3	ND	ND	ND	ND	ND	ND	ND	ND	ND	ND	ND
SnO_2	ND	ND	ND	ND	ND	ND	1.0	0.4	0.2	0.3	0.1
Sb_2O_3	ND	0.4	ND	ND	ND	0.1	ND	ND	ND	0.3	0.1
BaO	0.1	0.1	0.1	0.1	0.1	0.1	0.2	0.1	0.1	0.1	0.1
PbO	0.1	0.1	ND	ND	ND	ND	14.2	6.3	4.5	1.2	0.4

Table 2. Continued.

Sample number Colour	18 opaque yellow green	19 opaque red	20 opaque white	21 opaque yellow	22 opaque yellow	23 opaque yellow green	24 opaque yellow	25 opaque yellow	26 opaque yellow	27 opaque white	28 opaque white
Na_2O	12.9	15.3	15.5	14.2	10.9	8.4	6.5	10.7	9.1	19.3	16.7
MgO	0.6	1.2	1.3	1.4	0.9	0.5	0.5	1.2	1.0	1.3	1.5
Al_2O_3	2.3	2.5	2.6	2.4	2.2	2.4	2.0	2.3	1.6	2.6	3.2
SiO_2	57.1	60.9	61.7	50.6	40.6	40.6	26.8	57.9	33.4	64.8	60.0
P_2O_5	0.1	0.2	0.1	0.2	0.1	0.1	0.1	ND	0.1	0.2	0.2
SO_3	0.1	0.3	0.3	0.3	0.3	0.1	0.1	0.1	0.2	0.4	0.3
Cl	0.8	0.8	0.8	0.8	0.6	0.5	0.3	0.7	0.5	1.1	0.9
K_2O	0.8	0.9	0.9	0.9	2.6	0.5	0.5	0.3	0.8	0.8	0.9
CaO	5.7	6.8	7.1	5.4	3.8	4.5	2.5	6.4	3.1	7.1	6.1
TiO_2	0.1	0.1	0.1	0.1	0.1	0.1	0.1	0.1	0.1	0.1	0.2
Cr_2O_3	ND	ND	ND	ND	ND	ND	ND	ND	ND	ND	ND
MnO	0.4	1.6	1.6	0.3	0.7	ND	0.2	ND	0.1	0.3	0.5
Fe_2O_3	0.8	1.3	1.4	1.4	1.7	0.6	0.9	0.6	1.0	1.2	2.0
CoO	ND	0.1	ND	ND	0.1	ND	ND	ND	ND	ND	ND
NiO	ND	ND	ND	ND	ND	ND	ND	ND	ND	ND	ND
CuO	2.1	ND	ND	0.2	0.1	ND	ND	ND	ND	ND	ND
ZnO	ND	ND	ND	0.1	ND	ND	ND	0.1	ND	ND	ND
As_2O_3	ND	ND	ND	ND	ND	ND	ND	0.2	ND	ND	ND
SnO_2	0.8	3.6	4.8	1.3	4.2	4.6	6.7	2.6	4.6	0.3	7.6
Sb_2O_3	0.2	ND	ND	ND	ND	ND	ND	ND	ND	ND	ND
BaO	0.1	0.1	0.1	0.1	0.1	0.1	0.1	0.1	0.1	0.1	0.1
PbO	17.4	3.5	3.6	21.5	33.3	39.1	54.5	27.0	46.1	0.0	2.9

Sample number Colour	29 opaque white	30 opaque bluish white	31 trans-lucent turquoise	32 colour-less	33 trans-lucent olive	34 opaque white	35 colour-less	36 opaque red	37 opaque grey	38 opaque yellow
Na_2O	17.5	12.7	14.4	11.6	17.6	17.4	17.4	14.0	3.7	1.3
MgO	1.3	0.6	0.6	5.5	1.4	1.6	1.2	1.4	5.8	1.9
Al_2O_3	2.5	2.6	2.9	1.1	3.0	3.2	2.4	2.7	9.8	3.7
SiO_2	61.3	62.3	63.9	69.5	64.5	60.3	64.9	57.3	56.8	21.4
P_2O_5	0.2	0.1	ND	ND	0.1	0.2	0.1	0.3	4.0	0.1
SO_3	0.4	0.1	0.2	0.2	0.2	0.4	0.3	0.3	ND	0.1
Cl	0.9	0.8	0.9	0.7	1.1	0.9	0.9	0.6	ND	0.1
K_2O	1.0	0.7	0.4	2.2	0.7	1.1	0.7	1.5	2.4	0.7
CaO	6.9	7.1	8.7	6.4	6.2	6.6	8.8	7.1	11.3	3.4
TiO_2	0.1	0.1	ND	ND	0.3	0.1	0.1	0.2	0.5	0.2
Cr_2O_3	ND	ND	ND	ND	ND	ND	ND	ND	ND	ND
MnO	1.4	ND	ND	0.7	1.5	0.6	1.5	1.1	0.2	0.1
Fe_2O_3	1.8	0.5	0.5	0.4	3.2	2.1	0.9	1.7	4.0	1.7
CoO	ND	ND	ND	ND	ND	ND	ND	ND	ND	ND
NiO	ND	ND	ND	ND	ND	ND	ND	ND	ND	ND
CuO	ND	ND	3.3	ND	0.1	ND	ND	2.2	ND	0.1
ZnO	ND	ND	ND	ND	0.1	ND	ND	0.4	ND	0.1
As_2O_3	ND	ND	ND	ND	ND	ND	ND	ND	0.1	ND
SnO_2	1.6	2.1	0.4	ND	ND	4.2	ND	2.3	ND	4.6
Sb_2O_3	ND	ND	ND	ND	0.1	ND	ND	ND	ND	ND
BaO	0.1	0.1	0.1	ND	0.1	0.1	0.1	0.1	0.1	0.2
PbO	3.4	10.9	2.9	ND	0.1	2.5	ND	7.9	ND	63.5

ND = Not Detected.

date. Number 12, the earliest glass vessel analysed dating to the mid 4th-early 5th century of a yellowish-green colour, contains 0.3 titania which is unusually high for a glass of this period; the explanation is probably because it was associated with a manganese- or iron-rich impurity.

The translucent turquoise glass bead sample analysed (number 31) is coloured by 3.3% cupric oxide and the presence of lead and tin oxide may mean that the original colorant was a fragment of lead-rich bronze. The translucent deep green glass (number 33) is coloured by 3.2% iron oxide, probably in a reduced (ferrous) state; this glass also contains low impurity levels of copper, zinc and antimony oxides.

The three colourless glasses analysed (numbers 8, 32 and 35) differ quite widely in composition. The 0.4% antimony trioxide detected in analysis 8 is probably responsible for decolorizing the glass; the cobalt impurity detected has been effectively neutralized by it. Another colourless sample (number 32) contains an unusually high magnesia level which is correlated with a higher potassium oxide level; the glass also contains low aluminia and soda levels (see below). The other colourless sample (number 35) on the other hand contains impurity levels which are more typical of 'Roman' glasses with elevated manganese oxide levels. It is nevertheless very difficult to ascertain why these last two glasses are colourless in the absence of antimony trioxide (a glass clarifier and decolorizer). The former sample contains relatively low levels of potential colorants so it is probable that the atmosphere within the furnace was monitored and controlled in order to produce a colourless glass. The chemical composition of sample 32 is typical of Islamic glass and indeed the late date of the bead on the site helps to confirm this. The principal compositional features which show it to be different from the other translucent glasses are low soda, low aluminia, high magnesia and high potassium oxide. These can be clearly related to published post-800 AD Islamic glass compositions.[6] The same composition was found in metal-foil beads found at Rostock-Dierkow[7] and Hedeby[8]; the beads were made at Hedeby in the 9th century[9], presumably using glass imported from the Islamic world.

Overall, it is clear that the variation of the pale translucent glass colours from the common pale translucent green through yellowish green to bright yellow green in the glass vessels sampled have been achieved by variations in the gaseous atmosphere of the furnace.[10] One can claim this with some confidence because in most cases there are no *obvious and distinctive* differences in the levels of potential colorants in the glasses. However with more data of contemporary glasses the following characteristics might be noted: the yellowish-green colour of sample 5 (Carolingian) may be due to a higher relative level of manganese to iron, when compared to contemporary samples, a yellow-brown (number 9) dating to 550-650 has no detectable manganese oxide- though the same is true of sample 1 which is blue-green (an 8th century form).

6 THE OPAQUE GLASSES ANALYSED

Samples taken from beads of opaque red (6), orange (1), yellow-green (2), yellow (6), white (6) and grey (1) glasses were chemically analysed. The reason that the glasses are opaque, as opposed to translucent or transparent, is that masses of small crystals are present in the glass which impede and reflect light rays.[11] In the case of opaque red glasses the normal relatively high levels of ferrous oxide are present with the copper oxide also in a reduced form (cuprous oxide) producing the opaque dull red colour. The elevated levels of iron oxide are probably explained by their use as an internal reducing agent in the glasses which would help to bring about the formation of the cuprous oxide crystals. Most of the opaque red glasses analysed are further characterized by the presence of elevated levels of magnesia and potassium oxide when compared to translucent glasses (analysis numbers 14, 15, 16, 19 and 36). This is a characteristic of Roman opaque red-brown enamels and of opaque red-brown tesserae, so it can be assumed that a similar stock of glass was involved.[12] Another characteristic of the Wijnaldum opaque red-brown glasses is highly variable lead oxide levels with a range of 1.2% to 14.2%, though five out of the six analysed contain levels between 1.2% and 6.3%. Sample 16 contains unusually high calcium oxide: 11.7% compared with the rest which contain between 4.9% and 7.1%. This glass contains both tin and antimony oxide impurities, whereas the rest contain only tin; this latter feature suggests that a bronze was used as a source of copper colorant. Indeed in almost all respects, including levels of magnesia and potassium oxides which are even higher than the already elevated levels found in four other red glasses, analysis 16 is different from the other reddish glasses – it is also brighter in colour. Given these compositional differences, there is a possibility that the bead originated in the Islamic world, though more analyses of Islamic red glasses from secure archaeological contexts need to be published before greater confidence in this interpretation can be achieved. One other exception to the pattern of using opaque red glass produced in the 'Roman' tradition is analysis number 13 of a ribbed bead. This contains more than twice the lead oxide level found in the other red glasses analysed and contains impurity levels of potassium and magnesium oxides which are inconsistent with the use of the red glass produced using a 'Roman' tradition.[13]

The single opaque orange glass (number 17) analysed has one analytical feature in common with the red-brown glasses: elevated magnesia and potassium oxide levels. However the apparent presence of 21.1% cuprous oxide in the glass suggests that the electron beam was mainly focused on a high concentration of copper-rich opacifying crystals in the glass when analysed.

One of the two opaque yellow-green glasses analysed (numbers 18) has been produced by the presence of lead stannate crystals bathed in a translucent turquoise glass. The lead stannate crystals would have been developed by heat-treating the glass in which the lead and tin oxides had already been dissolved. The other opaque yellow-green example appears to be coloured by iron combined with lead stannate crystals. As with five samples of opaque yellow glasses analysed the glasses are opacified with crystals of lead antimonate, $Pb_2Sn_2O_7$.

Most of the opaque yellow glasses contain between 21.5% and 54.5% lead oxide. The levels of lead oxide of 54.5% and 46.1% in samples 24 and 26 respectively are relatively high, though not exceptional for early medieval glasses.[14] These high lead oxide samples from Wijnaldum have context dates of probably late 6th/early 7th century. In other respects the tin-opacified opaque yellow glasses are generally quite consistent in composition. Most of those analysed are apparently opacified by lead stannate crystals; in the case of sample 3 (catalogue 30), however, the lead antimonate crystals are evidently responsible for the opacity, though their density is relatively low, resulting in the detection of low levels of lead and antimony oxides.

The single opaque yellow tesserae from Wijnaldum which was analysed contained lead antimonate crystals which provided the colour and opacity. Together with the trailed on opaque yellow decoration of a cylindrical beaker (sample 3) of early to mid-8th century date the use of lead antimonate is entirely consistent with a 'Roman' technology and in direct contrast to the use of lead stannate in most (non-tesserae) Early Medieval yellow glass. It is notable that tesserae do not appear to have been used on 6th and 7th century bead making sites, but re-appear in the 8th century. When tesserae do appear in the early 8th century onwards at sites like Ribe and Åhus, they are opacified by antimonate compounds consistent with a Byzantine/Roman production technology.[15] On the other hand, other opaque yellow glasses used to make cables, millefiori rods and other forms of bead decoration from Ribe and Åhus are of the lead stannate composition; lead antimonate is therefore a distinct ('Roman') technology found in glass tesserae. Chemical analyses of opaque yellow lead-rich glasses from an early medieval cemetery at Griesheim, Germany apparently revealed the use of a lead oxide-lime-silica glass[16]; if this is the case (with unusually high levels of arsenic oxide of up to 1.98%), it constitutes the earliest lead oxide-lime-silica glass in a European context; minimal alkali levels were detected. However, analysis of unweathered contemporary opaque yellow glass from Maastricht has shown that the glasses are soda-lime-silica-lead oxide glasses[17], so it is likely that if unweathered samples of the Griesheim glasses were obtained and analysed they too would turn out to be of this composition.

The single opaque yellowish-grey sample of vitreous material attached to a crucible that was analysed (Table 1: no. 37, Table 2: analysis 37) is clearly not a by-product of glass-working or glass-making; it is a fuel-ash slag probably resulting from metal working.

7 THE OPAQUE YELLOW GLASS ATTACHED TO A CERAMIC MATERIAL

Scientific analyses of a thick deposit of yellow glass was carried out. The glass was attached to a fragment of ceramic material. It is difficult to be sure precisely what the fragment originally formed a part of. It is possible that it is the remains of a flat surface (tray or 'plate') on which opaque yellow glass was heated. Given that glass such as this opaque yellow glass contains a high lead oxide level it melts at a relatively low temperature of ca 550-650°C. so it would be possible to heat it within a rudimentary kiln. Another possible interpretation is that the artefact is itself a fragment of a small oven or furnace, and that the yellow glass has fallen onto the floor of the structure in some quantity. This is a less likely interpretation because there would have been a high degree of waste involved, though it can not be ruled out entirely.

Scientific analysis revealed that the yellow lead-rich glass had interacted with the ceramic material comprehensively, the results showing a high degree of contamination in the glass. This degree of interaction suggests that the glass had been heated for some time, though it does not help to decide which of the two artefact functions suggested above is most likely. An analysis of a discrete area of opaque yellow glass (Table 2: analysis 38) showed that this area of glass contained higher lead oxide levels (63.5%) than found in any of the beads or bead decoration. The enrichment may simply be a result of the high mobility of the lead. However, in other respects, including the use of tin for the manufacture of opaque yellow lead stannate crystals, the composition of the opaque yellow glass is basically very similar to the analyses for other opaque yellow glasses from the site. The main differences are the slightly higher levels of magnesia and aluminia both of which have probably migrated from the ceramic body into the glass and enriched it. There can be little doubt that the object constitutes evidence for opaque yellow glass-working at Wijnaldum. It can be added to the other examples of European early medieval sites where evidence for yellow glass working opacified by lead stannate have been documented.[18]

8 THE ARCHAEOLOGICAL SIGNIFICANCE OF THE CHEMICAL ANALYSES

It is interesting to note that many of the translucent glasses, irrespective of their colour, consistently contain relatively low levels of metallic impurities. Tin oxide impurities occur in translucent samples 2 and 4, and tin is also associated with the use of cuprous oxide as an opacifier in the opaque red glasses, suggesting that a copper-bronze raw material was used for colouring the red glass. Low impurity levels of antimony oxide occur in translucent samples (samples 2, 4-6, 8, 12, 33). There has been some discussion recently about the significance of the impurities of antimony and tin oxides in translucent greenish glasses[19] and it has become clear that these impurities are not unusual. They tend also to be associated with elevated impurity levels of lead and copper oxides. One finds very similar impurity levels in contemporary funnels found at Wijnaldum and Borg, Norway [20] for example and they occur in analysed glasses from Dorestad.[21]

If we examine this range of impurities in Wijnaldum pale green glasses (a glass colour where one avoids the possibility that the impurities are associated with deliberately added colorants and opacifiers) several interesting results are revealed. Firstly, copper oxide occurs in four translucent pale green vessel glasses which date to the 8th-9th centuries (Samples 1, 2 , 4 and 6) but not in any of the samples of pale green glass vessels analysed which dating to before 700 (Samples 9, 10, 11 and 12); this must result from the use of a different kinds of glass being in circulation at the time (i.e. a slight change in technology).

From about the late 8th century a new raw material (a plant ash alkali) was introduced for the manufacture of Islamic glass which should be reflected in the increase in the levels of magnesium and potassium oxides in these glasses. One bead, a silver foil bead, is clearly of this Islamic composition so, given the evidence for the production of this bead type at Hedeby, one can suggest that the glass was imported from the Islamic sphere of production to make beads. Compositional parallels for such metal foil beads have been found at Hedeby and published by Dekówna[22] so it is clear that north-western Europe was using glass of this distinctly higher magnesia composition from the late 8th century onwards. The elevated magnesia and potassium oxide levels detected in the red glass used for decorating the 6th-7th century millefiori bead (sample 36) – is consistent with the compositions detected in Roman opaque red glasses used for enamels and for the manufacture of opaque red glass tesserae.[23] If this is a type which was imported from Byzantine Egypt this is evidence for the continuity of the use of Roman glass in the Byzantine world. The same characteristics are found in the contemporary red glass in sample 14 (a cylindrical bead with white decoration) and also in a stray cylindrical red bead with white decoration (sample 19) of late 6th – early 7th century type, so it is clear that this recipe or, more likely, recycled red glass made in the 'Roman' tradition continued to be used into the 7th and 8th centuries. The exception to this trend in the chemical composition of an opaque red ribbed bead (sample 13) which displays very different compositional characteristics, as well as having a different perforation shape[24] and was probably imported to Wijnaldum.

If we turn to the levels of magnesium oxide in the rest of the glasses analysed, there is no obvious correlation between sample age and impurity level; one might expect the levels of magnesia to increase in later glasses if new alkali raw materials were used or formed part of the recycling process then, but this is not the case here. Indeed a number of 'late' samples contain low levels (e.g. sample numbers 1-4, 6-8, 26 and 31) and two of the earlier samples (samples 11 and 33) contain higher levels, the opposite to what might be expected. Magnesia and potassium oxide are invariably positively correlated in ancient glasses because of the nature of the alkalis used. However some Wijnaldum glasses contain elevated levels of magnesia which are *not* reflected in elevated potassium oxide levels. Although this might seem a technically trifling point to make, it is nevertheless an important one: large data bases of Roman glasses show that when magnesia levels are elevated, so are potassium oxide levels so this seems to be a characteristic distinction.

There is however a clearer relationship between the occurrence of lead oxide impurities and glass date. Apart from one probable 8th century vessel fragment (sample 1) only the translucent vessel glasses dating to between ca 700 and 900 contain a lead oxide impurity, suggesting, along with the copper impurity levels, that a slight change in the soda-lime glass technology occurred in the mid 8th century. The picture is therefore consistent and very interesting because at the same time 8th century palm funnels tend to be of a better quality than 7th century palm cups.[25]

Overall, it can be suggested that the technology used for the manufacture of Wijnaldum pale green vessel glass is very similar to Roman glass, in that it is all of a soda-lime-silica composition. This situation has been found on a variety of other early medieval sites.[26] However, as noted above, there are also variations in the impurity levels of most notably antimony, tin, copper and lead oxides in some early medieval glasses. In addition, much other (early) 8th century and later glass that has been analysed from Dorestad[27], Ribe (Jutland, Denmark; unpublished data), Åhus (Sweden)[28] and from Early Christian Ireland[29] contains elevated levels of these impurities, in glass beads, bichrome cables and glass vessels.[30] Matthes has even found an increase in impurity levels of the oxides of arsenic and antimony in opaque high lead glasses in beads dating to ca 600 at the Griesheim cemetery in Germany.[31] This compositional change probably relates to a change only in the lead source which would have been deliberately used, perhaps as suggested through the use of a crucible which had been used for melting a lead-rich metal. The change in impurity levels observed by Matthes in opaque glasses[32] therefore reflects a different technological change from that reflected in the changes in impurity levels detected in the soda-lime translucent glasses and appears to be an intriguing precursor to changes in translucent glass technology; lead isotope determinations of some of these Merovingian opaque yellow glasses from Germany show that a 'local' source of lead was used as a primary raw material.[33] If one examines the chemical analysis of 4th century Roman glass such impurities have not been detected[34], yet they have in Roman glasses of 1st-2nd century AD date, but at generally lower levels.[35] These compositional comparisons within Roman glasses are for those found in the Levant, but this does not detract from a comparison with Wijnaldum glass or western Roman glass compositions, since in other respects there is little compositional variation in Roman glass irrespective of its date and place of origin. By the 8th century, impurity levels in early medieval European translucent glasses do appear to increase and chemical analysis of the Wijnaldum vessel glasses reflect this change, suggesting a slight change in the production technology. Funnel beakers from Wijnaldum in particular contain very similar levels of lead, antimony and tin oxides to some contemporary funnels from Borg (Norway).[36] The only glass tesserae found on the site- of an opaque yellow colour- when analysed qualitatively using X-ray fluorescence spectroscopy- was found to contain the standard composition for a 'Roman' product being of a soda-lime-silica composition opacified by lead antimonate crystals. A contemporary 8th century trailed-on yellow vessel decoration is also opacified with lead antimonate.

This is in total contrast to the yellow (lead stannate) opacifier which was used for all locally produced opaque yellow bead glasses analysed, a distinctive early medieval glass production tradition. Tesserae of the same composition were found at Ribe and Åhus and it is thought that these originally derived from looted Byzantine or Roman mosaics made using a Roman technology.

From these analyses it can be suggested that we have evidence for the use of three distinct technological traditions for the manufacture of the glass used to make vessels and beads at Wijnaldum. The first is the characteristic 'Roman' tradition with low heavy metal impurity levels, and of which the tessera with a context date of ca 750-770 and the contemporary yellow vessel decoration of the same date are primary examples. The second is translucent glass which has characteristic (heavy metal) impurity patterns and which came into use especially (though not exclusively) after 700. It is clear that a proportion of the glass of 8th century and later dates in Europe was made independently of Roman glass technology, perhaps because the amount of Roman recycled glass available was starting to run out. Before ca 700 we can not discard the possibility that (even though the glass is of 'the same' composition as many Roman glasses), the same primary raw materials were fused in Europe between the 5th and 7th centuries producing the same glass chemical compositions as used in the Roman world. We can include the predominant use of tin as an opacifier after 700 as another characteristic 'early medieval' component to this second technological type. One geographical area outside the European world which produced the third distinct glass type, and in which one or possibly two of the beads found at Wijnaldum were made, is the Islamic world: Sample 32 a colourless metal foil bead was certainly made using this technology and there is evidence for production of the bead type at Hedeby at this time (the 9th century). The other is half a red bead (sample 16) which may also have been made in the Islamic world.

It is clear then that scientific analyses of Wijnaldum glass vessel fragments and beads have produced evidence for a complex picture of early medieval glass technology in Europe.

REFERENCES

AERTS, A., K. JANSSENS, F. ADAMS, et al. in press. Trace-level Microanalysis of Roman glass from Khirbet Qumrân, Israel. In: J. Henderson, H. Neff & T. Rehren (eds), Proceedings of the International Symposium on Archaeometry, Illinois, USA, *Journal of Archaeological Science.*

BRILL, R.H. 1988. Scientific Investigations of the Jalame Glass and Related Finds. In: G.D. Weinberg (ed.) *Excavations at Jalame site of a Glass Factory in Late Roman Palestine*, Columbia, 257-294.

CALLMER, J. & J. HENDERSON 1991. Glassworking at Åhus, S. Sweden (eighth century AD), *Laborativ Arkeologi* 5: 143-154.

CALLMER, J., M. HECK, P. HOFFMAN, et al. 1997. Glasperlenproduktion im Frühmittelalter, *Ethnographisch Archäologisches Zeitschrift* 38: 225-234.

DEKÓWNA, M. 1980. Szkło W Europie Wczesnośredniowiecznej, Warsaw.

GREEN, L.R. & A.F. HART 1987. Colour and chemical composition in ancient glass: an examination of some Roman and Wealden glass by means of ultra-violet-visible-infrared spectroscopy and electron microprobe analysis, *Journal of Archaeological Science* 14: 271-282.

Heck, M. 1999. Die spezielle chemische Zusammensetzung gelber merowingerzeitlicher Glasperlen, paper given at *Herstellungstechniken and Herstellungswerkstätten von frühmittelalterlichen Glasperlen aufgrund der farbgebenden Komponenten und Mineralien Arbeitstreffen*, Technical University of Darmstadt, School of Materials Science, 21-22.1.99.

HENDERSON, J. 1985. The raw materials of early glass production, *Oxford Journal of Archaeology* 4: 267-291.

HENDERSON, J. 1988a. Electron-probe microanalysis of mixed-alkali glasses, *Archaeometry* 30: 77-91.

HENDERSON, J. 1988b. Electron-microprobe investigation of Early Irish glass and glass-making practices. In: E.V. Sayre, P.B. Vandiver, J. Druzik, et al. (eds) Materials issues in art and archaeology, Pittsburgh (Materials Research Society Symposium Proceedings 123), 141-146.

HENDERSON, J. 1991. Chemical characterization of Roman glass vessels, enamels and tesserae. In: *Materials Issues in Art and Archaeology II*, Pittsburgh (Materials Research Society Symposium Proceedings, 185), 601-608.

HENDERSON, J. 1995a. An investigation of Early Islamic glass production at Raqqa, Syria. In: P. Vandiver et al. (eds) *Materials Issues in Art and Archaeology III*, 1995a (Materials Research Society Symposium Proceedings, 352), 433-444.

HENDERSON, J. 1995b. Le verre de Dorestad: continuité technologique ou innovation?. In: Le verre de l'antiquite tardive et du Haut Moyen Age, Association Française pour l'Archéologie du Verre, huitième rencontre, Guiry-en-Vevin 1993, Musée Archaeologique Departmental du val d'Oise, 51-56.

HENDERSON, J. & I. HOLAND 1992. The glass from Borg, an early medieval chieftain's farm in Northern Norway, *Medieval Archaeology* 36: 29-58.

HEYWORTH, M.P. 1991. An archaeological and compositional study of early medieval glass from North West Europe, Bradford (unpubl. Ph.D. thesis University of Bradford).

HUNTER, J.R. & M.P. HEYWORTH 1998. *The Hamwich Glass*, York (CBA Research report 116).

JACKSON, C.M. 1996. From Roman to early medieval glasses. Many happy returns or a new birth?. In: Proceedings of the 13th Congress of the International Association for the History of Glass, The Low Countries 28th August – 1st September 1995, Lochum, 289-302.

MATTHES, C. 1999. Zusammenfassung der Ergebnisse der Magisterarbeit: Die Glasperlen des merowingerzeitlichen Gräberfeldes Griesheim, paper given at *Herstellungstechniken and Herstellungswerkstätten von frühmittelalterlichen Glasperlen aufgrund der farbgebenden Komponenten und Mineralien Arbeitstreffen*, Technical University of Darmstadt, School of Materials Science, 21-22.1.99, 1999.

SABLEROLLES, Y., J. HENDERSON & W. DIJKMAN 1997. Early Medieval glass bead making in Maastricht (Jodenstraat 30), The Netherlands. In: *Proceedings of the Bead*

Conference, Mannheim, November 1994, Römisch-Germanische Kommission, 293-313.

SANDERSON, D.C.W., J.R. HUNTER, & S.E. WARREN 1984. Energy dispersive XRF analysis of first millennium AD glass from Britain, *Journal of Archaeological Science* 11: 53-69.

STEPPUHN, P. 1997. Bleiglasperlen des frühen und hohen Mittelalters in Nordeuropa. In: *Proceedings of the Bead Conference, Mannheim, November 1994, Römisch-Germanische Kommission*, 202-209.

STEPPUHN, P. 1998. *Die Glasfunde von Haithabu,* Neumünster (Berichte über die Ausgrabungen in Haithabu 32).

TURNER, W.E.S. 1956. Studies of ancient glass and glass-making processes. Part III: The chronology of glass-making constituents, *Journal of the Society of Glass Technology* 40: 39-52.

NOTES

1. Henderson & Holand (1992, p. 41 and Fig. 4b).
2. Hunter & Heyworth (1998, p. 1 and p. 44 and Fig. 22).
3. See Sablerolles (this volume).
4. Henderson (1988).
5. Turner (1956, p. 40) and Henderson (1985, pp 272-274).
6. Henderson (1995a, pp 436-437, Tables I and II).
7. Steppuhn (1997, p. 208).
8. Steppuhn (1998, pp 100-101).
9. Steppuhn (1998, pp 100-101, Table 4).
10. Green & Hart (1987).
11. Henderson (1985, p. 285).
12. Henderson (1991, pp 603-604, Table II).
13. See also Sablerolles (this volume).
14. Callmer et al. (1997, Table 4).
15. Callmer & Henderson (1991, p. 151 and unpubl. data).
16. Callmer et al. (1997, Table 4) and Matthes (1999).
17. Sablerolles et al. (1997, Table 2).
18. Sablerolles et al. (1997, pp 304-305).
19. Heyworth (1991, section 11.2), Henderson & Holand (1992, pp 36-37) and Jackson (1996, pp 294-295).
20. Henderson & Holand (1992, p. 41 and p. 43, Table I, Fig. 4).
21. Henderson (1995b, p. 53, Fig. 1) and Hunter & Heyworth (1998, p. 44, Fig. 22).
22. Dekówna (1980, P2, P4, P7 in Fig. 12 and Table 44: analysis numbers 4, 5 and 8).
23. Henderson (1991, pp 603-604).
24. See Sablerolles (this volume).
25. Sablerolles (this volume).
26. Sanderson et al. (1984, 53 ff).
27. Henderson (1995b, p. 53).
28. Callmer & Henderson (1991, p. 149).
29. Henderson (1988b, p. 144, Figs 1 and 2).
30. Henderson & Holand (1992, pp 36-37).
31. Matthes (1999, Fig. 10).
32. Matthes (1999, Fig. 10).
33. Heck (1999).
34. Brill (1988, Tables 9-1 to 9-8).
35. Aerts et al. (in press).
36. Henderson & Holand (1992, Table I and Fig. 4).

The leather finds

W. GROENMAN-VAN WAATERINGE

1 INTRODUCTION

Leather as wind- and watertight material had been indispensable for the entire prehistoric and most of the historic period. Besides clothing it was used for all kinds of equipment. That despite this, leather finds, with the exception of footwear, are rare in the archaeological record is caused in the first place by the way the hides and skins were dressed. Until the late Iron Age hides and skins were prepared by salting, smoking etc.[1] These processes are, in contrast with tanning, reversible. Leather dressed in this way keeps less well. Secondly, large objects will not be found often, because when worn they could be cut into smaller pieces and be used again. Worn shoes, however, cannot easily be made into something else. Thus experience has shown that 90% or more of the leather finds belong to footwear.

2 MATERIAL

Leather finds were rare. Only in three wells, dated 775-850, 800-850 and 850-900 small pieces of leather were found (see catalogue). Their preservation was rather bad and identification of the leather was difficult and only possible in a few cases. So far calfskin, calf/cowhide and sheep/goatskin were identified with any certainty, deerhide is possibly represented with a single fragment. The leather fragments are extremely small, making identification of objects practically impossible. As shoe finds always make up more than 90% of leather finds in general, the supposition that most fragments are from shoes seems reasonable. However, the fragments are too small to give a positive identification of any fragment. Rather uncommon are four small, more or less rectangular pieces, probably belonging together. Their size and shape make it difficult to interpret them as common shoe fragments.

2.1 *The four rectangular pieces from well 29* (Fig. 1)

From well 29 (feature 322), dated 775-850, four more or less rectangular pieces were excavated, measuring ca 9.5 × 4, 7.5 × 3.5, 7 × 3.5 and 5.5 × 3.5 cm. The leather was identified as sheep/goat skin. Two of these have seams at four sides, the two other ones only at the short sides. The stitching of all four pieces is rather irregular. The largest piece has thread impressions along one short and two long sides at both flesh- and grain-sides, the other short side shows a whip stitch; the second piece with stitching around the four sides has thread impressions along the long sides only at the grainside, but at the grain- and fleshside along the two short ends.

2.2 *The fragments from well 12* (Fig. 2)

From well 12 (feature 2957), to be dated 850-900, two leather fragments were excavated, which, in view of the seams did not belong to shoes or at least to a common type of shoes. These seams are so-called closed seams, made with flesh-grain stitching. Sofar known, seams in uppers were either made with flesh/edge stitching or with a whip stitch in the Carolingian Period.[2] Neither can these seams be seen as the closed seam part of a combination of a closed (in the upper) and butted seam (in the sole), as the stitching is too fine for a sole seam. Only in a specific type of shoes, i.e. shoes with a heel tongue is this kind of seam known.[3] The closed seam was used for sewing the two layers of thickness of the tongue together. Another possibility is an object, for example a purse, where again two pieces of leather are sewn together with a closed seam. As the fragments are so small, no definitive choice can be made for one or the other.

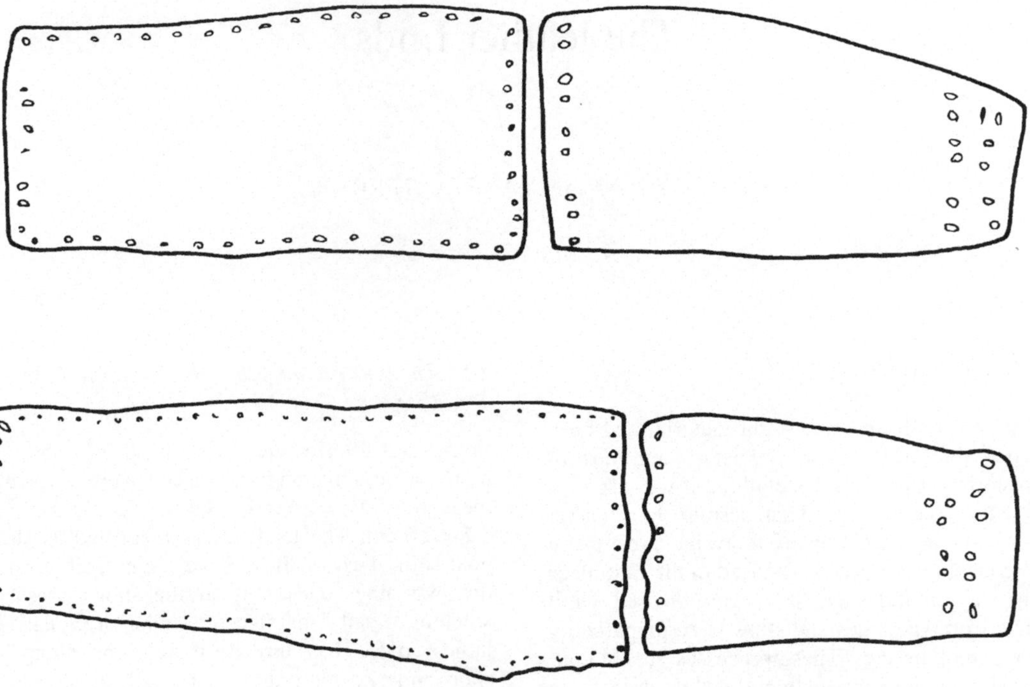

Figure 1. Wijnaldum, the four rectangular pieces of leather. Scale 1:2 (drawing AAC).

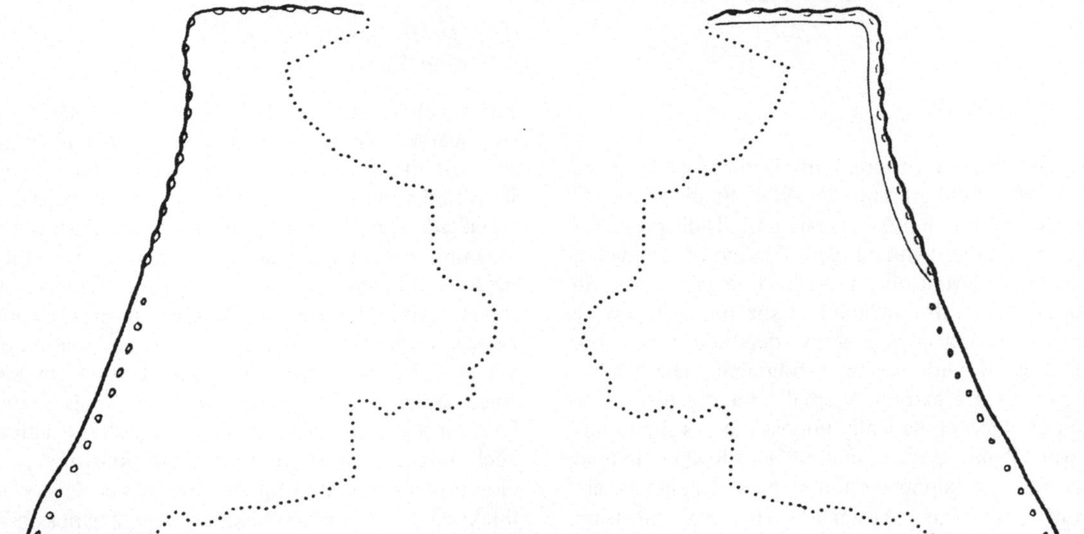

Figure 2. Wijnaldum, grain- and fleshside of the leather fragment with closed seam. Scale 1:2,5 (drawing AAC).

However, there are two arguments to make a choice for shoe fragments, although both are rather weak. There is in the first place the fact that shoe finds by far outnumber other leather finds in the medieval period.[4] In the second place there is the shape of one of the preserved pieces, that resembles not only the upper part of a heel with raised tongue, but also matches it in size (Fig. 3). But we have to be very careful with this criterion, because there is always the possibility of deformation with leather that has been preserved for such a long time in the soil. It is regrettable that the part that could have taken away any doubt of its purpose is not present, i.e. the heel part of the shoe itself or the lower half of the tongue. If this part had been preserved we would have been able to determine the presence or absence of slip stitches, used to sew the strengthening piece of the tongue into the heel. The absence of decoration is not a negative argument, as this is not always present.[5]

If the interpretation of shoe fragments belonging to a shoe with raised heel fragment is accepted the four pieces described in the preceding paragraph, found in another well and dated earlier, could well have been heel stiffeners of shoes with a raised heel tongue. Matching the four pieces two by two (see Fig. 1)[6] makes for two heel stiffeners, each consisting of a piece with stitching around four sides and one with stitching only along the short sides. The latter would be the part that was fixed in the heel of the shoe itself, the upper the part that was sewn fleshside to fleshside to the raised tongue.

Although wells 29 and 12 seem not to be contemporary, but succeed each other, one cannot be sure that the raised heel fragment and the four pieces interpreted as possible heel stiffeners ever belonged together.

3 CONSEQUENCES

Let us now consider the consequences of the find of a shoe with a raised heel tongue in Wijnaldum. As far as archaeological leather is concerned, the only parallels known from either the British Isles or the European Continent are those from the early medieval monastery of Iona, an Irish foundation. This notwithstanding the large number of leather

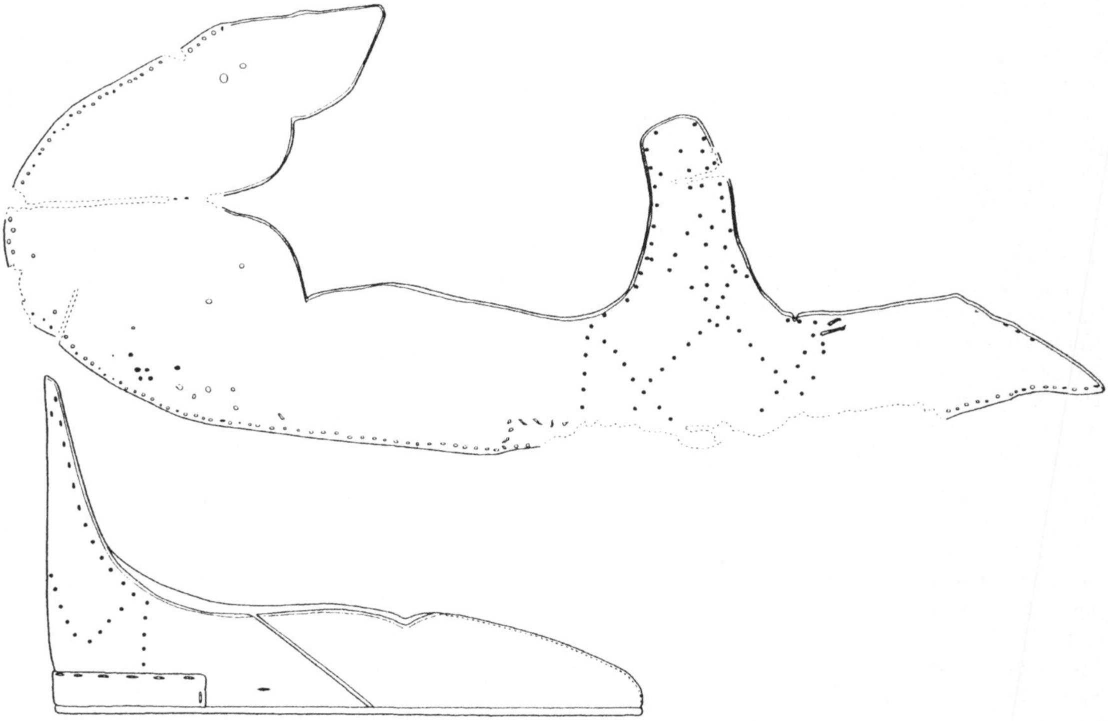

Figure 3. Iona, shoe upper and reconstruction of shoe with heel tongue (drawing AAC).

Figure 4. Shoes with heel tongue as illustrated in (from left to right) the Book of Mulling, the Book of Durrow and MS. 1395 of St. Gall (after Henry 1977, fig. 33).

objects from the period (7th-10th century) known to the author, in publications and from her own work.[7] The leather finds from Iona have to be dated between AD 600 and 650.[8] If the search for parallels encompasses manuscripts as well, however, we are more successful (Fig. 4).[9] Shoes with a heel tongue are clearly illustrated in several manuscripts, all of which are of Irish origin or have been made under strong British influence. They cover the feet not of ordinary people, but the feet of St. Matthew, for example, in the Book of Durrow (ca AD 680), and the feet of St. John in the Book of Mulling (ca AD 800?). Shoes with raised, tongue-shaped heels have never been found amongst the many Irish finds described by Lucas.[10] Thus we may assume that they were rare and made specifically for the (clerical?) elite from Ireland, or probably more generally, in the British Isles.

One is tempted to see in these shoe fragments a link with the earliest mission amongst the Frisians in the 7th-8th century, connected with two names, Willibrord (658-739) and Boniface (674-754). Both clerics came originally from the British Isles and were followed by a series of British monks and nuns. Willibrord became bishop of the Frisians and Boniface of Germany east of the Rhine. The latter undertook three journeys to Frisia; on his last, he was murdered with his entire escort by the heathen Frisians.[11] With the dating of the two fragments from Wijnaldum to the end of the 9th century, no direct connection seems to have been possible. However, many older finds were excavated from well no. 12 as well. On the other hand the shoe could have survived Boniface mission by many years, or the model could have been imitated later on.

The above given explanation is not necessarily the only one possible. Contacts between Ireland and the Continent are known from another context as well. It is known from written sources that in the Merovingian period shoes and textile were imported from Ireland towards the Continent, i.e. the Loire region, as by-products of the slave trade.[12] The role of Frisian merchants in overseas trade is not recorded before the middle of the 8th century.[13] As Merovingian trade was not so much characterised by bulk goods as well by luxury items the shoes probably belonged to the same category.[14] Not the common type shoes, made everywhere in large quantities, but something special like the shoes with raised heel tongues, must have been highly regarded by the continental elite.

CATALOGUE

Find number 4896, feature 403, filling of well 5, date 800-850 [3][15], with older finds: 13 offcuts (goatskin 7×, sheep/goat 1×, calfskin 1×, calf/cowhide 1×, unidentifiable 3×); three worn fragment (without seams, calf/cowhide 1×, probably goatskin 1×, unidentifiable 1×); two matching fragments with stitching (goatskin); one rather thick fragment with stitching (from a belt? possibly deerhide).

Find number 4921, feature 322, filling of well 29, date 775-850 [3], with older finds: one offcut (skin margin, calf/cowhide); one fragment with triangular (lace?) holes along the edge (calfskin); one fragment with three holes (sheep/goatskin); four rectangular fragments with seams (sheep/goatskin); two strips (unidentifiable).

Find number 11871, feature 2957, filling of well 12, date 850-900 [3], with much older (Merovingian) material: one sole margin; two fragments with seams, one of which with a closed seam (unidentifiable), scraps (unidentifiable).

REFERENCES

BARBER, J. 1981a. Excavations on Iona, 1979, *Proceedings of the Society of Antiquaries of Scotland* 111: 282-380.

BARBER, J. 1981b. Some observations on early Christian footwear, *Journal of the Cork Historical and Archaeological Society* 86: 103-106.

BENCARD, M. 1979. Wikingerzeitliches Handwerk in Ribe, eine Übersicht, *Acta archaeologica* 49: 113-138.

BLOCKMANS, W. 1970. Quellenanhang. In: A. Verhulst (ed.), Der Handel im Merowingerreich: Gesamtdarstellung nach schriftlichen Quellen, *Early medieval studies* 2: 2-54, 25-52.

DORGELO, A. 1961. Vondsten van middeleeuws schoenwerk te Deventer, *Berichten van de Rijksdienst voor het Oudheidkundig Bodemonderzoek* 10-11: 453-461.

GANSSER-BURCKHARDT, A. 1940. Die frühzeitliche Handwerkersiedlung am Petersberg im Basel, *Zeitschrift für schweizerische Archäologie und Kunstgeschichte* 2: 10-29.

GRENANDER-NYBERG, G. 1985. *Die Lederfunde aus der frühgeschichtlichen Wurt Elisenhof*, Frankfurt am Main etc. 219-248, 254-266 (Elisenhof, die Ergebnisse der Ausgrabung der frühgeschichtlichen Marschensiedlung beim Elisenhof in Eiderstedt 1957/58 und 1961/64, 5).

GROENMAN-VAN WAATERINGE, W. 1976. Schuhe aus Wijk bij Duurstede, *Berichten van de Rijksdienst voor het Oudheidkundig Bodemonderzoek* 26: 189-197.

GROENMAN-VAN WAATERINGE, W. 1981. The leather finds from ditch 1, *Proceedings of the Society of Antiquaries of Scotland* 111: 318-328.

GROENMAN-VAN WAATERINGE, W. 1984. *Die Lederfunde von Haithabu* 4, Neumünster (Berichte über die Ausgrabungen in Haithabu 21).

GROENMAN-VAN WAATERINGE, W. 1995. Pollenanalyse als Indikator für das Gerbeverfahren bei den Tierfellen des Mannes vom Tisenjoch. In: K. Spindler, E. Rastbichler-Zissernig, H. Wilfing, D. zur Nedden & H. Nothdurfter (eds), *Der Mann im Eis, Neue Funde und Ergebnisse*, Wien, 67-70 (Veröffentlichungen des Forschungsinstituts für Alpine Vorzeit der Universität von Innsbruck 2).

GROENMAN-VAN WAATERINGE, W., M. KILIAN & H. VAN LONDEN (submitted). The curing of hides and skins in European prehistory.

HENDRIKS, A. 1964. Karolingisch schoeisel uit Middelburg, *Berichten van de Rijksdienst voor het Oudheidkundig Bodemonderzoek* 14: 112-116.

HENRY, F. 1977. *The Book of Kells*, London.

JAMES, E. 1982. Ireland and western Gaul in the Merovingian period. In: D. Whitelock, R. McKitterick & D. Dumville (eds), *Ireland in early medieval Europe*, Cambridge, 362-386.

JURRIAANS-HELLE, G. 1998. *Kopten, christelijke cultuur in Egypte*, Amsterdam.

LUCAS, A.T. 1956. Footwear in Ireland, *The Journal of the County Louth Archaeological Society* 13: 309-394.

OJATEVA, E.I. 1965. Shoes and other leather objects from the prehistoric settlement at Staraja Ladoga, *Archeologiceskij sbornik* 7: 42-59.

PAOR, M. & L. DE 1958. *Early Christian Ireland*, London.

PREVITÉ-ORTON, C.W. 1971. *The shorter Cambridge medieval History*, 2 vols, Cambridge.

WIKLAK, K. 1969. Polish Early Medieval footwear of the 8th-13th centuries on the basis of excavations. *Materialy wczesnosredniowieczne* 6: 475-517.

NOTES

1. Groenman-van Waateringe (1995) and Groenman-van Waateringe, Kilian & Van Londen (submitted).
2. Groenman-van Waateringe (1984).
3. Groenman-van Waateringe (1981).
4. Groenman-van Waateringe (1976, 1981, 1984 with more literature).
5. Cf. Groenman-van Waateringe (1981).
6. Cf. Groenman-van Waateringe (1981, Fig. 22: 5).
7. E.g. Gansser-Burckhardt (1940), Lucas (1956), Dorgelo (1961), Hendriks (1964), Ojateva (1965), Wiklak (1969), Bencard (1979, pp 134-136), Grenander-Nyberg (1985) and Groenman-van Waateringe (1976, 1984).
8. Barber (1981a).
9. Henry (1977, p. 184) and Barber (1981a, b).
10. Lucas (1956). However, in the exposition 'Kopten, christelijke cultuur in Egypte' in the Allard Pierson Museum, Amsterdam 1998/1999 (Jurriaans-Helle 1998) two Coptic shoes (Cat. nos 90-91) show tongue-shaped heels. Although the Coptic connection of the early Irish church or at least the influence of Coptic art on early Christian Irish art (M. & L. de Paor 1958) is nowadays greatly denied (James 1982) shoes with tongue-shaped heels seem to form a tangible connection between the early Irish church and Coptic Christianity.
11. Previté-Orton (1971, pp 294-296).
12. Blockmans (1970), James (1982, pp 375-378, note 86: *Nec multum post Scothorum nauis diuersis mercimoniis plena ad litus adfuit, qui calciamenta ac uestamenta fratribus larga copia ministrauit*).
13. Verhulst (1970).
14. Verhulst (1970, pp 19-20).
15. [3] is an indication for the value of the dating: [1] finds and context/stratigraphy consistent = very reliable; [2] finds and context/stratigraphy non-consistent = unreliable; [3] finds and context/stratigraphy partly consistent = under certain conditions reliable; [4] no finds, date based on context/stratigraphy = reliable (see Gerrets & De Koning, this volume).

The human burials

A.G.F.M. CUIJPERS, C.M. HAVERKORT, J.M. PASVEER & W. PRUMMEL[1]

1 INTRODUCTION

Investigations of human graves and burials are not new in the terp region. Several excavations and studies have provided interesting information about burial customs through time.[2] At the terp of Wijnaldum-Tjitsma the excavators were already prepared for the discovery of human graves because of the well-known fibula, found some forty years ago (Schoneveld, this volume). This fibula was not interpreted as a direct indication of the presence of human burials, but it did suggest a possibility.

During a survey preceding the first excavation campaign in 1991, the surface of the terp was searched for the presence of human bones, or fragments thereof. These would have indicated the presence of a burial or multiple graves in the terp. The abundance of animal remains indicated that the preservation conditions for bone were good. However, of all the bone material that was recovered from the surface, nothing could be identified as human. This could either mean that there were no human burials, or, given the considerable erosion of the terp, that they had already disappeared. Another possibility was that they were located at a considerable depth below the surface of the terp.

In the course of successive excavation campaigns several human burials were indeed found on a somewhat deeper level: a single subadult inhumation, five infant burials and one cremation burial. Remarkably so, none of these was found in the context of a cemetery: the burials were found widely scattered across the terp (Fig. 1), with dates ranging from the 4th to the 9th century. The small size and rather unusual composition of the human skeletal sample of Wijnaldum severely limits the extent to which analysis of the skeletal remains can provide insight into the lives of the early inhabitants of the terp. Nevertheless, in this paper an attempt will be made to interpret the burials, by combining the osteological and archaeological data with evidence from proto-historic accounts and by comparing the burials in the terp of Wijnaldum with burials from similar terps in the region.

2 THE BURIALS AND THEIR SETTING

2.1 *The subadult inhumation*

This grave was dug into the natural salt-marsh just outside the contemporary terp (Figs 2 and 3). It gradually came to be covered by the terp as the latter expanded in the course of time. This not only resulted in an increased depth of the grave below the surface, it also protected the grave from the disturbing effects of agricultural activities in later centuries. The fact that this burial was located near the edge of the excavation area (see Fig. 1) leaves open the possibility of an association with a cemetery.

The individual (W'93 11702) was buried supine with extended legs, the right lower leg crossing the left lower leg. The left arm lay beside the body, the right arm across the abdomen. The body was oriented North-South, with the face turned to the West. The skeleton is fairly complete and very well preserved. In the marrow cavity of the right femur dried matter was found. After microscopic examination, this appeared to be bone material, probably lamellae from the marrow substance.[3] A sample of this matter has been used for DNA analysis.

Some small grave goods were found in association with different parts of the body: a chain of coloured beads of glass and amber was found around the neck; two bronze penannular brooches, one of which is badly eroded and fragmentary, near the shoulders, and a Donar amulet made of the ultimate tip of a red deer antler (*Cervus elaphus*) with a point-circle decoration[4] (Fig. 3) was found between the knees. These objects were probably associated with the clothes of the deceased, and are not inter-

Figure 1. Locations of the human burials at Wijnaldum within the excavation area. 1. Inhumed infant; 2. Cremation burial; 3. Inhumed subadult.

Figure 1. Continued.

Figure 2. Wijnaldum (Friesland), the inhumed subadult during the process of excavation (photograph C.M. Haverkort).

preted as gifts. Based on its stratigraphical position and the associated finds, the date of this burial ranges from AD 450 to 500.[5]

2.2 *The infant inhumations*

The skeletal remains of five newborn individuals were found in a settlement context. No grave goods were found with any of the burials. In general, the state of preservation of the remains is very good, as may be illustrated by the fact that even such small elements as immature teeth, ossicles and third phalanges of the fingers were recovered. The incompleteness of most of the skeletons is therefore likely to be the result of later disturbance such as ploughing.

It was not until they were washed and sorted that some small bones (W'92 7379) were recognized as being those of a human infant. Because the find number was associated with a position fairly close to the surface, and because the bones were ex-

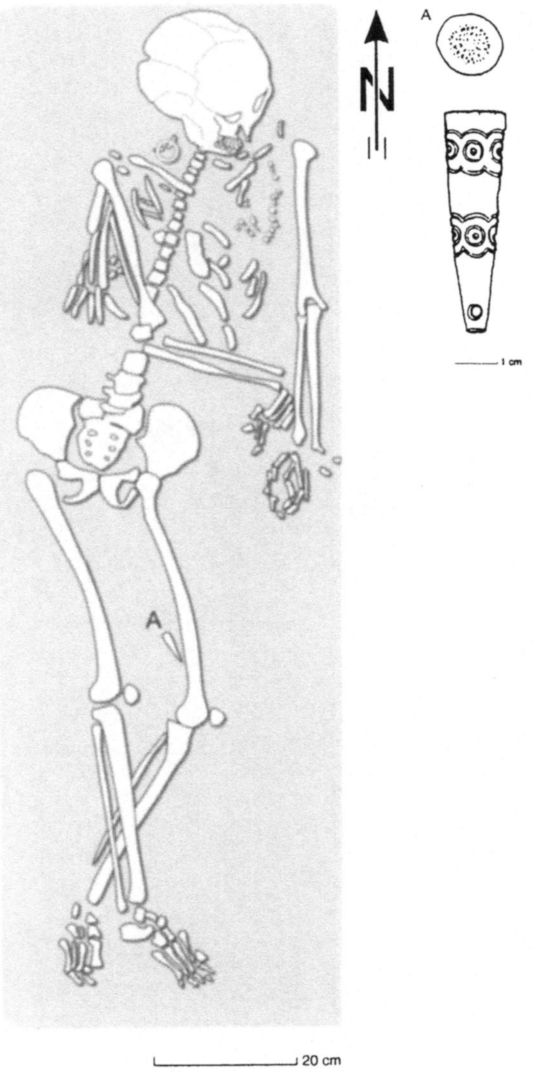

Figure 3. Wijnaldum (Friesland), inhumation burial W'93 11702. The Donar amulet of red deer antler (*Cervus elaphus*) (length of the amulet: 48.6 mm).

tremely well preserved, a recent date was at first suspected and the regional police as well as the Laboratory for Forensic Pathology (Rijswijk) were informed. However, the circumstances of the find could be reconstructed afterwards and its stratigraphical position indicated a date between AD 650 and 750.[6] This infant was buried in a small pit in a sod wall of a house. The skeleton is nearly complete, as opposed to those of the subsequently found infants.

Infant W'92 7946, found shortly after the first, was buried on its left side, facing North. Unfortunately the grave was disturbed. Only the trunk, upper extremities and skull were present, although the latter badly crushed by the weight of the soil. The grave was cut through the edge of a well. This and other stratigraphical evidence dates the child between 775 and 850.[7]

Buried on its right side, facing South, infant W'92 8260 was found a few weeks after the second, in a layer with the remains of sod houses and sunken huts (Fig. 4). This skeleton also lacks its lower extremities due to later disturbance of the grave. Because the position of the burial had been right underneath the ploughed topsoil, the features of its context were badly eroded and difficult to read. The infant was found very close to the burial of two foetal horse skeletons[8] which will be discussed in detail in Vol. 2. An association between the infant and the horses is suggested by the horizontal and stratigraphical position of the finds. The date of these burials is inferred from a collection of associated wheel-thrown pottery sherds between AD 450 and 500.[9]

The remains of the infants W'93 9021 and W'92 8008 were found in the same area. Both children were buried in a bronzecaster's workshop outside the associated house, amidst numerous fireplaces (Fig. 1). The skeleton of infant W'93 9021 was incomplete, with no pelvis or lower extremities other than the right tibia and fibula. The remains of infant W'92 8008 consisted of only both femora and the right tibia. The disturbance of their graves can probably best be explained by recent ploughing. The position of the body of W'93 9021 was

Figure 4. Wijnaldum (Friesland), one of the inhumed newborns (W'92 8260) (photograph C.M. Haverkort).

prone and its orientation was approximately Northeast-Southwest with the head Southwest, facing Southeast. The position of the W'92 8008 infant is unknown. Based on stratigraphy, both burials have been dated between AD 300 and 350.[10]

2.3 *The cremation burial*

The cremated bones (W'91 4570, 4840 and 4671) were deposited in an earthenware Anglo-Saxon jar. Cut by a Carolingian ditch (Fig. 1), the urn was slightly damaged. The different find numbers were treated separately until it was clear that there were no intrusive elements from the Carolingian ditch. The results presented below refer to the combined find numbers.

With regard to its location, this burial is in some aspects comparable to that of the subadult inhumation. The jar with the cremated remains was located along the edge of the excavation area and was dug into the natural salt-marsh outside the contemporary dwelling mound. In a later period the terp extended over the burial. Eventually the Carolingian ditch was dug along, and partly through the jar and its contents.

The stratigraphical position of the find as well as the associated Anglo-Saxon jar date these cremated remains in the second quarter of the 5th century AD. This date associates the burial with the beginning of the second period of habitation, which started around the middle of the 5th century.[11]

The total weight of the human bones is 360 g. This is rather low, considering the fact that after the cremation of an adult usually between 1.6 and 3.6 kg of bone fragments is left.[12] The overall colour of the fragments varies from brown-blue to white.

Among the grave goods contained in the jar were some fragments of rock crystal, glass-like beads, parts of a bronze ornament, a decorated antler spindle whorl and another decorated piece of antler. Particles of melted glass and metal were stuck to some bone fragments, and the other goods were burnt, indicating that these objects had been burnt together with the deceased.

The calcined spindle whorl (W'91 4671) is made from the basal part of a red deer antler, as is revealed by its compact structure. The whorl is rounded, and is decorated with at least nine concentric circles (three near each top, three or more over the widest part). Its diameter is 24.9 mm, its height 17.4 mm; the diameter of the perforation is 5.8 mm (Fig. 5).

The other calcined worked piece of antler is a

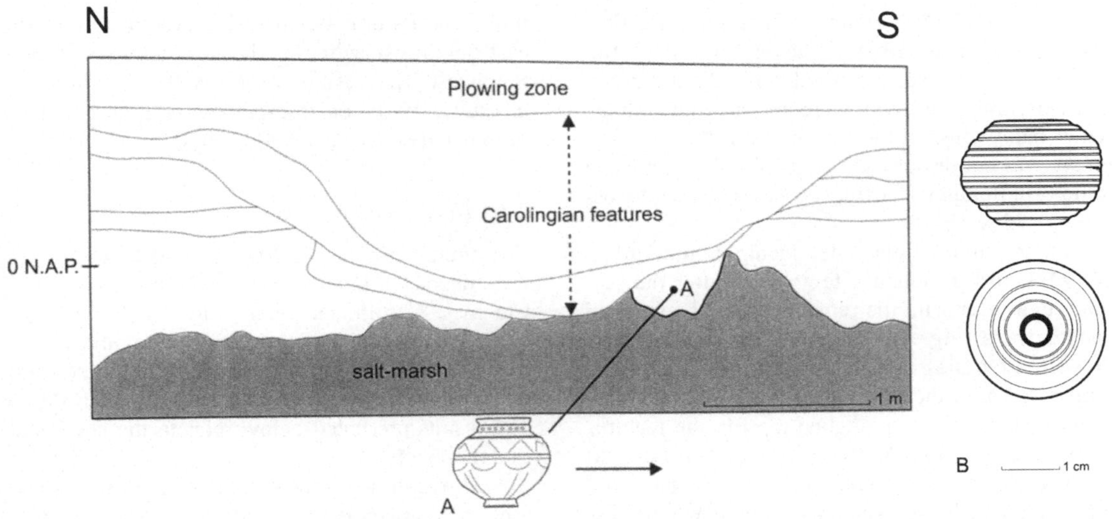

Figure 5. Wijnaldum (Friesland), cremation burial W'91 4570/ 4840/4671. The calcined spindle whorl of red deer antler (*Cervus elaphus*) (diameter of spindle whorl: 24.9 mm).

small fragment (26.5 mm in length, 16 mm in width, 3 mm in height) of a larger plate of antler. On one side it shows two rows of a fine decoration of small triangular notches. No traces of sawing for making teeth or rivets are present, such as in the coverplates of combs. The fragment may come from a broken comb case.

In addition to these grave goods some animal skeletal remains were recognized among the urn's contents. These bones will be described and discussed in section 3.3.

2.4 *Single finds of human bone*

Single human bones, bone fragments and teeth without any association with other finds have also been found during the excavations. Further details will be discussed in Vol. 2.

3 WIJNALDUM'S INHABITANTS

The study of human skeletons, representing the actual human beings that once inhabited a settlement or area, can provide a valuable addition to other archaeological data. A burial site typically provides information on a number of population characteristics, such as age and sex ratios and mortality profiles. In addition, frequencies of trauma and diseases may provide an indication of the overall health patterns, whereas differential mortuary practices might indicate social differences within a population. In order to obtain these types of information, a basic physical anthropological study of the remains is a prerequisite. However, any skeletal sample – i.e. from a burial site – is a *biased* sample rather than an accurate representation of the once living population. In the case of Wijnaldum this is quite obvious and our interpretation will therefore necessarily be restricted to the isolated individuals.

3.1 *The inhumed individuals*

3.1.1 *Methods of investigation*

Sex determination

Most methods that have been developed for a sex determination are based on either a visual or a statistical evaluation of morphological characteristics of the *adult* skeleton. The characteristics used in this study concern the morphology of the skull and the pelvis, as they are defined by the Workshop of European Anthropologists.[13]

Although all populations are characterized by a certain (population-specific) degree of sexual dimorphism, there is a considerable overlap in skeletal morphology between the sexes, which can make sexing extremely difficult. Moreover, the fact that in subadults sexually dimorph characteristics may not yet be fully expressed complicates the classification of such individuals as either male or female.

The skeletons of neonates and infants form a specific category for sex determination. Not only do their skeletons consist of many more elements due to the presence of unfused epiphyses, but also the shape of the immature bones differs from that of their adult counterparts. Although several studies have attempted to find morphological features to distinguish between male and female neonates or infants,[14] the results must still be considered inconclusive.[15]

However, recent developments in the study of ancient DNA have led to promising results with regard to sexing of individuals on the basis of Y-chromosome linked sequences as well as the amplification of part of the amelogenin gene (Richards et al., this volume). The application of such techniques to infant remains will be of immense value for paleodemographic studies, since it would provide information which is as yet very difficult to retrieve in any other way. Some preliminary results from the DNA-analyses are included here. For a further discussion of the DNA study see Richards et al. (this volume).

Age determination
Age determination is based on a combination of skeletal and dental *development*, in the case of subadult individuals, or, for adult individuals, on the basis of both skeletal *degeneration* and dental *wear*. In general the physiological age is estimated, rather than the chronological age. These two ages do not necessarily correspond, since both maturation and degeneration of bones and dentition are subject to environmental factors. The degree of degeneration of bones, for example, is strongly influenced by factors such as physical activity and nutritional status, while degree of tooth wear is to a large extent determined by the abrasiveness of the food that is consumed. No remains of adults were found at Wijnaldum, but the degree of tooth wear of the subadult individual was evaluated on the basis of Brothwell's diagram.[16]

Because of the variable effects of environmental factors, the age of adult individuals can only be determined within relatively broad ranges. This is in contrast to subadult individuals, where the immature bones consist of separate parts that will eventually fuse to form a single, full-grown bone. Individual bones follow a specific developmental pattern, so that the age range in which the epiphyses fuse is relatively narrow. However, this growth pattern is slightly different between girls and boys, and when the sex is unknown the total age range of epiphyseal fusion for both sexes should be considered.

Dental development, which can be evaluated independently of the skeletal age, is more uniform for both sexes. The eruption of teeth also follows a specific pattern and diagrams have been developed to estimate age on the basis of the eruption of both deciduous and permanent dentition.[17] Generally, dental age shows a closer relationship with chronological (or calendar) age than does skeletal age, since the latter is more subject to environmental influences.[18]

Although many epiphyses are not yet present in neonates, skeletal development can also be used as an indicator of age for this category. It has been shown that in neonates a strong relationship exists between age and body length. Series of foetal and infant skeletons of known age up to two months after birth have been used to construct tables and regression formulas. These formulas, based on the lengths of many of the immature bones, relate developmental stage of the skeleton to age[19] and provide an accurate age estimation. Developmental stage of the tooth crowns[20] and other skeletal parts,[21] however small, may provide additional information about the age.

Archaeological material is often incomplete, damaged or eroded. Age determination is therefore usually based on a combination of the information from as many of the available skeletal and dental elements as possible. In the case of very fragmentary material it may be impossible to determine age using any of the above mentioned techniques. In such cases, the microscopic analysis of the internal bone structure may be used for age determination. This method is especially important for the study of cremated remains and will be discussed later in further detail.

3.1.2 *Stature*
An estimate of stature can be obtained by measuring the long bones of the individual. Trotter and Gleser[22] and Trotter[23] have developed regression formulas based on a sample of 'American Whites' and 'American Blacks', both males and females. These formulas are widely used and generally considered as being reliable.

3.1.3 *Results*

The subadult
The subadult was a nearly full-grown individual, judging from the stages of epiphyseal fusion. This

means that the sex-specific features of the bones are sufficiently developed to allow for a reliable sex diagnosis. The pelvis as well as the skull show clear and unambiguous female characters: the pelvis has low and broad innominates with a wide subpubic angle; the skull is round and lacks significant development of supraorbital ridges, it has a nearly vertical forehead and large orbits. The diagnosis as 'female' is supported by the overall gracility of the skeleton. On the basis of DNA analysis this individual was also identified as female (Richards *et al.*, this volume).

The age at death of this young woman was determined on the basis of skeletal development and on the degree of dental wear. Unfused epiphyses of the skeleton are: the acromion of the scapula, the crista iliaca of the pelvis, the distal ulna and the medial clavicula. Using the age ranges for epiphyseal fusion in girls,[24] her age must have been 18 or 19 years.

The minimum of tooth wear indicates an age of 17-19 years, using Brothwell's diagram.[25] The third molars are absent. Although the time of eruption as well as the congenital presence of these elements is highly variable, their absence in this case does not conflict with a young age.

Since all epiphyses of the major long bones have fused, we may consider her bodylength to have attained its maximum. The most reliable estimate of stature can be obtained using the regression formula based on a combination of the measurements of femur (37.4 cm – left) and tibia (29.6 cm – left), since these bones contribute significantly to body length. Application of these numbers to the appropriate formula gives:

$$1.39\,(\text{femur} + \text{tibia}) + 53.20 = 146.33 \pm 3.55 \text{ cm}$$

In general, traumatic and pathological affections may become visible in the skeleton, depending on the nature and the gravity of the disorder. Chronic diseases are most likely to result eventually in visible bone lesions. Theoretically, this individual has lived long enough for the bones to show physical manifestations of a disease. However, morphological investigation of the bones provided no indications of traumatic lesions or disease.

Thus, the only non-infant inhumation found at Wijnaldum concerns the burial of a young woman. She died at an age of 18 or 19 years and was rather short, between about 143 and 150 cm. The skeletal remains do not provide any clues with regard to her health status or cause of death for this individual.

The infants

Given the difficulties with sexing infant remains, the sex of the newborn individuals could not be established by means of morphological methods. Tentative results from the DNA-analyses (Richards et al., this volume) include the determination of two female infants (W'92 7379 and W'92 7946) and three male infants (W'92 8260, W'93 9021 and W'92 8008). The age at death of the infants could be determined fairly accurately by means of body length, form and size of the bones and tooth buds. However, it must be emphasized that in infants, as in older individuals, the developmental characteristics will show a certain degree of variation.

Infant W'92 7379 from the 7th/8th century, with a body length of 52.7 cm was the youngest, having died at birth or a few weeks after birth. Infant W'92 7946, dated in the 8th/9th century, is comparable in body length (53.9 cm) and therefore age, but may be a few weeks older considering the size and development of the bones. The other three individuals (W'93 9021, W'92 8260 and W'92 8008) have reached the age of about two months after birth, having a body length of 56.5 cm, 56.9 cm and 57.8 cm respectively.

None of the bones show signs of trauma or pathological features and no cause of death could be established.

3.2 *The cremated human remains*

3.2.1 *Methods of investigation*

Only one cremation burial was found in Wijnaldum. The study of such remains, which are characterized by various effects of burning such as fragmentation, shrinkage and distortion, is a challenge for the anthropologist. Whether cremated remains can be studied successfully not only depends on the completeness of recovery of the remains from the pyre after the cremation, but also on the characteristics of the burial environment and the care exercised during the excavation and cleaning of the material.

Although cremated remains differ from inhumed remains, the objectives of the studies are the same and involve a determination of the minimum number of individuals, their sex and an estimation of age. Due to shrinkage and warping of the bones and their fragmentary state, however, methods for age and sex determination that have been developed for unburnt and relatively complete skeletal remains (WEA)[26] are not readily applicable to cremated remains. For example, the sexing criteria used by WEA should be applied with caution, since they are based upon the shape and robusticity of several skeletal elements. These factors can be altered by the burning process.

The cremated remains are first sorted using sieving of different mesh widths, in this case 10 and 3 mm,[27] yielding different fractions which are analysed separately. The <3 mm fraction is searched for middle ear ossicles or small bronze fragments The 3-10 mm fraction is more important, since it may contain crucial skeletal elements such as epiphyseal fragments, tooth fragments or other bone fragments indicative of age or sex. Among the >10 mm fraction are the more recognizable bone elements, which are sorted into the categories neurocranium, viscerocranium, axial, diaphyseal, epiphyseal, residue and animal. Because the total weight of the cremated remains will be calculated on the basis of the fractions 3-10 mm and >10 mm, other archaeological artefacts such as small parts of bronze, glass or charcoal fragments, as well as stones and other intrusive soil elements are removed. The different categories of the >10 mm fraction are measured and weighed to get an idea of the degree of fragmentation and of the proportional representation of the different parts of the skeleton in the cremated remains. Also the colour of the different categories is noted, since it provides an indication of the degree of burning as well as the temperature. The initial sorting is followed by an assessment of the minimum number of individuals (MNI), the determination of age and sex (in this order, because a sex determination is restricted to adults) and possible evidence of trauma or disease.

MNI can be determined on the basis of e.g. double skeletal elements or clear age differences in the dentition and other skeletal parts. The presence of skeletal elements of more than one individual need not necessarily imply an intentional double or multiple burial, since mixture at the place of burning or intrusion later in the ground could have occurred.

In the presence of e.g. epiphyses or dentition, age determination of the remains of individuals belonging to the two infans groups, 0-6 years (infans 1) and 7-12 years (infans 2), and the juvenile group of 13-18/19 can be relatively easy. For individuals older than 24 years, however, the age determination is more problematic. The most reliable results are obtained by using the so-called complex method, which gives an estimate of age based on a combination of four characteristics according to WEA standards.[28] Unfortunately, in cremated remains only one of the four features that are used in this method is regularly present, i.e. the sutures of the skull. Since suture obliteration is the least reliable of the four characteristics and because only parts of the sutures are present in cremated remains, it only permits a rough estimate of age. A morphological age determination can also be very difficult or even impossible on cremated remains consisting of only a few bone fragments.

In response to these problems, the histological age assessment method of thin sections of cremated bone was developed during the last decade. Histological age determination is based on age-related changes in the bone structure. To study the bone structure, thin sections (30-80 μm) of long bones are made with a microtome and studied with a light microscope.

For long bones, there are principally two histological ageing methods: the quantitative and the qualitative method. The former is based upon counts of, for example, the number of osteons (complexes of concentric bone lamellae around a central canal) per unit, e.g. mm^2. The age estimate is obtained by inserting this value in a regression formula. Following Kerley's[29] work on femur, tibia and fibula, methods and associated regression formulas have been published for unburnt bones (e.g. for the femur,[30] the tibia,[31] mandibula,[32] ribs and clavicle)[33] as well as for burnt femur fragments.[34]

In the qualitative method, various microstructures in bone are examined for their age-dependent changes and are compared with a reference series.[35] Living bone is characterized by a constant rearrangement of the internal structures, and the specific stages in this remodelling process provide an indication of age. However, some caution should be exercised in the application of the histological methods since a number of factors may affect the internal organization of bone structure, such as nutrition, sex, and metabolical diseases.[36] For many factors it is not yet known how and to what extent they may act upon bone tissue.

In this study, the qualitative method was used. Research carried out by Hummel and Schutkowski showed that with histological age determination of cremated remains the qualitative method provides better results than the quantitative, given sufficient experience of the observer.[37]

The most important microstructures (see Figs 6 and 7) for the qualitative histological age determination are:

– *The outer and inner circumferential lamellae.* These layers on the outer/cortex side and inner/marrow cavity side decrease with age.

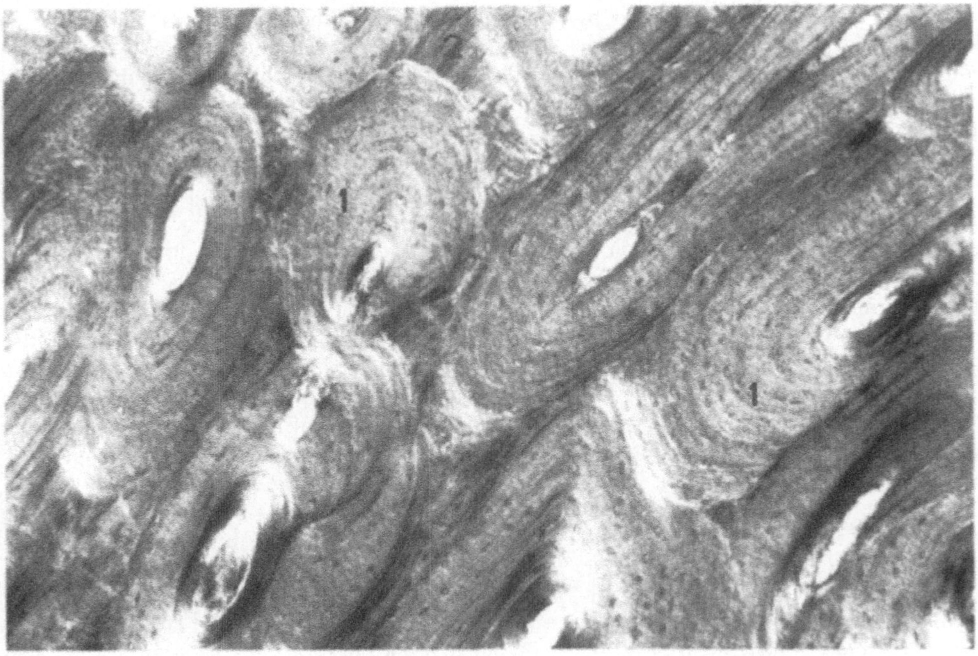

Figure 6. Osteons (1) and interstitial lamellae (2) of the adult individual (125×) (photograph A.G.F.M. Cuijpers & R. de Man).

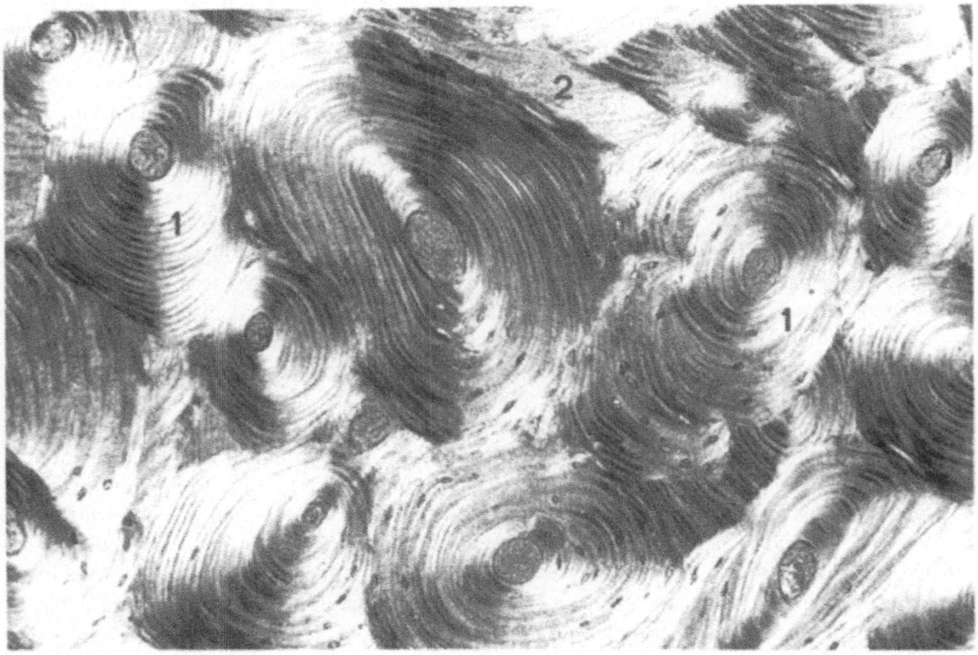

Figure 7. Osteons (1) and interstitial lamellae (2) of an individual from the reference series, age 28 years (125×) (photograph A.G.F.M. Cuijpers & R. de Man).

– *The osteons or Haversian systems.* These complexes of concentric bone lamellae around a central canal change in shape, size and density with age. In addition the area that is occupied by osteons and osteon fragments serves as an age-indicator.

– *The resorption canals.* These canals, formed by osteoclastic activity, vary in terms of number and the position in bone tissue with the age of the individual.

A small fragment of one of the long bones, preferably femur, is embedded in an epoxy resin,[38] in order to prepare thin sections with a Leitz saw microtome.[39] The thin sections are examined under a microscope with differential contrast (Nomarski), and compared with sections of individuals from a reference series (in this case the reference series of Uytterschaut[40]). The histological age determination takes place blindly from the morphological results.

In morphologically unidentified diaphyseal fragments histological research can also be applied to distinguish between human and animal bone. Generally this method gives conclusive results, since the bone structure of an animal, like pig, sheep or cattle, differs from that of a human.

As mentioned before, sex determination on cremated remains is not an easy procedure, due to the distortion of the bone material. As a complementary method to the standards of the Workshop (WEA), measurements of the pars petrosa[41] can be applied to cremated remains with good results.

3.2.2 *Results*

In the >10 mm fraction fragments of all skeletal parts – skull (24.7%), spine (22.3%), long bones (18.2%) and epiphyseal (6.5%) – were represented. Also, some animal bone fragments (28.3%) were identified. The overall colour of the fragments varied from brown-blue to white. This indicates a burning temperature between 550 and 800°C.[42]

Some skull and dentition fragments were identified as those of a child, whereas other skull, long bone and tooth fragments were definitely from an older, juvenile or adult, individual. Thus, it could be established that the MNI was two.

On the basis of dentition, a morphological age determination was possible for the child. Fragments of a deciduous molar and some of the other teeth indicated an age between 3 years ± 12 months and 5 years ± 16 months old, according to Ubelaker's diagram for dental development.[43] Using this same diagram the age of the second individual could not be specified more accurately than as 'older than 11 years ± 30 months'.

The histological age method was applied to both individuals. The complete cross section of a bone fragment of the child shows non-Haversian systems and some osteons. There are a lot of lamellae present, especially in the inner area. Qualitative histological age determination confirms the morphological age determination of between three and six years.

Even more important is the histological age determination of the second individual, for which the morphological investigation could only give a minimum age of 11 years ± 30 months. The bone structure of this individual is characterized by a lot of inner basic lamellae and some outer basic lamellae. There are also lamellae present between the osteons. The osteons represent several generations. Non-Haversian fragments in the outer area of the thin sections are indicative of a 'young age'. After consulting the reference series, the age was eventually determined as between 20 and 34 years old. This confirms the presence of at least two individuals in the cremation burial.

No primary sex indicating bone features, as defined in the standards of the WEA were discovered among the remains. A secondary feature, the processus marginalis, appeared feminine. This, however, on its own is not sufficient to provide a sex diagnosis. There was no pars petrosa present, by which the sex could have been established. Close examination of the fragments did not reveal any evidence for trauma or disease.

Thus, the only cremation burial found at Wijnaldum up to now contained two individuals. The child was between 3 years ± 12 months and 5 years ± 16 months old and the adult was between 20 and 34 years old. The sex of the individuals could not be determined. Apart from the human remains some animal remains are also present.

3.3 *The animal remains in the cremation burial*

Thirty cremated animal bone fragments, with a total weight of 53.5 g, were found among the cremated human remains in the urned burial. All of them are calcined (white). They are:

– A shaft fragment of a metacarpus and two proximal fragments of possibly the same right metatarsus of adult cattle (*Bos taurus*), weighing together 20.7 g,

– A shaft fragment of a right metatarsus of a roe deer (*Capreolus capreolus*) weighing 1.1 g,

– Six rib fragments of sheep/goat (*Ovis aries*/*Capra hircus*) weighing together 5.4 g,

– A spindle whorl (W'91 4671, discussed earlier) made from red deer antler (*Cervus elaphus*), weighing 7.8 g,

– A fragment of an implement made from red deer antler (*Cervus elaphus*) weighing 1.0 g (discussed earlier),

– Five unidentified remains from cattle/horse size mammal(s), weighing together 7.2 g,

– 13 unidentified remains from sheep/goat/pig size mammal(s), weighing together 10.3 g.

The human or animal origin of ten other calcined bone remains, weighing together 10.1 g, could not be morphologically established.

4 FRISIAN INHUMATION AND CREMATION TRADITIONS

The human skeletal remains that were uncovered at Wijnaldum – one subadult woman, five infants and one cremation burial of an adult and a child – form a rather unusual sample. It is unlikely that this sample is the result of differential preservation at the site, given the excellent condition of all bone material found in the terp. It is theoretically possible that the erosion of the topsoil caused the destruction of burials. However, because no loose human bones were found in the topsoil, we conclude that no more individuals were buried within the excavated area. One or several cemeteries may be hidden outside the excavated area of the terp Tsjitsma, or on neighbouring terps of the Wijnaldum row of terps.

Few inhumation and cremation burials are known from the terp region of Groningen and Friesland that date from the period before AD 400. The prevailing funeral custom during the first four centuries AD was possibly cremation without the burial of the ashes, or cremation followed by an unurned burial. No grave goods were given. Around the beginning of the 5th century the burial custom changes. There are now formal cemeteries, with cremation and inhumation burials. Isolated burials, however, still occur. Grave goods are often given to the dead.[44] After the 8th century, inhumation is the predominant rite of burial under influence of Christianity. Grave goods are no longer given.

At Wijnaldum, remains of the inhumation as well as the cremation rite were discovered: a single inhumation burial of a young woman and the cremated remains of an adult and a child found in the same urn. Each of these burials may be associated with a separate cemetery, or they were isolated burials. The five newborns were inhumed in the settlement, away from a formal cemetery.

Except for W'93 9021 and W'92 8008, all of these burials are of different date, none of them overlapping in time. Because of the small sample for each period (a maximum of two), we cannot consider any of these burials to represent the prevailing burial custom of its period. It is therefore impossible to conclude any change in burial tradition through time.

The Wijnaldum cremation burial, dated AD 425-450, and the subadult inhumation burial, dated AD 450-500, fit in the mixed burial tradition of cremation as well as inhumation which existed in the terp region during the period of the 5th to the 9th century.

The infant inhumations of Wijnaldum date from the 4th to the 9th century. Neonates buried in a settlement context are all inhumations. That cremated remains of infants under the age of one may be recovered in the terp area, is demonstrated by the early medieval cemetery of Oosterbeintum. None of these infants, however, was identified as being newborn.[45]

4.1 *Inhumation*

The inhumed individual was a young woman who had reached the age of 18 or 19. She was rather short. The calculated stature of between 143 and 150 cm is verified by the length of the skeleton measured *in situ*, which was roughly 140 cm. The average stature of early medieval Frisian women was estimated at 160 cm (range: 150-173 cm, n = 30).[46] So, even to contemporaneous standards, the Wijnaldum young woman was of short stature.

One of the grave goods found in the young woman's grave was a Donar amulet. It was found between the knees and it is possible that is was attached to a belt. Donar amulets were found across Europe from England to Russia, exclusively in female or infant graves. In Friesland they are known from cemeteries at Hogebeintum (5th/6th century) and Dokkum-Berg Sion (7th/8th century). Although the amulet is usually interpreted as a magic object[47] it cannot be excluded that it was purely used as an ornament.[48]

4.2 *Infant burials*

Inhuming infants in the settlement was not unusual in the past. Saunders[49] mentions that, as appeared from ethnohistoric sources, 'infants are frequently buried far away from cemeteries, in house floors, entrances, or in other contexts'. The underrepresentation of infants in cemeteries may thus be better explained by differential burial practices than by differential preservation. The question that arises, then, is *why* the infants were given a different treatment from that given to other age groups.

Infant burials were found at five other Frisian terps:

1. Elisenhof (8-9th century AD),
2. Hessens (6-7th century AD),
3. Tofting (4-5th century AD),
4. Jelsum (4th century AD), and
5. Hatsum (4th century AD).

At Elisenhof[50] the remains of seven neonates without any grave goods were found inhumed in the floors of houses. It was concluded that the infants were almost full-grown foetuses or neonates, with a maximum age of one month after birth.[51] Whether they were stillborn or not could not be determined on the basis of the skeletons.[52] In this case infanticide was considered, referring to Frisian customs of killing infants on the grounds of 'family planning', 'solange das Kind noch keine der Erde erwachsene Nahrung zu sich genommen hatte'.[53]

During the excavation at Hessens, an infant inhumation was discovered beneath a hearth.[54] The infant was wrapped in a woollen cloth which was tied to the body with a woollen cord. Near the head a small earthenware bowl was found. The cervical vertebrae showed evidence of strangulation. The presence of a woollen cord tied around the neck leaves little doubt as to the way the infant was killed. Fractures of the 7th, 8th and 9th ribs suggest a stab with a knife in the area of the heart. The age of the infant was estimated at six months after birth.

The infant inhumation discovered at Tofting[55] shows some similarities to the burial at Hessens. It has been interpreted as a 'building sacrifice'. An infant of about 10 months after birth was buried in a cradle together with an earthenware bowl underneath the pathway in the stable of a farmhouse. The position of the cervical vertebrae suggested that the infant's head had been severed from the spine.[56]

At Jelsum a foetus of seven months old was found buried in the settlement. Unfortunately, no more details about the context of this burial could be identified in the narrow trench.[57] At Hatsum a neonate was found.[58] The circumstances of this burial are not described.

The infants of the first three sites have been interpreted as sacrifices, which implies that they were killed before being buried. Their burial inside a house indicates that it may have been a building sacrifice. A building sacrifice was supposed to have a protective influence on the inhabitants of the house in which the corpses were buried.[59] That the children were killed is quite certain in the case of the Hessens and Tofting infants. Wijnaldum infant W'92 7379 was buried in the sod wall of a house dated between 650 and 750. This is the only Wijnaldum infant that could be interpreted as a building sacrifice, although there is no evidence for an intentional death. The context of the other Wijnaldum infants could not be established in detail.

In many cultures and periods stillborns and newborns were not buried in the cemetery, but along pathways,[60] under the fire place[61] or under the threshold of the entrance.[62] The infants were buried in these locations to avert disease or misfortune. Some authors, however, consider the inhumation of infants in the house as a normal part of the burial custom, without any protective connotation.[63]

That an excavated newborn died a natural death is very difficult to establish. Skeletal evidence of physical weakness in an infant is highly unlikely to be found,[64] since the child did not live long enough for any effects to become visible in the bones. Other causes of neonatal death, such as complications during birth, trauma or malformations caused by lethal congenital disorders may be recognized in skeletal remains. The death of unhealthy infants may have been natural, or provoked intentionally.

Both ethnohistoric and ethnographic literature show that infanticide is widespread among human populations.[65] As was mentioned before, these infants are mostly buried within or between the houses. Killing of an unwanted infant usually takes place at an early age, in most cases before the individual is considered to be fully 'human'. The definition of the 'beginning of human life' differs widely between cultures, for example[66]: after the infant has been nursed, after the first year of life, after the child has been named or after the first three years of life. From the *Lex Frisionum*,[67] we learn that the Frisians did not consider an infant as a human being until it had been nursed. Apparently newborns were looked upon as different from the rest of the population. This was perhaps the reason why it was allowed to kill infants, and it may also explain why infants were buried away from the rest of the population.

4.3 *Cremation*

The cremation burial contained the remains of two

individuals. Most of the remains are of the adult individual. All skeletal parts are represented. From the child, fragments of the skull, teeth and long bones are present. There are not many bone remains left, considering the fact that after the cremation of an adult usually between 1.6 and 3.6 kg of bone fragments remain.[68] Both individuals are only represented by 360 g of cremated bone. This small amount can be the result of two factors. First and foremost in this case are the postdepositional processes. The cremation burial was cut by a Carolingian ditch, which caused the disappearance of bone fragments. One fourth of the urn itself was lost. The second reason that so few cremated remains were found, was presumably that the people who cremated the individuals did not collect all of the bone fragments for the burial.

The question arises if this was an intentional burial of two individuals whose bodies were cremated. Intrusion of cremated bone fragments from another grave can be excluded on archaeological grounds: the cremated remains of both individuals were buried in the same urn. The second possible way in which remains of the two individuals ended up in the same grave, is that remains of an earlier cremation of a child which were left on the burning place, were mixed with the remains of the adult. Because of the fact that the skull and the extremities of the child are represented, it is more likely that the Wijnaldum cremation grave was an intentional burial of two cremated individuals. The remains of the individuals show the same degree of burning, so could have been cremated together.

4.4 *Animals in the cremation ritual*

Cattle and roe deer are only represented by metapodia; bones that, being foot bones, are not covered by meat. The metapodia found in this burial may have been parts of feet that were left attached to skins in butchering. At least two feet, a front foot and the right hind foot, were attached to the cowhide. All cow and roe deer feet were probably attached to the skins when they were deposited on the pyre. Only parts of the largest elements of the feet, the metapodia, were collected to be buried in the urn together with the human ashes.

Recovered foot bones in other early medieval burials reveal that the deceased was cremated or inhumed while lying on, covered by or wrapped in an animal skin from which the feet were not removed. So the 'skins' found in the cremation burial might have been used as wraps, or shrouds, to cover the dead bodies, and to transport them to the pyre. The feet of the animals functioned as natural handles.

Humans were cremated or inhumed in or on skins of brown bear (*Ursus arctos*) in Scandinavia[69] and Britain,[70] as is revealed by the presence of brown bear third phalanges in their graves. An inhumed 4 to 6 years old Romano-British child in Asthall (Oxfordshire, Britain), dating from the 4th century AD, was wrapped in a dog skin, as is shown by the bones of the fore feet and one of the hind feet of a dog.[71] A cremated woman in the Anglo-Saxon cemetery at Illington (Britain) was laid on or wrapped in a sheepskin, as is revealed by the cremated remains of three sheep feet in the grave (no. 5).[72] Cattle hides in inhumations or cremations have not been recorded so far.

Up to now, no other early medieval inhumation or cremation burial in the Frisian/Groningen terp region is known in which the body was laid on or wrapped in a skin. Any skin, however, from which the feet had been removed, is unrecognizable in cremation and inhumation graves. The only animal foot bones belonging to a skin found in a burial in this region so far, are three metatarsi of a hind foot of cat (*Felis domesticus*) in a male tree trunk inhumation grave at Lutje Lollum presumably from the early Middle Ages.[73] The cat bones were found near the man's hands.[74] They possibly belonged to the clothing (sleeve, glove?) of the man.

The number of two possible skins coincides with the number of humans in the urned burial. This is an argument to the contrary that the skins were a cover or bed on the pyre on which the deceased were laid out. A cowhide would offer enough room for the adult and the child. The cow hide may have been used as a wrap or shroud for the adult, the roe deerskin for the child. Whether the skins were tanned or not, is unknown.

The spindle whorl and the implement of deer antler may have been grave gifts or goods originating from the clothes of the deceased. If the spindle whorl belonged to the adult person of the urned burial, this person was presumably a woman. Four out of six spindle whorls in the cemetery of Oosterbeintum were found with humans that were identified osteologically as women (three inhumation burials and one cremation burial). The fifth was found in a cremation burial of unknown sex and age. The sixth spinning whorl, however, was found in the unurned cremation burial, *Brandgrube* 527, of a 2-5 year-old child (AD 500 to 700).[75] Therefore it is not excluded that the spindle whorl belonged to the child.

The rib fragments of a sheep or goat represent a trunk, or part of a trunk, that most likely served as a grave gift. Parts of sheep/goat, cattle, pig (*Sus domesticus*) and sandpipers (little or Temmincks stint, *Calidris minuta* or *C. temminckii*, and dunlin, *C. alpina*) were put on the cremation pyres at the early-medieval cemetery of Oosterbeintum (AD 400 to 750). A lamb or kid and a teal (*Anas crecca*) were cremated together, obviously apart from a human, after which their cremated remains were put in a separate burial.[76] The deposition of complete animals or parts of animals on cremation pyres or in inhumation burials was widespread in Europe in the Early Middle Ages. Sheep, goat, cattle and pig were mostly cremated or inhumed in parts, horses and dogs as complete animals.[77]

5 CONCLUSIONS

The skeletal sample from Wijnaldum did not allow for any conclusions on the population level. Still, the data correspond with the early medieval burial customs in the terp region known so far. In the terp the inhumation of a young adult female individual was found. The grave did not contain any grave gifts other than decorational artefacts such as fibulae, beads, possibly from a necklace, and a 'Donar amulet'. Since the grave was located at the edge of the excavated part of the terp (Fig. 1) the association with a more extended cemetery cannot be excluded.

Regarding its location the same applies to the only cremation burial that was found during the excavations. This burial contained two individuals, an adult and a child, who were most likely wrapped in cow and deer skins respectively. Some glass, rock crystal, bronze and antler objects were burnt with the deceased.

The five newborn individuals were buried at various locations on the terp. Only two of them are of comparable date. It is possible that one of the infants served as a building sacrifice, for the grave was associated with a sod wall of a house. Although contemporaneous infant burials in similar circumstances suggest an intentional death, the skeletal remains of this child do not give positive proof for killing. In fact, all five Wijnaldum infants may have died a natural death. They were buried in a domestic setting, possibly because of a perceived benefit for their relatives. But the reason for their burial outside a cemetery may indicate that they were considered different from older human beings.

ACKNOWLEDGEMENTS

We would like to express our gratitude to Prof. Dr G.N. van Vark (Dept. of Anatomy, University of Groningen), Dr E. Knol (Groninger Museum), drs E. Kramer (Fries Museum), Prof. Dr W.M. Molenaar (Dept. of Pathology, University of Groningen), Dr C.E. Le Poole, Dr H.T. Uytterschaut, R. de Man (ROB), drs F.J. Laarman (ROB), the Department of Anatomy, University of Leiden, the ROB for the permission to use their facilities and K.M. Clark for permission to refer to a paper in press.

REFERENCES

AHLQVIST, M.D. & O.D. DAMSTEN 1969. A modification of Kerley's method for the microscopic determination of age in human bone, *Journal of Forensic Sciences* 14: 205-212.

BANTELMANN, A. 1975. *Die frühgeschichtliche Marschensiedlung beim Elisenhof in Eiderstedt*, Bern etc. (Studien zur Küstenarchäologie Schleswig-Holsteins, Serie A. Elisenhof Band 1).

BOND, J.M. 1994. The cremated animal bone. In: J.I. McKinley (ed.), *The Anglo-Saxon cemetery at Spong Hill, North Elmham Part VIII: The cremations*, Norfolk (EAA Report 69), 121-135.

BOOTH, P., K.M. CLARK & A. POWELL in press. A dog skin from Asthall, *International Journal of Osteoarchaeology*.

BROEKE, P.W. VAN DEN 1977. Bouwoffer en archeologie. Een literatuurstudie van het bouwoffer en identificatie van bouwofferresten in de archeologie van Noordwest-Europa van vóór 1300 AD, Leiden (MA thesis Leiden).

BROTHWELL, D.R. 1981. *Digging up Bones: The excavation, treatment and study of human skeletal remains*, Oxford.

ECKHARDT, K.A. & A. ECKHARDT (eds) 1982. *Lex Frisionum*, Hannover (Monumenta Germaniae Historica, Fontes Iuris Germanici Antiqui 12).

EL-NOFELY, A.A. & M.Y. I'CAN 1989. Assessment of age from the dentition in children. In: M.Y. I can (ed.), *Age markers in the human skeleton*, Illinois, 237-254.

GIFFEN, A.E. VAN 1923/1924. Voortgezet onderzoek in de terp te Hatsum bij Dronrijp, *Jaarverslag van de Vereeniging voor Terpenonderzoek* 7/8: 9-47.

GREEN, B., & B. MILLIGAN 1993. Catalogue of cremations and inhumations. In: A. Davison, B. Green & B. Milligan (eds), *Illington: A study of a breckland parish and its Anglo-Saxon cemetery*, Norfolk (EAA Report 63), 18-46.

HALBERTSMA, H. 1982. Frieslands Oudheid, Groningen (thesis Groningen).

HAVERKORT, C.M., M. HOPMAN, J.M. PASVEER ET AL. 1993. De jongste bewoners van Wijnaldum (Fr), *Paleo-aktueel* 4: 123-126.

HERRMANN, B., G. GRUPE, S. HUMMEL et al. 1990. *Prähistorische Anthropologie*, Berlin etc., 186-207.

HOLCOMB, S.M.C. & L.W. KONIGSBERG 1995. Statistical study of sexual dimorphism in the human fetal sciatic notch, *American Journal of Physical Anthropology* 97: 113-125.

HUMMEL, S. & H. SCHUTKOWSKI 1993. Approaches to the histological age determination of cremated human remains. In: G. Grupe & N. Garland (eds), *Histology of Ancient Human Bone: Methods and Diagnosis*, Berlin, 111-123.

KERLEY, E.R. 1965. The microscopic determination of age in human bone, *American Journal of Physical Anthropology* 23: 149-163.

KNOL, E. 1988. Magische voorwerpen in vroeg-middeleeuwse graven in Friesland. In: M. Bierma, A.T. Clason, E. Kramer et al. (eds), *Terpen en Wierden in het Fries-Groningse kustgebied*, Groningen, 117-128.

KNOL, E. 1993. De Noordnederlandse kustlanden in de Vroege Middeleeuwen. Groningen (Dissertation VU Amsterdam).

KNOL, E., W. PRUMMEL, H.T. UYTTERSCHAUT et al. 1996. The early medieval cemetery of Oosterbeintum (Friesland), *Palaeohistoria* 37/38.

KÓSA, F. 1989. Age estimation from the fetal skeleton. In: M.Y. I can (ed.), *Age Markers in the Human Skeleton*, Illinois, 21-54.

LUBBE, H.B.M. VAN DER, C.P.A.T. KLEIN et al. 1988. A simple method for preparing thin (10 μm) histological sections of undecalcified plastic embedded bone with implants, *Stain Technology* 63: 171-176.

MAAT, G.J.R. 1985. A selection method of human cremations for age and sex determination. In: *Abstracts XII International Anatomical Congress*, London, 419.

MARTIN, R. & R. KNUßMANN (eds) 1989. *Anthropologie*, I/1, Stuttgart.

MCKINLEY, J.I. 1989. Cremations: expectations, methodologies and realities. In: C.A. Roberts & F. Lee (eds), *Burial Archaeology: Current Research, Methods and Developments*, Oxford (BAR British Series 211), 65-76.

MÜLLER-KARPE, H. 1968. *Handbuch der Vorgeschichte*, II, München.

PITZ, E. 1987. Ökonomische Determinanten der Bevölkerungsentwicklung im Mittelalter. In: B. Herrmann & R. Sprandel (eds), *Determinanten der Bevolkerungsentwicklung im Mittelalter*, Weinheim, 155-171.

PRUMMEL, W. & E. KNOL 1991. Strandlopers op de brandstapel, *Paleo-aktueel* 2: 92-96.

RALSTON, W.R.S. 1872. *The songs of the Russian people, as illustrative of Slavonic mythology and Russian social life*, London.

REICHSTEIN, H. 1994. Die Säugetiere und Vögel aus der frühgeschichtlichen Wurt Elisenhof. In: *Die Ergebnisse der Ausgrabung der frühgeschichtlichen Marschensiedlung beim Elisenhof in Eiderstedt 1957/58 und 1961/64*, Frankfurt/M. (Studien zur Küstenarchäologie Schleswig-Holsteins, Serie A. Elisenhof Band 6), 1-214.

SAUNDERS, S.R. 1992. Subadult Skeletons and Growth Related Studies. In: S.R. Saunders & M.A. Katzenberg (eds), *Skeletal Biology of Past Peoples: Research Methods*, New York, 1-20.

SCHLABOW, K. 1953. Leichtvergängliche Stoffe aus der Wurtengrabung Hessens, *Probleme der Küstenforschung im Gebiet der südlichen Nordsee* 5: 26-43.

SCHLABOW, K. 1955. Textilfunde und Hausopfer von Tofting. In: A. Bantelmann (ed.), *Tofting – Eine vorgeschichtliche Warft an der Eidermündung*, Neumünster, 94-97.

SCHUTKOWSKI, H. 1983. Über den diagnostischen Wert der Pars petrosa ossis temporalis für die Geschlechtsbestimmung, *Zeitschrift für Morphologie und Anthropologie* 74: 129-144.

SCHUTKOWSKI, H. 1986. Geschlechtsdifferente Merkmale an kindlichen Skeletten – Kenntnisstand und diagnostische Bedeutung, *Zeitschrift für Morphologie und Anthropologie* 76: 149-168.

SCHUTKOWSKI, H. 1987. Sex determination of fetal and neonate skeletons by means of discriminantanalysis, *International Journal of Anthropology* 2: 347-352.

SCHUTKOWSKI, H. 1989. Beitrag zur Alters- und Geschlechtsdiagnose am Skelett nicht erwachsener Individuen, *Anthropologische Anzeiger* 79: 1-9.

SCRIMSHAW, S.C.M. 1984. Infanticide in human populations: Societal and individual concerns. In: G. Hausfater & S.B. Hrdy (eds), *Infanticide: Comparative and Evolutionary Perspectives*, New York, 439-462.

SIGVALLIUS, B. 1994. Funerary pyres: Iron Age cremations in North Spånga, Stockholm (Theses and papers in Osteology 1).

SINGH, J.J. & D.L. GUNBERG 1970. Estimation of age at death in human males from quantitative histology of bone fragments, *American Journal of Physical Anthropology* 33: 373-381.

STEN, S. & M. VRETEMARK 1988. Storgravsprojektet – osteologiska analyser av yngre järnålderns bernika brandgravar, *Fornvännen* 83: 145-156 (also published as: Osteologische Analysen knochenreicher Brandgräber der jüngeren Eisenzeit in Schweden, *Zeitschrift für Archäologie* 26, 1992, 87-103).

STOUT, S.D. 1989. The use of cortical bone histology to estimate age at death. In: M.Y. I°can (ed.), *Age markers in the human skeleton*, Illinois, 195-207.

STOUT, S.D. & R.R. PAINE 1992. Histological age estimation using rib and clavicle, *American Journal of Physical Anthropology* 87: 111-115.

TROTTER, M. 1970. Estimation of stature from intact long limb bones. In: T.D. Stewart (ed.), *Personal identification in mass disasters: Report of a seminar held in Washington, D.C., 9-11 December 1968*, Washington, 71-84.

TROTTER, M. & G.C. GLESER 1952. Estimation of stature from long-bones of American Whites and Negroes, *American Journal of Physical Anthropology* 10: 463-514.

UBELAKER, D.H. 1989. *Human skeletal remains: Excavation, analysis, interpretation*, Washington DC (Manuals on archaeology 2).

UYTTERSCHAUT, H.T. 1985. Determination of skeletal age by histological methods, *Zeitschrift für Morphologie und Anthropologie* 75: 331-340.

WAHL, J. 1982. Leichenbranduntersuchungen. Ein Überblick über die Bearbeitung und Aussagemöglichkeiten von Brandgräbern, *Prähistorische Zeitschrift* 57: 1-180.

WERNER, J. 1964. Herkuleskeule und Donar-Amulett, *Jahrbuch des Römisch-Germanischen Zentralmuseums Mainz* 11: 176-197.

WERNER, J. 1972. Zwei prismatische Knochenanhänger ('Donar-Amulette') von Zlechov, *Acta Musei Moraviae* 57: 133-140.

WORKSHOP OF EUROPEAN ANTHROPOLOGISTS 1980. Recommendations for age and sex diagnoses of skeletons, *Journal of Human Evolution* 9: 517-549.

NOTES

1. The four authors are responsible for the study of the human cremation, of the human inhumations (Nos 2 and 3) and of the animal remains and antler grave goods respectively.
2. Knol (1993) and Knol et al. (1996).
3. Molenaar (pers. comm.).
4. Knol (1993, pp 67-68).
5. Gerrets (pers. comm.).
6. Gerrets (pers. comm.).
7. Gerrets (pers. comm.).
8. Haverkort et al. (1993).
9. Gerrets (pers. comm.).
10. Gerrets (pers. comm.).
11. Gerrets (pers.comm.).
12. McKinley (1989, p. 66).
13. Workshop of European Anthropologists (1980, pp 517-527).
14. Kósa (1989, p. 24) and Schutkowski (1986, 1987, 1989).
15. Holcomb & Konigsberg (1995).
16. Brothwell (1981, pp 71-72).
17. Ubelaker (1989, pp 63-65).
18. Saunders (1992, p. 7).
19. Kósa (1989).
20. El-Nofely & I°can (1989).
21. Kósa (1989).
22. Trotter & Gleser (1952).
23. Trotter (1970).
24. Martin & Knußmann (1989, p. 424).
25. Brothwell (1981, pp 71-72).
26. Workshop of European Anthropologists (1980).
27. Maat (1985).
28. Workshop of European Anthropologists (1980).
29. Kerley (1965).
30. Ahlqvist & Damsten (1969).
31. Uytterschaut (1985).
32. Singh & Gunberg (1970).
33. Stout & Paine (1992).
34. Hummel & Schutkowski (1993).
35. Hummel & Schutkowski (1993).
36. Stout (1989, pp 202-204).
37. Hummel & Schutkowski (1993, pp 122).
38. Herrmann et al. (1990, p. 188).
39. Van der Lubbe et al. (1988).
40. Uytterschaut (1985).
41. Schutkowski (1983).
42. Wahl (1982, p. 21).
43. Workshop of European Anthropologists (1980, pp 528-529).
44. Knol (1993, pp 155-156).
45. Knol et al. (1996, pp 45-46).
46. Knol (1993, thesis 9).
47. Werner (1964) and Werner (1972).
48. Knol (1988, pp 122-124).
49. Saunders (1992, pp 2-3).
50. Bantelmann, (1975, p. 79).
51. Bantelmann (1975, p. 79).
52. Reichstein (1994, p. 136).
53. Pitz (1987, p. 164).
54. Schlabow (1953, p. 42).
55. Schlabow (1955, p. 97).
56. 'Die Todesursache dürfte in der Lage der Halswirbel, die scharf gekrümmt sind, zu suchen sein. Es mu angenommen werden, da die Halswirbel vom Kopf getrennt waren' (Schlabow 1955, p. 97).
57. Kramer (pers. comm).
58. Van Giffen (1923/1924, p. 37).
59. Van den Broeke (1977).
60. Saunders (1992, pp 2-3).
61. Schlabow (1953, p. 42).
62. Ralston (1872, p. 136) and Van den Broeke (1977, p. 74).
63. Müller-Karpe (1968, p. 442).
64. Saunders (1992, p. 3).
65. Scrimshaw (1984).
66. Scrimshaw (1984, p. 441).
67. Eckhardt & Eckhardt (1982, pp 46-47) and Halbertsma (1982, p. 589).
68. McKinley (1989, p. 66).
69. Sten & Vretemark (1988) and Sigvallius (1994).
70. Bond (1994, p. 134).
71. Booth et al. (in press).
72. Green & Milligan (1993, p. 22).
73. Fries Museum Leeuwarden, Inventory Number 66A.
74. Knol & Prummel (pers. comm).
75. Knol et al. (1996, pp 406-407).
76. Prummel & Knol (1991) and Knol et al. (1996, p. 327).
77. Sten & Vretemark (1988), Green & Milligan (1993), Sigvallius (1994) and Bond (1994).

DNA analysis of the human remains[1]

M.B. RICHARDS, I.B. COLSON, J.F. BAILEY, R.E.M. HEDGES & B.C. SYKES

1 INTRODUCTION

Two kinds of DNA analysis which could assist in the investigation of archaeological sites such as the Tjitsma terp at Wijnaldum currently look promising: the analysis of population markers such as mitochondrial DNA (mtDNA), and sex determination. A third, more remote, possibility is kinship analysis. They all depend critically on the invention in the 1980s of DNA amplification using the polymerase chain reaction,[2] and the recent discovery that trace amounts of the genetic material sometimes survive in archaeological remains, which can be analysed after amplification by PCR.[3]

DNA has now been shown to survive in a variety of organic remains, although a picture of its extent in archaeological material is only now beginning to become clear.[4] For bone remains it seems that trace amounts of very fragmentary DNA (often less than 0.1% of concentrations in living tissues) survive when the general preservation of the bone is good, but poorer samples rarely contain amplifiable material. Our experience to date suggests that poorer samples may have sufficient collagen to be radiocarbon-dated by AMS and yet not enough DNA for analysis. This effectively means they contain no DNA of the appropriate region since under optimal conditions PCR can amplify from a single copy.

Aside from lack of surviving DNA, two problems have plagued studies of archaeological remains. The first is that the DNA often copurifies with humic and fulvic acids, which inhibit the Taq enzyme that is used to amplify the DNA in the PCR. This is widely obviated by the use of a silica/guanidine thiocyanate purification step[5]; but the effectiveness of this depends on the initial concentration of the inhibitors and several such steps may be needed to remove all traces of inhibition. Secondly, there is the difficulty of contamination of the extracts by modern excavators, museum staff and laboratory workers. An explanation of this requires a brief description of the amplification process.

PCR uses two short oligonucleotide primers to target and amplify a specific part of the genome. The double-stranded DNA is heated to about 94°C to separate the strands, and the primers are then annealed to their target by recooling. The Taq polymerase then copies the region of DNA between them, which may be several thousand base-pairs for modern DNA but is rarely more than a few hundred in fragmentary ancient DNA. The cycle is then repeated about forty times so that the number of copies increases exponentially (at least in theory), eventually leading to enough of the amplified DNA for sequence analysis to be carried out. Unfortunately, the great sensitivity of the technique means that the reaction is very prone to contamination from the DNA of modern people: a single cell falling into the reaction, for example, might contain about 800 copies of the mtDNA molecule.[6] There is also a great risk of carryover from one reaction, which produces hundreds of millions of copies, into the next, unless very strict isolation procedures are enforced. These problems are exacerbated when working with ancient DNA since the target DNA is at levels which may well be lower than that of potential contaminants (such as cells or aerosol droplets) and is, additionally, usually damaged and therefore less efficiently amplified than modern contaminants.

Sex determination using PCR rests on the amplification of parts of the human genome which differ between males and females. This is a technique which could be valuable in cases such as Wijnaldum, where the burials are of mainly of infants, which are impossible to sex using current osteological methods. Currently two different methods are being pursued: the use of primers covering a part of the Y chromosome, which will amplify in males and not females[7] and the amplification of part of the amelogenin gene, which is present on a non-recombining part of both Y and X chromosomes and has a different length in each.[8] The latter method is preferable since there is a different pattern for males and females: males yield two fragments of slightly different size, one from the Y chromosome copy of the gene and one from the X; females yield a single, double-strength fragment from the two X chromosome copies. There is therefore an internal control. However, the amelogenin gene is present at a level of only one copy per cell, which in practice makes it very difficult to recover from archaeological samples at present. The DYZ1 region of the Y chromosome, on the other hand, is a repeated region which is present at a level of around 500-1000 copies per cell, and survives in a correspondingly greater range of material.[9] The disadvantage of this method is that it is not controlled for lack of DNA in the female case: a negative result could indicate either a female or a male with no detectable surviving Y-chromosome DNA, although a parallel amplification of mtDNA can help to overcome this problem. Both methods are also at risk from contamination, and should only be performed in conjunction with other more variable loci, which can more readily be authenticated on phylogenetic grounds; most studies using osteologically sexed archaeological remains so far have suffered from this difficulty. However, the use of other loci as cross-checks for contamination can reduce this problem. At present, the locus from which the most confidence can be gained is the control region of mtDNA.

mtDNA is a particularly valuable tool in the investigation of ancient DNA because it not only has a high copy number but also a high mutation rate and therefore has diversified in different human populations.[10] This means that, on the continental scale at least, different populations often carry distinctive mtDNA control region sequences (or haplotypes), which can therefore be used as 'markers' or labels for particular populations. Since mtDNA is also inherited maternally, the relations between the sequences can also be traced phylogenetically to reconstruct maternal genealogies. Given an acceptable calibration of the rate of mutation,[11] lineage clusters can even be assigned ages, albeit with rather large standard errors. Marker sequences, however, are informative only at the population and not at the individual level. For example, the discovery of an African marker sequence in an individual of north European descent does not make that person an African; but it does indicate that at some time in the past a lineage which had arisen in sub-Sahara Africa subsequently migrated into northern Europe.

Whether or not mtDNA can be used as a marker at the much more local level of, say, different language communities is presently not known, and would in any case depend on the contingencies of history. However, certain populations may be distinctive in certain respects. We have studied mtDNA samples from modern human populations at the fringes of Europe, in southern Scandinavia, Celtic Britain, Finland, Iceland, the Basque country and northern Spain, and have divided the haplotype variation into five clusters or haplogroups.[12] The largest group of lineages in Europe is derived from a single sequence, referred to as the Cambridge Reference Sequence (CRS) as it was also (coincidentally) the first mtDNA haplotype to be sequenced and is used as a standard. Haplotypes radiate from that sequence in a star pattern, indicating an expansion within Europe which we tentatively date to within the last 25,000 years ago: perhaps reflecting demographic expansions after the last glacial maximum. Four other haplogroups also appear to radiate from the CRS, but more detailed analysis shows them to have been derived much earlier, before the Late Upper Palaeolithic. We have found that some lineage clusters are very widespread across Europe (and also, in some cases, western Asia), but that some occur more commonly in some of the populations within Europe than others. Of particular relevance here is a cluster defined by a transitional mutation at position 16,189 in the control region (which we call 189 for short). This cluster occurs at a frequency of around 25% in north-western Germany (delimited by the limes saxoniae in the east, the Danish and Dutch borders to the north and west, and extending approximately as far south as Hannover) but at less than 5% in other European populations (except a small sample of Finns, where it is intermediate). Since it is virtually restricted to this region at appreciable frequency, it is possible that it may serve as a marker for the north-west German population.

The material culture of the Wijnaldum area changes dramatically at the second half of the fourth and beginning of the 5th century, when the Anglo-Saxon complex takes over.[13] It has been suggested that this may have been due to the replacement of the old Frisian population by Anglo-Saxons, and possibly other southern Scandinavian peoples, at this time (perhaps after emigration of the earlier population); alternatively the change may have been the result of settlement and population mixing, lite dominance or new patterns of trade. Although the possibilities are complex, and the human remains from the region at present sparse, mtDNA sequencing might lend support to one or other of these, if DNA proved to have survived in the skeletal remains. The site has an advantage over others in that it was excavated with DNA analysis in mind, and bone samples were retrieved using disposable gloves to minimise the possibility of on-site contamination. In such a case, any DNA recovered is much more likely to be authentic than DNA recovered from museum specimens. Furthermore, the preservation of the remains at Wijnaldum is quite exceptional, as a result of the fine anoxic chalk environment, so that DNA is more likely to have survived.

2 MATERIALS AND METHODS

The samples are listed in Table 1. The crucial factor in the processing of these samples was their recovery from the excavation using disposable gloves, thereby minimizing the risk of modern DNA contamination. Two indicators of the diagenetic status of the bone were investigated, histological condition and nitrogen survival.[14] Approximately 200 mg samples of humerus or femur (tooth and bone lamellae from the marrow cavity in the case of W93-11702) were transferred from Groningen to Oxford in sterile plastic bags. They were then powdered using a hypochlorite-sterilized coffee grinder and extracted as described[15] but scaled down to one-tenth volumes. Phage lambda DNA (1.35g/ml) was included in the extracts as a carrier to enhance the likelihood of recovery. 221 and 223 base pair (bp) fragments were amplified using overlapping primer pairs (conL3/H3-B and conLC/H2-B), and the first hypervariable segment of the mtDNA control region sequenced directly using the biotin-Streptavidin method as described.[16] As an authentication check, amplification of a third fragment 513 bp long was also attempted for five of the samples.[17] Furthermore, the DNA sequences of two of the principal excavators were also determined for comparison as described.[18] Sex determination was performed using the DYZ1 primers Y1.1. and Y1.2, which amplify a 149-base pair fragment of Y chromosome DNA, also as described. It was also attempted with the amelogenin system.[19]

3 RESULTS

Bone diagenesis

The histological preservation of the infant bones could not be assessed directly using the histological index of Hedges et al.[20] as the structure of infant bone differs from that of adults, for which the criteria were developed. However, the bones lacked destructive foci and appeared very similar to modern infant bone structure, and were estimated to have values of around 4-5, except in one case (W92-Excavated soil) for which a value of approximately 3 was estimated. This good preservation was reflected in the nitrogen content, which were around 3%, except again the excavated soil sample which was 1.98%. Modern adult bones give values of 4-5%.

Amplification

All 14 extracts of the seven samples yielded PCR products with the closer mitochondrial primers but none of those tested (all except W92-7946 and W92-Excavated soil) yielded the 513-bp PCR product.

Table 1. Summary of the results of the DNA analysis. The sequences extend from 16,090 to 16,400, numbered as per Anderson et al. (1981) less 16,000. The table shows the number of PCR products that were sequenced from each of two extractions, and the variants from the CRS. Note that position 391 is excluded from the network as it is outside the region sequenced for modern samples. The sex of each individual, as determined by DNA analysis, is also given.

Sample no.	Sample type	Date	1st ext.	2nd ext.	Variants	Sex
W93-11702	Young adult	550-600	2	1	192	F
W92-7946	Infant	775-850	1	1	129 223 304 391	F
W92-8260	Infant	450-500	2	2	263	M
W92-excavated soil	Loose infant bone	Undated	2	1	391	M
W92-7379	Infant	650-750	2	1	145 231 261	F
W93-9021	Infant	300-350	2	1	93 126 294 296 304	M
W92-8008	3 bones of 1 infant	Roman	2	1	189 270	M

Sex determination
We assigned sex to the individuals using the DYZ1 system. The DYZ1 product was reproducibly amplified from the W92-8260, W93-9021 and W92-8008 extracts after one round of PCR, and from W92-Excavated soil after two rounds. We therefore assign these as male. W93-11702, W92-7946 and W92-7379 did not yield product even after two rounds. We therefore assign these as female, since mtDNA was amplifiable from these extracts indicating that DNA was present. None of the samples yielded PCR products with the amelogenin primers. The assignment of W93-11702 is supported by osteological sex determination[21] but since the remaining individuals are infants, it is not possible to cross-check the results with those of physical anthropology.

Mitochondrial sequence data
We have obtained complete direct sequences for the region 16,090 to 16,400[22] from duplicate extracts of all seven samples (Table 1). Each sequence is distinctive, none exactly matches the Cambridge Reference Sequence, none of the sequences matches those of two of the principal excavators at the terp site, and the five samples tested did not yield PCR products with primers that are further apart, suggesting the high degree of fragmentation expected of ancient DNA. This latter test is particularly appropriate in the present case since the samples were excavated so recently that it is unlikely that there has been sufficient time for modern contamination to have degraded, unlike the case of museum specimens. We therefore have reason to feel confident that they are authentic.

Phylogenetic analysis
The sequences can be visualized by presentation as a reduced median network,[23] which illustrates the principal phylogenetic relationships in the data. Data from bone samples, which are often rather sparsely represented due to the difficulties inherent in the isolation and analysis of ancient DNA, can be presented by incorporation into networks constructed from modern populations data. In Figure 1, we have added the seven terp sequences to a network of haplotypes from modern Frisian speakers from the North Frisian islands and the East Frisian and West Frisian regions of the German and Dutch mainland. The terp data fits closely into the patterns of variation we have identified European populations. One of the seven sequences, W92-Excavated soil, is identical to the Cambridge reference sequence (CRS) for the segment for which we make the comparison with modern data (16,090-16,365, abbreviated to 90-365) although it includes a variant outside this region at position 391. The CRS itself occurs at about 20% in modern European populations. It is the ancestral haplotype of Group 1, a cluster which we believe may have expanded throughout Europe during the Upper Palaeolithic. Samples W93-11702 and W92-8260 also appear to be members of Group 1, with single transitions at positions 192 and 263 respectively.

Sample W92-7379 falls into Group 2A, a cluster which we believe to be associated with the Neolithic colonization of Europe, as it is relatively rare and lacking in diversity in Europe but common and diverse in the Middle East. In fact, we have previously suggested that one branch of this cluster, characterized by transitions at positions 126, 145, 231 and 261 in this region, was specifically involved in the colonization of central Europe by early farming communities, perhaps associated with the LBK material culture complex.[24] This is corroborated by the sequence obtained here, which (in the region 90-400) has each of these variants except that at position 126. This position would therefore seem to have reverted.

Sample W93-9021 falls into Group 2B, which we believe to be an Upper Palaeolithic founder group with its origins in the Middle East. Sample W92-7946 is a member of Group 3, an apparently ancient cluster with its origins in western Asia or Africa. Sample W92-8008 has transitions at positions 189 and 270, making it a member of Group 5, a cluster whose origin we have not been able to trace.[25] We have therefore found in the seven terp samples representatives of four of the five major lineage groups which characterize European populations, including both sub-groups of Group 2.

4 DISCUSSION

If the calibration of the mutation rate for mtDNA is accepted, these major groups describe a time-scale of perhaps more than 60,000 years, encompassing the first settlement of Europe by anatomically modern humans and subsequent dispersals and divergence. In the present context, however, their interest would lie with whether they can be used to trace more recent population movements, such as occurred during the European Migration Period. In this regard, although on the whole European populations are rather homogeneous in their mtDNA

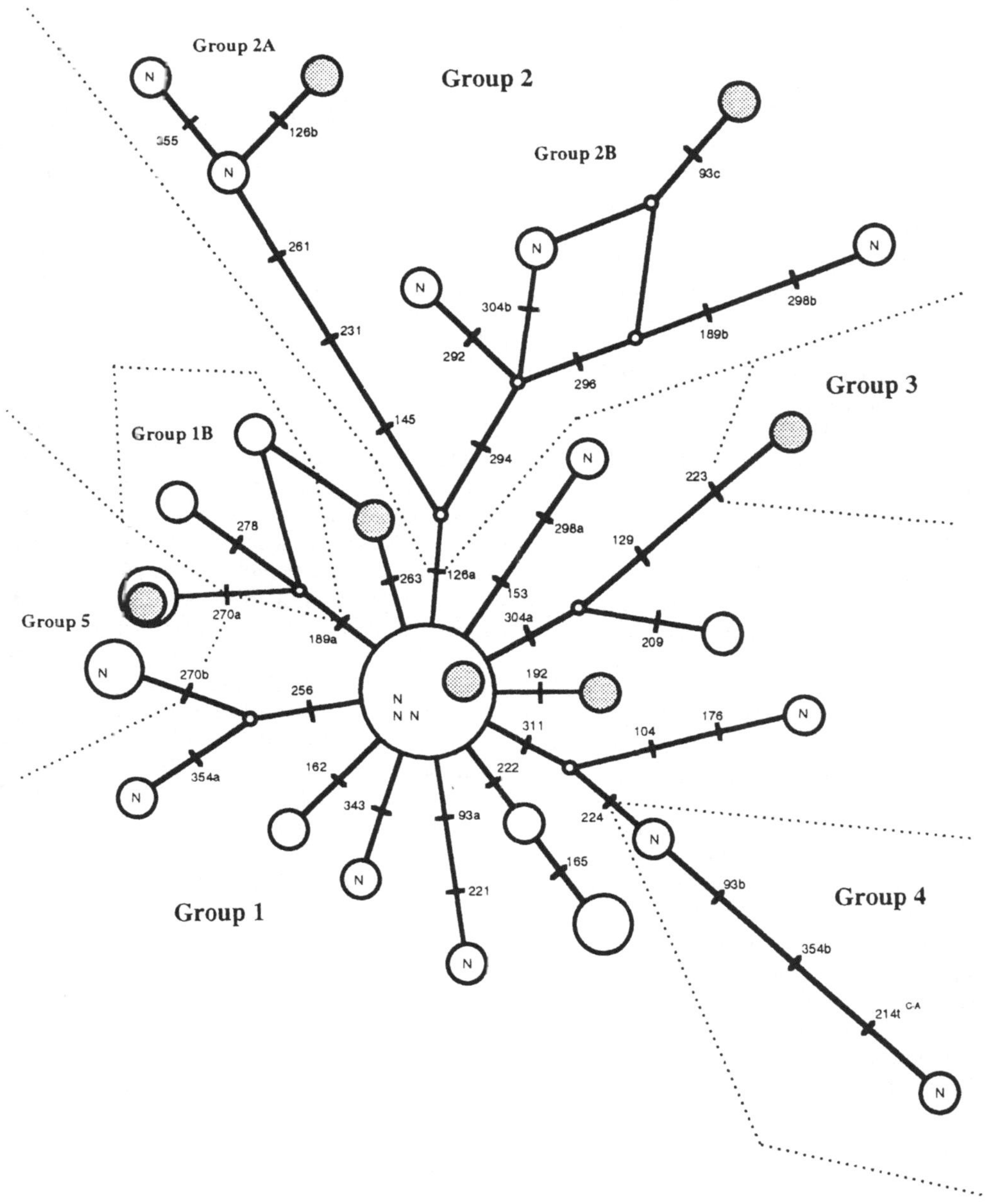

Figure 1. Reduced median network of seven sequences from the terp at Wijnaldum and 28 modern Frisian speakers (16 North Frisian islanders, four German East Frisians, three German Saterlanders and three Dutch West Frisians). Circles represent sequence haplotypes for the mtDNA control region 16,090-16,365. The area of the circle is proportional to the frequency of the haplotype; small circles represent hypothetical intermediate nodes. Lines represent mutational events between haplotypes and are labelled with the positions of transitional mutations using the number system of Anderson et al. (1981) less 16,000. Transversional mutations are labelled 't' and further specified. Parallel mutation events which have been resolved by the reduction process are labelled a, b etc. Extant Frisian speakers from Germany and the Netherlands are represented by unfilled circles and Wijnaldum samples by shaded circles. Extant North Frisian islanders are marked 'N'. The central haplotype, marked *, is the CRS. Haplogroups 1 through 5 are marked. Note that Group 5, distinguished by a variant at position 270, can be more closely clustered with additional information (Richards et al., submitted); here the network method separates 270 into two events. The topology of the network is very tree-like, despite the high number of parallel mutations; the squares indicate regions of ambiguity in the branching structure.

structure (probably reflecting post-glacial re-expansions and Neolithic immigration as well as historic and prehistoric migration events), we have identified a sub-cluster of Group 1, characterized by a transition from the CRS at position 189, which is very much more frequent (at 25%) in the present-day north-west German population than other populations we have sampled, where its frequency is generally below 5%.[26] A possible explanation for this is a local effect of the post-glacial expansion into the north European plain, which was probably uninhabited in the preceding Early Upper Palaeolithic.[27] This is fortuitous insofar as the geographical range of high frequency of this marker corresponds approximately to the region historically thought to comprise the 'Saxon' area.[28] This sub-cluster, which we refer to as Group 1B, might therefore be considered as a molecular marker for the north-west German population, and by implication may be useful in investigating possible migrations out of this region in the past.

Using the phylogenetic diagram, we can compare the Wijnaldum haplotypes with those of extant Frisian speakers from the North Frisian islands and the mainland. We should be cautious about drawing conclusions at this stage since the sample sizes are very small. However, it is interesting that amongst the mainland Frisian speakers (six from north Germany and three from the Netherlands), only Group 1 is represented (including two individuals with the 1B marker) with the exception of a single member of Group 5. By contrast, amongst the Frisian islanders (16 individuals, mainly from Føhr), all of the main European groups, with the exception of Group 3, are represented, making the islanders correspondingly much more diverse.[29] By comparison, the seven Wijnaldum samples include members of all of the groups except Group 4, and are similarly diverse (the mean number of mutational differences between them is 4.2, in comparison with 4.1 for modern Frisians and 3.3 for modern north-west Germans.[30] This may provide a first indication of a shared history between the extant islanders and the inhabitants of the terp.

The site at Wijnaldum is notable for the replacement of old Frisian by Anglo-Saxon material culture during the 5th century, and the change is sufficiently great to implicate a population replacement.[31] However, the molecular evidence provides, as yet, no support for this suggestion. Group 1B is absent from the seven individuals present at the terp, two of which (W93-9021 and W92-8008) were in any case dated stratigraphically to before the hiatus.[32] However, the probability, given by the binomial theorem, that Group 1B would be absent from the terp sample even if it were part of the north-west German population as we see it today, is 0.24. Clearly one would desire a probability of less than 0.05 in order to reject the replacement hypothesis with confidence. Moreover, a further individual (W92-Excavated soil) was represented only by a loose find of a single infant bone and has not yet been dated.[33] If it were necessary to exclude this material from the post-Saxon sample, the probability would rise to 0.32. Therefore, the sample size is too low to reject the population replacement suggestion, even given all the assumptions we have made.

Whilst we have to rely on interpretations based on modern German data, the suggestion that Group 1B can be associated with populations which were once ethnically 'Saxon' remains an assumption. Work with Saxon cemeteries from north-west Germany may help to strengthen or refute this idea. If this assumption is maintained, however, we would only need to increase our post-hiatus sample size to eleven individuals, all lacking the 1B marker, in order to lend the hypothesis of replacement at Wijnaldum statistical support. For example, our modern sample from the North Frisian islands meets these criteria and may therefore be taken to imply that these islands were settled by a population of non-northwest-German origin.[34] By contrast, the extant mainland Frisian speakers carry the marker and are therefore indistinguishable from the bulk of the north-west German population. Of course, much larger sample sizes would be necessary if, for example, different levels of population admixture were to be tested. Nevertheless, we are optimistic that it may indeed be possible to address questions about the history of this region using mitochondrial DNA using a combination of modern and prehistoric populations.

This is the first study of ancient DNA in which human remains have been deliberately excavated with gloves in order to avoid contamination. There are good grounds for thinking that the results presented here genuinely represent the survival of endogenous DNA, in stark contrast to others we have obtained from museum specimens.[35] It therefore seems that, if they are reasonably well preserved, human remains that are freshly excavated under clean conditions are likely to provide a much more accessible source of ancient DNA than museum specimens, and that this is the best, and possibly even the only, avenue for human ancient DNA research to take in the future.

ACKNOWLEDGMENTS

We thank Danny Gerrets and Juliëtte Pasveer for the bone samples, Kate Smalley and Sue Tetzner for modern DNA sequencing, Peter Forster and Antje Arfsten for modern samples, and Peter Forster and Hans-Jürgen Bandelt for analytical advice.

REFERENCES

ANDERSON, S., A.T. BANKIER, B.G. BARRELL et al. 1981. Sequence and organization of the human mitochondrial genome, *Nature*, 290: 457-65.

BANDELT, H.-J., P. FORSTER, B.C. SYKES et al. 1995. Mitochondrial portraits of human populations using median networks, *Genetics* 141: 743-753.

FISCHER, A. 1991. Pioneers in deglaciated landscapes. The expansion and adaptation of Late Palaeolithic societies in Southern Scandinavia. In: N. Barton, A.J. Roberts & D.A. Roe (eds), *The Late Glacial in north-west Europe. Human adaptation and environmental change at the end of the Pleistocene*, London (Council for British Archaeology, Research Report 77).

GAMBLE, C. 1986. *The Palaeolithic Settlement of Europe*, Cambridge.

GERRETS, D.A. 1995. The Anglo-Frisian relationship seen from an archaeological point of view. In: F.F. Faltings, A.G.H. Walker & O. Wilts (eds) *North-Western European language evolution*, Supplement vol. 12, Odense, 141-163.

HAGELBERG, E., L.S. BELL, T. ALLEN et al. 1991. Analysis of ancient bone DNA: techniques and application, *Philosophical Transactions of the Royal Society of London* (Series B) 333: 399-407.

HAGELBERG, E., B. SYKES & R. HEDGES 1989. Ancient bone DNA amplified, *Nature* 342: 485.

HANDT, O., M. RICHARDS, M. TROMSDORF et al. 1994. Molecular genetic analyses of the Tyrolean Ice Man, *Science* 264: 1775-1778.

HARPENDING, H.C., S.T. SHERRY, A.R. ROGERS et al. 1993. The genetic structure of ancient human populations, *Current Anthropology* 34: 483-496.

HEDGES, R.E.M., A.R. MILLARD & A.W.G. PIKE 1995. Measurements and relationships of diagenetic alteration of bones from three archaeological sites, *Journal of Archaeological Science* 22: 201-211.

HSS, M. & S. PBO 1993. DNA extraction from Pleistocene bones by a silica-based purification method, *Nucleic Acids Research* 21: 3913-3914.

HUMMEL, S., & B. HERRMANN 1991. Y-chromosome-specific DNA amplified in ancient human bone, *Naturwissenschaften* 78: 266-7.

KOGAN, S.C., M. DOHERTY & J. GITSCHIER 1987. An improved method of prenatal diagnosis of genetic diseases by analysis of amplified DNA sequences, *New England Journal of Medicine* 317: 985-990.

MULLIS, K.B. & F.A. FALOONA 1987. Specific synthesis of DNA in vitro via a polymerase catalyzed chain reaction, *Methods in Enzymology* 155: 335-350.

PBO, S., R.G. HIGUCHI & A.C. WILSON 1989. Ancient DNA and the polymerase chain reaction. The emerging field of molecular archaeology, *Journal of Biological Chemistry* 264: 9709-9712.

RICHARDS, M.B., H. CORTE-REAL, P. FORSTER et al. Palaeolithic and neolithic lineages in the European mitochondrial gene pool, *American Journal of Human Genetics* 59: 185-203.

RICHARDS, M.B., P. FORSTER, H. CORTE-REAL et al. (in prep.) Mitochondrial DNA variation in north-west Europe.

RICHARDS, M., P. FORSTER, S. TETZNER et al. 1995. Mitochondrial DNA and the Frisians. In: F.F. Faltings, A.G.H. Walker & O. Wilts (eds) *North-Western European Language Evolution*, Supplement vol. 12, Odense, 141-163.

RICHARDS, M.B., R.E.M. HEDGES & B.C. SYKES 1995. Authenticating DNA extracted from ancient skeletal remains, *Journal of Archaeological Science* 22: 291-299.

ROBIN, E.D. & R. WONG 1988. Mitochondrial DNA molecules and virtual number of mitchondria per cell in mammalian cells, *Journal of Cellular Physiology* 136: 507-513.

SULLIVAN, K.M., A. MANNUCCI, C.P. KIMPTON et al. 1993. A rapid and quantitative DNA sex test: Fluorescence-based PCR analysis of X-Y homologous gene amelogenin, *Biotechniques* 15: 636-641.

TODD, M. 1993. *The early Germans*, Oxford.

WILKINSON-HERBOTS, H.M., M.B. RICHARDS, P. FORSTER, et al. (submitted). Site 73 in hypervariable region II of the human mitochondrial genome and the origin of European populations, *Genome Research*.

WILSON, A.C., R.L. CANN, S.M. CARR et al. 1985. Mitochondrial DNA and two perspectives on evolutionary genetics, *Biological Journal of the Linnaean Society* 26: 375-400.

NOTES

1. Note added in proof: This manuscript was submitted in 1995, during the prehistoric phase of European mitochondrial research. In the intervening years, the picture has been considerably refined (see Richards, et al., *Annals of Human Genetics* 62: 241-260 and Macaulay et al., *American Journal of Human Genetics* 64: 232-249) and indeed is still evolving. One consequence of this is that it is no longer clear that there might be significant differences in the mitochondrial patterns between northwest Germany and other parts of Europe, making the kind of high-resolution migration studies suggested here much more difficult to envisage. We would no longer like to suggest that our data could allow us to distinguish between immigration and acculturation on a site such as Wijnaldum.
2. PCR: Mullis & Faloona (1987).
3. Pbo, Higuchi et al. (1989) and Hagelberg et al. (1989).
4. Hagelberg et al. (1991) and Richards et al. (1995).
5. Hss & Pbo (1993).
6. Robin & Wong (1988).
7. Kogan et al. (1987) and Hummel & Herrmann (1991).
8. Sullivan et al. (1993).
9. Kogan et al. (1987).

10. Wilson et al. (1985).
11. E.g. Harpending et al. (1993).
12. Richards et al. (1996.)
13. Gerrets (1995).
14. Hedges et al. (1995).
15. Richards et al. (1995).
16. Richards et al. (1995) and Richards et al. (1995.)
17. Cf. Handt et al. (1994).
18. Richards et al. (1995).
19. Sullivan et al. (1993).
20. Hedges et al. (1995).
21. Cuijpers et al. (this volume).
22. Anderson et al. (1981).
23. Bandelt et al. (1995).
24. Richards et al. (1996).
25. Note that the 189 event is distinct from that characterizing Group 1B, although this is not clear at the level of resolution in the network herem (see Richards et al. 1996).
26. Richards et al. (1995) and Richards et al. (in prep.).
27. Gamble (1986) and Fischer (1991).
28. Todd (1993) and see above.
29. Bandelt et al. (1995).
30. Richards et al. (1995).
31. Gerrets (1995).
32. Cuijpers et al. (this volume).
33. J.M. Pasveer (pers. comm.).
34. Richards et al. (1995).
35. Richards et al. (1995).

Conclusions

D.A. GERRETS

In his introductory outline of the Wijnaldum-Tjitsma excavation within the wider perspective of current research taking place on political centralization in early medieval Frisia, Heidinga refers to the Helgö excavations in central Sweden and the different opinions among archaeologists about the economic basis, the socio-political and ideological context and on the place of the site on the evolutionary ladder (proto-urban or not).[1] Reviewing these opinions Kristina Lamm stated: 'So I am convinced that we shall never reach a definite conclusion on the question what Helgö really was and we must simply accept, that however rich the material, archaeological evidence cannot give a true picture of life at that time'.[2]

The excavations at Wijnaldum-Tjitsma also gave rise to different opinions among archaeologists about the significance of the site and they will probably do so in the future. But what picture do we get when we fit all the jigsaw pieces together? First we have to conclude that the outer pieces of the puzzle are missing as we have only excavated slightly more than seven percent of the whole terp. Furthermore, many pieces are missing due to site formation processes and erosion, while also part of the material is still under study.[3] But as we will see much can already be said about the life of the inhabitants of Wijnaldum-Tjitsma. And certainly, the Wijnaldum-Tjitsma excavations have contributed to a better understanding among the archaeologists involved in the project, of the society of which it formed part. The excavations started as an independent project, but finally appeared to be a starting point for new archaeological research into the *Dark Ages of Frisia*.

In these conclusions the present information provided by the contributors will be summarized and evaluated, after which an attempt will be made to draw some preliminary conclusions on the nature of Wijnaldum-Tjitsma. What do the facts tell us compared with the expectations we had of the site as expressed in Heidinga's contribution?

1 THE LANDSCAPE

In the beginning was the landscape, in this particular case a relatively recent and very dynamic one. The history of the settlement and the possibilities of its habitants were very much connected with the landscape, so knowledge of its development, which is provided by Peter Vos in his contribution, is indispensable for understanding Wijnaldum.

Vos points out that the salt-marshes in Westergo started to expand over the sandy tidal flat deposits from ca 1000 BC onwards. The first colonists moved into the salt-marsh about 700 BC in the southernmost part of Westergo where no salt-marsh ridges occur and where a settlement pattern of scattered terps came into being. At this time small local rivers belonging to the Boorne system and draining the hinterland of southern Friesland discharged into a funnel-shaped estuarine-like, tidal basin in the centre of present-day Westergo. Gradually this remnant of the former mouth of the Boorne river silted up, due to high energetic circumstances in the North Sea. During storms sandy material was eroded from channels and tidal flat areas further to the north and transported to the salt-marsh hinterland. Because the environment of Wijnaldum-Tjitsma was protected by the former Grienderwaard this western part of Westergo silted up to a lesser extent and to a lower level than the eastern part. The hinterland will have drained mainly via this western depression.

After 600 BC the first salt-marsh ridges were formed, because sand was deposited along the margins of the salt-marsh and along creeks. These rather small and low salt-marsh ridges are one of the most characteristic morphological features in

Westergo. The terps are located on these natural elevations. Each time a new salt-marsh ridge developed, new terp settlements were founded on this ridge. About 300 BC the former Boorne estuary had silted up to such an extent that the first terp settlements were founded in the former mouth of the estuary and during the following centuries a linear settlement pattern developed.

During the 1st century AD the erosion of the Grienderwaard caused increased sedimentation during storms in the western tidal basin and the formation of a sandy cap ridge started along the whole western part of Westergo. At this time a lower marsh had developed in the environment of Wijnaldum-Tjitsma, which was suitable for grazing cattle, but not yet for founding permanent settlements. In the next century, when the cap ridge in the north gradually closed and the salt-marsh had developed into a middle marsh, the ridge of Wijnaldum-Voorrijp and further to the east the ridge of Dongjum-Ried-Berlikum were occupied.

Of particular interest is the changing system of gullies and streams, not only because of the drainage of the region but also because they largely determined communication. Due to filling up of the channels in Westergo after the late-Roman Period, the hinterland south of Westergo started to drain via the Middelzee and Marne systems. With the continuous rise in sea-level rise tidal action penetrated these systems, possibly enhanced by artificial drainage. At the end of the 9th century the Middelzee system rapidly expanded and in the 10th century reached as far inland as Sneek where it made contact with the Marne system.

The Ried must have been one of the main tidal channels of the western depression in northern Westergo in the early Roman Period. According to Vos the ridge on which the cluster of terps at Wijnaldum is situated should be considered as a levee of this Ried tidal channel and not as a levee along the seaward margin of the salt-marsh. At the time that the terp settlement was founded, Wijnaldum-Tjitsma could probably still be reached by boat. It is not clear whether the Ried was still navigable after AD 250, when a large cap ridge completely blocked off the western depression. Vos does not want to exclude the possibility that the Ried was kept navigable artificially. The main part of Westergo, however, then drained via the Middelzee and the Marne tidal systems.

Let us now turn to the landscape where Wijnaldum was to be founded. It is generally assumed that middle marshes could be exploited as such when people thought the time was right, but this was not the case here. The first human activity at the site starts with the construction of a dike-like structure. The structure cannot be dated precisely because no datable finds were discovered, but it must have been constructed shortly before the oldest occupation phase which started about AD 175. Recent excavations at Dongjum-Heringa revealed a similar dike-like structure protecting small agricultural fields against the sea.[4] Closer inspection of the documented sections at Wijnaldum-Tjitsma showed that the dike-like structure here also served to protect an agricultural field. Contrary to Dongjum, at Wijnaldum several storm layers covering the partly eroded construction phases of the structure show that it was sufficient protection against normal high waters, but that it was not high enough to protect against the few extreme storm surges.

It has often been assumed that the North Sea littoral of the northern Netherlands was abandoned in the Late Roman Period because of the Dunkirk II transgression phase.[5] No evidence could be found, however, for the supposed cyclicity of the Dunkirk transgression phases in the research area. The salt-marshes in Westergo show a continuous expansion (regressive trend), whereas other areas show contemporaneous marine inundations (transgressive trend). According to Vos the coastal evolution in the northern Netherlands has to be considered a long-term marine infilling process of the Pleistocene valley systems which was a consequence of the Holocene rise in sea-level.

What did the landscape actually look like, which plants were growing in it and how usable was it for man, in view of the impact of occasional floods? Questions which are answered in contributions of Groenman-van Waateringe and Pals. After the arrival of the first inhabitants, the site was situated in a brackish marshy environment which was, as shown by diatom research, not flooded frequently any more. The surrounding salt-marsh was well drained by tidal creeks. The high salinity of the soil, however, made it impossible for trees to grow. As Pals has indicated, the natural grassland of the middle marsh was pre-eminently suited for grazing. The presence of plant remains which are characteristic of gradients ranging from high salt-marsh to low salt-marsh and for a brackish and even fresh-water environment show the differentiation of this in many other respects very homogeneous landscape.

2 SETTLEMENT DEVELOPMENT

The biography of the main subject of this book, the settlement (in fact, initially we are probably dealing with two settlements which grew together), as described in the contribution of Gerrets and De Koning, is especially focused on terp building phases, changing settlement patterns and on building traditions. It is, so to speak, the scene of human activities which will be dealt with later on.

In the second half of the 2nd century the settlement started with the construction of one and perhaps two house-terps on the site. Pottery suggests that the founders of the settlement came from within the region, probably from the more southern terp row between Midlum and Herbayum. Surveying on the neighbouring fields suggest that more contemporaneous house-terps were situated at short distance from each other in the unexcavated part of the terp. Apparently, there was still a threat from the sea at this stage of the occupation, as a building was constructed on a 1.25-m-high platform, while a second lower extension of this platform served as an activity area for e.g. copper working. The erection of buildings with a wooden, roof-supporting construction on an elevated platform is characteristic of settlements of the Roman Period in the terp region.[6]

After a demographic peak in the 2nd/3rd centuries, most terps in the Westergo region were abandoned during the 2nd half of the 3rd century. According to Taayke, the small number of terps with Eddelak pottery, which is characteristic of the latter part of the 3rd century, are an indication for the demographic decline in the region.[7] Wijnaldum-Tjitsma belongs to the few settlements where habitation continued until the end of the 3rd or the beginning of the 4th century.[8] At the same time the number of inhabitants in the region decreased to a level where they become almost invisible archaeologically. Only few finds suggest that Westergo was not completely abandoned.[9]

The abandonment of Wijnaldum-Tjitsma for some generations can be derived from the radiocarbon datings, the almost complete absence of finds from the 4th/early 5th century, the discontinuity in the development of indigenous handmade pottery and other categories of material culture, the change in settlement lay-out, the change in house building tradition and the introduction of sunken huts at the site.

Immediately after the site was re-occupied habitation expanded and the different house-terps grew together quickly to form one big terp. House plans from this period are best preserved compared with house remains from later occupation phases. The main buildings lack any visible remains of a wooden roof-supporting construction. The roof was probably supported by the massive sod walls. Besides the relatively small N-S orientated main buildings were some E-W orientated sunken huts. Due to erosion it is hard to say how many main buildings were standing contemporaneously on the excavated site but it will probably have been a maximum of five.

At the later post-Roman stages of the settlement the threat from the sea became less as no intentional heightening of any significance of the terp occurred any more and activity areas regularly extended across the outer limits of the terp. The pollen evidence shows that the environmental conditions towards the Carolingian Period definitely seem to have changed towards a more fresh-water environment.[10] Heightening occurred in the early Middle Ages more as the result of occupation itself e.g. by garbage-dumping, levelling of house remains etc.

The northern limit of the settlement remained at the same place rather constantly. Occupation gradually expanded in a southerly direction (away from the sea). About 500 the occupation will have moved temporarily to an adjacent part of the terp, when most of the excavated area was in use as an agricultural field. The first indications for the reintroduction of three-aisled buildings with roof-supporting poles date about 700. With the introduction of wooden poles, the thickness of the sod walls decreased, while constructive elements dating in the 8th century suggest that the roof-supporting function of the walls was gradually taken over by these poles. Regarding the number of wells dating to the Carolingian Period the number of buildings must have increased considerably, but due to erosion only small parts have been preserved.

3 THE SIGNIFICANCE OF AGRARIAN PRODUCTION

In a society with a low degree of specialization it may be expected that each settlement, whatever its status, was involved somehow in agriculture. However, this does not mean that every household was. As we will see, the small part of Wijnaldum that was excavated may provide a deviant picture. As the animal bones of Wijnaldum are still under study, it is hard to say whether animal husbandry, which is generally so important in the salt-marsh area, played an important role within the Tjitsma settlement. This means that we have to rely on indirect evidence here.

The number of buildings that can be reconstructed is very small, considering the total number of buildings that must have stood on the terp. But all the remarkable features of the three-aisled longhouses which are so characteristic of the coastal plain of the northern Netherlands and North Germany, such as cattle boxes in the byre and the pavement in the central aisle of the byre with ditches on either side for the dung were completely absent at Tjitsma. Dung was found only in a very small number of features. The clearest indication for animal husbandry is the possible drinking place, found at the outer limits of the terp, dating in Period IV (550-650), which was filled up with dung at a later stage. Very small numbers of imprints of cows were observed but certainly not as many as at e.g. Elisenhof, where large numbers of imprints of cows showed that cows followed fixed routes to and from the stable.[11]

According to Groenman-van Waateringe, the low percentage of *Plantago lanceolata* (ribwort plantain) in the pollen diagrams cannot be regarded as an indication that animal husbandry was of no importance because this species germinates poorly and with difficulty in compacted soils. Perhaps the analysis of the animal bone can show whether we are dealing with a consumer or producer site at Tjitsma.[12]

More can be said about tillage. As the recent excavation at the terp settlement of Dongjum has shown the agricultural fields in the Middle Roman Period (ca AD 175-300) located on the salt-marsh were surrounded by small dikes and this also seemed to be the case at Wijnaldum-Tjitsma although not recognized as such in the field. At Wijnaldum-Tjitsma these agricultural fields were threatened by extreme high water during storm surges. This perhaps explains the complete dominance of barley (*Hordeum vulgare*), which is more salt resistant. The presence of a second dike, contemporaneous with the Roman settlement, suggests that agricultural fields were still located on the natural salt-marsh, surrounded by small dikes. The small platforms on which the main buildings together with wells and metal working activities were located probably did not offer enough space for agricultural activities on a substantial scale.

As Pals shows, macroscopic remains of plants of economic importance show more variation during the Migration Period (Period III A/B, ca 425-550) compared with the preceding Middle Roman Period. Agricultural fields are now located on the terp itself. About 500 a large surface of the excavated part of the terp was in use for agricultural production. Also in later periods smaller plots on the terp were in use as agricultural fields. There were no indications that in the early Middle Ages agriculture extended beyond the borders of the terp. The presence of a few small granaries indicates the presence of storage facilities, but does not necessarily say anything about agricultural production on the site. As Groenman-van Waateringe points out in her contribution, we have to be careful with the interpretation of the analysis of pollen due to preservation circumstances. The pollen analysis, however, shows that cultural indicators are low, suggesting that no important agricultural production took place on or in the immediate environment of the terp. It is not until the last stage of the occupation during Period VIII that a considerable increase of cultural indicators seems to occur. At least it is clear that rye (*Secale cereale*) was never cultivated at Tjitsma. The small number of pollen from rye found at Tjitsma can be regarded as an indication for the import of rye from elsewhere.

Although the pollen analysis seems to suggest that no agriculture of any importance took place, the number of chaff remains of barley like rachis internodes, culm nodes and occasional awn fragments can be regarded as an indication for local cultivation and processing of this crop, which would have been sown in spring. Perhaps the fact that the agricultural fields were located on the terp also explains the presence of spikelet forks of emmer wheat (*Triticum dicoccum*), as this cereal is not especially suited for cultivation in the salt-marsh. Besides barley and perhaps emmer wheat also celtic beans (*Vicia faba var. minor*), field peas (*Pisum Sativum*) and flax (*Linum usitatissimum*) will have been cultivated.

4 CRAFT PRODUCTION

A supposed central place should produce ample evidence of craft production and Wijnaldum-Tjitsma indeed did so. The presence of small-scale craft production and especially metal-working from the beginning of the settlement is remarkable.[13] In the Roman Period both iron and copper are worked, but the importance of copper working seems to decrease in favour of iron working during the Early Middle Ages. Copper alloys are transported in the shape of ingots to the site, but scrap was also gathered especially in the Roman Period. We have the impression that copper working is a much more common feature in the region than we thought. According to amateurs looking for metal objects with a metal-detector, melted pieces of lead, melted pieces of copper and other remains of smelting are regularly found. However, their value is not always estimated correctly and they are often kept with other scrap-finds. When we asked for them, they say that these kind of finds can be found almost everywhere in the region. It is remarkable that pieces of moulds, which are so characteristic of large production sites like Ribe and Birka are nearly absent, compared with the number of other remains related to copper working.

From the Migration Period onwards, remains of small-scale craft production like spinning, weaving, bone- and antler-working, gold- and silver-working and glass- and possibly amber bead-production are represented, but at a much smaller scale than at the large production sites mentioned earlier.[14] As Henderson points out in his contribution, working of opaque yellow glass could be confirmed by chemical analysis, adding Wijnaldum-Tjitsma to the few sites in Northwest Europe with primary evidence for glass working. This probably does not mean that it did not occur at other sites, but this category of finds was probably not discovered before, because of the excavation methods used. This underlines the importance, for reasons of comparability, of excavation methods just as water screen sieving and systematic metal detection, together with the need for a growing awareness among archaeologists, for this category of finds.

The sod buildings as excavated at Tjitsma are often associated with crafts and craft production.[15] The importance of craft production decreases after 650. Only iron is still worked on a small scale.

5 CAREFUL USE OF RESOURCES

It was also observed at Tjitsma that especially during the Roman Period and the Migration Period scrap metal was collected to be re-used for the production of new metal objects. The re-use of older material was a widespread phenomenon as has been generally recognized. Its implication for the archaeological interpretation of artefact distribution patterns and chronology, however, are not yet fully understood. The stratigraphically collected data from Tjitsma clearly show the traps into which we can fall without such contextual information of high quality.

Secondary reuse especially of Roman artefacts such as copper money, Roman militaria, terra sigillata sherds, glass vessel fragments and fragments of tiles, is a striking feature within the settlement.[16] Careful stratigraphical analysis made it clear that many Roman objects (not all of them) were not dug up by later human activities on the site, but must have been brought to the settlement at a (sometimes much) later stage. This conclusion is underlined by several glass and copper finds from the 1st and first half of the 2nd century, when the site was not yet inhabited. The finds from Wijnaldum also show what at least some of the material was used for. Terra sigillata was transformed into a pendant and playing discs, one fragment was probably used for needlework, while two other fragments had rounded off edges.[17] Two late-Merovingian spindle whorls were made of Roman tiles.[18] Henderson also indicates in his contribution that there is increasing evidence from chemical analysis for the recycling of Roman glass in the production of early-medieval glass beads.

Although there is some indication for the use of Roman copper money for remelting, most other Roman coins from an early medieval context are not found in relation with metal working activities. Van der Vin justly stresses in his contribution, that the Roman coins must have had a special meaning within Germanic society because they were preserved for decades and sometimes even for centuries without disappearing into the melting pot. This was not only the case with silver coins but also with copper money. It is possible that they were still in use as coins.[19] The need for smaller denominations is reflected in the presence of halved and quartered coins.

It is interesting to see that not all objects from the Roman Period were collected at a post-Roman date. In his contribution Erdrich shows that a distinction must be made between finds of Germanic origin and finds of Roman origin.[20] Many of the Roman finds were badly damaged by cutting or hacking. This is in contrast with the Germanic finds, which showed only traces of normal wear. A similar observation is made by Galestin, who shows how Roman tiles and part of the Roman imported pottery were brought to the site in the early Middle Ages, while many sherds of the terra nigra-like pottery were found in a Roman context. She assumes that this latter category of wheelthrown pottery has an origin outside the *limes* in Germania Magna.

The analysis of charcoal shows that much construction wood was re-used as firewood.[21] The great number of rivets from the Migration Period and the early Middle Ages found at Tjitsma shows that this will at least partly have been wreckage.[22] Dung will probably also have been used for heating. This can very well be explained by the almost complete absence of trees in this salt-marsh landscape. Altogether we get the impression that the inhabitants were very economical in the use of the available resources.

6 WIJNALDUM-TJITSMA AND ITS CONTACTS WITH THE OUTSIDE WORLD

It was suggested in Heidinga's contribution that the people of Wijnaldum belonged to a maritime society, not only having contacts by ship with neighbouring 'nuclear regions', but also with peoples living beyond more remote horizons. What is the evidence? Though no ships or harbour installations were found, dozens of clench nails were discovered, which shows that there was a close relation with the sea. The amount of imported wheelthrown pottery and the number of glass vessel fragments does not say much about the social status of the terp inhabitants at Wijnaldum, but it does tell us something about its participation in international networks of communication. During the 6th/7th century wheelthrown pottery forms about 70% of the total amount of pottery. For the southern part of the Netherlands these numbers are not unusual, but for the northern part of the Netherlands these numbers are very high. Wheelthrown pottery in the province of Drenthe is almost completely absent during the same period. Also many other finds show that intensive contacts were maintained with the Anglo-Scandinavian world, especially England, and with the Middle Rhine area.

Apart from pottery other finds also show that Wijnaldum-Tjitsma formed part of an international network of long-distance trade. Chemical analysis has shown that at least one, but probably two Carolingian beads found at Tjitsma must have been produced in Islamic workshops. Also an Arab dirhem of Harun al Rashid transformed into a ornament is evidence for indirect contacts with the Arabian world. These finds probably reached Frisia via Russia and Scandinavia as was recently shown by the find of the Westerklief hoard.[23]

The amount of imported pottery had already decreased in the second half of the 7th century and began to increase again in the second half of the 8th century, but it never reached the same numbers as in the 6th/7th century, although the latter development can also be explained by the success of the *Kugeltopf*. It is interesting to note that vessel glass, which originated almost exclusively from the Rhineland, decreases in number during the (second half of the) 6th century to reach a low during the 7th century after which it increases again during the 8th century, reaching a 'peak' during the (first half of the) 9th century.[24]

Apart from pottery, the raw materials for the production of red garnet and other precious jewellery, copper alloys, iron, tin and lead used for the production of clothing accessories, tools, weapons, utensils for personal body care, glass in the form of a *tessera* and glass rod fragments for the production of beads, etc. also refer to intensive exchange of goods and perhaps even trading contacts.

Groenman-van Waateringe suggests that the somewhat higher proportion of rye pollen (*secale*) in the Carolingian samples may point to the import of rye in the form of grains.[25]

7 BURIAL AND OTHER RITUAL PRACTICES

The part of the Wijnaldum community beyond the boundary of the final *rite de passage* is less familiar to us than the world of the living. Information on funeral or other practices is even more scarce. No burials of adults from the Roman Period are found at Wijnaldum-Tjitsma.[26] This is a characteristic phenomenon for the terp region and Knol has posed the hypothesis that cremation without an archaeologically visible context was the rule at that time.[27] He assumed that in the beginning of the 5th century a dramatic change in burial customs took place when a mixed burial practice with cremations and inhumations in formal cemeteries is introduced. At Wijnaldum-Tjitsma only one cremation burial containing two persons, namely a child and an adult, dating in the early 5th century was excavated. Both individuals were probably wrapped in skins when they were cremated. An inhumation of a 18/19-year-old woman was initially dated in the second half of the 6th century, but in her contribution Sablerolles proposes a somewhat earlier date in the second half of the 5th century. That both burials were excavated at a considerable distance from each other seems to contradict the earlier mentioned formal cemetery, which is characteristic of the terp region in the early Middle Ages. This can be explained, however, by the location of the excavated trenches. Both burials were excavated near the limits of the excavation in natural salt-marsh deposits, south of the contemporary settlement. It is very well possible that other burials can be excavated in this part of the terp, but since the settlement expanded over this part of the salt-marsh during the later habitation phases, many cubic metres of soil will have to be excavated first. All together five newborn infants were excavated, one of which was a building sacrifice. All these infant burials come from a settlement context. It is still difficult to distinguish male from female neonates or infants on morphological characteristics. The sex of the Wijnaldum infants, however, could be determined by DNA-analysis showing that two infants were female and three were male.[28] It is hard to establish whether the children died naturally or were killed instantly. As Cuijpers et al. describe in their contribution, infanticide was widespread among human populations. These infants were mostly buried between the houses. Killing of an unwanted child took place before the child was considered to be fully 'human'. From the *Lex Frisionum*, we learn that the Frisians did not consider an infant a human being until it had been nursed.

One of the infant burials must be mentioned separately. It was excavated immediately beneath the ploughing zone where features are hardly readable, which makes the context somewhat uncertain. In the immediate environment two horse foetuses were excavated together with 17 rim fragments of completely 'new' pots (almost exclusively pots of the Alzey 32/33 type). All rim sherds belonging to different pots, were lying together in a small pit and were never used as cooking pots, which is generally the case with these kinds of pots. It seems that these pots were ritually destroyed. All finds can be dated in the 5th century.

8 THE MAJOR ISSUES

Two major issues are to be discussed: the discontinuity of habitation at Tjitsma, which also raises the question of the origin of the inhabitants of the second occupation, and the socio-political importance of the site, especially in the early Middle Ages. It is obvious that these issues can only be dealt with in a wider regional and supra-regional perspective. As we will see, the re-colonization in the 5th century heralded Wijnaldum's heyday, in which also traffic overseas and international exchange (of which the evidence is described above) was intensified.

9 CONTINUITY OR DISCONTINUITY IN THE OCCUPATION OF THE TERP REGION: A HOT DEBATE IN DUTCH ARCHAEOLOGY

As we have seen, Wijnaldum was abandoned for some generations during the 4th-early 5th century and the situation on the terp changed with the arrival of new inhabitants. Wijnaldum was not an exception in the terp region. The question of what happened in Frisia during the Migration Period belongs to the most hotly debated issues in the archaeological research of the terp-region.[29] The then curator of the Frisian Museum, Boeles, already supposed in 1906 that an invasion of Angles and Saxons had taken place in the coastal region of the Northern Netherlands on their way to England.[30] He came to his conclusion because of the sudden appearance of cruciform brooches and Anglo-Saxon pottery in certain terp layers. The close philological relationship between Old English and Old Frisian was seen by him as a confirmation that Frisia was involved in the Anglo-Saxon expansion to the West.[31]

The first person to criticize the idea of an extensive Anglo-Saxon migration was the historian Gosses who argued in 1929 that the dispersal of a material culture with Anglo-Saxon characteristics could very well be explained by acculturation.[32] Gosses was remarkably modern in his criticism of the way archaeologists explained changes in material culture by the migration of whole populations. The debate culminated in 1953 after the publication of the second edition of Boeles' *Friesland tot de 11e eeuw* in 1951 where he reconfirmed his hypothesis of an Anglo-Saxon invasion.[33] Acculturation of Anglo-Saxon material culture together with the continuity of the Frisian tribal name from the Roman Period into the early Middle Ages were the most important arguments of Boeles' opponents to refute his migration hypothesis.

In the 1960s and 1970s migrationist models were generally criticized and most Dutch archaeologists became convinced of long-term continuity. The demographic decline of the late 3rd and 4th century was now generally explained by the Late Roman Dunkirk II-transgression, when a rise in sea level made the possibilities for habitation in the terp region less favourable.[34]

Critical analysis of the archaeological data, however, made it clear that Wijnaldum-Tjitsma was abandoned at least for some generations, but there was no single piece of evidence that environmental deterioration was the reason for abandonment. The excavation showed no indication whatsoever for extraordinary sedimentation. The surrounding saltmarsh at Wijnaldum was already too high to be seriously threatened by the sea or by a rising groundwater table, while the sea shore was protected by the newly developing cap ridge.

We do not know where the people living on the site went to nor why they left. So far no traces have been found for Frisians elsewhere in Northwest-Europe during the mid-/late-Roman Period, except for the Housestead Ware found near Hadrian's Wall in England, but this 'Frisian' pottery must probably be explained by the presence of Frisian auxiliary forces in the Roman army. Some historians assumed that (some of) the Frisians migrated south and integrated in Frankish society, because Constantius Chlorus meets Frisii and Chamavii in the central river area or at the mouth of the river Scheldt.[35]

We have pointed out before that the lay-out of the settlement changed together with the tradition in house construction and the appearance of sunken huts. We know from elsewhere in the coastal region of the northern Netherlands that burial customs changed also from *Brandgruben* to a mixed burial custom of inhumation and cremation after which the burnt bones were buried with the additional gifts in jars. From then on cremations and inhumations are buried in formal cemeteries.[36] Recently Blok pointed out that also from the angle of place-names discontinuity is most likely for the Frisian coastal region.[37]

10 THE ORIGIN OF THE NEW INHABITANTS

It is still difficult to say where the new inhabitants of Wijnaldum-Tjitsma came from. The new material culture they brought with them, shows an origin in the eastern North Sea region. Decorated pottery shows many similarities with pottery from the *Elbe-Weser Dreieck*, but there are metal finds showing that there are influences from southern Scandinavia.[38]

During the excavation we still believed that the analysis of mtDNA could help to answer the question whether migration or acculturation had taken place at Wijnaldum, and if migration occurred, where the immigrants came from. Recent developments is this branch of science, however, made it clear that the differences in mtDNA between populations in Northwest Europe are not as great as expected at the time, making this kind of high resolution migration studies much more difficult to envisage.[39]

The origin of the new inhabitants of Wijnaldum-Tjitsma and the Westergo region is the subject of further study at the moment.[40]

11 SOCIO-POLITICAL IMPORTANCE OF WIJNALDUM-TJITSMA AND ITS INHABITANTS

The main research problem to be answered with the excavation of Wijnaldum-Tjitsma was the meaning of the settlement within the historical perspective of the socio-political and economic growth of early medieval Frisia.[41] It will be clear that this question is not easily answered. As already stated, we have excavated slightly more than 7 percent of the present terp. Furthermore, the site is very much affected by erosion and site formation processes.

We are also dealing with a research bias. Crafts like metal-working have only received the attention they deserve during the last decade. Previously, we sometimes have seen mentioned in excavation reports that crucibles are found, but the other remains of the production process have generally been neglected. Now that archaeologists are becoming more aware of the importance of the traces of crafts-production, traces of e.g. copper working are showing up at many sites in the Netherlands and abroad.

There is also a problem of comparability. At the Tjitsma excavation some new methods were introduced in terp research, such as water-screen sieving and systematic metal-detection, producing large numbers of finds, which were previously almost unknown in the region. And here we touch upon the more general problem that it is still hard to say anything about the meaning of the settlement within its regional context, because more systematic research in Westergo has only just started. Nevertheless, the excavations at the Tjitsma terp have provided us with important new information about the settlement, its inhabitants and their material culture.

The indication that the main occupation of the

inhabitants of Wijnaldum-Tjitsma was craft production and the apparent secondary importance of agricultural production is in remarkable contrast with most excavated terp settlements.

Perhaps the presence of crafts as such and the scale of production at Tjitsma is not so extraordinary, but the variety of crafts represented is remarkable. Especially the indications for the presence of a gold- and silversmith as evidenced by finds like gold and silver drops, silver ingots, a fragment of a crucible with gold inclusions, touchstones with gold traces, a small hammer, a punch, weights, fragments of balances, etc. show that Wijnaldum-Tjitsma was not an ordinary rural settlement.[42] That also the most precious red garnet jewellery belonged to the products of this goldsmith is shown by the find of a die, meant for making the cross hatched gold foil under the garnets and the find of a small fragment of unworked garnet.[43] The circulation of gold in early-medieval society was so much interwoven with the foundation of political power of the ruling elite, that the gold- and silversmith must have worked under its direct patronage.

Schoneveld and Zijlstra have pointed out that the disc-on-bow brooch from Wijnaldum must be regarded as a symbol of power due to the materials which were used in its production and the symbolic content of its decoration, which is embedded in Germanic ritual cosmology. Because of the similarities between the Wijnaldum brooch and jewellery from the royal ship burial at Sutton Hoo and some other graves with a royal signature, they assume that the woman who owned this brooch belonged to the leading aristocracy within Frisian society. The Wijnaldum brooch, like many other pieces of precious jewellery from the littoral of the northern Netherlands, reflects characteristics which are regarded as typically Frisian, suggesting that they were produced in workshops within the region itself.[44] This shows that Frisian jewellery was not produced for exchange in international networks but that it must probably be regarded as a category of goods which were inalienable.[45] It was handed down from one generation to the next and the longer it remained within the possessions of an aristocratic family, the more it's worth increased. As pointed out before, there are good reasons to believe that the disc-on-bow brooch from Wijnaldum may have belonged to the regalia of a royal family and as such it will have belonged to the category of inalienable goods.

The date of production is still a matter of discussion. Schoneveld and Zijlstra propose for art-historical reasons a date at the end of the first half of the 7th century, while Nijboer and Van Reekum suggest a date at the beginning of the 7th century because of the high gold content.[46] Recently, they received support from Nicolay, who pleaded on stylistic grounds for a date at the end of the 6th or in the first two decades of the 7th century.[47]

The find of additional pieces of the large disc-on-bow brooch on the terp by amateur archaeologists with a metal-detector and during the excavation confirm that the fibula originates from Wijnaldum-Tjitsma. But what does this piece of jewellery, related with the highest elite or even royalty in society, mean? All finds come from the ploughing zone, distributed over a large area of ca 600 m^2, and do not clarify its presence. No grave was excavated indicating that somebody with a higher status was buried here, neither did we excavate buildings indicating that the owner of the brooch lived here. Nijboer and Van Reekum suggest that the brooch was found at Tjitsma because it was in possession of the local gold- and silver smith for re-use or repair.[48] Regarding the die and the raw red garnet this may have been the same goldsmith who had made this disc-on-bow brooch. The missing central disc of the brooch may have been re-used for a new piece of jewellery or may not yet have been excavated.

Considering the degree of re-use of raw materials at Tjitsma the small number of gold finds reflect only a small proportion of the amount of gold that will have circulated within the settlement. Thus it seems reasonable to assume that a small number of craftsmen were living within the excavated area who were affiliated with a high ranking elite. This close relationship to the highest elite is perhaps also reflected by the find of a fragment of a shoe with a raised heel tongue. Unfortunately, the context of the find is uncertain as the contents of the well, from which it originates, covers almost the complete early Middle Ages. According to Groenman-Van Waateringe, however, the origin of this shoe must probably be sought among the clerical elite from Ireland or more generally the British Isles.[49] Therefore this shoe perhaps reflects a diplomatic gift from a British missionary to a member of the Frisian aristocracy. This would mean that the shoe, or what remained of it, arrived about a century after the presumed heyday of the terp settlement.

Also other finds like brooches, fragments of glass vessels, horse-gear, weaponry, excavated burials of humans and animals etc. suggest that it was not the highest aristocracy itself who was living here, but people who belonged to their clientele. We are referring here to evidence from the excavated area, but we have to keep in mind that the settlement as a whole was much larger. Probably not only the 11.2 ha of the Tjitsma terp belonged to it, but also the other terps of the Wijnaldum cluster. Therefore, it is quite well possible that the residence of this aristocracy must be sought on the unexcavated part of the Tjitsma terp or on one of the other terps. In his contribution Pol underlines the special importance of Wijnaldum-Tjitsma and its immediate environment when he discusses the relatively large number of golden tremisses from Wijnaldum, compared with other sites in Friesland. Regarding the number of sceattas and silver denarii from the late 7th and first half of the 8th century found at Wijnaldum this special position of Wijnaldum appeared to be maintained well into the 8th century.

Prosperity in the 6th/7th century is not confined to the Wijnaldum cluster of terps. Egge Knol has shown that, regarding the number of gold- and silver finds, Westergo was the richest region in the North Sea littoral of the northern Netherlands during this period. Dorestad and Utrecht are generally regarded by archaeologists and historians as the political centres of early medieval Frisia. We should not forget, however, that at the moment that the first Frankish written sources shed some light on Dark Age Frisia, competition of Frisians and Franks is focusing on the control over the central river area. The excavation at Wijnaldum-Tjitsma, together with Egge Knol's study, suggest that processes of political centralization were already going on further to the north in a period when the Frankish authors paid no attention to the northern Netherlands.[50] It seems likely that the significance of Wijnaldum decreased with the Frankish conquest of this part of Frisia after the battle at the Boorne river in 734, when Charles the Hammer defeated the Frisian duke Bubo and his army. Exactly when the number of people living at Wijnaldum-Tjitsma reached its maximum in the second half of the 8th and 9th century, the general impoverishment of Central Frisia as reflected by the rapid decrease of the number of coins in circulation also seemed to have affected the former prosperity of Tjitsma's inhabitants. According to Knol, precious metal finds and imported pottery are more evenly distributed over the whole coastal area of the northern Netherlands in the Carolingian Period.[51] This development seems to be reflected in changes taking place at Wijnaldum-Tjitsma. The possible return of the three-aisled long-house could be an indication for the increasing importance of cattle breeding. Metal working which was so obviously present during the preceding centuries becomes less important. Also Sablerolles states that in contrast with the preceding periods no luxury glasses are represented among the Carolingian glass finds from Wijnaldum-Tjitsma. She suggests that during the late Merovingian/Carolingian Period a change in Germanic drinking ceremonies takes place when beer is replaced by wine which also becomes available for 'commoners' in society.[52] This appears to be underlined by the find of two wooden barrels, generally used for the transportation of wine, in wells on the Wijnaldum-Tjitsma terp.

Summarizing it can be said that the excavated settlement area at Wijnaldum-Tjitsma probably did not distinguish itself from other terp settlements in the North Sea littoral of the Netherlands in the mid-Roman Period. It developed into a more specialized settlement after the arrival of new immigrants in the 5th century. Craft production and especially metal working were the most important activities of the inhabitants of the excavated part of the settlement until the middle of the 8th century. As many of the objects related with the production of precious jewellery are coming from undefinable, sometimes even secondary contexts it is hard to say exactly when goldsmiths were active on the terp. If we take the waste material from other metal working activities such as copper and iron, with which some of the objects are related, as an indication the late 6th/early 7th century was probably the period that goldsmithing was of greatest importance. This is also the period in which the majority of the Frisian gold finds must be dated. The importance of craft production seems to decline somewhere in the middle of the 8th century, when Frisia had been conquered by the Franks. Together with the decrease of craft production, the first indications of an increasing importance of agricultural activities on the site were excavated.

12 FUTURE RESEARCH

During the excavations at Wijnaldum we became more and more aware of the need for further research in Westergo and in Frisia in the Early Middle Ages. The excavations should not be seen as an aim in itself, but as a new starting point. It was unique for Dutch archaeology that three of the main archaeological institutions active in the North Sea littoral and the central river area decided to join forces in the *Frisia project*.[53] Part of the project is to continue research in Westergo to get a better understanding of Wijnaldum-Tjitsma in its regional context.

REFERENCES

BANTELMANN, A. 1975. *Die frühgeschichtliche Marschensiedlung beim Elisenhof in Eiderstedt. Landschaftsgeschichte und Baubefunde*, Bern etc. (Studien zur Küstenarchäologie Schleswig-Holsteins, Serie A, 1).

BAZELMANS, J. 1996. *Eén voor allen, allen voor één. Tacitus' Germania, de Oudengelse Beowulf en het ritueel-cosmologische karakter van de relatie tussen heer en krijger-volgeling in Germaanse samenlevingen*, Amsterdam (thesis University of Amsterdam).

BAZELMANS, J. (in press a). Het laat-Romeinse bewoningshiaat in het West- en Noord-Nederlanse kustgebied en het voortbestaan van de Friezennaam, Jaarverslagen van de Vereniging voor terpenonderzoek.

BAZELMANS, J. (in press b). By weapons made worthy. Lords, retainers and their relationship in Beowulf, Amsterdam (Amsterdam Archaeological Studies 5).

BAZELMANS, J., D.A. GERRETS & P.C. VOS 1998. Zoden aan de dijk. Kleinschalige dijkbouw in het Friesland van de Romeinse tijd, *Noorderbreedte* 22/6: 18-22.

BESTEMAN, J.C. 1997. De vondst van Westerklief, gemeente Wieringen: een zilverschat uit de Vikingperiode, *Oudheidkundige Mededelingen uit het Rijksmuseum van Oudheden te Leiden* 77: 199-226.

BLOK, D.P. 1996. Das Alter der friesischen Wurtnamen. In: Looyenga, T. & A. Quak (eds), *Frisian Runes and Neighbouring Traditions. Proceedings of the First International Symposium on Frisian Runes at the Fries Museum, Leeuwarden 26-29 January 1994*, Amsterdam etc. (Amsterdamer Beiträge zur Alteren Germanistik 45), 25-35.

BOELES, P.C.J.A. 1906. De opgravingen in de terp Hoogebeintum, *De Vrije Fries* 20: 391-430.

BOELES, P.C.J.A. 1927. *Friesland tot de elfde eeuw: zijn oudste beschaving en geschiedenis*, 's-Gravenhage.

BOELES, P.C.J.A. 1951. *Friesland tot de elfde eeuw, zijn vòòr- en vroege geschiedenis*, 2nd edition, Den Haag.

BOERSMA, J.W. 1988. Een voorlopig overzicht van het archeologisch onderzoek van de wierde Heveskesklooster (Gr.). In: M. Bierma, A.T. Clason, E. Kramer en G.J. de Langen (eds), *Terpen en wierden in het Fries-Groningse kustgebied*, Groningen.

BOONE, W.J. DE 1954. *De Franken. Van hun eerste optreden tot de dood van Childerik*, Amsterdam.

CARVER, M.O.H. 1994. Environment and Commodity in Anglo-Saxon England. In: Rackham, J. (ed.), *Environment and Economy in Anglo-Saxon England. A review of recent work on the environmental archaeology of rural and urban settlements in England* (Council for British Archaeology Research Reports 89), 1-6.

ES, W.A. VAN 1968. Friezen en Romeinen. In: Kalma, J.J., J.J. Spahr van der Hoek & K. de Vries (eds), *Geschiedenis van Friesland*, Drachten, 48-94.

GOSSES, I.H. 1929. Friese jubileumliteratuur, *Tijdschrift voor geschiedenis* 44: 14-39.

GOSSES, I.H. 1936. De grootte volksverhuizing, *Tijdschrift voor geschiedenis* 51: 5-42.

HALICI, H. 1997. Gebruiksvoorwerpen van been en gewei uit Tjitsma, Wijnaldum, MA thesis, Groningen University.

HEERINGEN, R.M., P.A. HENDERIKX & A. MARS (eds) 1995. *Vroeg-Middeleeuwse ringwalburgen in Zeeland*, Goes etc.

HEIDINGA, H.A. 1997. *Frisia in the First Millenium. An outline*, Utrecht.

KNOL, E. 1993. De Noordnederlandse kustlanden in de Vroege Middeleeuwen, Groningen (thesis Vrije Universiteit, Amsterdam).

LAMM, K. 1988. Helgö – international trading centre, local market or farming community? In: A. Lundström & J. Karlernäs (eds), *Thirteen Studies on Helgö*, Stockholm, 89-100 (The Museum of National Antiquities Studies 7).

MAZZO KARRAS, R. 1985. Seventh-century jewellery from Frisia: A re-examination, *Anglo-Saxon Studies in Archaeology and History* 4: 159-177.

NICOLAY, J.A.W. 1998. Goudvondsten uit het Noordnederlandse Terpengebied (450/500-650 n. Chr.). Politieke, religieuze en sociale aspecten van import, circulatie en depositie van goud, Groningen (MA thesis Groningen university).

NICOLAY, J.A.W. (in press). Gold finds from the Northern Netherlands: some hypothesis on the development of kingship in the Frisian area (450/500-650 AD). In: J. Bazelmans, A. Pol & F. Theuws (eds), *The circulation and deposition of valuables in the late Roman period and the early Middle Ages*.

SCHMID, P. 1994. Oldorf – eine frühmittelalterliche friesische Wurtsiedlung, *Germania* 72: 231-267.

SIEBS, TH. 1889. *Zur Geschichte der englisch-friesischen Sprache*, Halle.

SIPMA, P. 1953. Een Angelsaksische invasie in Friesland? *It Beaken*, 162-192.

TAAYKE, E. 1988. Terpenaardewerk uit de IJzertijd en de Romeinse tijd. In: M. Bierma (red.), *Terpen en wierden in het Fries-Groningse kustgebied*, Groningen, 50-60.

TAAYKE, E. 1992. Die einheimische Keramik der nördlichen Niederlande 600 v. Chr. bis 300 n. Chr., Teil I: Westergo (Friesland), *Berichten van de Rijksdienst voor Oudheidkundig Bodemonderzoek* 40: 109-222.

TULP, C. 1996. *Metaalbewerking op de terp Tjitsma*, Groningen (MA-thesis Groningen University).

TULP, C. & N. MEEKS (in press). The Tjitsma (Wijnaldum) die: a tool for making a cross-hatched pattern on gold foil, *Journal of the Historical Metallurgy Society*.

WATERBOLK, H.T. 1979. Siedlungskontinuität im Küstengebiet der Nordsee zwischen Rhein und Elbe, *Probleme der Küstenforschung in südlichen Nordseegebiet* 13: 1-21.

NOTES

1. Heidinga (this volume).
2. Lamm (1988, p. 98).
3. Especially the large amount of animal bones is now

being studied by Kristin Bosma (Groningen Institute of Archaeology), which will probably reveal more about the character of the site. The final analysis of the medieval pottery can still slightly alter the chronological framework.

4. Bazelmans et al. (1998).
5. E.g. Boersma (1988, p. 77) and Waterbolk (1979, p. 17).
6. Van Es (1968, p. 57).
7. Taayke (pers. comm.).
8. See also Erdrich & Sablerolles (this volume).
9. e.g. Van Es (1968, p. 89), Taayke (1988, p. 59; 1992, pp 191-192) and Knol (1993, p. 110).
10. Groenman-van Waateringe (this volume).
11. Bantelmann (1975, p. 55 and figs 38,1; 38,2 and 38,3).
12. Carver has suggested that differences between producer and consumer sites and sites with a subsistence economy can be observed in the spectra of animal bones (Carver 1994, pp 3-5). See also note 3.
13. Tulp (1996) and next volume.
14. Halici (1997, and next volume), Tulp (next volume) and Sablerolles (this volume).
15. Bantelmann (1975, 139ff) Van Heeringen et al. (1995, p. 133) and Schmid (1994, 246ff).
16. See the contributions of Van der Vin, Erdrich, Volkers & Sablerolles (this volume).
17. Volkers (this volume).
18. See the contributions of Volkers & Galestin (this volume).
19. See also Van der Vin (this volume).
20. Erdrich (this volume).
21. See Stuijts (next volume).
22. Reinders & Neef (next volume).
23. Besteman (1997).
24. Sablerolles (this volume).
25. Groenman-Van Waateringe (this volume).
26. All human remains and their context are discussed by Cuijpers et al. (this volume).
27. Knol (1993, pp 155/156) and Cuijpers et al. (this volume).
28. Colson et al. (this volume).
29. e.g. Boeles (1951) and Sipma (1953). Bazelmans (in press a) recently made a study of the debate regarding continuity or discontinuity in the occupation of the terp region.
30. Boeles (1906), Boeles has repeated his opinion several times: e.g. 1927 and 1951.
31. Boeles followed Siebs (1889).
32. Gosses (1929, 1936), see also Bazelmans (in press a).
33. Boeles (1951, 207 ff).
34. See also note 5 and the contribution of Peter Vos (this volume).
35. See e.g. De Boone (1954, p. 17).
36. Knol (1993, p. 156).
37. Blok (1996, 25ff).
38. Taayke (next volume) and Nicolay (1998, and in prep.).
39. Richards (this volume) and note 1.
40. Westergo is one of the regions of early medieval Frisia, which is being studied at the moment in the scope of a project financed by the Dutch Foundation for Scientific Research (NWO) with the title 'The relationship between landscape/land use and the process of socio-political centralization in Frisia (3rd-9th century AD).
41. See Heidinga (this volume).
42. Caroline Tulp (next volume). A second small hammer from Tjitsma is in the private collection of J. Zijlstra, Leeuwarden.
43. Kars & Kars (next volume).
44. Mazzo Karras (1985, 170 ff.) and Nicolay (1998, and in prep.).
45. Bazelmans (1996, and in press b).
46. Schoneveld & Zijlstra (this volume) and Nijboer & Van Reekum (this volume).
47. Nicolay (1998, p. 42).
48. Nijboer & Van Reekum (this volume).
49. Groenman-van Waateringe (this volume).
50. Knol (1993, pp 218-243).
51. Knol (1993, pp 238 and 243).
52. Sablerolles (this volume).
53. Heidinga (1997).

List of contributors

B.A.M. Baardman, Fugro Engineers, P.O. Box 63, 2260 AB Leidschendam, Netherlands.

J.F. Bailey, Department of Cellular Science, Institute of Molecular Medicine, John Radcliffe Hospital, Headington, Oxford OX3 9DU, UK.

J.C. Besteman, Amsterdam Archaeological Centre, University of Amsterdam, Nieuwe Prinsengracht 130, 1018 VZ Amsterdam, Netherlands.

J.M. Bos, Groningen Institute of Archaeology, University of Groningen, Poststraat 6, 9712 ER Groningen, Netherlands.

I.B. Colson, Department of Zoology, University of Oxford, Oxford OX1 3PS, UK.

A.G.F.M. Cuijpers, State Service for Archaeological Investigations (ROB), Kerkstraat 1, 3811 CV Amersfoort, Netherlands.

M. Erdrich, Amsterdam Archaeological Centre, University of Amsterdam, Nieuwe Prinsengracht 130, 1018 VZ Amsterdam, Netherlands.

M. Galestin, Groningen Institute of Archaeology, University of Groningen, Poststraat 6, 9712 ER Groningen, Netherlands.

D.A. Gerrets, Groningen Institute of Archaeology, University of Groningen, Poststraat 6, 9712 ER Groningen, Netherlands.

W. Groenman-van Waateringe, Amsterdam Archaeological Centre, University of Amsterdam, Nieuwe Prinsengracht 130, 1018 VZ Amsterdam, Netherlands.

C.M. Haverkort, 15 Church Street, Great Shelford CB2 5EL, UK.

R. Hedges, University of Oxford, Research Laboratory for Archaeology and the History of Art, 6 Keble Road, Oxford OX1 3QJ, UK.

H.A. Heidinga, Amsterdam Archaeological Centre, University of Amsterdam, Nieuwe Prinsengracht 130, 1018 VZ Amsterdam, Netherlands.

J. Henderson, Department of Archaeology, University of Nottingham, University Park, Nottingham, NG7 2RD, UK.

J. de Koning, Amsterdam Archaeological Centre, University of Amsterdam, Nieuwe Prinsengracht 130, 1018 VZ Amsterdam, Netherlands.

A.J. Nijboer, Groningen Institute of Archaeology, University of Groningen, Poststraat 6, 9712 ER Groningen, Netherlands.

J.P. Pals, Amsterdam Archaeological Centre, University of Amsterdam, Nieuwe Prinsengracht 130, 1018 VZ Amsterdam, Netherlands.

J.M. Pasveer, Western Australian Museum, Francis Street, Perth WA 6000, Australia.

A. Pol, The Royal Coin Cabinet, P.O. Box 11208, 2301 EA Leiden, Netherlands.

W. Prummel, Groningen Institute of Archaeology, University of Groningen, Poststraat 6, 9712 ER Groningen, Netherlands.

J.E. van Reekum, Marnixkade 51 III, 1018 XT Amsterdam, Netherlands

M. Richards, Department of Cellular Science, Institute of Molecular Medicine, John Radcliffe Hospital, Headington, Oxford OX3 9DU, UK.

Y. Sablerolles, Department of Archaeology, University of Nottingham, University Park, Nottingham, NG7 2RD, UK.

B. Sykes, Department of Cellular Science, Institute of Molecular Medicine, John Radcliffe Hospital, Headington, Oxford OX3 9DU, UK.

J. Schoneveld, Archaeological Research and Consultancy Centre, University of Groningen, Poststraat 6, 9712 ER Groningen, Netherlands.

J.P.A. van der Vin, The Royal Coin Cabinet, P.O. Box 11028, 2301 EA Leiden, Netherlands.

T.B. Volkers, Bewaarschaoolstraat 9, 8748 AN Witmarsum, Netherlands.

P.C. Vos, Netherlands Institute of Applied Geoscience TNO – National Geological Survey, District North and East, P.O. Box 511, 8000 AM Zwolle, Netherlands.

J. Zijlstra, Jan Piebengawei 51, 8915 DC Leeuwarden, Netherlands.